Spherical Geometry in the Eighteenth Century I

Renzo Caddeo • Athanase Papadopoulos
Editors

Spherical Geometry in the Eighteenth Century I

Euler, Lagrange and Lambert

Springer

Editors
Renzo Caddeo
Dipartimento di Matematica e Informatica,
Palazzo Delle Scienze
Università degli Studi di Cagliari
Cagliari, Italy

Athanase Papadopoulos
Institut de Recherche Mathématique
Avancée
Université de Strasbourg et CNRS
Strasbourg, France

ISBN 978-3-032-12465-4 ISBN 978-3-032-12466-1 (eBook)
https://doi.org/10.1007/978-3-032-12466-1

Mathematics Subject Classification: 01A30, 01A50, 00A30, 01A20, 03A05

Cover illustration: The Astronomer, by Johannes Vermeer (1668). Musée du Louvre, Paris.

This Springer imprint is published by the registered company Springer Nature Switzerland AG
The registered company address is: Gewerbestrasse 11, 6330 Cham, Switzerland

[. . .] With silent worship of the Great of old,–
The dead, but sceptred, Sovereigns, who still rule
Our spirits from their urns. Lord Byron

*(*Manfred, act 3, scene 4.*)*

Preface

Spherical geometry, as a mathematical field, started in Ancient Greece where several fundamental treatises were written on this subject. Among these works, we mention, in historical order, the *Moving sphere* by Autolycus of Pitane (fourth century BC), the *Spherics* by Theodosius of Tripoli (first century BC), and the *Spherics* by Menelaus of Alexandria (first–second centuries AD). The next major developments in this domain are due to the Arab mathematicians between the tenth and the thirteenth century, and one has to mention here their important work on spherical trigonometry. The Arabs made the connection between the Greek world and the Renaissance in Europe, as far as mathematics is concerned. The next significant progress took place in the eighteenth century and it is due to a large extent to Leonhard Euler, Joseph-Louis Lagrange, and Johann Heinrich Lambert, in the works that are presented in the present volume. Spherical geometry received renewed attention in the nineteenth century by Nikolai Ivanovich Lobachevsky, who lectured intensively on this topic and who realized important analogies between the formulae of spherical trigonometry and those of the new geometry he discovered, which he called "imaginary geometry." As a matter of fact, Lobachevsky used spherical geometry in his foundational work on hyperbolic geometry. In the twentieth century, spherical geometry was studied as one of the two geometries of constant nonzero curvature, a subject which became at the forefront of low-dimensional geometry and topology, where spherical and hyperbolic (three-dimensional) geometries were recognized by Thurston as two of the eight fundamental geometries in dimension three.

The present volume contains the complete works on spherical geometry by Euler, Lagrange, and Lambert, translated into English from the Latin, French, and German. The translators are mathematicians working in this field. From the purely literary point of view, each of these memoirs carries the style of a master, and reading them can be a real pleasure for today's geometers. These translations are accompanied by explanatory footnotes and by several synthetic essays that insert the original memoirs in the adequate mathematical, historical, and philosophical background. This is the first time that the important works on spherical geometry by the great mathematicians of the eighteenth century are collected in a unified way, and that the relationships between them are highlighted and analyzed.

The editors would like to thank Marie-Pascale Hautefeuille for her help in the proof-reading.

Cagliari, Italy
Strasbourg, France
May 2026

Renzo Caddeo
Athanase Papadopoulos

Chronologies

Year	**Some major events in Euler's life and publication dates of his main works mentioned in the present volume**
1707	On April 15, Euler was born in Basel
1720	Euler enters the University of Basel
1723	Euler obtains a degree in philosophy and starts a theology curriculum at the University of Basel
1724	Euler delivers his public lecture at the University of Basel on the comparison between the philosophical systems of Descartes and Newton
1726	Euler completes his studies in mathematics
1727	Euler leaves Basel, to which he will never return. He arrives to Saint Petersburg
1731	Euler is appointed adjunct professor of physics at the Imperial Academy of Sciences at Saint Petersburg
1733	Euler is appointed to the Chair of Mathematics of the Imperial Academy of Sciences
1735	Euler solves the Basel problem, that is, the problem of finding the sum of the infinite series of the inverses of squares of integers. He starts his collaboration with the geographer Joseph-Nicolas Delisle
1736	Euler publishes his *Mechanica*
1737	Using the theory of continued fractions, Euler proves that e and e^2 are irrational
1739	Euler publishes his famous treatise on music, the *Tentamen novae theoriae musicae ex certissimis harmoniae principiis dilucide expositae*
1741	Euler becomes the head of the geography department at the Imperial Academy of Saint Petersburg. The same year he quits Saint Petersburg for Berlin.
1744	Publication of Euler's *Methodus inveniendi lineas curvas maximi minimive proprietate gaudentes, etc.*, inaugurating the calculus of variations
1746	Publication of the *Introductio in analysin infinitorum*
1750	In a letter to Goldbach, Euler announces the formula $V - E + F = 2$
1753	Publication of the *Atlas Geographicus omnes orbis terrarum regiones in XLIV tabulis exhibens*

1755	Euler discovers the Euler equation of fluid mechanics. Publication of the memoirs *Principes de la trigonométrie sphérique tirés de la méthode des plus grands et plus petits* and *Élémens de la trigonométrie sphéroïdique tirés de la méthode des plus grands et plus petits*
1766	Euler returns to Saint Petersburg. Publication of the first volume of the *Institutiones calculi integralis*
1768	First edition of Euler's *Lettres à une princesse d'Allemagne*
1781	Publication of the memoir *De mensura angulorum solidorum*
1782	Publication of the memoir *Trigonometria sphaerica universa, ex primis principiis breviter et dilucide derivata*
1783	Publication of the memoir *Problematis cuiusdam Pappi Alexandrini constructio*
1783	Euler dies on September 7 in Saint Petersburg, at age 76
1797	Publication of the memoir *Variae speculationes super area triangulorum sphaericorum*
1815	Publication of the memoir *Geometrica et sphaerica quaedam*

Year	**Some major events in Lagrange's life and publication dates of his main works mentioned in the present volume**
1736	January 25, Lagrange was born in Turin
1754	June 28, Lagrange, for the first time, writes to Euler. The same year, he sends his first work in the form of a letter to the Italian mathematician Giulio Fagnano
1755	Lagrange is appointed professor of mathematics at the Turin Royal Artillery School. On August 12, in a letter to Euler, he outlines the principles of the calculus of variations as he conceives it. On September 6, Euler congratulates Lagrange for "having brought this theory to its highest degree of perfection"
1756	At the proposition of Euler, Lagrange is elected foreign member of the Berlin Academy of Sciences
1757	Lagrange is co-founder of the *Società Scientifica Privata Torinese*, a learned society which became later the Turin Academy of Sciences. Most of Lagrange's works were published in the Memoirs of this Society
1764	Lagrange studies the three-body problem for the Earth, Sun, and Moon. The same year, he receives the Grand Prix of the Académie des Sciences de Paris
1766	On the recommendation of Euler and d'Alembert, Lagrange becomes Euler's successor as director of the Mathematics section at the Prussian Academy of Sciences
1768	Lagrange presents his first memoir on number theory
1775	Lagrange initiates the theory of quadratic forms
1776	Publication of Lagrange's *Solution algébrique d'un problème de géométrie*
1779	Lagrange becomes senator in Paris. Publication of the two memoirs *Sur la construction des cartes géographiques, I and II*
1783	Lagrange's wife dies, which leads him to a deep depression that causes him to stop his research for a few years
1787	After the death of King Frederick II, Lagrange leaves Berlin for Paris, where he becomes member of the Académie des sciences
1788	Publication of the *Mécanique analytique*

1794	Lagrange becomes the first professor of mathematical analysis at the newly founded École Polytechnique
1795	Lagrange becomes professor of mathematics at the newly founded École Normale, at the same time as Laplace and Monge
1799	Publication of Lagrange's *Solution de quelques problèmes relatifs aux triangles sphériques avec une analyse complète de ces triangles*
1797	Publication of Lagrange's *Théorie des fonctions analytiques*, which W. R. Hamilton describes as a "scientific poem"
1801	Publication of the *Leçons sur le calcul des fonctions*
1813	Lagrange dies in Paris on April 10, at age 77

Year	Some major events in Lambert's life and publication dates of his main works mentioned in the present volume
1728	August 26 or 28, Lambert was born in Mulhouse
1740	Lambert leaves school definitively and becomes an apprentice tailor at his father's shop
1744	Lambert works as a clerk in a foundry. The same year, he is intrigued by the appearance of the six-tailed comet that appeared on the nights of March 7 to 8 and 8 to 9, which stirred his vocation as an astronomer
1746	Lambert works as a secretary at a bi-weekly newspaper in Basel
1748	Lambert works as a private tutor for the children of the von Salis family in Chur
1750	Lambert starts a series of systematic observations of the starry sky
1753	Lambert becomes member of the Helvetic Society of Basel
1756	Lambert performs a grand tour of Europe
1758	Lambert publishes *Les propriétés remarquables de la route de la lumière par les airs, et en général par plusieurs milieux réfringens, sphériques et concentriques*. On July 1st, 1758, he writes his first letter to Euler
1759	Lambert publishes his treatise on perspective, *Die freye Perspektive, oder, Anweisung, jeden perspektivischen Aufriss von freyen Stücken und ohne Grundriss zu verfertigen*
1760	Lambert publishes his *Photometria sive de mensura et gradibus luminis, colorum et umbrae*. The same year, Euler recommends Lambert as a professor of astronomy at the Imperial Academy of Sciences at Saint Petersburg
1761	Lambert publishes his *Cosmologische Briefe über die Einrichtung des Weltbaues* in which the world is represented as an ordered compound of galaxies, supergalaxies, and higher systems of stars
1764	Lambert arrives in Berlin, at the invitation of Euler. Publication of his treatise on logico-mathematical philosophy, the *Neues Organon oder Gedanken über die Erforschung und Bezeichnung des Wahren und dessen Unterscheidung vom Irrthum und Schein*

1765	Lambert is appointed member of the Prussian Academy of Sciences of Berlin, class of Physics
1765–1772	Lambert writes his largest mathematical publication, the *Beiträge zum Gebrauche der Mathematik und deren Anwendung* in 4 volumes. The third volume is his treatise on spherical trigonometry, *Anmerkungen und Zusätze zur Trigonometrie*
1766	Lambert writes his *Theorie der Parallellinien*, for which he is considered as a forerunner of hyperbolic geometry
1767	Lambert presents to the Berlin Academy his proof that π is irrational
1771	Publication of Lambert's *Anlage zur Architectonic*, in which philosophy is presented as a deductive science
1772	Publication of Lambert's treatise on cartography *Anmerkungen und Zusätze zur Entwerfung der Land- und Himmels-Charten*
1777	Lambert dies in Berlin on September 25, at age 49

Contents

Contributors

Annette A'Campo-Neuen Departement Mathematik und Informatik, Universität Basel, Basel, Switzerland

Vincent Alberge Lycée Jean Macé Vitry-sur-Seine, France

Renzo Caddeo Faculty of Science, University of Cagliari, Cagliari, Italy

Charalampos Charitos Laboratory of Mathematics and Theoretical Mechanics, Agricultural University Athens, Athens, Greece

Stelios Negrepontis Mathematics Department, National and Kapodistrian University of Athens, Athens, Greece

Athanase Papadopoulos Institut de Recherche Mathématique Avancée, Université de Strasbourg et CNRS, Strasbourg, France

Guillaume Théret Laboratoire Interdisciplinaire Carnot de Bourgogne (ICB), Dijon Cedex, France

Vladimir Turaev Department of Mathematics, Indiana University, Bloomington, IN, USA

Alena Zhukova Saint Petersburg Electrotechnical University, St. Petersburg, Russia

Chapter 1
Introduction

Renzo Caddeo and Athanase Papadopoulos

Abstract This is the introduction to the volume *Spherical geometry in the eighteenth century, I: Euler, Lagrange and Lambert*. The introduction contains a summary of the book, and at the same time it gives a detailed overview of the important works on spherical geometry done in the eighteenth century.

Keywords Mathematics in the eighteenth century · Leonhard Euler · Joseph-Louis Lagrange · Johann Heinrich Lambert · Spherical trigonometry · Spherical geometry · Solid angle · Greek mathematics · Euclid · Proclus · Pythagorean mathematics · Plato · Anthyphairesis · Pappus problems · Menelaus theorem · Discovery of the sine rule · Geometry in the Middle-ages · Ibn 'Irāq · Naṣīr al-Dīn al-Ṭūsī · Biographies · Foundations of mathematics

AMS Classification 01A20, 01A30, 01A50, 00A30, 03A05

This volume is a commented edition of new English translations of the writings on spherical geometry by three prominent mathematicians of the eighteenth century, Leonhard Euler, Joseph-Louis Lagrange and Johann Heinrich Lambert.

The book is divided into two parts. The first part contains introductory essays on the works of Euler, Lagrange and Lambert that are translated in this volume. The essays have mathematical, historical and philosophical characters. The second part is devoted to the English translations of the original memoirs, made from the Latin, French and German. These translations are new and are made by mathematicians

R. Caddeo
Dipartimento di Matematica e Informatica, Università di Cagliari, Cagliari, Italy
e-mail: caddeo@unica.it

A. Papadopoulos (✉)
Institut de Recherche Mathématique Avancée, Université de Strasbourg et CNRS, Strasbourg, France
e-mail: athanase.papadopoulos@math.unistra.fr

R. Caddeo, A. Papadopoulos (eds.), *Spherical Geometry in the Eighteenth Century I*, https://doi.org/10.1007/978-3-032-12466-1_1

familiar with the subjects of these memoirs. Footnotes by the editors and the translators are appended to the translations.

Beyond spherical geometry, we find in this volume some important pieces of eighteenth-century geometry, explained and introduced in a way that highlights their Greek roots and the developments they underwent. In particular, in addition to our three main authors, we shall encounter on several occasions the names of Euclid, Apollonius, Pappus, Ibn ʻIrāq, Naṣīr al-Dīn al-Ṭūsī and others, as well as names of more recent mathematicians such as Lobachevsky, Gauss, Chasles and Poncelet.

In the rest of this introduction, we present in some detail the content of the volume.

Chapters 2 to 10 contain the introductory essays and Chaps. 11 to 19 contain the translations of the original memoirs.

Chapter 2, by Athanase Papadopoulos and Vladimir Turaev, is titled *Euler's work on spherical geometry: An overview with comments*. In this chapter, the authors start by explaining the importance of the role played by trigonometry in any given geometry, Euclidean or non-Euclidean. They then provide a detailed overview of the complete writings of Euler on spherical geometry, pointing out relations between this work and his works on other topics such as astronomy, geography and geomagnetism. They also review other memoirs of Euler which are not properly on spherical geometry but which involve spheres, including a memoir on the three-dimensional Apollonius problem and others concerning algebraic curves on the sphere. Thus, beyond spherical geometry, the reader is invited to discover in this chapter several interesting facets of Euler's work on geometry.

Chapter 3, by Charalampos Charitos, is an exposition of some fundamental results in spherical geometry, based on Lagrange's memoir *Solution of some problems relative to spherical triangles with a complete analysis of these triangles* translated in Chap. 18 of the present volume. At the same time, Charitos points out relations between Lagrange's and Euler's writings on this subject, putting the two works in perspective and mentioning their impact on modern mathematics. A proof of the spherical Pythagorean theorem, from which the other formulae of spherical trigonometry follow, is presented in detail, using the methods of the calculus of variations, as Euler did in one of his memoirs on this subject. A duality (or polarity) theorem, used by both Euler and Lagrange in their works on spherical geometry, is also established. The question, stressed by Lagrange, of finding a minimal set of formulae from which all the others follow easily, and that of getting the rectilinear trigonometric formulae as a limit of the spherical ones, are addressed.

In Chap. 4, Guillaume Théret reviews the approaches of several mathematicians, ancient and modern, to solve the problem that Euler studies in his memoir *A construction relative to a problem of Pappus of Alexandria* (Chap. 16 of the present volume). The problem is to construct a triangle which is inscribed in a given circle and whose three sides, when they are produced, pass through three given points. A certain number of mathematicians, some of which are among the greatest, have given their own solution to this problem and have sought to generalize it, placing it in different frameworks (modification of the hypothesis in the Euclidean framework, solution in the spherical framework, generalization to the case of projective geometry, etc.). Théret simultaneously provides a mathematical and historical overview of this problem, mentioning works of Cramer, Fuss, Viète, Descartes, Castillon, Gergonne, and, closer to us, Marcel Berger and Daniel Perrin. In particular, Pappus' original proof of this result is presented in detail, as well as a proof of a "generalized Pappus theorem", within the framework of projective geometry. Théret also mentions a connection with the famous Apollonius problem concerning the construction of circles tangent to three given objects in the plane each of which may be a point, a line or a circle.

Chapter 5, by Stelios Negrepontis and Athanase Papadopoulos, is titled *Notes on angles and solid angles, in relation with Euler's memoir "De mensura angulorum solidorum"*. This chapter is motivated by Euler's memoir mentioned in the title. The authors discuss the notion of solid angle, but also that of plane angle, from the mathematical, historical and philosophical points of view. They make an overview that starts from Ancient Greece, where the Pythagoreans, using the notion of anthyphairesis, established a relationship between, on the one hand, the quality of an angle to be acute, straight or obtuse, and on the other hand, the philosophical concepts of finite and infinite.

Several difficulties related to the notion of angle have been pointed out throughout history, and going through them shows the complexity of this notion. We mention in particular the problems that arise, since Euclid's times and even before, in dealing with angles made by curved lines, and the question of how to include the notion of angle in one of the Aristotelean categories (quantity, quality, relation, etc.). These questions were natural questions in those times. The definition of solid angle as an area leads to difficulties, starting with the question of what is "area" in general, and how one can define area without measure theory. (We recall incidentally that Euclid, in the *Elements* does not define area and never computes area, but he deals only with the comparison between areas and the concept of "two figures having the same area".)

Thus, Chap. 5, while providing a historical context to Euler's work on solid angles, highlights at the same time the intermingling between mathematics and philosophy. It also illustrates the fact that questions raised some 2500 years ago are still relevant and concern today's mathematics.

Chapter 6, by Athanase Papadopoulos, is titled *On the life and work of Johann Heinrich Lambert*. Lambert is the typical example of the complete autodidact who later became one of the most famous, original and versatile scholars of his times. Born in a poor family and forced to quit school at twelve, he spent the last twelve years of his life at the Royal Prussian Academy of Sciences as one of its most eminent members. His writings on mathematics, perspective, pyrometry, cosmology, astronomy, philosophy and other topics are reviewed in this chapter, which is also an occasion to quote some excerpts from Lambert's interesting correspondence with Euler. Besides Lambert's relation with Euler, his connections with several thinkers of his time, including Johann III and Daniel Bernoulli, Joseph-Louis Lagrange, Emmanuel Kant and others are highlighted.

Chapter 7, by Athanase Papadopoulos and Guillaume Théret, is titled *Johann Heinrich Lambert's Memoir Theorie der Parallellinien: A Review with Commen-*

tary. As the title indicates, this chapter is an overeview of Lambert's memoir *Theorie der Parallellinien*, written in 1766, which the author conceived as an attempt to prove the parallel postulate of Euclid's *Elements*, and consequently, the non-existence of the geometry known today as hyperbolic geometry. With this in mind, Lambert, in a substantial part of his memoir, developed a geometry obtained from Euclidean geometry by replacing the parallel postulate by its negation while keeping Euclid's other postulates unchanged. His hope was to reach a contradiction, which is a way of proving that the parallel postulate is a consequence of the others. Needless to say, Lambert failed in his goal. But it turns out that his attempts contain fundamental results of hyperbolic geometry, the discovery of which, by Lobachevsky, Bolyai and Gauss, was not to take place only the following century. Thus, Lambert's memoir (which he did not intend to publish but which was eventually published in 1895, that is, 117 years after his death) constitutes one of the founding texts of non-Euclidean geometry.

We have included this chapter in the present volume because, as the reader may know, spherical geometry is one of the three geometries of constant curvature, the other two being Euclidean geometry and hyperbolic geometry. In this sense, along with hyperbolic geometry, spherical geometry constitutes one of the two non-Euclidean geometries. In fact, Lambert, like Lobachevsky and others after him, understood the deep relationship between spherical and hyperbolic geometry, in particular the formal and the more profound analogies between the trigonometric formulae of the two geometries, the properties of birectangular isosceles quadrilaterals and of trirectangular quadrilaterals, and the monotonicity properties, which can be formulated in terms of convexity properties which hold, with the inequalities in reverse direction, in these geometries.

These relations and others, between the three geometries, highlighted in the work of Lambert, led us to include in this volume this chapter which gives an overview on the important memoir by Lambert, underlining its major ideas, and making connections with the works of several mathematicians on the parallel postulate covering a period of two millennia.

Chapter 8, by Annette A'Campo-Neuen, is titled *The* pentagramma mirificum *from Napier to Lambert: Notes on Lambert's memoir on spherical trigonometry*. The important relations between the elements of a right-angled spherical triangle are summarized in the so-called Napier mnemonic rules, associated with a diagram that first appeared in the work *Mirifici logarithmorum canonis descriptio etc.* by John Napier, published in Edinburgh in 1614. This diagram consists of a closing chain of five right-angled triangles enclosing a self-polar spherical pentagon. It gives a convenient way of presenting ten relations between any three of the five parts of the triangle other than the right angle. In his memoir on spherical trigonometry, Lambert clarified the meaning of this diagram and gave a proof of Napier's rules in terms of cyclic transformations of a given triangle. In her commentary, Annette A'Campo-Neuen focuses on Lambert's elegant proof of Napier's rule and the properties of the pentagonal star of triangles. She also recalls that the configuration of five triangles with the enclosed pentagon was called *pentagramma mirificum* (an expression which means "marvelous pentagon") by Gauss, who found further properties of this pentagon.

Chapter 9, by Athanase Papadopoulos, is titled *Spherical trigonometry before the modern era: The treatise of Naṣīr al-Dīn al-Ṭūsī*. This chapter consists in an overview of an invaluable thirteenth century treatise by Naṣīr al-Dīn al-Ṭūsī on spherical geometry which was translated into French in 1891 under the name *Traité du quadrilatère* (Treatise of the quadrilateral). The treatise contains a large amount of historical information on the discovery of the spherical sine rule and the other spherical trigonometric formulae by the Arabs of the Middle-Ages and the transformation of the field of spherical geometry that this discovery gave rise to. The aim of this chapter is to give an idea of the state of the art of spherical geometry before the modern era.

Chapter 10, by Athanase Papadopoulos, titled *The polar triangle, from Ibn 'Irāq to Euler and Lagrange*, contains some landmarks on the history of the notion of polar triangle. In their memoirs contained respectively in Chaps. 12 and 18 of the present volume, Euler and Lagrange refer to this triangle when they notice that we can obtain some formulae of spherical trigonometry starting from others by interchanging the letters for sides and angles and replacing the values of the new angles and sides by their supplement to two right angles. The overview in Chap. 10 covers some important landmarks concerning the polar triangle and its use, from its discovery by the Arab mathematician Ibn Irāq (tenth century) until the modern period.

With Chap. 11, we start the second part of this volume, which contains the English translations of memoirs written originally in French, German or Latin.

We start this part by Euler's two memoirs on spherical trigonometry, *Principles of spherical trigonometry deduced from the method of maxima and minima* (Chap. 11) and *General spherical trigonometry, deduced from fundamental principles in a brief and clear manner* (Chap. 12).

In the first memoir, Euler uses what he calls the *method of maxima and minima*, that is, the calculus of variations, to establish a set of formulae of spherical trigonometry. He writes that the formulae he gives "are very handful for making the calculation with the aid of logarithms." In the last section, he uses the techniques and results obtained in the preceding sections to study the area of spherical triangles, recovering analytically the so-called Girard formula which gives the area of such a triangle in terms of its angle excess.

In the second memoir (Chap. 12), Euler establishes the trigonometric formulae of spherical geometry using more classical methods, namely, he works in the 3-dimensional ambient space of the sphere, deducing the spherical trigonometric formulae from the Euclidean ones.

Chapter 13 consists of Euler's memoir *Various speculations on the area of a spherical triangle*. In this memoir, Euler starts by giving several formulae for the area of a spherical triangle in terms of its side lengths. This is an analogue, in the setting of spherical geometry, of the well-known Euclidean Heron formula. The formulae that Euler obtains involve trigonometric functions of the sides of the triangle. Presumably, Euler's formulae are the oldest known analogues of Heron's formula in spherical geometry. Lagrange, in his memoir *Solution of some problems relative to spherical triangles with a complete analysis of these triangles*, published 18 years after Euler's and translated in Chap. 18 of the present volume, establishes such formulae (without mentioning the work of Euler). Euler's main object in his memoir is to give a solution to a problem which he attributes to his young colleague Anders Johan Lexell, a Swedish astronomer and geometer who was working like him at the Russian Imperial Academy of Sciences at Saint Petersburg. The problem is to find the locus of the vertices of a spherical triangle with a given base and fixed area. This is the spherical analogue of a result in Euclid's *Elements*, which is the subject of Propositions 37 and 39 of Book I. We note that Lexell's work on this problem and on other questions related to spherical geometry are translated and discussed in a sequel to the present volume, carrying the title *Spherical geometry in the eighteenth century, II: Lexell, Schubert and Fuss*.

Chapter 14 of the present volume consists of Euler's memoir *On the measure of solid angles*, whose aim is to obtain numerical values of the solid angles of the five regular polyhedra. We recall that the measure of a solid angle is the area of the region intercepted by this angle on a sphere of radius one. Thus, Euler starts his memoir by giving a proof of Girard's formula for the area of a spherical triangle. The proof he gives goes back to Bonaventura Cavalieri. It is intrinsic to the sphere, and it is based on the area of lunes, that is, regions bounded by two half-great circles. In the same memoir, Euler gives several formulae for the area of a spherical triangle in terms of its side lengths and he uses them in his computations of the values of the solid angles of the five regular polyhedra.

In the next two memoirs, Euler obtains results on spherical geometry that are motivated by theorems of Euclidean geometry.

Chapter 15 is Euler's memoir *Some questions of the geometry of the plane and of the sphere*. In this memoir, Euler solves the following problem. Consider a triangle in the Euclidean plane and three segments joining each vertex to the side opposite to it, and assume that these three segments intersect at a common point. Euler obtains an equation satisfied by the six parts cut on these three segments by the intersection point. He then shows that conversely, if we start with such an equation, we can construct a triangle and three lines joining the vertices of this triangle to the opposite sides such that the six segments that arise in this construction satisfy the given equation. After the Euclidean case, Euler proves the spherical analogue of this result.

Chapter 16 consists of Euler's memoir *A construction relative to a problem of Pappus of Alexandria*. In this memoir, Euler solves the problem of constructing a triangle inscribed in a given circle and whose three sides are contained in three lines that are required to pass through three given points. He solves the problem, first in the Euclidean case, and then in the spherical case. A solution of this problem, in the Euclidean case, is contained in Pappus' *Collection*, in the special case where the three points are aligned. Pappus' solution of the problem is reviewed in the essay by Guillaume Théret contained in the present volume (Chap. 4). Another construction, in the Euclidean case, is given in Lagrange's memoir *Algebraic solution of a problem in geometry* translated in Chap. 17 of this volume.

Chapter 18, also by Lagrange, is titled *Solution of some problems relative to spherical triangles with a complete analysis of these triangles.* Lagrange starts this memoir by recalling that in a Euclidean triangle, the sides are proportional to the sines of the opposite angles, and he remarks that the constant of proportionality is equal to the diameter of the circumscribed circle and can be easily expressed in terms of the area of the triangle and of its sides. Lagrange then passes to spherical geometry, considering similar questions in this setting. He finds analogous formulae for the ratios of the sides to the sines of the angles opposite to them. He obtains several formulae involving radii of circumscribed circles and areas of spherical triangles. After some considerations on the volume of the triangular pyramid formed by the three radii of the sphere that connect the center to the three angles, Lagrange makes several remarks on Girard's formula, and establishes a formula for the area of a triangle in terms of its side lengths, that is, a spherical Heron formula, of the kind Euler gave in his memoir *On the measure of solid angles* that we reviewed above (Chap. 14). After this, Lagrange establishes a complete set of spherical trigonometric formulae. At the end of his memoir, he points out a relation between spherical and rectilinear triangles, obtained by making the radius of the sphere infinitely large, thus transforming the sphere into a plane, and the spherical triangles into rectilinear ones. He shows that in this way, the formulae of rectilinear trigonometry can be deduced from those of spherical trigonometry.

The last chapter (Chap. 19), consists of an English translation of Lambert's memoir *Notes and Comments on Trigonometry*. The memoir is extracted from a larger work titled *Contributions to mathematics and its applications*. In his memoir, Lambert starts with the so-called Napier mnemonic rules which summarize the relations between any three parameters (angles and side lengths) of a right-angled spherical triangle. These rules are recorded in a diagram representing a closing chain of five right-angled triangles enclosing a spherical pentagon. Lambert clarifies the meaning of this diagram and gives a proof of Napier's rules in terms of cyclic transformations of a given triangle. He reduces the formulae of spherical trigonometry to four basic ones, being careful to making these formulae suitable for their use with logarithms. Lambert's *Notes and Comments on Trigonometry* constitutes a valuable treatise on spherical trigonometry that complements the ones of Euler and Lagrange that are published in the present book.

In conclusion, in this volume, we have collected important geometrical essays by three of the main founders of modern mathematics. The reader may find there the ideas that are at the foundation of spherical geometry and techniques that are at the basis of modern geometry. At the same time, we have tried to communicate to the reader our interest in reading texts written by the masters, as well as our respect and admiration for the great mathematicians of the past.

Part I
Essays

Chapter 2
Euler's Work on Spherical Geometry: An Overview with Comments

Athanase Papadopoulos and Vladimir Turaev

Wasser den Rhein hinunterfließen, bis die Schule endlich merkt, daß die Mathematik eine Geisteswissenschaft sein kann und daß die Schüler EULER *so gut verstehen können wie* PLATO *oder* GOETHE. *(Andreas Speiser,* Introduction to Euler's *Introductio in analysin infinitorum. Tomus secundus. In: Leonhardi Euleri opera omnia Vol. 9, Lipsiae 1945,* p. XIX.*)*

Abstract We review Euler's work on spherical geometry. After an introduction concerning the general place that trigonometric formulae occupy in geometry, we start by the two memoirs of Euler on spherical trigonometry, in which he establishes the trigonometric formulae using different methods, namely, the calculus of variations in the first memoir, and classical methods of solid geometry in the other. In a third memoir, Euler gives several formulae for the area of a spherical triangle in terms of its side lengths (these are "spherical Heron formulae"). He uses this in the computation of numerical values of the solid angles of the five regular polyhedra, which is his goal in his memoir. We then review memoirs in which Euler systematically starts by establishing a theorem or a construction in Euclidean geometry and then proves an analogue in spherical geometry. We point out relations between Euler's memoirs on spherical trigonometry and his works in astronomy, on the problem of drawing geographical maps, and in geomagnetism. We also review some other works of Euler involving spheres, including a memoir on the three-dimensional Apollonius problem and others concerning algebraic curves on the sphere. Even though these works are not properly on spherical geometry, they show Euler's interests in various questions related to spheres and we think that

A. Papadopoulos (✉)
Institut de Recherche Mathématique Avancée, Université de Strasbourg et CNRS, Strasbourg, France
e-mail: athanase.papadopoulos@math.unistra.fr

V. Turaev
Department of Mathematics, Indiana University, Bloomington, IN, USA
e-mail: vtouraev@indiana.edu

R. Caddeo, A. Papadopoulos (eds.), *Spherical Geometry in the Eighteenth Century I*, https://doi.org/10.1007/978-3-032-12466-1_2

they are worth highlighting in such an overview. Beyond spherical geometry, the reader is invited to discover in this chapter an important facet of the work of the great Leonhard Euler.

Keywords Leonhard Euler · Spherical geometry · Spherical trigonometry · Spherical triangle · Polar triangle · Euclidean geometry · Eighteenth century · History of geometry · Applications to geography · Applications to astronomy · Applications to geomagnetism · Algebraic curves on the sphere · Pappus problem · Apollonius problem · Generalized Apollonius problem

AMS Classification 01A50, 51-03, 53-03

2.1 Introduction

In this chapter, we review Euler's memoirs on spherical geometry together with other works that include this work on spherical geometry in a broad perspective. The plan of the chapter is as follows.

We start (Sect. 2.2) with an excursus on the importance of trigonometric formulae, in any geometry, mentioning in particular the work of Lobachevsky and the paramount importance of the formulae of spherical trigonometry that he established in his lectures and in several writings.

After this digression which provides one motivation for studying Euler's work on spherical trigonometry, we review (Sect. 2.3) Euler's memoirs on this subject, namely, the *Principes de la trigonométrie sphérique tirés de la méthode des plus grands et plus petits* (Principles of spherical trigonometry deduced from the method of maxima and minima) [27], published in 1755, and *Trigonometria sphaerica universa, ex primis principiis breviter et dilucide derivata* (General spherical trigonometry, deduced from fundamental principles in a brief and clear manner) [34], published in 1782. In each of these two memoirs, Euler establishes a set of formulae of spherical trigonometry. In the first memoir, he uses, in proving these formulae, methods of differential geometry and the calculus of variations, whereas in the second he relies on more classical proofs that are based on the geometry of the 3-dimensional Euclidean space: cutting by planes the solid angles centered at the origin defined by a spherical triangle and applying to the resulting Euclidean triangles the formulae of plane trigonometry.

In Sect. 2.4, we review Euler's memoir *De mensura angulorum solidorum* (On the measure of solid angles), in which he gives a geometrical proof of Girard's formula for the area of spherical triangles which is intrinsic to the sphere. The proof begins with a remark on the areas of lunes on the sphere[1] that goes back to Bonaventura Cavalieri. In the same memoir, Euler obtains several formulae for the area of a spherical triangle as a function of its side lengths, using analytical tools.

[1] A *lune* on the sphere is a figure bordered by two half-great circles.

He then uses these formulae to obtain approximate values of the solid angles of the five regular polyhedra.

In Sect. 2.5, we review Euler's memoir *Variae speculationes super area triangulorum sphaericorum* (Various speculations on the area of spherical triangles) [39], in which he proves a result in spherical geometry which is the analogue of a well-known result in Euclidean geometry, namely, a description of the locus of the vertices of triangles having a fixed base and a fixed area. In the Euclidean case, this locus is a straight line, and the result is contained in Euclid's *Elements*. In the spherical case, the locus is not a geodesic, and the result is more complicated to prove. Euler attributes the problem and the result to his young collaborator Anders Johann Lexell, who wrote a memoir on this subject [61].

Section 2.6 is an overview of several memoirs in which Euler, like in the one which is the subject of the preceding section, proves some results in Euclidean geometry and then their analogues in spherical geometry. This includes the memoir *Problematis cuiusdam Pappi Alexandrini constructio* (A construction relative to a problem of Pappus of Alexandria) [35] in which he obtains a generalization of a Euclidean construction from Pappus' *Collection*, before considering the spherical case. This problem (in the Euclidean case) has attracted the attention of several mathematicians, in particular Euler's young collaborator at the Russian Academy, Nicolaus (better known at the Saint Petersburg Academy as Nikolai Ivanovich) Fuss. The other mathematicians who worked on this problem include Lagrange, Castillon, Carnot, Brianchon and many others which we discuss in Sect. 2.6. In another chapter of the present volume, Guillaume Théret gives a detailed commentary on Pappus' problem. In the same section, we review Euler's memoir *Geometrica et sphaerica quaedam* (Some questions of the geometry of the plane and of the sphere) [41], in which he starts again by proving a result in Euclidean geometry and then its spherical analogue. This also gives us the occasion to review several works of Euler on Euclidean geometry.

In Sect. 2.7, we discuss some of the spherical geometric arguments that appear in Euler's works on geography, astronomy and geomagnetism.

In Sects. 2.8 and 2.9, we survey some work of Euler on the sphere (even though this work is not, strictly speaking, on spherical geometry), including his memoir on algebraic curves on the sphere (and other surfaces), a generalization to dimension three of Apollonius' problem of kissing circles (which becomes then a problem of kissing spheres), making relations with recent problems, some of which are still part of current mathematical research.

In the memoirs we review in this chapter, beyond Euler's work on spherical geometry, it is Euler's work itself that the reader will discover.

We start with some thoughts on spherical trigonometry.

2.2 Why Spherical Trigonometry?

The expression *spherical trigonometry* refers to a set of formulae that make relations between the side lengths and the angles of a spherical triangle. Such formulae constitute a non-negligible part of Euler's mathematical heritage. In fact, even before writing his memoirs on this topic, Euler used extensively spherical trigonometry in his works on geography, astronomy and geomagnetism. But in our opinion, the importance of spherical trigonometry goes beyond these applications, as we shall explain later in this chapter, after including the question in a broader context.

First of all, let us state a general fact, namely, that the set of trigonometric formulae relative to a given geometry is an important aspect of that geometry, perhaps the most important. Indeed, geometers know that the basic features of a geometry are contained in the properties of its triangles: the properties of the small (and infinitesimal) triangles tell us about the local properties of the geometry, and the properties of large triangles tell us about its large-scale properties.

John Napier, in his *Mirifici logarithmorum canonis descriptio, ejusque usus, in utraque trigonometria; ut etiam in omni logistica mathematica, amplissimi, facillimi, & expeditissimi explicatio* (1614) [65], highlighted the importance of the trigonometric formulae. The following quote by Napier is borrowed from Annette A'Campo's chapter in the present volume [1]:

> Since geometry is the art of measuring proposed quantities with complete accuracy, and as a figure [..] can be decomposed into the sides of triangles, or by further triangulation into smaller triangles, then a figure is composed of triangles with some of its angles and sides measured, and all the other triangular parts are to be found, and from which the extent of the figure can be found. Therefore it is clear that the arithmetical solution of any geometrical question depends on the principles by which triangles are solved.

It is interesting to remember in this respect that the great Lobachevsky, when he started establishing the bases of the geometry that carries now his name, was very careful in establishing a complete set of trigonometric formulae. Less well known is that Lobachevsky, who was very much concerned about the question of the non-contradiction of his geometry, argued that if there were a contradiction in his geometry, then there would be a contradiction stemming from the associated trigonometric formulae, for the reason we have just explained. But he had proved his (hyperbolic) trigonometric formulae using solely the geometry of the Euclidean plane and of the sphere. More precisely, he used the geometry of the horospheres of three-dimensional hyperbolic space (having shown beforehand that these horospheres naturally carry a Euclidean geometry) and that of the geometric spheres of his geometry (having shown beforehand that they bear the same geometry as those of the ordinary spheres of the three-dimensional Euclidean space). But then, because of the way he produced the trigonometric formulae of hyperbolic geometry, he was able to claim that if there were a contradiction stemming from these hyperbolic trigonometric formulae, then the contradiction would be already inherent in the formulae of Euclidean or spherical geometry. Obviously, the question of the non-contradiction of spherical geometry was not addressed in those days, since this was not a geometry defined axiomatically, but a geometry that was part of Euclidean space geometry. Thus, Lobachevsky, without having a Euclidean model of hyperbolic geometry, reduced the non-contradiction problem of his geometry to that of Euclidean geometry. This fact is little known. Let us quote the remarks that he made on this question, at the end of his *Elements of geometry* (the translation of the following text is from [73, p. 223]):

> After we have found Equations (17)[2] which represent the dependence of the angles and sides of a triangle; when, finally, we have given the general expressions for elements of lines, areas and volumes of solids, all else in the Geometry is a matter of analytics, where calculations must necessarily agree with each other, and we cannot discover anything new that is not included in these first equations from which must be taken all relations of geometric magnitudes, one to another. Thus if one now needs to assume that some contradiction will force us subsequently to refute the principles that we accepted in this geometry, then such contradiction can only hide in the very Equations (17). We note, however, that these equations become Equations (16) of spherical trigonometry as soon as, instead of the sides a, b, c we put $a\sqrt{-1}, b\sqrt{-1}, c\sqrt{-1}$; but in ordinary Geometry and in spherical Trigonometry there enter everywhere only ratios of lines; therefore ordinary Geometry, Trigonometry and the new Geometry will always agree among themselves.

It is also good to know that the trigonometric formulae (spherical and hyperbolic) are already contained in Lobachevsky's first memoir, the *Elements of geometry* [62], and they appear in his later works, such as the *Geometrische Untersuchungen* [63, §35 and 36 respectively].

Talking about the importance of the geometry of triangles, let us also recall that a large portion of the geometrical part of Euclid's *Elements* and of Menealus' *Spherics* [72] is a study of Euclidean, respectively spherical triangles (even if none of these treatises contains trigonometric formulae). Let us also recall that each of Herbert Busemann and A. D. Aleksandrov and M. Gromov developed geometric theories that are based on a property of triangles, see [10], [3] and [49]. Finally, let us recall that W. Thurston, in his *Geometry and topology of Three-manifolds* [79], which arguably is the most influential twentieth century work on geometry and topology, starts by establishing a set of trigonometric formulae for spherical and hyperbolic geometry.

2.3 On Euler's Work on Spherical Trigonometry

Euler wrote two memoirs on spherical trigonometry, *Principes de la trigonométrie sphérique tirés de la méthode des plus grands et plus petits* (Principles of spherical trigonometry deduced from the method of maxima and minima) [27] and *Trigonometria sphaerica universa, ex primis principiis breviter et dilucide derivata* (General spherical trigonometry, deduced from fundamental principles in a brief and clear manner) [34]. They are translated in Chaps. 11 and 12 respectively of the present volume. Regarding these memoirs, we start by quoting Rosenfeld, from his *History of non-Euclidean geometry* [73, p. 31]:

[2] Equations (16) and (17) of Lobachevsky's treatise to which he refers in the present passage are respectively sets of trigonometric formulae of spherical and hyperbolic geometry he established in his text, before this passage.

> The modern form of spherical trigonometry, as well as of all trigonometry, is due to the great Leonhard Euler (1707–1783), a native of Basel, who worked in Petersburg and Berlin. Whereas trigonometry before Euler was concerned with trigonometric *lines*, Euler's trigonometry dealt with trigonometric *functions*, which he linked to the exponential function by means of the well known *sinus totus*,[3] the complete sine, that is, the radius of the circle, and replaced it with unity.

To describe in some detail Euler's two memoirs, let us first recall that a spherical triangle is a figure on the sphere consisting of three points (the vertices) together with three shortest segments (the sides) joining them pairwise. Thus, a spherical triangle gives rise to six quantities: the three angles at the vertices and the three side lengths. By the usual criteria of equality of spherical triangles, the knowledge of any three of these six quantities determines the triangle. A trigonometric formula is typically a relation between four or more of these quantities. In each of his two memoirs, Euler derives a set of trigonometric formulae that can be useful in various situations, involving a wealth of variables like sums or differences of pairs of sides or of angles, etc.

We first consider Euler's memoir *Principes de la trigonométrie sphérique tirés de la méthode des plus grands et plus petits* [27], published in 1755.

In the introduction to this memoir, Euler addresses the question of why one should use the *methods of maxima and minima*, that is, the calculus of variations, to establish trigonometric formulae, while simpler methods exist. He answers this question in three points:

(1) First, the methods of maxima and minima, he says, acquire a new domain of application in the field of spherical trigonometry. Indeed, it seems that this is the first time that trigonometric formulae are established using this method.[4]
(2) Euler declares that it is always useful to have different proofs of the same result, and that this will help for a better understanding of this result.

[3] In ancient sine tables, the *sinus totus* (the Latin expression for "total sine") appeared as the radius of the circle that was used to construct this table. It corresponds there to the value of $\sin 90^{\circ}$.

[4] It should be recalled that the methods of the calculus of variations, since their introduction by Euler and then Lagrange, have been especially useful in the solution of problems of physics, even if these methods were formulated as part of mathematical analysis. Euler's memoir *Methodus inveniendi lineas curvas maximi minimive proprietate gaudentes, sive solutio problematis isoperimetrici latissimo sensu accepti* (A method for finding curved lines enjoying properties of maximum or minimum, or solution of isoperimetric problems in the broadest accepted sense), published in 1744 [24] and in which he gives a general presentation of the problems that are dealt with using the calculus of variations as well as an exposition of the general methods for solving them, contains, by way of illustration, a list of one hundred special cases to which the general methods apply. These problems include a solution to the brachistochrone problem, others related to the principle of least action, others in geometric optics, and many more, all originating in physics. Talking about the calculus of variations, Euler writes in his memoir *Principes de la trigonométrie sphérique tirés de la méthode des plus grands et des plus petits* [27] that "[. . .] since we have shown that most of the mechanical and physical problems can be resolved very promptly with the aid of this method, it cannot be but very pleasant to see that the same method brings such a great help for the resolution of problems of pure Geometry."

(3) Euler states that the methods of maxima and minima, unlike the previously known methods used for deriving the trigonometric formulae, are very general and can be applied to arbitrary surfaces.

Concerning the third point, one may recall that in the memoir *Éléments de la trigonométrie sphéroïdique tirés de la méthode des plus grands et des plus petits* (Elements of spheroidal trigonometry drawn from the method of the maxima and minima) [28],[5] published the same year (1755) as the memoir *Principes de la trigonométrie sphérique tirés de la méthode des plus grands et plus petits*, Euler studies triangles on a spheroid,[6] and that he even starts there by considering triangles on an arbitrary surface. Let us quote him from the introduction to that memoir (translation from [11, p. 206]):

> Having established the Elements of spherical trigonometry on the principle of the maxima and minima, my main goal was to fix such a general principle, from which one could extract the resolution of triangles formed not only on a spherical surface, but in general on an arbitrary surface. Since the sides of a spherical triangle are arcs of great circles which, being the shortest paths that one can draw on the surface of a sphere from one point to another, it is on an equal footing that I consider the sides of a triangle described on an arbitrary surface, in such a way that they become the shortest paths that lead from an angle to another on this surface.

Let us return to Euler's memoir *Principes de la trigonométrie sphérique etc.* [27]. Euler starts by recalling the definition of a spherical triangle. The idea of using the calculus of variations in this setting is based on the fact that an edge of such a triangle is a geodesic on the sphere, that is, a shortest path between its endpoints. Finding the length of a side of a spherical triangle becomes a variational problem. The first result in Euler's memoir is the solution of this problem in the case where the side whose length is sought is opposite to a right angle. More precisely, he gives a solution to the following problem:

Problem Given the arc AP on the equator and the arc PM on a meridian, to find on the spherical surface the shortest line AM that can be drawn from point A to point M.

The use of geographical terminology in this statement is not unexpected, since Euler had in mind the example of the Earth, generally identified with a sphere in the problems of drawing geographical maps in which he had been heavily involved; see [11], and also Sect. 2.7 of the present chapter.

---

[5] The memoir is translated into English in the volume [11] which is concerned by mathematical geography in the eighteenth century.

[6] A spheroid is a surface obtained by the rotation of an ellipse along one of its axes. The study of the trigonometry of the spheroid was motivated by questions of geography. Indeed, at Euler's time, it was realized, based on Newton's work, that the Earth's shape was spheroidal rather than spherical. Therefore, spheroidal trigonometry became useful in cartography, an activity in which Euler was heavily involved. Euler writes in the same memoir [27]: "[...] Since the surface of the Earth, being not spherical, but spheroidal, a triangle formed on the surface of the Earth will belong to the species of which I have just talked about [that is, on the spheroid]." The memoir [28] also contains explicit computations of distances between points on a spheroid.

As in several other memoirs of Euler, the main results in [27] are presented in the form of problems and corollaries. Euler starts by establishing a list of 30 formulae involving only three variables, valid for the case of a right triangle: Given any two quantities (sides or angles), he gives three formulae giving the values of the remaining ones (excluding the right angle, which is known). Using this, he obtains a large collection of trigonometric formulae for general (not necessarily right) triangles, involving sides, angles and various combinations of them (half-sums of sides, differences of sides, etc.). Among them are formulae for the value of a side of a spherical triangle in terms of the three angles. Obviously, such formulae cannot exist in Euclidean geometry, where the angles do not determine the sides. The memoir concludes with a proof using differential and integral calculus of the formula attributed to Albert Girard[7] which gives the area of a spherical triangle in terms of its angle excess.

Beyond the obtention of trigonometric formulae, Euler developed in the memoir [27], using the calculus of variations, a version of the intrinsic differential geometry of surfaces embedded in 3-space, a theory that Gauss and, after him, Riemann, continued to develop in the century that followed. Let us also mention in this respect that in the memoir *De linea brevissima in superficie quacunque duo quaelibet puncta iungente* (On the shortest line joining two points on a surface) [21] published in 1729, Euler (who was 22 years old) established the equation of a geodesic on an arbitrary surface.

In the memoir *Trigonometria sphaerica universa, ex primis principiis breviter et dilucide derivata* (General spherical trigonometry, deduced from fundamental principles in a brief and clear manner), published in 1782 [34]. Euler recovers the trigonometric formulae in a more classical manner, by methods that use the ambient three-dimensional space and with no differential or integral calculus. He starts by establishing the following three basic formulae,

$$\frac{\sin C}{\sin c} = \frac{\sin A}{\sin a},$$

$$\cos A \sin c = \cos a \sin b - \sin a \cos b \cos C,$$

and

$$\cos c = \cos a \cos b + \sin a \sin b \cos C,$$

[7] Albert Girard (1595–1632) was a French-born mathematician who studied and spent all his life in the Netherlands, where his family was exiled for being protestant. He worked on algebra and number theory. Like several mathematicians of his times, he was also an engineer. He wrote treatises on fortifications. He translated into French works of the Flemish mathematician and music theorist Simon Stevin (1548–1620) and of the Dutch astronomer and mathematician Willebrord Snellius (1580–1626). His work is situated, historically, between those of François Viète and René Descartes. Girard died at age 37. Tannery wrote a biographical note on Girard [77], in which he declares that the latter was one of the most remarkable geometers of the beginning of the seventeenth century.

from which he derives the rest of the trigonometric formulae by performing algebraic transformations. Right at the beginning of the memoir, Euler notes that the choice of the notation A, B, C and a, b, c for the angles and sides opposite to them leads to a set of symmetric formulae. He highlights a rule for obtaining new formulae out of known ones, and this rule turns out to be a form of duality theory in spherical geometry, or the theory of the polar triangle. This discovery was attributed by Michel Chasles, the famous mathematician and historian of mathematics, in [15], to François Viète (1540–1603), but it is known now that this discovery was already known to the Arab mathematicians of the tenth century; see the historical exposition in Chap. 10 of the present volume [67].

Several authors after Euler worked out, like him, a complete system of trigonometric formulae based on very few starting formulae. We mention in particular Lagrange, in his memoir *Solution de quelques problèmes relatifs aux triangles sphériques avec une analyse complète de ces triangles* (Solution of some problems relative to spherical triangles with a complete analysis of these triangles), published in 1799 and translated in the present volume [58], and de Gua [50], whose work on the subject came shortly after that of Euler and before that of Lagrange. We end this section by quoting Delambre, from his *Rapport historique sur les progrès des sciences mathématiques depuis 1789* (Historical report on the progress of mathematical sciences since 1789) [19, p. 45], who includes Euler's work in a broader historical context:

> Euler, in a memoir printed in 1779,[8] had reduced spherical trigonometry to a fully analytical form. Since everything is determined in a triangle of which we know, for example, two sides and the included angle, it follows that the whole of trigonometry is contained in the equation which serves to solve this particular case. To extract it is a purely analytical problem. However, it requires some skill; it is not even useful to know in advance what one is looking for. This problem, moreover, was only a game for Euler. Mr. Lagrange, in treating the same subject in the sixth *Cahier* of the *Journal de l'École Polytechnique*, added some curious propositions concerning the area of the spherical triangle and the pyramid for which this surface serves as a base. It should be noted, however, that this new way of considering trigonometry added almost nothing to the formulae we had in our possession, and we should not be surprised: the astronomers who made continual use of these methods had turned them upside down in every way; they had even arrived at solutions which are astonishing in their symmetry and elegance, when we consider the elementary means by which they had reached there. The research of Euler and Lagrange made the analytical method as elementary as the old one; and since that time several geometers, Messrs. Bertrand, Lacroix, Puissant and some others have varied this development, which in fact can be explained in several ways, starting with whichever of the particular formulae one wishes: given only one, the rest will follow from it.

For the works of Bertrand, Lacroix and Puissant mentioned by Delambre, see [7, 56, 71].

[8] Delambre refers here to the memoir *Trigonometria sphaerica universa, ex primis principiis breviter et dilucide derivata*, published in 1782. The volume of the *Acta* in which it appeared carries the mention "pro anno 1779, pars prior", which explains Delambre's assertion on dates.

To end this section, let us note that Euler, Lagrange and Lambert, in their memoirs on spherical trigonometry that are published in the present volume, all worked out sets of formulae that are suitable for computations with logarithms, that is, involving multiplications rather than additions. This was important for computations and for the establishment of tables that were used in astronomy and geography.

2.4 On the Measure of Solid Angles

Spherical area is a central theme of Euler's memoir *De mensura angulorum solidorum* (on the measure of solid angles) (1781) [33]. In this memoir, Euler's goal is to compute the numerical values of the solid angles of the five regular polyhedra. But the measure of a solid angle is the area of the region of the sphere of radius 1 centered at the vertex of this solid angle. This is the 3-dimensional analogue of the fact that the measure of an angle in the plane whose vertex is at the center of a circle of radius 1 is the length of the arc of circle subtended by this angle. Thus, the computation of the solid angles becomes a question of computation of area in spherical geometry. The simplest case is that of a solid angle formed by joining pairwise along their edges three Euclidean angles so that a solid angle appears as a vertex of a pyramid. In this case, the value of the solid angle is equal to the area of a spherical triangle. Euler recalls in the memoir considered that, according to Girard's formula, the area of a spherical triangle is equal to its angle excess with respect to two right angles. He starts by giving a very simple proof of Girard's formula, based on the computation of areas of lunes.[9] A similar proof was already given by John Wallis (1616–1703), who attributes it to Bonaventura Cavalieri (1598–1647). See Figs. 2.1 and 2.2, extracted respectively from Wallis' and Euler's memoirs. The drawings summarize the main ideas of the proofs and their similarity is an indication of the fact that the two proofs are the same.[10] The excerpt from Wallis' proof, reproduced in Fig. 2.1, mentions the names of Girard and Cavalieri.

Regarding the so-called Girard formula, let us note additionally that Lagrange, in his memoir *Solution de quelques problèmes relatifs aux triangles sphériques avec une analyse complète de ces triangles*, which is translated in the present volume, Chap. 18 [58], writes the following: "[...] we should rather attribute this theorem to Cavalieri who gave it in the *Directorium generale uranometricum*, printed in

[9] We recall that a *lune*, in spherical geometry, is a region on the sphere bounded by two half-great circles, that is, a lune is bounded by two geodesic segments joining a pair of antipodal points, see Fig. 2.1 or Fig. 2.2.

[10] This proof with intersections of lunes can be adapted to hyperbolic geometry to prove the analogous result, namely, that the area of a hyperbolic triangle is equal to its angle deficit. Such a proof, in the hyperbolic case, was given by Gauss in [48, Vol. VIII, p. 221]. Thurston, in his book [80, p. 83 ff], gives the same proof. The analogy between the Cavalieri–Wallis and the Gauss–Thurston proofs is accounted for in the article [2].

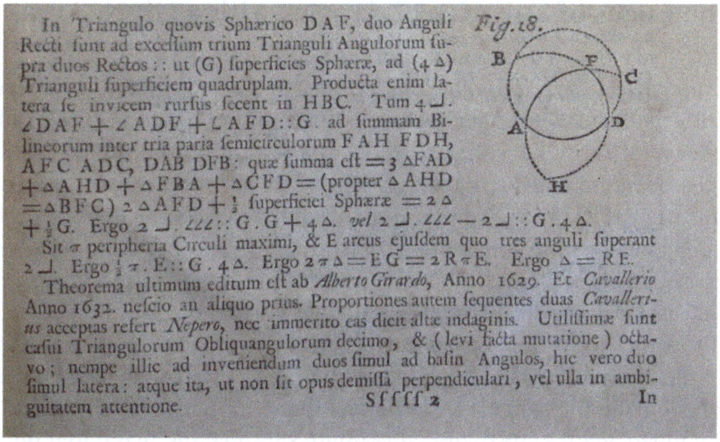

In Triangulo quovis Sphærico DAF, duo Anguli Recti ſunt ad exceſſum trium Trianguli Angulorum ſupra duos Rectos : : ut (G) ſuperficies Sphæræ, ad (4 Δ) Trianguli ſuperficiem quadruplam. Producta enim latera ſe invicem rurſus ſecent in HBC. Tum 4 ∟. ∠DAF + ∠ADF + ∠AFD : : G. ad ſummam Bilineorum inter tria paria ſemicirculorum FAH FDH, AFC ADC, DAB DFB: quæ ſumma eſt = 3 ΔFAD + ΔAHD + ΔFBA + ΔCFD = (propter ΔAHD = ΔBFC) 2 ΔAFD + ½ ſuperficiei Sphæræ = 2 Δ + ½ G. Ergo 2 ∟. ∠∠∠ : : G. G + 4 Δ. *vel* 2 ∟. ∠∠∠ − 2 ∟ : : G. 4 Δ.

Sit π peripheria Circuli maximi, & E arcus ejuſdem quo tres anguli ſuperant 2 ∟. Ergo ½ π. E : : G. 4 Δ. Ergo 2 π Δ = E G = 2 R π E. Ergo Δ = R E.

Theorema ultimum editum eſt ab *Alberto Girardo*, Anno 1629. Et *Cavallerio* Anno 1632. neſcio an aliquo prius. Proportiones autem ſequentes duas *Cavallerius* acceptas refert *Nepero*, nec immerito eas dicit altæ indaginis. Utiliſſimæ ſunt caſui Triangulorum Obliquangulorum decimo, & (levi facta mutatione) octavo; nempe illic ad inveniendum duos ſimul ad baſin Angulos, hic vero duo ſimul latera: atque ita, ut non ſit opus demiſſa perpendiculari, vel ulla in ambiguitatem attentione.

Sſſſſ 2 In

Fig. 2.1 From Wallis' proof of Girard's Theorem ([82], Vol. 2, p. 875), the construction for the proof of Girard's formula. The region FAHDF is a lune

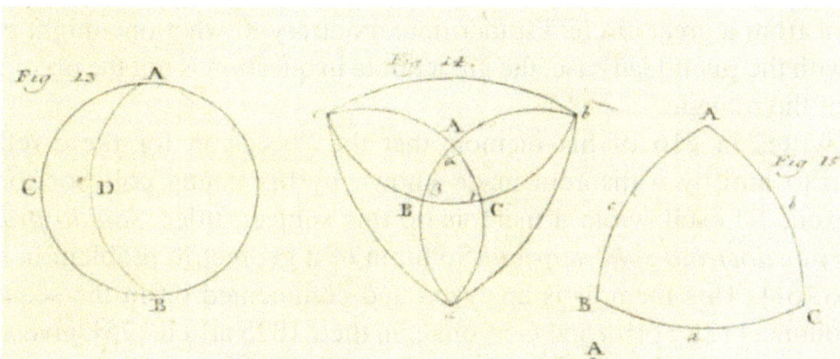

Fig. 2.2 From Euler's memoir [33], the construction for the proof of Girard's formula. The region ACBDA (left) is a lune. In the middle figure, we have three lunes

Bologna in 1632, with the beautiful proof reported on by Wallis and inserted since then in most of the Trigonometries."

After the proof he gives in [33] of the so-called Girard formula, Euler establishes several formulae for the area of a spherical triangle in terms of its side lengths. This is a spherical analogue of the Euclidean Heron formula. This allows Euler to give numerical values for the solid angles of all five regular polyhedra, after he solves the following problem:

> Given a solid angle consisting of n planar angles which are (all) equal to a certain quantity a and which are equally inclined among themselves, to find the measure of the solid angle.

It is conceivable that this is the first time in history where numerical values of the solid angles of the five regular polyhedra are computed. Incidentally, after Euler gives the spherical Heron formula, he shows that it reduces to the usual (Euclidean) Heron formula when the three sides of the spherical triangle become infinitely small.

2.5 The Locus of Spherical Triangles with a Given Area

In the memoir *Variae speculationes super area triangulorum sphaericorum* (Various speculations on the area of spherical triangles) [39], published in 1797, Euler starts by giving several formulae for the area of a spherical triangle as a function of its side lengths. He then considers the problem of finding the locus of the vertices of a spherical triangle with a given base and fixed area. The analogous problem, in the Euclidean case, is the subject of Propositions 37 and 39 of Book I of Euclid's *Elements*. In this case, the desired locus is a straight line. At the same time, Euclid proves that any triangle in this family of triangles can be cut into pieces that can be joined together to form a square having the same area as the initial triangle. We recall that the *Elements* do not contain any definition of area, but only the notion of "two figures having the same area". The fact that two figures can be cut into pieces that can be glued again to form the same figure says that the two initial figures have the same area.

Unlike the Euclidean case, in spherical geometry, the sought locus is not a straight line (i.e. a great circle), but a small circle, that is, a locus of points equidistant from a great circle. Furthermore, contrary to what one might expect by analogy with the Euclidean case, the great circle in question is not the one containing the base of the triangle.

Euler writes in §16 of his memoir that the "occasion for these reflections" was given to him by a theorem made known by his young collaborator Anders Johan Lexell.[11] Lexell wrote a memoir on this subject, titled *Solutio problematis geometrici ex doctrina sphericorum* (Solution of a geometric problem in spherical geometry) [61]. This memoir is analyzed and commented on in the sequel of the present volume, [12]. Sorlin and Gergonne, in their 1825 article [75], give a solution to what is called now Lexell's problem, without quoting Euler, but quoting Lexell and Legendre.

In 1855, Lebesgue published a paper titled *Démonstration du théorème de Lexell* (A proof of Lexell's theorem) [59] in which he gives a proof of that theorem which in fact is Euler's proof. At the end of this paper, the editor of the journal adds a comment saying that Eugène Catalan, in his *Éléments de Géométrie* (Book VII, Problem VII), gives a proof of Lexell's theorem. Catalan does not mention Euler. The only original contribution of Lebesgue in this paper consists in two remarks that he makes after the proof. The first one says that there is an analogous result for surfaces of revolution which admit an equator. The second remark says that by taking the stereographic projection of the figure on the sphere, we get a family of curved triangles in the Euclidean plane that have a given base, whose angle sum is constant, and whose vertices are on the same circle.

[11] Anders Johan Lexell (1740–1784) started working at the Saint Petersburg Academy of Sciences under Euler's guidance and he eventually became an important astronomer and geometer at the Academy. The reader may refer to the articles [83] by Alena Zhukova and [66] by the first author of the present chapter for an account on Lexell's work on spherical geometry.

2.6 Other Memoirs: From Euclidean to Spherical Geometry

We have already noted that the so-called Lexell problem, which Euler studies in spherical geometry, is the spherical version of the result contained in Euclid's *Elements* concerning the locus of the vertices of the triangles having the same base and a fixed area. We also said that in his memoir on the measure of solid angles [33], Euler gives a spherical analogue of the Euclidean Heron formula for the area of a triangle in terms of its side lengths. There are two other memoirs by Euler, translated in the present volume, in which Euler obtains results in spherical geometry that are analogues of results in Euclidean geometry. We shall now discuss them.

In the memoir *Problematis cuiusdam Pappi Alexandrini constructio* (A construction relative to a problem of Pappus of Alexandria), published in 1783 [35], Euler solves, in the Euclidean and then in the spherical setting, a problem inspired by a construction of Pappus,[12] from the *Collection* [69, Proposition 117 of Book VII]. The problem is to construct a triangle which is inscribed in a given circle and whose three sides are contained in three lines that are required to pass through three given points. The construction is non-trivial. Pappus gives this construction in the Euclidean setting and only in the special case where the three given points are collinear.[13]

[12] Pappus of Alexandria was a prominent mathematician who lived around the end of the third century and the first half of the fourth century. His major work, a substantial part of which has survived, is the *Collection* (Synagoge, Συναγωγή), which contains mathematical results, some of which are difficult to prove, covering much of geometry. His name is attached to a theorem concerning the alignment of three points (and which is therefore not the one Euler is talking about), which is at the foundations of projective geometry. Pappus also introduced the notion of anharmonic ratio. His *Collection* contains a host of theorems with detailed proofs (sometimes several proofs), some of which are difficult, which are not due to him, but which would not have come down to us without him. Some are taken from works by Euclid, Archimedes, Heron, Apollonius and others which are completely lost. The *Collection* contains results in Euclidean geometry, spherical geometry, analytic geometry, astronomy, and mechanics, as well as problems of the type that would today be called "recreational mathematics." Jean-Étienne Montucla in [64, t. 1, p. 329] writes that Pappus "gives in his *Mathematical Collection* marks of an uncommon intelligence in geometry, and in several places one finds traces of genius." Among the mathematicians who have immersed themselves in the study of Pappus, at least René Descartes, Thomas Simpson and Michel Chasles should be mentioned. The latter published a work, which he called *Les trois livres des porismes d'Euclide, rétablis pour la première fois d'après la notice et les lemmes de Pappus et conformément au sentiment de R. Simson sur la forme des énoncés de ces proposition* (The three books of Euclid's porisms recovered for the first time according to the notice and lemmas of Pappus and in accordance with the feeling of R. Simson on the form of the statements of these propositions) [16]. The work is entirely drawn from Pappus's *Collection*. Pappus's *Collection* is both a great source of inspiration for mathematicians and an invaluable document on the history of mathematics. The Introduction to Book VII of that work contains a significant amount of unique information on ancient mathematicians and their results.

[13] Pappus' *Collection* contains sections on spherical geometry, but the problem which Euler solves in both the Euclidean and the spherical case was considered by Pappus only in the Euclidean case.

This problem has attracted the attention of several prominent mathematicians. In the same volume of the *Acta* in which Euler's memoir was published, another solution of the same problem, in the spherical setting as well, was given by Nicolaus Fuss,[14] another young collaborator of Euler, in a memoir titled *Solutio problematis geometrici Pappi Alexandrini* (Solution of a geometrical problem of Pappus of Alexandria) [46].[15] An elementary solution by Lagrange of the same problem, based on trigonometric formulae, appeared in the memoir *Solution algébrique d'un problème de géométrie*, whose English translation is contained in Chap. 17 of the present volume, see [57]. Michel Chasles, in his *Aperçu historique sur l'origine et le développement des méthodes en géométrie* [15, pp. 328–329] has a note on the history of this so-called Pappus problem (in the Euclidean setting). According to this account, Euler's proof (in the Euclidean setting) can be traced back to a theorem attributed to the Scottish mathematician Matthew Stewart concerning four points taken arbitrarily on a straight line [76]. We also learn from the same source that in 1742, the Swiss mathematician Gabriel Cramer (1704–1752) proposed this problem in the general (Euclidean) setting of non-collinear points to Johann Castillon (1704–1791) and that the latter published a solution, in 1776, in the *Nouveaux mémoires de l'Académie royale des sciences et belles-lettres de Berlin*. It is interesting to note that both Cramer and Castillon were correspondents of Euler, the first one between 1743 and 1751, and the second one from 1745 to 1765. This correspondence is published in [60]. Chasles also notes that Annibale Giordano (1769–1835), a 16 year old Napolitan, and Gian Francesco Malfatti (1731–1807)[16] solved the more general problem in which a triangle is replaced by a polygon with an arbitrary number of sides passing through an equal number of points. Their works are published in Volume IV of the *Memorie della Societa Italiana delle scienze detta dei XL*. In the *Berlin Memoirs* of 1796, the Swiss mathematician Simon l'Huillier (1750–1840) published a modified algebraic solution of the same general problem. Lazare Carnot (1753–1823), in his *Géométrie de position* [13, p. 383], gave a modified form of Lagrange's solution and applied it to the case of an arbitrary polygon, a case already considered by Giordano and Malfatti. Chasles also mentions solutions to generalizations of Pappus' problem by Charles-Julien Brianchon (1783-1864) in the case where a conic section replaces the circle, and where the three points are aligned, by Joseph Diez Gergonne (1771–1859) in the general case considered by Brianchon and allowing a solution only with straightedge (and no compass), and by Jean-Victor Poncelet (1788–1867), who, in his *Traité des propriétés projectives*

---

[14] Nicolaus (or Nikolai Ivanovich) Fuss (1755–1826), originating from Basel, was sent to Saint Petersburg by Daniel Bernoulli to assist Euler in his works. He later became one of the most brilliant geometers there, with important contributions to spherical geometry. Fuss became later Euler's dear friend and he married his granddaughter, Albertine Benedikte Philippine Luise Euler, the daughter of his son Johann Albrecht, who was also a mathematician.

[15] An English translation of this memoir by Fuss together with commentaries is published in the forthcoming volume [12].

[16] Malfatti's name is attached to the so-called *Malfatti problem* which asks for the construction of three circles inside a triangle that are tangent to each other and to the sides of the triangle.

[70] considers an extension to the case of a polygon with an arbitrary number of sides and where the circle is replaced by an arbitrary conic. A detailed mathematical commentary on Euler's result presented in this memoir, with various approaches and generalizations by other mathematicians, is provided by Guillaume Théret in Chap. 4 of the present volume [78].

Now we pass to the memoir *Geometrica et sphaerica quaedam* (Some questions of the geometry of the plane and of the sphere) [41], which also contains results in Euclidean plane geometry followed by their analogues in spherical geometry. In this memoir, Euler starts by proving the following theorem:

Let ABC be a triangle in the plane and let a, b, c be points on the sides BC, AC, AB respectively. If the lines Aa, Bb, Cc intersect at a common point O, then we have

$$\frac{AO}{Oa} \cdot \frac{BO}{Ob} \cdot \frac{CO}{Oc} = \frac{AO}{Oa} + \frac{BO}{Ob} + \frac{CO}{Oc} + 2.$$

He also gives the following converse:

In the Euclidean plane, given three segments AOa, BOb, COc meeting at a common point O and satisfying the above equation, we can construct a triangle ABC such that the points a, b, c are on the sides BC, AC, AB respectively.

In the Euclidean case, Euler derives several different proofs of the theorem and its converse. He also considers the case where the point O lies outside the triangle. In the spherical case, he obtains an analogous result. The new condition is:[17]

$$\frac{\tan AO}{\tan Oa} \cdot \frac{\tan BO}{\tan Ob} \cdot \frac{\tan CO}{\tan Oc} = \frac{\tan AO}{\tan Oa} + \frac{\tan BO}{\tan Ob} + \frac{\tan CO}{\tan Oc} + 2.$$

Thus, to pass from the Euclidean to the spherical case, the equation satisfied is the same up to replacing the lengths by their tangents. Mathematicians are fond of such analogies, and they seek to understand the reasons for them. Euler gives two proofs of this spherical result, one of them based on spherical trigonometry, and the second one using a projection of the triangle onto a plane which is tangent to the sphere at the point O, reducing the problem to the Euclidean one, which he considered before.

It seems that the questions studied in this memoir have not been studied before Euler, neither from the Euclidean nor from the spherical point of view. In the paper [68] written with Weixu Su, the first author of the present chapter has worked out the hyperbolic case, where a similar condition holds, with the tangent function replaced by the hyperbolic tangent, that is, the condition reads:

$$\frac{\tanh AO}{\tanh Oa} \cdot \frac{\tanh BO}{\tanh Ob} \cdot \frac{\tanh CO}{\tanh Oc} = \frac{\tanh AO}{\tanh Oa} + \frac{\tanh BO}{\tanh Ob} + \frac{\tanh CO}{\tanh Oc} + 2.$$

[17] To avoid dealing with special cases, we are assume that the spherical triangle is contained in a quarter sphere.

To end this section, and since we are talking about Euclidean geometry, let us mention that Euler wrote several memoirs in which he obtained a large number of results in this domain. Among these memoirs, we mention the *Solutio problematis geometrici circa lunulas a circulis formatas* Solution of a geometric problem on the lunules made by circles) [23], in which he gives a solution to a problem formulated by Daniel Bernoulli in his *Exercitationes quaedam mathematicae* (Some mathematical exercises) [6] (1724),[18] related to the question of dissecting circles into pieces that have equal area. It is interesting to read Euler's comments on the solution he presents:

> [The solution] is not only presented geometrically without analysis, but analysis is also considered less suitable for its solution. This disadvantage of analysis, although commonly resorted to in many geometric problems, nevertheless seems to me to be attributable less to analysis than to the analyst. In this problem, I clearly prove that not only is analysis unsuitable for its solution, but that the geometric method is far preferable. Using this method, I will present a general solution to this problem, which can be obtained by a geometric method.

One may compare this statement with the fact that a few years later, and concerning other kinds of problems, Euler would seek to replace geometric arguments with analytical ones, and it was in the person of Lagrange that he found someone who could do it. The latter's ideas simplified Euler's approach in that they avoided tricky geometric arguments by replacing them with analytical ones, which led directly to what later came to be called the Euler–Lagrange equation. Euler expressed this in a letter dated September 6, 1755, in which he wrote to Lagrange:

> After reading your last letter, in which you seem to have brought the theory of *maxima* and *minima* almost to its highest point of perfection, I cannot sufficiently express my admiration for the extraordinary sagacity of your mind. For not only had I called for in my treatise on this subject a purely analytical method which would allow the rules set out therein to be deduced, but I had subsequently devoted much effort to the search for such a method: and you have truly given me great joy by having the kindness of sharing with me your very profound and very solid reflections on these questions; this is why I acknowledge my debt to you.

Among Euler's other memoirs on Euclidean geometry, we mention the *Solutio facilis problematum quorundam geometricorum difficillimorum* (Easy solutions to some difficult geometrical problems) [30] and *Variae demostrationes geometriae* (Various geometric demonstrations) [25]. In the latter, he proves several elementary results involving the perimeter, area and radius of inscribed circles and other characteristics of Euclidean figures. For example, one result says that the area of a Euclidean triangle is equal to the product of half the sum of its side lengths with the radius of the inscribed circle.

In the memoir *Solutio problematis geometrici* (Solution of a problem of geometry) [26], Euler presents his solution to another problem of Pappus, namely, given two conjugate diameters of an ellipse, to find the conjugate axes. (This is Problem

[18] Euler says that this problem was proposed by Christian Goldbach.

XVII of Book VIII of the Collection, p. 851 of Ver Eecke's edition [69].) The problem is also quoted in T. L. Heath's, *History of Greek Mathematics* [54, Vol. II, p. 437–438]. Talking about Heath, let us also point out a note in his edition of Euclid's *Elements* [55, Vol. 1 p. 402] in which he quotes the following theorem of Euler: *In any quadrilateral the sum of the squares on the sides is equal to the sum of the squares on the diagonals and four times the square on the line joining the middle points of the diagonals.* Euler published several other elementary (but not necessarily easy to prove) results on Euclidean geometry and it would be interesting to work out the analogous statements on the sphere and in the hyperbolic plane.

2.7 Geography, Astronomy and Geomagnetism

Spherical geometry has applications in geography, astronomy and other fields. Chasles writes in his *Aperçu historique* [15, p. 235] that since the works of the ancients (he mentions Theodosius, Menelaus and Ptolemy), "if the theory [of spherical geometry] was extended and if it has attained, in the hands of our most celebrated geometers, a high degree of perfection, this was always done by almost preserving the same framework, because the goal was always the same: the computation of triangles in order to fit the service of the astronomer and the navigator, and for the great geodesic operations which showed the true form of the Earth." Euler was interested in all domains of applications of mathematics, in particular in geography, astronomy and geomagnetism, which all involve spherical trigonometry. We recall incidentally that at some point, Euler was the head of the geography department at the Russian Academy of Sciences in Saint Petersburg.

As already noted in Sect. 2.3, Euler, in his memoir *Principes de la trigonométrie sphérique tirés de la méthode des plus grands et plus petits* [27], used geographical terminology while he established his trigonometric formulae. In the introduction of this memoir, he talks about using the same methods for the study of the spheroid, "the surface of the Earth, being not spherical, but spheroidal", and announcing his forthcoming memoir *Élémens de la trigonométrie sphéroïdique tirés de la méthode des plus grands et plus petits* [28] in which he establishes, by the same methods, the trigonometric formulae for the surface of the spheroid and from which he deduces explicit distances between locations on the Earth.

The main question in the problem of drawing geographical maps is to find maps of least distortion (with an appropriate meaning for this word) from a subset of the sphere into the plane. The volume [11] contains memoirs by Euler, Lagrange and Lambert on this theory, translated into English and with commentaries. In Euler's memoirs on this subject, the geometry of the sphere, and in particular spherical trigonometry, plays a major role. Astronomy is also linked to spherical geometry, because the celestial vault is assimilated to a sphere, and making a map of the sky and studying the movements of the stars amount to questions of spherical geometry. In any case, Euler's memoirs on astronomy involve some amount of spherical trigonometry. For example, his memoir *Solutio problematis astronomici ex datis tribus stellae fixae altitudinibus et temporum differentiis invenire elevationem poli et declinationem stellae* (Solution to problems of astronomy: given the altitudes and time differences for three fixed stars, to find the elevation of the pole and the declination of the star) [22], published long before his memoirs on spherical trigonometry, consists almost entirely of considerations of spherical trigonometry, with some numerical examples in the last part. The goal of this memoir is to solve the problem announced in the title. Euler starts by recalling the following result which he attributes to Friedrich Christoph Mayer:[19] In any spherical triangle ABC, we have

$$\cos A = \frac{\cos BC - \cos AB \cdot \cos AC}{\sin AB \cdot \sin AC}.$$

He then proves the following statement, quoted as a theorem.

In any spherical triangle ABC, *we have*

$$\cos BC = \frac{\cos(AB + AC) - \cos(AB - AC)}{2} + \frac{\cos A \cdot \cos(AB - AC) - \cos A \cdot \cos(AB + AC)}{2}.$$

Likewise, the memoir *De inventione longitudinis locorum ex observata lunae distantia a quadam stella fixa cognita* (On finding the longitude of a place by observing the distance between the moon and a known fixed star) [32] contains a substantial part on spherical trigonometry. There are many other examples.

On geomagnetism, we mention Euler's memoir *Recherches sur la déclinaison de l'aiguille aimantée* (Researches on the declination of the magnetized needle) [29]. The (magnetic) declination at a given point on the Earth's surface is the angle between the magnetic north and true north. In this memoir, Euler mentions the work of Edmund Halley, who proposed a new idea for showing the declination on geographical maps, see [53]. The memoir [29] starts by the following: "Since this research, as well as the following ones, requires the analytical resolution of spherical triangles, it will be good to put the formulae before our eyes, so that we do not need to look for them elsewhere. I will therefore begin with right triangles;

[19] Friedrich Christoph Mayer (1762–1841) was a mathematician who, like Euler, was working at the Imperial Academy of Sciences in Saint Petersburg. From the time of Euler's arrival to this Academy, in 1727, Mayer helped him with various problems of celestial mechanics applied to the determination of the sun's orbit, the movement of the planets and the calculation of lunar eclipses. In addition, Mayer worked with Jakob Hermann and Daniel Bernoulli on a theory of the moon based on Delisle's extensive programme of astronomical observation, which included the occultations of stars and planets by the moon, as well as lunar and solar eclipses. At that time, Delisle was the major astronomer and geographer of the Academy. Mayer's works on spherical trigonometry include the following: *Trigonometrica*, published in the *Commentarii academiae scientiarum Petropolitanae* 2, 1729, p. 12–30, and *Problema trigonometricospahericum*, published in the *Commentarii academiae scientiarum Petropolitanae* 4, 1735, p. 31–37.

etc." The memoir, which is 77 pages long, contains a complete set of formulae of spherical trigonometry (without proofs), that are used all along.

Finally, let us mention that spherical geometry is the basis of an argument that is the subject of a letter sent by Euler to Maupertuis dated July 4, 1744 (letter No. 7, p. 49 of [42]). The argument is analyzed in Chap. 4, titled *On the Duration of the Passage of a Star from an Almucantar to Another* in volume [11], see [14].

2.8 Curves on the Sphere, the Spheroid, etc.

The works mentioned in this section are not on spherical geometry, but we have collected them here because they involve the sphere. Indeed, Euler studies in these memoirs curves on a sphere embedded in the 3-dimensional Euclidean space.

In the memoir *De curva rectificabili in superficie sphaerica* (On rectifiable curves on the surface of the sphere) [31] (1771), Euler studies curves on the sphere that are expressible by algebraic equations.[20] He says that this problem arises from what he calls the "great Florentine problem [. . .] raised on a spherical surface a century ago." He finds such a curve, namely, the "spherical epicycloid", obtained by the motion on a great circle of a small circle whose diameter makes a rational ratio with the diameter of the sphere.

From this question regarding the surface of the sphere, he moves on to more general surfaces.

In the memoir *De lineis rectificabilibus in superficie sphaeroidica quacunque geometrice ducendis* (On the rectifiable lines that have to be drawn geometrically on a spheroidal surface) [37] published the same year, Euler gives a simpler proof of the result of the preceding one [31], at the same time generalizing it from the case of the sphere to the case of a spheroid. In the introduction, he points out the difficulty of the question of drawing geometrically algebraic lines on a surface. He writes: "A most difficult question is when it is required to draw geometrically lines on the surface of any body, with all of whose arcs defined algebraically; since in cylinders, besides straight lines parallel to the axis, no other rectifiable curves can be drawn."

In the memoir *De curvis rectificabilibus in superficie coni recti ducendis* (On rectifiable curves drawn on the surface of a right cone) [36], published in 1781, Euler finds a family of rectifiable curves on a right cone whose ratio of side-length to the diameter of the base is rational. Let us quote from the introduction:

> Since we can certainly affirm that no rectifiable line can be drawn on a cylindrical surface except the straight lines themselves parallel to the axis of the cylinder, the same seems to be true of conical surfaces, because a cylinder can be viewed as a sort of cone, while the height is increased infinitely. However, I have observed that there are such right cones on the surface of which, in addition to the straight lines drawn from the vertex of the cone to the

[20] Euler calls these curves "rectifiable", a word used today in a different meaning.

> circumference of the base, innumerable other curved lines can be geometrically described, admitting rectifications; but this does not seem to be possible on all right cones, much less on oblique ones.

2.9 Another Problem on Spheres

In this section, we consider Euler's memoir *Solutio facilis problematis, quo quaeritur sphaera, quae datas quatuor sphaeras utcunque dispositas contingat* (An easy solution to the problem of finding a sphere that touches four given spheres, arranged in an arbitrary way) [40]. This is a three-dimensional analogue of the famous Apollonius' problem in the plane.[21] Like the problems considered in the preceding section, strictly speaking, Euler's memoir is not on spherical geometry, nevertheless it concerns spheres. Euler presents two solutions to the problem of constructing a sphere that is tangent to four given spheres. The proofs rely on trigonometric formulae. Towards the end of this memoir, Euler begins using astrological signs as variables, see Fig. 2.3.[22]

At the end of the memoir, Euler writes:

> In this way, we find a pair of solutions to a very difficult problem, which at first sight, seemed to require the most abstruse stereometric investigations, problems of this kind usually requiring both the most intricate figures and the most troublesome calculations, while the solutions given here can be obtained with the help of not too lengthy calculations. The problem itself is not new, but was once found to have been solved by the great geometer Fermat; but since at that time the calculation of angles was almost completely unknown, it is not surprising that our solution is found to be much more convenient.

There is a relationship between the two-dimensional Apollonius' problem, which is at the origin of the problem discussed in this memoir, and the problem of Pappus, which is considered by Euler in the memoir [35] and on which we commented in Sect. 2.6. Guillaume Théret mentions this relationship in his commentary on Euler's memoir, [78], Chap. 4 of this volume. The reconstruction of Apollonius' proof and

[21] The reference is to Apollonius of Perga (ca. 262–190 BC). The problem, in its simplest form, is to construct (using straightedge and compass) a circle that is tangent to three given circles in a plane. Apollonius' problem is part of his work ἐπαφαί ("Contacts", or "Tangencies"), which does not survive but which is reported on by Pappus in his *Collection*. A detailed content of what is known about this work of Apollonius, and a history of the attempts made for its reconstruction, are given by Ver Eecke, the Belgian editor of Pappus' *Collection*, see [69, Introduction, p. LXVI ff.] We learn from there that Apollonius' work on this problem presumably contained twenty-one lemmas, sixty theorems and eleven problems, which Pappus, in his *Collection*, summarizes in a unique proposition saying that given three arbitrary elements which could be points, lines or circles, to construct a circle which passes by the points (in the case where the objects are points) or is tangent to the lines or circles. Pappus enumerates ten cases in which Apollonius solved the problem.

[22] Lambert used these signs in his memoir on taxeometry (the science of measure of order), see Chap. 6 of the present volume, and the figure reproduced there.

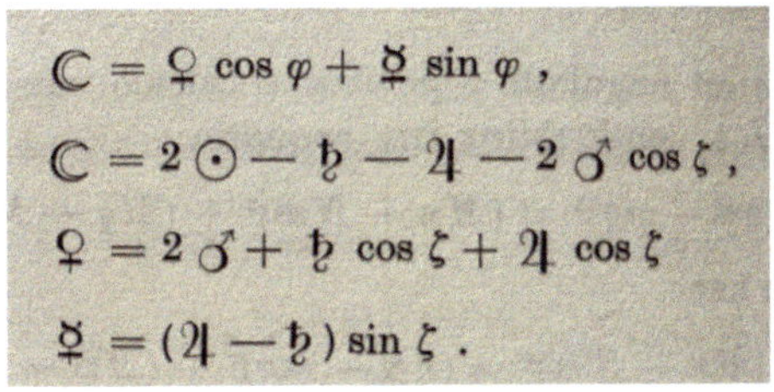

$$☾ = ♀ \cos\varphi + ☿ \sin\varphi\,,$$
$$☾ = 2☉ - ♄ - ♃ - 2♂ \cos\zeta\,,$$
$$♀ = 2♂ + ♄ \cos\zeta + ♃ \cos\zeta$$
$$☿ = (♃ - ♄) \sin\zeta\,.$$

Fig. 2.3 A plate from Euler's memoir *Solutio facilis problematis, quo quaeritur sphaera, quae datas quatuor sphaeras utcunque dispositas contingat*

more generally the work on this problem attracted the attention of the greatest geometers such as Viète, Descartes, Newton, Poncelet, Gergonne, Cauchy and, as we said, Euler in his memoir *Solutio facilis problematis, quo quaeritur circulus, qui datos tres circulos tangat* [38]. The three-dimensional version, treated by Euler in this memoir, was also considered by Fermat in his *De contactibus sphaericis*, see [43, p. 52–69].

On this higher-dimensional Apollonius problem, besides Euler's memoir [38], we mention the works of Carnot [13], Hachette [51, 52], Binet [8], Battaglini [5], Français [44, 45], Dupin [20], Serret [74], Coaklay [17] and Alvord [4]. Chasles, in his *Aperçu historique sur l'origine et le développement des méthodes en géométrie* [15, p. 66], writes about Euler's work on the three-dimensional Appollonius problem:

> The question of the sphere tangent to four others is one in which Geometry has long had the upper hand over Analysis. In 1779, Euler presented two analytical solutions of it to the Academy of Petersburg, which did not appear until the beginning of this century, in the Volume of Academy for the years 1807–1808 (printed in 1810). Carnot had already indicated an analytical solution in his *Géométrie de position* (p. 416), but without carrying out the developments that would have led to a second-degree equation. Today, it is M. Poisson who was the first to completely resolve this question by calculation (Bulletin de la Société philomatique, ann. 1812, p. 141.) Shortly afterwards, MM. Binet and Français also gave different analytical solutions (17th issue of the Journal de l'école polytechnique, and tome III of the Annales de Mathématiques).

Naturally, Apollonius' problem in the plane has been extended to a similar problem on the sphere and on other quadratic surfaces. The question on the sphere is to construct all spherical circles that are tangent to three given circles, see [13, 47, 81].

To conclude this section, let us say that Euler has studied the classical (2-dimensional) Apollonius problem in his memoir *Solutio facilis problematis, quo quaeritur circulus, qui datos tres circulos tangat* (An easy solution of a problem, in which a circle is searched for, tangent to three given circles) [38], written in 1779 and published in 1810. This problem has been generalized to the case of $n > 3$ circles and to the case where some of the circles are replaced by lines or by conics, to the case of intersecting circles, etc. There is a very large bibliography on generalized Apollonius problems. Among the mathematicians who worked on this topic, we mention again Pappus, then Descartes, Newton, Gauss, Cauchy, Poncelet, Gergonne, Laguerre, and there are many others.

Jean-Alfred Serret, in his paper *De la sphère tangente à quatre sphères données* (On the sphere tangent to four given spheres) [74], solves the problem of a sphere tangent to four given spheres using the notion of power of a point with respect to a sphere (the square of the distance from the center minus the square of the radius). Gaston Darboux, in his paper *Sur les relations entre les groupes de points, de cercles et de sphères dans le plan et dans l'espace* (On the relationships between groups of points, circles and spheres in the plane and in space) [18] solves the problem of constructing a circle intersecting three given circles at given angles. In his paper, Darboux says that the problem has been proposed by Jakob Steiner.

Among the modern works around generalized Apollonius' problems, we cannot help but think of the theorem of the Koebe–Andreev–Thurston theorem, see [79, Chapter 13] and the review in [9].

Finally, talking about Euler's work on spherical geometry, we should mention the works on this topic done by the young mathematicians of the Saint Petersburg Academy of Sciences who worked under his direction or who were inspired by him. We have devoted a forthcoming volume to this subject, [12]. The first author of the present chapter has also published a short overview of this topic [66].

References

1. A. A'Campo-Neuen, The pentagrammma mirificum from Napier to Lambert. Notes on Lambert's memoir on spherical trigonometry, in *Spherical Geometry in the Eighteenth Century, I: Euler, Lagrange and Lambert*, ed. by R. Caddeo, A. Papadopoulos, Chap. 8 (Springer, Berlin, 2026)
2. N. A'Campo, A. Papadopoulos, Area in non-Euclidean geometry, in *Eighteen Essays on Non-Euclidean Geometry*, ed. by V. Alberge, A. Papadopoulos (European Mathematical Society, Helsinki, 2019), pp. 3–25
3. A.D. Aleksandrov, V.A. Zalgaller, *Intrinsic Geometry of Surfaces* (The Academy of Sciences of the USSR, Moscow, 1962). English translation by J. M. Danskin, AMS Translations of Mathematical Monographs, Amercian Mathematical Society, Providence (1967)
4. B. Alvord, The intersection of circles and intersection of spheres. Am. J. Math. **5**(1), 25–44 (1882)
5. G. Battaglini, Soluzione di un problema di geometria a tre coordinate: Descrivere una sfera in modo che intersechi quattro altre sfere ad angoli dati. Annali di Scienze Matematiche e fisiche, Roma **2**, 373–380 (1851)
6. D. Bernoulli, *Exercitationes Quaedam Mathematicae* (Domenico Louvisa, Venice, 1724)
7. J. Bertrand, *Calcul Différentiel et Calcul Intégral* (Gauthier-Villars, Paris, 1864–1870)
8. J.P.M. Binet, Sur la détermination analytique d'une sphère tangente à quatre autres. J. de l'École Polytech. Paris **10**, 113–128 (1815)
9. P.L. Bowers, Combinatorics encoding geometry: the legacy of Bill Thurston in the story of one theorem, in *In the Tradition of Thurston. Geometry and Topology*, ed. by K. Ohshika, A. Papadopoulos (Springer, Cham, 2020), pp. 173–239
10. H. Busemann, Spaces with non-positive curvature. Acta Math. **80**, 259–311 (1948)
11. R. Caddeo, A. Papadopoulos, *Mathematical Geography in the Eighteenth Century: Euler, Lagrange and Lambert* (Springer, Cham, 2022)
12. R. Caddeo, A. Papadopoulos, *Spherical Geometry in the Eighteenth Century, II: Fuss, Lexell and Schubert* (Springer, Berlin, 2026)

13. L.N.M. Carnot, *Géométrie de Position* (J. B. M. Duprat, Paris, 1803)
14. Ch. Charitos, On the duration of the passage of a star from an almucantar to another, in *Mathematical Geography in the Eighteenth Century: Euler, Lagrange and Lambert*, ed. by R. Caddeo, A. Papadopoulos (Springer, Cham, 2022), pp. 81–88
15. M. Chasles, Aperçu historique sur l'origine et le développement des méthodes en géométrie, in *Mémoires couronnés par l'Académie royale des sciences et belles-lettres de Bruxelles*, tome XI (Hayez, Bruxelles, 1837)
16. M. Chasles, *Les trois livres de porismes d'Euclide, rétablis pour la première fois d'après la notice et les lemmes de Pappus et conformément au sentiment de R. Simson sur la forme des énoncés de ces propositions* (Mallet-Bachelier, Paris, 1860)
17. G.W. Coaklay, Analytical Solutions of the Ten Problems in the Tangencies of Circles; and also of the Fifteen Problems in the Tangencies of Spheres. Math. Month. **2**, 116–126 (1859–1860)
18. G. Darboux, Sur les relations entre les groupes de points, de cercles et de sphères dans le plan et dans l'espace. Ann. Sci. l'É.N.S. 2e Série, t. **1**, 323–392 (1872)
19. J.-B. Delambre, *Rapport historique sur les progrès des sciences mathématiques depuis 1789 et sur leur état actuel* (Courcier, Paris, 1810)
20. C. Dupin, Mémoire sur les sphères. Correspondance sur l'école Impériale Polytechnique **2**(5), 423 (1813)
21. L. Euler, De linea brevissima in superficie quacunque duo quaelibet puncta jungente. Comment. Acad. Sci. Petrop. **3**, 110–124 (1732); *Opera Omnia* Series 1, vol. XXV, p. 1–12
22. L. Euler, Solutio problematis astronomici ex datis tribus stellae fixae altitudinibus et temporum differentiis invenire elevationem poli et declinationem stellae. Comment. Acad. Sci. Petrop. **4**, 98–101 (1735). Opera Omnia, Series 2, Vol. 30, pp.1–4
23. L. Euler, Solutio problematis geometrici circa lunulas a circulis formatas. Comment. Acad. Sci. Petropol. **9**, 207–221 (1737). Opera Omnia Series 1, Volume 26, pp. 1–14
24. L. Euler, *Methodus inveniendi lineas curvas maximi minimive proprietate gaudentes, sive solutio problematis isoperimetrici latissimo sensu accepti.* Première édition, 1744; Opera Omnia Series 1, vol. XXIV (Apud Marcum-Michaelem Bousquet & Socios, Geneva, 1744)
25. L. Euler, *Variae demostrationes geometriae.* Novi Commentarii academiae scientiarum Petropolitanae, vol. 1 (1950), pp. 49–66. Opera Omnia, Series 1, Volume 26, p. 15–32
26. L. Euler, *Solutio problematis geometrici.* Novi Commentarii academiae scientiarum Petropolitanae, vol. 3 (1753), pp. 224–234. Opera Omnia, Series 1, Volume 26, pp. 60–70
27. L. Euler, *Principes de la trigonométrie sphérique tirés de la méthode des plus grands et des plus petits.* Mémoires de l'Academie des Sciences de Berlin, vol. 9, 233–257 (1755). Opera Omnia, Series 1, vol. XXVII, pp. 277–308
28. L. Euler, *Élémens de la trigonométrie sphéroïdique tirés de la méthode des plus grands et des plus petits.* Mémoires de l'Academie des Sciences de Berlin, vol. 9 (1755), pp. 258–293; Opera Omnia, Series 1, vol. XXVII, pp. 309–339
29. L. Euler, Recherches sur la déclinaison de l'aiguille aimantée. Mémoires de l'académie des sciences de Berlin, vol. 13 (Birkhäuser, Basel1757), pp. 175–251. Opera Omnia, Series 3, volume 10, pp. 261–343
30. L. Euler, Solutio facilis problematum quorundam geometricorum difficillimorum. Novi Comment. Acad. Sci. Imp. Petropol. **11**, 103–123 (Birkhäuser, Basel, 1767). Opera Omnia, Series 1, Volume 26, pp. 139–157
31. L. Euler, De curva rectificabili in superficie sphaerica. Novi Comment. Acad. Sci. Imp. Petropol. **15**, 195–216 (1771). Opera Omnia: Series 1, Volume 28, pp. 142–160
32. L. Euler, De inventione longitudinis locorum ex observata lunae distantia a quadam stella fixa cognita. Acta Acad. Sci. Imper. Petropol. **2**, 301–307 (1780). Opera Omnia, Series 2, Volume 30, pp. 334–338
33. L. Euler, De mensura angulorum solidorum. Acta Acad. Sci. Imper. Petropol. **2**, 31–54 (1781); Opera Omnia, Series 1, vol. XXVI, pp. 204-223. English translation, this volume, Chapter 14
34. L. Euler, Trigonometria sphaerica universa, ex primis principiis breviter et dilucide derivata. Acta Acad. Sci. Imper. Petropol. **3**, 72–86 (1782); Opera Omnia Series 1, vol. XXVI, pp. 224–236

35. L. Euler, Problematis cuiusdam Pappi Alexandrini constructio. Acta Acad. Sci. Imper. Petropol. **4**, 91–96 (1783); Opera Omnia Series 1, vol. XXVI, pp. 237–242
36. L. Euler, De curvis rectificabilibus in superficie coni recti ducendis. Acta Acad. Sci. Imper. Petropol. **1781**, 60–73 (1784); Opera Omnia: Series 1, Volume 28, pp. 336–347
37. L. Euler, De lineis rectificabilibus in superficie sphaeroidica quacunque geometrice ducendis. Nova Acta Acad. Sci. Imper. Petropol. **3**, 57–68 (1788). Opera Omnia: Series 1, Volume 29, pp. 112–125
38. L. Euler, Solutio facilis problematis, quo quaeritur circulus, qui datos tres circulos tangat. Nova Acta Acad. Sci. Imper. Petropol. **6**, 95–101 (1790). Opera Omnia, series 1, volume 26, pp. 270–275
39. L. Euler, Variae speculationes super area triangulorum sphaericorum. Nova Acta Acad. Sci. Imper. Petropol. **10**, 47–62 (1797); Opera Omnia, Series 1, vol. XXIX, pp. 253–266
40. L. Euler, Solutio facilis problematis, quo quaeritur sphaera, quae datas quatuor sphaeras utcunque dispositas contingat. Mém. de l'académie Impér. de St-Pétersbourg **2**, 17–28 (1810). Opera Omnia, Series 1, Volume 26, pp. 334–343
41. L. Euler, Geometrica et sphaerica quaedam. Mém. de l'académie Impér. de St-Pétersbourg **5**, 96–114 (1815); Opera Omnia, Series 1, vol. XXVI, pp. 344–358
42. L. Euler, *Opera Omnia*, Ser. 4, Vol. VI, Correspondance d'Euler avec P.-L. M. de Maupertuis et Frédéric II (Birkhäuser Verlag, Basel, 1986)
43. P. de Fermat, *Œuvres de Fermat, Texte établi par Paul Tannery*, vol. 1 (Gauthier-Villars, Paris, 1891), pp. 52–69
44. J. Français, De la sphère tangente à quatre sphères données. Correspond. sur l'école Impér. Polytech. **2**(2), 63–66 (1810)
45. J. Français, Solution analytique du problème de la sphère tangente à quatre sphères données. Correspond. sur l'école Impér. Polytech. **2**(5), 409–410 (1813)
46. N. Fuss, Solutio problematis geometrici Pappi Alexandrini. Acta Acad. Sci. Imper. Petropol. **1**, 97–104 (1783)
47. J.D. Gergonne, Recherche du cercle qui en touche trois autres sur une sphère. Ann. Math. Pures Appl. **4**, 349–359 (1813–1814)
48. C.F. Gauss, *Werke, Königliche Gesellschaft der Wissenschaften, Göttingen* (The Bavarian State Library, Munich, 1900)
49. M. Gromov, Hyperbolic groups, Essays in group theory. Publ. Math. Sci. Res. Inst. **8**, 75–263 (1987)
50. Gua (Abbé de), Trigonométrie sphérique, déduite très brièvement et complètement de la seule solution algébrique du plus simple de ses problèmes, au moyen de diverses transformations dont les rapports des sinus et cosinus, tangentes et cotangentes, sécantes et cosécantes d'un même arc ou d'un même angle-plan rendent cette solution susceptible, et comprenant quelques formules et observations qu'on croit utiles et neuves. Mém. l'Académie R. Sci. Paris 291–343 (1783)
51. J.P.N. Hachette, Sur le contact des sphères; sur la sphère tangente à quatre sphères données; sur le cercle tangent à trois cercles donnés. Correspond. l'école Polytech. **1**(2), 27–28 (1808)
52. J.P.N. Hachette, Déterminer le centre et le rayon d'une sphère qui touche quatre sphères données (solution analytique). J. l'école Polytech. Paris **10**, 129–136 (1815)
53. E. Halley, An account of the cause of the change of the variation of the magnetic needle; with an hypothesis of the structure of the internal parts of the earth. Philos. Trans. R. Soc. London **16**, 179–191, 470-478 (1692)
54. T.L. Heath, *History of Greek Mathematics*, 2 volumes (Cambridge University Press, Cambridge, 1921). Reprint, Dover, New York, NY, 1981
55. T.L. Heath, *The Thirteen Books of Euclid's Elements*, 3 volumes, translated from the text of Heiberg with introduction and commentary (Cambridge University Press, Cambridge, 1926). Reprint, Dover
56. S.-F. Lacroix, *Traité élémentaire de trigonométrie rectiligne et sphérique et d'application de l'algèbre à la geometrie* (Courcier, Paris, 1807)

57. J.-L. Lagrange, *Solution algébrique d'un problème de géométrie*. Nouveaux Mémoires de l'Académie royale des Sciences et Belles-Lettres de Berlin (1776). Translated in the present volume, Chapter 17
58. J.-L. Lagrange, Solution de quelques problèmes relatifs aux triangles sphériques avec une analyse complète de ces triangles. J. l'école Polytech. Paris **2**, 270–297 (1800). English translation, Chapter 18, this volume
59. H. Lebesgue, Démonstration du théorème de Lexell. Nouvelles Ann. Math. **14**(1), 24–26 (1855)
60. Leonhardi Euleri Opera Omnia IVA/7, in *Commercium Epistolicum*, ed. by S. Bodenmann, V. Hug, M. Ilić, A. Kleinert. Correspondence with French speaking scientists from Switzerland (Basel, Birkhäuser, 2017)
61. A.J. Lexell, Solutio problematis geometrici ex doctrina sphaericorum. Acta Acad. Sci. Imper. Petropol. **5**(1), 112–126 (1781)
62. N.I. Lobachevsky, On the elements of geometry (in Russian). Kazansky Vestnik **25**, 178–187 (1829). 228–241; 27 (1829), 227–243; 28 (1830), 251–283, 571–683. Reproduced in Lobachevsky's Collected geometric works, Vol. I, 1-67. Complete works [104], Vol. I, 185-261, with a introduction and a commentary by A. P. Kotelnikov An abridged Russian version was published in St. Petersburg, 1908. German translation by F. Engel in Nikolaj Iwanowitsch Lobatschefskij. Zwei geometrische Abhandlungen aus dem Russischen übersetzt mit Anmerkungen und mit einer Biographie desVerfassers. Teubner, Leipzig, 1898
63. N.I. Lobachevsky, Geometrische Untersuchungen zur Theorie der Parallellinien (Geometrical researches on the theory of parallels). Kaiserl. russ. wirkl. Staatsrathe und ord. Prof. der Mathematik bei der Universität Kasan, in der Finckeschen Buchhandlung, Berlin, 1840. French translation by J. Hoüel, Etudes géométriques sur la théorie des parallèles, Mémoires de la Société de Sciences Physiques et Naturelles de Bordeaux IV (1866), pp. 83–182, and Gauthier-Villars, Paris, 1866; reprinted by Hermann, Paris, 1900. Russian translation by A. V. Letnikov, Matematicheski i Sbornik (Mathematical Collection) III (1868), 78–120; also published as a one-volume Russian edition in 1945. English translation by G. B. Halsted, Geometrical researches on the theory of parallels, first published by the University of Texas at Austin, 1891. Reprinted in Bonola, La geometria non-euclidea. Esposizione storico-critica del suo sviluppo. First edition, Ditta Nicola Zanichelli editore, Bologna, 1906. English translation by H. S. Carslaw, Non-Euclidean geometry, A critical and historical study of its development. First edition, Chicago, 1912, reprinted by Dover, New York, 1955
64. J.-F. Montucla, *Histoire des mathématiques*, 4 volumes (H. Agasse, Paris, 1799–1802)
65. J. Napier, Mirifici logarithmorum canonis descriptio, ejusque usus, in utraque trigonometria; ut etiam in omni logistica mathematica, amplissimi, facillimi, & expeditissimi explicatio, Edinburgi, Ex officinâ Andreæ Hart (1614)
66. A. Papadopoulos, On the works of Euler and his followers on spherical geometry, *Gaṇita Bhāratī* (Indian Mathematics). Bullet. Ind. Soc. History Math. **36**(1–2), 237–292 (2014)
67. A. Papadopoulos, The polar triangle, from Ibn 'Irāq to Euler and Lagrange, in *Spherical Geometry in the Eighteenth Century, I: Euler, Lagrange and Lambert*, ed. by R. Caddeo, A. Papadopoulos, Chap. 10 (Springer, Berlin, 2026)
68. A. Papadopoulos, W. Su, On hyperbolic analogues of some classical theorems in spherical geometry, in *Hyperbolic Geometry and Geometric Group Theory*, ed. by K. Fujiwara, K. Ohshika, S. Kojima. Advanced Studies of Pure Mathematics, vol. 73 (Mathematical Society of Japan, Tokyo, 2017), pp. 225–253
69. Pappus, La *Collection mathématique de Pappus d'Alexandrie, traduit du grec en français avec une introduction et des notes par P. Ver Eecke* (Desclée de Brouwer, Paris/Bruges, 1933); repr. Paris, A. Blanchard, 1982
70. J.-V. Poncelet, *Traité des propriétés projectives des figures: ouvrage utile à ceux qui s'occupent des applications de la géométrie descriptive et d'opérations géométriques sur le terrain* (Gauthier-Villars, Paris, 1865–1866)

71. L. Puissant, *Traité de geodésie, ou exposition des méthodes trigonométriques et astronomiques, applicables à la mesure de la terre et à la construction du canevas des cartes topographiques* (Bachelier, Paris, 1842)
72. R. Rashed, A. Papadopoulos, *Menelaus' Spherics: Early Translation and al-Māhānī/al-Harawī's Version.* Critical edition and mathematical commentary with English translation, De Gruyter, Series: Scientia Graeco-Arabica, vol. 21 (De Gruyter, Berlin, 2017), 890pp.
73. B.A. Rosenfeld, *History of Non-Euclidean Geometry.* Translated by A. Shenitzer, Studies in the History of Mathematics and Physical Sciences, vol. 12 (Springer, NewYork, 1988)
74. J.-A. Serret, De la sphère tangente à quatre sphères données. J. für die reine und angewandte Math. **37**, 51–57 (1848)
75. A.N.J. Sorlin, J.D. Gergonne, Trigonométrie, Recherches de trigonométrie sphérique. Ann. Math. Pures Appl. **15**, 273–304 (1824–1825)
76. M. Stewart, Some general theorems of considerable use in the higher parts of mathematics. Edimbourg, 1746, in - 8°
77. P. Tannery, Albert Girard de Saint-Mihiel. Bullet. Sci. Math. Astron. **7**(1), 358–360 (1883)
78. G. Théret, On Euler's memoir A construction relative to a problem of Pappus of Alexandria, in *Spherical Geometry in the Eighteenth Century, I: Euler, Lagrange and Lambert*, Chap. 4 (Springer, Berlin, 2026)
79. W.P. Thurston, *The Geometry and Topology of Three-Manifolds.* Lecture Notes (Princeton University, Princeton, 1976). Published in the Collected Works of William P. Thurston with Commentary: IV. Ed. B. Farb, D. Gabai and S. P. Kerckhoff. Providence, RI: American Mathematical Society, 2022
80. W.P. Thurston, *Three-Dimensional Geometry and Topology*, ed. by S. Levy, vol. 1 (Princeton University Press, Princeton, 1997)
81. F.P.G. Vansson, Contact des cercles sur la sphère, par la geométrie. Nouvelles Ann. Math. **14**, 55–71 (1855)
82. J. Wallis, *Operum Mathematicorum* (Oxford University Press, Oxford, 1693)
83. A. Zhukova, On the contribution of Anders Johan Lexell in spherical geometry, Gaṇita Bhāratī (Indian Mathematics). Bullet. Ind. Soc. History Math. **41**(1–2), 127–149 (2019)

Chapter 3
An Introduction to Spherical Geometry Based on Fundamental Works of Euler and Lagrange

Charalampos Charitos

Abstract This chapter consists of an exposition of some fundamental footsteps of spherical geometry, based on Lagrange's memoir *Solution de quelques problèmes relatifs aux triangles sphériques avec une analyse complète de ces triangles,* making relations with Euler's work on the same subject and putting these works in perspective and establishing relations with modern works.

Keywords Spherical geometry · Lagrange · Euler · Spherical Pythagorean theorem · Spherical law of cosines · Duality · Polar triangle · Dual triangle

AMS Classification 01A50, 51-03, 53A35

3.1 Introduction

Spherical geometry is the geometry of the unit sphere S^2 equipped with the "standard" metric induced from the metric $ds^2 = dx^2 + dy^2 + dz^2$ of the Euclidean space E^3. The lines of S^2, i.e. its geodesics, are the great circles which are the intersection of the planes passing through the center of the sphere with the sphere. The angles between two lines are the dihedral angles of planes that contain these lines. The consideration of spherical geometry as a new geometry which is at the same level as Euclidean geometry is usually attributed to Riemann [17], but this fact is not correct. Indeed, clearly, spherical geometry, as conceived by Menelaus of Alexandria in his *Spherics*, was considered as a geometry *per se*, different but on a par with Euclid's geometry, and this was long before Riemann; see the recent English translation of the *Spherics* and the editors' comments there [16]. The reader will also see that Johann Heinrich Lambert, in his memoir titled *Theorie der Parallellinien* [11], written in 1766 and thoroughly reviewed in Chap. 7 of

C. Charitos (✉)
Laboratory of Mathematics and Theoretical Mechanics, Agricultural University Athens, Athens, Greece
e-mail: bakis@aua.gr

R. Caddeo, A. Papadopoulos (eds.), *Spherical Geometry in the Eighteenth Century I*, https://doi.org/10.1007/978-3-032-12466-1_3

the present volume [13], considers the three geometries, the Euclidean, spherical and hyperbolic (the latter being still, for him, purely hypothetical), as being three geometries which may be put in parallel. Still, in works of the late nineteenth and beginning of the twentieth century, the expression "Riemannian geometry" refers to the geometry of the sphere. If we denote by (t, u) the geographical coordinates of S^2, where t is the latitude and u is the longitude, then, with respect to the coordinates (t, u), the Riemannian metric on S^2 has the form

$$ds^2 = du^2 + dt^2 \cos^2 u. \tag{3.1.1}$$

In the modern theory of differential geometry of surfaces, this formula follows easily from the fact that the metric on S^2 is induced by the Euclidean metric of E^3. However, Euler obtained this formula as a corollary of one of his constitutive principles of spherical geometry which states that, infinitesimally, spherical geometry behaves like Euclidean geometry and thus infinitesimally the Euclidean Pythagorean theorem holds [6].

According to Rosenfeld [18] the modern form of spherical geometry and trigonometry is due to Euler who performed calculations using the metric ds^2 given by (3.1.1). Also, Euler was the first to give intrinsic proofs of the spherical trigonometric formulae using methods of the calculus of variations, a field which he discovered.

On the other hand, according to Bonola [3], p. 82, the recognition of spherical geometry as a geometric system of axioms in which the angle sum of triangles is greater than π is an eighteenth century discovery and is basically due to Lambert. The geometry of the sphere can be studied imposing a system of axioms, like Euclidean and hyperbolic geometry. This system of axioms is substantially different from the system of axioms in Euclidean geometry and the parallel axiom is replaced by the axiom saying that the angle sum in any triangle is greater than π. It seems that Menelaus in his emblematic work *Sphaerics* was the first who studied the geometry of the sphere using intrinsic methods, see [1, 16] (especially, the paragraphs 2.6 and 6.7) and [12] for an introduction to spherical geometry and a comparative study of it regarding the Euclidean and hyperbolic geometry.

The impact of spherical geometry on modern geometry and topology in general is of great significance. In dimension two, thanks to the uniformization theorem, spherical geometry is one of the three geometries, along with the Euclidean and hyperbolic. Each closed surface admits one of these geometries. In three dimensions, spherical geometry is one of the eight distinct geometric structures that closed 3-manifolds may admit under Thurston's geometrization theorem.

Finally, the geometry and the metric relations of spherical triangles play an important role in the comparison geometry and more precisely in the study of Alexandrov spaces of curvature greater than or equal to $k, k > 0$ (see for instance [2, 4]). Alexandrov spaces of curvature bounded below are locally compact, complete length spaces whose geodesic triangles are comparable with the spherical triangles of the sphere of radius $1/\sqrt{k}$. They are defined via axioms similar to those given by Euclid and Alexandrov's axioms replace certain equalities by inequalities. The study of these spaces started in the nineteen fifties by A. D. Alexandrov and his school and continued by his students. Nowadays the study of these spaces constitutes an important research area of metric geometry.

In the present chapter, we review and comment on Lagrange's work [10] on spherical geometry. The starting point of his work, as it is presented in [10], is the sine law for rectilinear triangles. Lagrange starts by giving a proof for spherical triangles. After this he deals with the problem of finding a single property of spherical triangles from which all the formulae of spherical trigonometry result. Finally, he compares spherical and rectilinear triangles and he obtains results for rectilinear triangles starting from results on spherical triangles. Besides, in the present exposition, we explain Euler's approach, who, using the methods of the calculus of variations he invented, obtained a set of formulae of spherical trigonometry. Our purpose is also to compare the approaches of Lagrange and Euler.

The present chapter is organized as follows:

In Sect. 3.2, basic trigonometric identities of Euclidean geometry are recalled and proved.

In Sect. 3.3, using solid geometry, the laws of cosines and sines for spherical triangles are obtained. Some formulae for the volume of rectilinear pyramids will be derived. Finally, the polar or supplementary triangle of a spherical triangle is introduced, since it is used in the next section in an essential way in the proofs of the formulae.

In Sect. 3.4, following [10], we derive all the trigonometric formulae of spherical geometry starting from the law of cosines and using the duality theorem for spherical triangles.

Section 3.5 is devoted to the area of spherical triangles and to formulae giving the area of these triangles.

In Sect. 3.6, based on [10], a comparison between rectilinear and spherical triangles is made.

In Sect. 3.7, Euler's approach to spherical geometry is discussed. In particular, by means of Euler's method of the calculus of variations, we obtain, using a modern mathematical language, the Pythagorean theorem for right spherical triangles. After that, we show that every trigonometric formula of spherical trigonometry can be obtained from the spherical Pythagorean theorem, in particular the law of cosines and the duality theorem. In this sense, the Pythagorean theorem can be considered as the fundamental equation of spherical trigonometry.

Sections 3.2–3.6 are based on Lagrange's work [10].

3.2 The Euclidean Law of Sines

Below we will prove the well-known property of rectilinear (Euclidean) triangles which asserts that the ratio of the sides to the sines of the opposite angles is constant and equal to the diameter of the circumscribed circle and that it is also equal to the product of the three sides divided by the double of the area. We shall recall these proofs because Lagrange in [10] began his study of spherical geometry by first reporting on these results of rectilinear triangles and then proved the corresponding identities for spherical triangles.

A rectilinear triangle with vertices A, B, C will be denoted by ABC. If no ambiguity arises, we shall denote by the same letters A, B, C the angles of the triangle ABC. Otherwise, the angles of ABC will be denoted by $\widehat{BAC}, \widehat{ABC}, \widehat{ACB}$. Finally, we shall denote by a, b, c the lengths of the sides that are opposite to the vertices A, B, C.

By the well-known rectilinear law of cosines, we have

$$a^2 = b^2 + c^2 - 2bc\cos A. \tag{3.2.1}$$

Therefore,

$$\cos A = \frac{b^2 + c^2 - a^2}{2bc}$$

and hence

$$\sin A = \frac{\sqrt{4b^2c^2 - (b^2 + c^2 - a^2)^2}}{2bc} = \frac{d}{2bc},$$

where $d = \sqrt{4b^2c^2 - (b^2 + c^2 - a^2)^2}$.

Therefore,

$$\frac{a}{\sin A} = \frac{b}{\sin B} = \frac{c}{\sin C} = \frac{2abc}{d}. \tag{3.2.2}$$

Considering the circumscribed circle to the triangle ABC, let us denote by K its center and by r its radius. Then, by the law of cosines applied to the triangle KBC we have

$$a^2 = 2r^2 - 2r^2\cos 2A.$$

Therefore,

$$a^2 = 2r^2(1 - \cos 2A) = 4r^2\sin^2 A$$

and hence

$$\frac{a}{\sin A} = 2r. \tag{3.2.3}$$

Let s be the area of the triangle ABC. We have

$$d = \sqrt{4b^2c^2 - (b^2 + c^2 - a^2)^2}, \tag{3.2.4}$$

thus,

$$\begin{aligned} d &= \sqrt{(2bc + b^2 + c^2 - a^2)(2bc - b^2 - c^2 + a^2} \\ &= \sqrt{[(b + c)^2 - a^2][a^2 - (b - c)^2]} \end{aligned}$$

and hence

$$d = \sqrt{(a + b + c)(a - b + c)(a + b - c)(b + c - a)}. \tag{3.2.5}$$

On the other hand,

$$s = \frac{1}{2}(base) \cdot (altitude) = \frac{1}{2}bc \sin A = \frac{1}{2}bc\frac{d}{2bc} = \frac{d}{4}. \tag{3.2.6}$$

Relation (3.2.6), which connects s and d, is the well-known Heron Theorem. Combining the relations (3.2.2), (3.2.3), (3.2.6) we get

$$\frac{a}{\sin A} = \frac{b}{\sin B} = \frac{c}{\sin C} = 2r = \frac{abc}{2s}. \tag{3.2.7}$$

The previous formula includes all the assertions at the beginning of this section and is known as the (rectilinear, or Euclidean) *law of sines*.

Remark 3.2.1 The law of cosines (3.2.1) can be obtained from the Pythagorean theorem by dividing an arbitrary triangle into right triangles by drawing the altitudes. Below, in Sect. 3.6, we will show that all the formulae of spherical trigonometry can be derived from the spherical Pythagorean theorem.

3.3 Applications of Solid Geometry to Spherical Trigonometry

In this section we will prove first the spherical law of cosines as an application of solid geometry, that is, of the Euclidean geometry of E^3. Subsequently, following Lagrange's method as it is explained in [10], we will derive a sine law for spherical triangles. Finally, following always [10], some formulae expressing the volume of rectilinear pyramids will be derived.

The Law of Cosines for Spherical Triangles

The spherical law of cosines is reported in § 3 of [10]. Its proof is not mentioned in [10] but we will recall it because this law is fundamental in the sense that Lagrange derived all trigonometric identities of spherical geometry from it.

Let ABC be a spherical triangle on the unit sphere S^2. The angles of ABC will be denoted by the letters A, B, C and the lengths of the sides opposite to A, B, C will be denoted by a, b, c. As for the rectilineal triangles, if there is any ambiguity, the angle A will be denoted by $\widehat{BAC}$ and similarly for the angles B and C.

For a spherical triangle ABC, we consider the points B', C' in E^3, see Fig. 3.1(ii), such that the point B' (resp. C') is defined as the intersection point of the line defined by the geodesic ray OB (resp. OC) and the tangent line to the side AB (resp. AC) at A. We set $h = B'C'$. Using Fig. 3.1(i), which depicts a unit circle of E^2 and a special triangle with its sides, and since $\widehat{AOB'} = c$ and $\widehat{AOC'} = b$, we have $AB' = \tan c$, $AC' = \tan b$, $OB' = \sec c$ and $OC' = \sec b$.

Now, from the rectilinear law of cosines in the triangle $AB'C'$ we have

$$h^2 = \tan^2 b + \tan^2 c - 2 \tan b \tan c \cos A,$$

and from the same law in the triangle $OB'C'$ we have

$$h^2 = \sec^2 b + \sec^2 c - 2 \sec b \sec c \cos a.$$

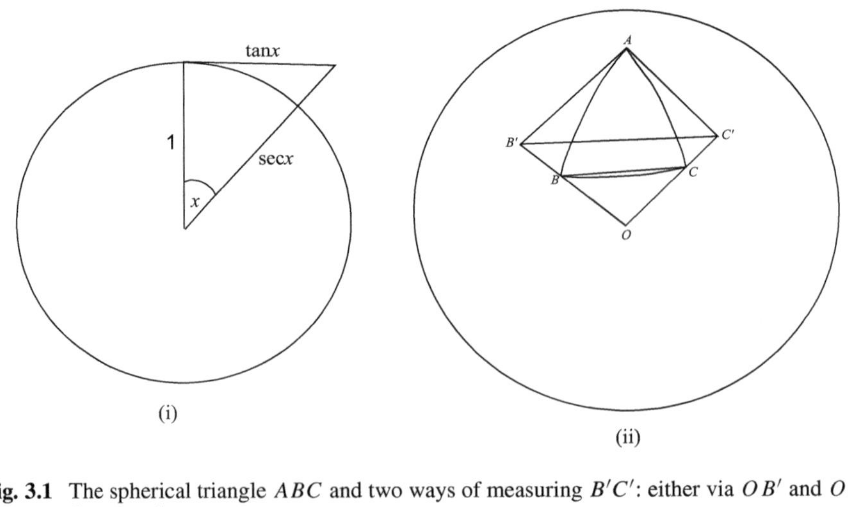

Fig. 3.1 The spherical triangle ABC and two ways of measuring $B'C'$: either via OB' and OC', or via AB' and AC'

Therefore,

$$\tan^2 b + \tan^2 c - 2\tan b \tan c \cos A^2 = \sec^2 b + \sec^2 c - 2\sec b \sec c \cos a.$$

Using the identities $\sec^2 b - \tan^2 b = 1$ and $\sec^2 c - \tan^2 c = 1$, the previous relation becomes

$$\sec b \sec c \cos a = 1 + \tan b \tan c \cos A$$

and setting $\sec b = \frac{1}{\cos b}$, $\sec c = \frac{1}{\cos c}$, $\tan b = \frac{\sin b}{\cos b}$, $\tan c = \frac{\sin c}{\cos c}$ we get the following fundamental relation referred to as the *spherical law of cosines*

$$\cos a = \cos b \cos c + \sin b \sin c \cos A. \tag{3.3.1}$$

Similarly we have

$$\cos c = \cos a \cos b + \sin a \sin b \cos C$$

and substituting the value of $\cos c$ in (3.3.1) we have

$$\cos a = \cos a \cos^2 b + \sin a \sin b \cos b \cos C + \sin b \sin c \cos A.$$

Dividing the previous relation by $\sin b$ we get

$$\cos a \sin b = \sin a \cos b \cos C + \sin c \cos A. \tag{3.3.2}$$

Relation (3.3.2) will be used in the next paragraph in order to derive a variation of the law of cosines for angles.

The Law of Sines for Spherical Triangles

In this paragraph, following [10], we will prove for a spherical triangle ABC relations analogous to the relations of Sect. 3.2. That is, we will show that the sines of the sides are proportional to the sines of the opposite angles and we will examine how this ratio depends on the radius of the circumscribed circle and on the area of the rectilinear triangle with vertices A, B, C.

From (3.3.1) we have

$$\cos A = \frac{\cos a - \cos b \cos c}{\sin b \sin c} \tag{3.3.3}$$

and thus

$$\sin A = \frac{\sqrt{\sin^2 b \sin^2 c - (\cos a - \cos b \cos c)^2}}{\sin b \sin c}.$$

Setting, for short,

$$f = \sqrt{\sin^2 b \sin^2 c - (\cos a - \cos b \cos c)^2},$$

we get

$$\sin A = \frac{f}{\sin b \sin c}, \quad \text{or} \quad \frac{\sin a}{\sin A} = \frac{\sin a \sin b \sin c}{f}$$

and since a, b, c appear cyclically we have

$$\frac{\sin a}{\sin A} = \frac{\sin b}{\sin B} = \frac{\sin c}{\sin C} = \frac{\sin a \sin b \sin c}{f}. \tag{3.3.4}$$

Therefore, substituting $\sin^2 b$ and $\sin^2 c$ by $1 - \cos^2 b$ and $1 - \cos^2 c$ we have

$$f = \sqrt{1 - \cos^2 a - \cos^2 b - \cos^2 c + 2 \cos a \cos b \cos c},$$

where we can see that the three sides a, b, c enter equally.

Similarly to the Euclidean case we may factorize f and get an expression convenient for the logarithmic calculus, as follows:

$$\sin^2 b \sin^2 c - (\cos a - \cos b \cos c)^2$$

$$= (\sin b \sin c + \cos a - \cos b \cos c)(\sin b \sin c - \cos a + \cos b \cos c)$$

$$= [\cos a - \cos(b + c)][\cos(b - c) - \cos a].$$

But we have in general

$$\cos x - \cos y = 2 \sin \frac{x + y}{2} \sin \frac{x - y}{2},$$

therefore

$$f = 2\sqrt{\sin \frac{a + b + c}{2} \sin \frac{a + c - b}{2} \sin \frac{a + b - c}{2} \sin \frac{b + c - a}{2}}. \tag{3.3.5}$$

Now, considering the circle circumscribed to the spherical triangle, this circle is also circumscribed to the rectilinear triangle formed by the chords corresponding to the spherical arcs a, b, c. The lengths of these chords are $2\sin\frac{a}{2}, 2\sin\frac{b}{2}, 2\sin\frac{c}{2}$ respectively. Thus, if we denote by h the quantity corresponding to the quantity d of (3.2.2) and if we denote by R the radius of the planar circumscribed circle, we get from (3.2.2) and (3.2.3)

$$R = \frac{8\sin\frac{a}{2}\sin\frac{b}{2}\sin\frac{c}{2}}{h}.$$

From (3.2.4) we may compute h and we have

$$\begin{aligned} h^2 = {} & 32(\sin\frac{a}{2}\sin\frac{b}{2})^2 + 32(\sin\frac{a}{2}\sin\frac{c}{2})^2 + 32(\sin\frac{b}{2}\sin\frac{c}{2})^2 \\ & - 16(\sin\frac{a}{2})^4 - 16(\sin\frac{b}{2})^4 - 16(\sin\frac{c}{2})^4 \end{aligned}$$

and substituting $2(\sin\frac{a}{2})^2 = 1-\cos a$, $2(\sin\frac{b}{2})^2 = 1-\cos b$, $2(\sin\frac{c}{2})^2 = 1-\cos c$, we get

$$h^2 = 4f^2 + 8(1-\cos a)(1-\cos b)(1-\cos c) = 4f^2 + 64(\sin\frac{a}{2}\sin\frac{b}{2}\sin\frac{c}{2})^2.$$

Thus, from (3.2.6), (3.2.7) and the above expression of R we get

$$R = \frac{4\sin\frac{a}{2}\sin\frac{b}{2}\sin\frac{c}{2}}{\sqrt{f^2 + 16(\sin\frac{a}{2}\sin\frac{b}{2}\sin\frac{c}{2})^2}}.$$

Denoting by ϕ the spherical arc that measures the distance from the pole of the circumscribed circle (that is, the center of this circle on the sphere), to the circumference of this circle, we have $R = \sin\phi$. Therefore,

$$\sin\phi = \frac{4\sin\frac{a}{2}\sin\frac{b}{2}\sin\frac{c}{2}}{\sqrt{f^2 + (4\sin\frac{a}{2}\sin\frac{b}{2}\sin\frac{c}{2})^2}},$$

hence

$$\cos\phi = \frac{f}{\sqrt{f^2 + (4\sin\frac{a}{2}\sin\frac{b}{2}\sin\frac{c}{2})^2}},$$

thus

$$\tan\phi = \frac{4\sin\frac{a}{2}\sin\frac{b}{2}\sin\frac{c}{2}}{f}. \tag{3.3.6}$$

Therefore, from (3.3.4) we have

$$\frac{\sin a}{\sin A} = \frac{\sin b}{\sin B} = \frac{\sin c}{\sin C} = 2\tan\phi \cos\frac{a}{2}\cos\frac{b}{2}\cos\frac{c}{2}. \tag{3.3.7}$$

The double equality $\frac{\sin a}{\sin A} = \frac{\sin b}{\sin B} = \frac{\sin c}{\sin C}$ that appears in Relations (3.3.4) and (3.3.7) constitutes the *law of sines* in spherical trigonometry.

Let us denote by S the area of the rectilinear triangle ABC. By the last equality of Relation (3.2.7) and setting $2\sin\frac{a}{2}, 2\sin\frac{b}{2}, 2\sin\frac{c}{2}$ instead of a, b, c and R instead of r, we have

$$S = \frac{2\sin\frac{a}{2}\sin\frac{b}{2}\sin\frac{c}{2}}{R} = \frac{2\sin\frac{a}{2}\sin\frac{b}{2}\sin\frac{c}{2}}{\sin\phi} \tag{3.3.8}$$

since $R = \sin\phi$.

Finally, from the law of sines, we substitue $\sin c$ by $\frac{\sin a \sin C}{\sin A}$ in Relation (3.3.2). Dividing the resulting relation by $\sin a$ we get

$$\cot a \sin b = \cot A \sin C + \cos b \cos C. \tag{3.3.9}$$

From (3.3.2), changing a into b and A into B we have

$$\sin a \cos b = \cos a \sin b \cos C + \sin c \cos B.$$

From the previous relation, substituting $\sin a \cos b$ in (3.3.2) we have

$$\cos a \sin b = \cos a \sin b \cos^2 C + \sin c \cos B \cos C + \sin c \cos A,$$

that is,

$$\cos a \sin b \sin^2 C = \sin c(\cos B \cos C + \cos A).$$

Setting $\sin c = \frac{\sin b \sin C}{\sin B}$, then dividing by $\sin b \sin C$ and then multiplying by $\sin B$, we get

$$\cos a \sin B \sin C = \cos B \cos C + \cos A,$$

or,

$$\cos A = -\cos B \cos C + \sin B \sin C \cos a. \tag{3.3.10}$$

This law consists of a variation on the law of cosines, also called the *cosine rule for angles*.

The Volume of Rectilinear Pyramids

The volume of a pyramid whose base is a rectilinear triangle ABC of area S, with vertices A, B, C lying on the sphere S^2 and whose apex is at the center of the sphere is

$$\frac{S\cos\phi}{3},$$

since the height of the pyramid is equal to $\cos\phi$. Recall that ϕ is the spherical arc measuring the angular distance from the pole of the circumscribed circle of the triangle ABC to its circumference.

Therefore from (3.3.6) and (3.3.8) the volume of the pyramid will be

$$\frac{2\sin\frac{a}{2}\sin\frac{b}{2}\sin\frac{c}{2}}{3\tan\phi}=\frac{f}{6}. \tag{3.3.11}$$

Now, a formula for the volume of an arbitrary rectilinear pyramid will be derived.

We have seen that the volume of the triangular pyramid, whose apex is the center of the sphere and having as edges the three radii of the sphere which form the angles a, b and c is equal to $\frac{f}{6}$.

These radii are produced in order that they become p, q, r and we will compute the volume of this new pyramid, see Fig. 3.2. If the angle between p and q is a then the area of the triangular face containing the edges p, q is $\frac{pq\sin a}{2}$. Thus, if we denote by H the perpendicular drawn from the extremity of r on the plane Π of the face formed by p and q, the volume of the pyramid will be $\frac{Hpq\sin a}{6}$.

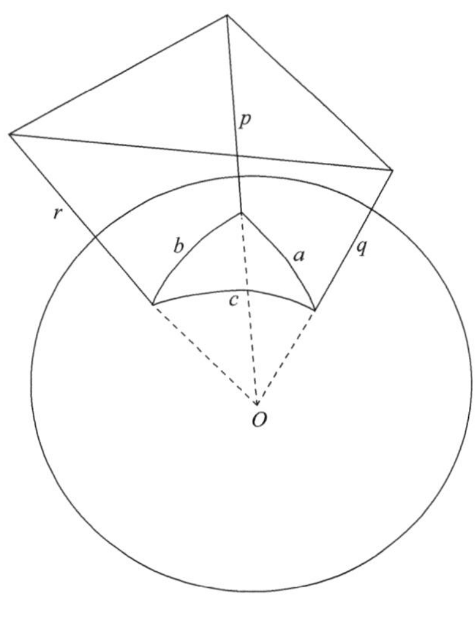

Fig. 3.2 A rectilinear pyramid whose apex is the center of the sphere

Let us call θ the angle that r makes with the plane Π. Since $H = r\sin\theta$, it follows that the volume V of the pyramid is

$$V = \frac{pqr\sin a\sin\theta}{6}. \tag{3.3.12}$$

In the special case where $p = q = r = 1$, we deduce from Formulae 3.3.11 and 3.3.12 that $\sin a\sin\theta = f$ which implies that

$$\sin\theta = \frac{f}{\sin a}. \tag{3.3.13}$$

Therefore, by (3.3.12) and (3.3.13) the volume of an arbitrary pyramid in terms of the lengths p, q, r of the three sides and the angles a, b contained by these three sides and c is equal to

$$V = \frac{pqrf}{6}$$

where f is given by (3.3.5).

Polar Triangle

Considering a spherical triangle ABC, its *polar* or *dual triangle* is another spherical triangle $A'B'C'$ defined as follows: Considering the line BC, we also consider the great circle containing it. The *pole* A' corresponding to BC is the point of the sphere which is farthest away from this great circle and which is also on the same side of the great circle than the point A. Similarly the poles B' and C' corresponding to the lines AC and AB are defined. Geometrically, considering lines perpendicular to BC at the vertices B and C, as in Fig. 3.3, the pole A' can be defined as intersection point of these lines. In a similar way, the poles B', C' can also be defined. The polar/dual triangle is by definition the spherical triangle with vertices A', B', C'. The polarity or duality theorem was discovered by the Arab mathematicians of the tenth century,

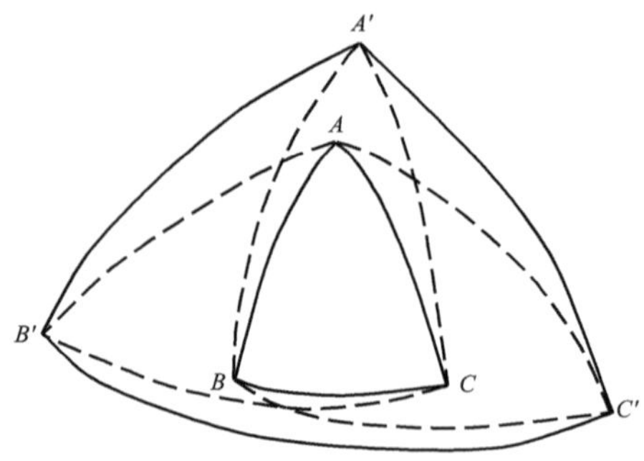

Fig. 3.3 The polar/dual triangle $A'B'C'$ of the spherical triangle ABC

see [16], p. 795, [14, 15] and Chap. 10 of the present volume. It can be stated as follows:

Theorem 3.3.1 *The sides (resp. the angles) of a polar triangle are equal to the supplements of the angles (resp. the sides) of the original triangle.*

Theorem 3.3.1 can be proved using solid geometry and it was used by Euler and Lagrange for obtaining new trigonometric formulae from known ones, as we will see below. In the final section of this chapter, we will explain how this theorem can be proved without using solid geometry.

3.4 Formulae of Spherical Trigonometry

Lagrange declared in § 13 of [10] that an interesting problem was to find a minimal set of fundamental propositions, so that all the formulae of spherical trigonometry can be deduced from these propositions. He noted that the most elegant effort in this direction was Euler's work [7] where all the trigonometric formulae were deduced from only three equations. Lagrange asked if the number of fundamental trigonometric equations, from which all other trigonometric equations follow, could be further reduced and he stated that de Gua had the idea of making all spherical trigonometry dependent on a single equation. But he claimed that the memoir that de Gua gave on this subject in the 1783 volume of the Academy of Sciences [5] contained very complicated computations that were not clear and consequently his methods could not be accepted.

In this section, following [10], all the trigonometric formulae for spherical triangles will be derived from the law of cosines and the polarity theorem for spherical triangles.

The Basic Equations

Each spherical triangle ABC has six elements, the three angles and the three sides, and three of these elements suffice to determine all the others. Therefore, it is clear that the simplest relations will combine four elements. Lagrange in [10] proved that all the different combinations that can be done with the six elements taken four-by-four are reduced to four basic equations (I)–(IV) which will be given below.

Of course, all equations hold with a circular permutation of sides a, b, c and of angles A, B, C.

The fundamental equation, in the sense that it is the starting equation, is the law of cosines, which henceforth will be labelled by (I), that is,

$$\cos a = \cos b \cos c + \sin b \sin c \cos A. \tag{I}$$

This is a relation between three sides and one angle.

As we saw, the law of sines can be derived from (I) and this equation will be labelled by (II), that is,

$$\frac{\sin a}{\sin A} = \frac{\sin b}{\sin B} = \frac{\sin c}{\sin C}. \tag{II}$$

These are relations between two sides and two angles.

Relation (3.3.9) will be labelled by (III), that is,

$$\cot a \sin b = \cot A \sin C + \cos b \cos C \tag{III}$$

and is again a relation between two sides and two angles.

Finally, Relation (3.3.10) will be labelled by (IV), that is,

$$\cos A = -\cos B \cos C + \sin B \sin C \cos a, \tag{IV}$$

which is a relation between three angles and one side.

Equations (I), (II), (III), (IV) are considered as basic by Lagrange because they contain the solution of all the questions in spherical trigonometry.

Right Triangles

Setting $A = \frac{\pi}{2}$ in relations (I)–(IV) we have

$$\cos a = \cos b \cos c, \tag{3.4.1}$$

$$\sin a = \frac{\sin b}{\sin B},$$

$$\cot a = \cot b \cos C, \tag{3.4.2}$$

$$\cos a = \cot B \cot C.$$

Setting $C = \frac{\pi}{2}$ in Relations (III)–(IV) we have

$$\cot A = \cot a \sin b$$

$$\cos A = \sin B \cos a.$$

Equation (3.4.1) is known as the Pythagorean theorem in spherical geometry. The above six equations provide the solution of all the cases of right triangles. Additionally, since they are under a product form which is very useful for the use of logarithms, they are also used to decompose all triangles into right triangles by drawing perpendiculars.

However, as we will see in the next paragraph, Lagrange goes one step further by factorizing the four basic equations in order to be able to make calculations and solve all cases in an arbitrary triangle.

Other Spherical Trigonometric Identities

- Determination of A from a, b, c.

  From Relation (I) we have

$$\cos A = \frac{\cos a - \cos b \cos c}{\sin b \sin c},$$

thus

$$1 + \cos A = 2(\cos \frac{A}{2})^2 = \frac{\cos a - \cos(b + c)}{\sin b \sin c} = \frac{2 \sin \frac{b+c-a}{2} \sin \frac{b+c+a}{2}}{\sin b \sin c},$$

$$1 - \cos A = 2(\sin \frac{A}{2})^2 = \frac{\cos(b - c) - \cos a}{\sin b \sin c} = \frac{2 \sin \frac{a-b+c}{2} \sin \frac{a+b-c}{2}}{\sin b \sin c},$$

therefore

$$\tan \frac{A}{2} = \sqrt{\frac{\sin \frac{a+b-c}{2} \sin \frac{a-b+c}{2}}{\sin \frac{a+b+c}{2} \sin \frac{b+c-a}{2}}}. \tag{3.4.3}$$

- Determination of a from b, c and A

  From (I) we have

$$\cos a = \cos b \cos c + \sin b \sin c \cos A.$$

Since it is not possible to factorize this equation we will use a subsidiary variable φ. In this way the resulting equation is in product form and thus appropriate for the usage of logarithms.[1]

More concretely, by drawing a perpendicular from B onto the side AC, let us denote by φ the segment on the side AC adjacent to A. That is, if P is the projection of B on AC then φ is the length of AP. Thus, from (3.4.2) we have

$$\tan c \cos A = \tan \varphi \tag{3.4.4}$$

[1] The use of logarithms in the trigonometric formulae is attributed by Lambert to John Napier, who is usually considered as one of the inventors of logarithms, see §1 of Lambert's memoir *Notes and comments on trigonometry*, translated in Chap. 19 of the present volume.

which implies that

$$\sin c \cos A = \cos c \tan \varphi.$$

By substituting the above term $\sin c \cos A$ in (I), that is, in the equation $\cos a = \cos b \cos c + \sin b \sin c \cos A$, we have

$$\cos a = \cos c(\cos b + \sin b \tan \varphi).$$

But

$$\cos b + \sin b \tan \varphi = \frac{\cos b \cos \varphi + \sin b \sin \varphi}{\cos \varphi} = \frac{\cos(b - \varphi)}{\cos \varphi},$$

thus

$$\cos a = \frac{\cos c \cos(b - \varphi)}{\cos \varphi}. \tag{3.4.5}$$

- Determination of A from a, b, B and a from b, A, B.

  From Relation (II) we have

$$\sin A = \frac{\sin a \sin B}{\sin b}, \quad \sin a = \frac{\sin b \sin A}{\sin B}.$$

- Determination of A from a, b, C and a from b, A, C.

  From Relation (III) we have

$$\cot A = \frac{\cot a \sin b}{\sin c} - \cot C \cos b, \quad \cot a = \frac{\cot A \sin C}{\sin b} + \cot b \cos C.$$

Additionally, using a proper subsidiary angle, (see §23 in [10]), we obtain

$$\cot A = \frac{\cot C \sin(b - \varphi)}{\sin \varphi}, \quad \cot a = \frac{\cot b \cos(C - \varphi)}{\cos \varphi}.$$

- Determination of a from A, B, C.

  From (3.4.3) and Theorem 3.3.1 we obtain

$$\cot \frac{a}{2} = \sqrt{\frac{\cos \frac{A+B-C}{2} \cos \frac{A-B+C}{2}}{-\cos \frac{B+C+A}{2} \cos \frac{B+C-A}{2}}}.$$

- Determination of A from a, B, C.

As before, by drawing the perpendicular BP from B onto the side AC of the triangle ABC, let us denote by φ the length of AP. If Φ' is the angle opposite to φ in the triangle ABP and $\Phi = \pi/2 - \Phi'$, then from (3.4.4), (3.4.5) and Theorem 3.3.1 we have

$$\tan C \cos a = \cot \Phi$$

$$\cos A = \frac{\cos C \sin(B - \Phi)}{\sin \Phi}.$$

Finally, if we want to find two sides from the third side and the two opposite angles, or two angles from the third angle and the two opposite sides, we use the following formulae attributed to Napier.

From (3.4.3) and Theorem 3.3.1, we have

$$\tan \frac{A}{2} = \sqrt{\frac{\sin \frac{a+b-c}{2} \sin \frac{a-b+c}{2}}{\sin \frac{b+c+a}{2} \sin \frac{b+c-a}{2}}}.$$

Similarly, changing A into B and a into b,

$$\tan \frac{B}{2} = \sqrt{\frac{\sin \frac{a+b-c}{2} \sin \frac{b-a+c}{2}}{\sin \frac{b+c+a}{2} \sin \frac{a-b+c}{2}}}.$$

Therefore

$$\tan \frac{A}{2} \tan \frac{B}{2} = \frac{\sin \frac{a+b-c}{2}}{\sin \frac{a+b+c}{2}}.$$

Changing b into c and B into C,

$$\tan \frac{A}{2} \tan \frac{C}{2} = \frac{\sin \frac{a-b+c}{2}}{\sin \frac{a+b+c}{2}}$$

and changing a into b and B into A, we get

$$\tan \frac{B}{2} \tan \frac{C}{2} = \frac{\sin \frac{b+c-a}{2}}{\sin \frac{a+b+c}{2}}.$$

Therefore

$$(\tan \frac{A}{2} + \tan \frac{B}{2}) \tan \frac{C}{2} = \frac{\sin \frac{a-b+c}{2} + \sin \frac{b+c-a}{2}}{\sin \frac{a+b+c}{2}} = \frac{2 \sin \frac{c}{2} \cos \frac{a-b}{2}}{\sin \frac{a+b+c}{2}},$$

$$(\tan\frac{A}{2}-\tan\frac{B}{2})\tan\frac{C}{2}=\frac{\sin\frac{a-b+c}{2}-\sin\frac{b+c-a}{2}}{\sin\frac{a+b+c}{2}}=\frac{2\cos\frac{c}{2}\sin\frac{a-b}{2}}{\sin\frac{a+b+c}{2}}.$$

On the other hand,

$$1+\tan\frac{A}{2}\tan\frac{B}{2}=\frac{\sin\frac{a+b+c}{2}+\sin\frac{a+b-c}{2}}{\sin\frac{a+b+c}{2}}=\frac{2\sin\frac{a+b}{2}\cos\frac{c}{2}}{\sin\frac{a+b+c}{2}},$$

$$1-\tan\frac{A}{2}\tan\frac{B}{2}=\frac{\sin\frac{a+b+c}{2}-\sin\frac{a+b-c}{2}}{\sin\frac{a+b+c}{2}}=\frac{2\sin\frac{c}{2}\cos\frac{a+b}{2}}{\sin\frac{a+b+c}{2}},$$

and since $\tan\frac{A\pm B}{2}=\frac{\tan\frac{A}{2}\pm\tan\frac{B}{2}}{1\mp\tan\frac{A}{2}\tan\frac{B}{2}}$ we have

$$\tan\frac{A+B}{2}\tan\frac{C}{2}=\frac{\cos\frac{a-b}{2}}{\cos\frac{a+b}{2}} \tag{3.4.6}$$

$$\tan\frac{A-B}{2}\tan\frac{C}{2}=\frac{\sin\frac{a-b}{2}}{\sin\frac{a+b}{2}}. \tag{3.4.7}$$

By Theorem 3.3.1 we also have

$$\tan\frac{a+b}{2}\cot\frac{c}{2}=\frac{\cos\frac{A-B}{2}}{\cos\frac{A+B}{2}} \tag{3.4.8}$$

$$\tan\frac{a-b}{2}\cot\frac{c}{2}=\frac{\sin\frac{A-B}{2}}{\sin\frac{A+B}{2}}. \tag{3.4.9}$$

Relations (3.4.6)–(3.4.9) were originally discovered by Napier (see § 25 in [10]).

3.5 The Area of a Spherical Triangle

Starting from the well-known and beautiful theorem, which is commonly attributed to Albert Girard and which asserts that, up to normalization, the area of a spherical triangle ABC with angles A, B, C is equal to $A+B+C-\pi$, Lagrange derived formulae which relate the area of the triangle with its sides.

Considering a spherical triangle ABC whose sides b, c are smaller than a quarter of circle, we may produce a triangle $AB'C'$ by extending AB and AC until AB' and AC' become equal to a quarter of circle, see Fig. 3.4. This triangle $AB'C'$ has area equal to A. If we subtract from it the triangle ABC, we obtain a quadrilateral

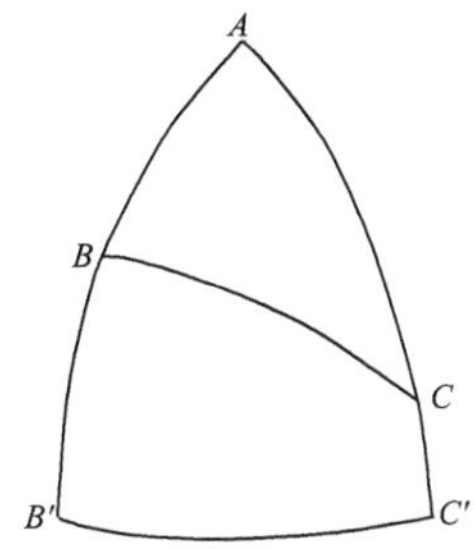

Fig. 3.4 The triangle ABC whose sides b, c are produced until a quarter of a circle

$BCC'B'$ with area $\pi - (B + C)$. The side $B'C'$ is equal to A and the two sides which are perpendicular to it are equal to $\beta = \frac{\pi}{2} - b$ and $\gamma = \frac{\pi}{2} - c$.

By the well-known Napier relation (3.4.6), we have

$$\tan \frac{B+C}{2} = \frac{\cos \frac{b-c}{2}}{\cos \frac{b+c}{2}} \cot \frac{A}{2}. \tag{3.5.1}$$

Thus, if the area of $BCC'B'$ is denoted by σ, we have

$$\tan \frac{\sigma}{2} = \cot \frac{B+C}{2} = \frac{1}{\tan \frac{B+C}{2}},$$

therefore

$$\tan \frac{\sigma}{2} = \frac{\cos \frac{b+c}{2}}{\cos \frac{b-c}{2}} \tan \frac{A}{2} = \frac{\sin \frac{\beta+\gamma}{2}}{\sin \frac{\beta-\gamma}{2}} \tan \frac{A}{2}.$$

Set $\Sigma = A + B + C - \pi$ for the area of ABC. Then

$$\cot \frac{\Sigma}{2} = -\tan \frac{A+B+C}{2} = -\frac{\tan \frac{A}{2} + \tan \frac{B+C}{2}}{1 - \tan \frac{A}{2} \tan \frac{B+C}{2}}.$$

From (3.5.1) we have

$$\cot \frac{\Sigma}{2} = -\frac{\tan \frac{A}{2} \cos \frac{b+c}{2} + \cot \frac{A}{2} \cos \frac{b-c}{2}}{\cos \frac{b+c}{2} - \cos \frac{b-c}{2}},$$

a formula which can be transformed into

$$\cot \frac{\Sigma}{2} = \frac{\cos \frac{b}{2} \cos \frac{c}{2} + \sin \frac{b}{2} \sin \frac{c}{2} \cos A}{\sin \frac{b}{2} \sin \frac{c}{2} \sin A}. \tag{3.5.2}$$

In the relation (3.5.2), we substitute the values of $\cos A$ and $\sin A$ as they result from (3.3.3) and (3.3.4) respectively. So, after dividing the top and the bottom by $\sin\frac{b}{2}\sin\frac{c}{2}$, we have

$$\cot\frac{\Sigma}{2} = \frac{4(\cos\frac{b}{2}\cos\frac{c}{2})^2 + \cos a - \cos b\cos c}{f}. \tag{3.5.3}$$

But, $2(\cos\frac{b}{2})^2 = 1 + \cos b$ and $2(\cos\frac{c}{2})^2 = 1 + \cos c$. Making these substitutions in the previous relation (3.5.3), we get finally the following nice equation

$$\tan\frac{\Sigma}{2} = \frac{f}{1 + \cos a + \cos b + \cos c}.$$

3.6 Rectilinear Geometry as the Limit of Spherical Geometry

In this section, considering spherical triangles on a family of spheres of radius r, with r tending to infinity, and using power expansions of the sine and cosine functions, some results for rectilinear triangles will be derived.

We consider a sphere of radius r and center O and a triangle $A'B'C'$ on it whose sides have lengths α, β, γ. Thus, if we consider the rays OA', OB', OC' then $\widehat{B'OC'} = \frac{\alpha}{r}$, $\widehat{A'OB'} = \frac{\gamma}{r}$, $\widehat{A'OC'} = \frac{\beta}{r}$. If A, B, C are the intersection points of these rays with the unit sphere, then the lengths of the sides of ABC will be $a = \frac{\alpha}{r}, b = \frac{\beta}{r}, c = \frac{\gamma}{r}$. Obviously, the angles of the spherical triangles ABC and $A'B'C'$ are equal.

Now, if the radius r becomes infinitely large, then the sphere changes into a plane and the spherical triangle $A'B'C'$ becomes rectilinear. Thus, if $V = 0$ is an equation between the sines and cosines of a, b, c and A, B, C of the triangle ABC, we may substitute $\frac{\alpha}{r}, \frac{\beta}{r}, \frac{\gamma}{r}$ by a, b, c. If we further assume that r is infinitely large and we substitute $\sin a$ with $\frac{\alpha}{r} - \frac{\alpha^3}{2\cdot 3r^3} + \cdots$ and $\cos a$ with $1 - \frac{\alpha^2}{2r^2} + \frac{\alpha^4}{2\cdot 34r^4} + \cdots$ and similarly for $\sin b, \cos b, \sin c, \cos c$, the equation $V = 0$ is transformed into an equation in which we may delete the terms which vanish under the assumption $\frac{1}{r} = 0$. In this way we may deduce the trigonometric rules of rectilinear triangles from the fundamental equations of spherical trigonometry.

More specifically, after the substitution described above, the equation $V = 0$ is transformed into a power series equal to zero according to the decreasing powers of r of the form

$$V = \frac{P}{r^m} + \frac{Q}{r^{m+1}} + \frac{R}{r^{m+2}} + \cdots = 0.$$

Thus, multiplying by r^m and setting $\frac{1}{r} = 0$ as $r \to \infty$, we get $P = 0$.

This is an interesting, rather than useful, method to get new formulae because it is not common to prove simplest relations from more complicated ones. Nevertheless Lagrange in [10] gave an example that we quote below in order to show how the method works. The interest of this example lies in the fact that it shows how Euclidean geometry is obtained as a limit of spherical geometry.

As $r \to \infty$ the side $a = \frac{\alpha}{r} \to 0$ and hence $\cos a = 1$. Therefore, from (IV) we have

$$\cos A = \sin B \sin C - \cos B \cos C.$$

Hence

$$\cos A = -\cos(B - C),$$

whence we deduce

$$A = \pi - B - C \Leftrightarrow A + B + C = \pi,$$

which is the fundamental relation between the angles of a rectilinear triangle.

In the following, we present a proof that Lagrange gave of a beautiful theorem of Legendre which gives an approximation of the area of a spherical triangle lying on a sphere of large radius r by the area of the rectilinear triangle whose sides have the same lengths as those of the spherical one. We will see below that if the spherical triangle is drawn on a sphere of very large radius, as is the Earth, the approximation is quite reliable for calculating the area of a spherical triangle [10].

Let ABC be a spherical triangle. Starting from the cosine rule, we have

$$\cos A = \frac{\cos a - \cos b \cos c}{\sin b \sin c}.$$

Now, substituting sines and cosines by their expansions and stopping to the term of order 4, we have

$$\cos A = \frac{\frac{\beta^2+\gamma^2-\alpha^2}{2r^2} + \frac{\alpha^4-\beta^4-\gamma^4}{2\cdot 3\cdot 4r^4} - \frac{\beta^2\gamma^2}{4r^4}}{\frac{\beta\gamma}{r^2}(1 - \frac{\beta^2+\gamma^2}{2\cdot 3r^2})}. \tag{3.6.1}$$

Then, first, we multiply the top and the bottom by r^2, subsequently the factor $1 - \frac{\beta^2+\gamma^2}{2\cdot 3r^2}$ in the denominator is substituted by the factor $1 + \frac{\beta^2+\gamma^2}{2\cdot 3r^2}$ and finally we neglect the terms divided by powers higher than r^2. In this way, from (3.6.1), we get

$$\cos A = \frac{\beta^2 + \gamma^2 - \alpha^2}{2\beta\gamma} + \frac{\alpha^4 - \beta^4 - \gamma^4}{2 \cdot 3 \cdot 4\beta\gamma r^2} - \frac{\beta^2\gamma^2}{4\beta\gamma r^2} + \frac{(\beta^2 + \gamma^2 - \alpha^2)(\beta^2 + \gamma^2)}{3 \cdot 4\beta\gamma r^2},$$

therefore

$$\cos A = \frac{\beta^2+\gamma^2-\alpha^2}{2\beta\gamma} + \frac{\alpha^4+\beta^4+\gamma^4-2\alpha^2\beta^2-2\alpha^2\gamma^2-2\beta^2\gamma^2}{2\cdot 3\cdot 4\beta\gamma r^2}. \quad (3.6.2)$$

Setting $\frac{1}{r}=0$, the angle A becomes the angle opposite to the side α in a rectilinear triangle, whose sides are α, β, γ. If we denote this angle by A' we get

$$\cos A' = \frac{\beta^2+\gamma^2-\alpha^2}{2\beta\gamma}.$$

Therefore

$$\sin^2 A' = \frac{2\alpha^2\beta^2+2\alpha^2\gamma^2+2\beta^2\gamma^2-\alpha^4-\beta^4-\gamma^4}{4\beta^2\gamma^2}$$

and from (3.6.2) we have

$$\cos A = \cos A' - \frac{\beta\gamma\sin^2 A'}{2\cdot 3r^2}.$$

If we denote by Θ the area of the rectilinear triangle whose sides are α, β, γ then $\Theta = \frac{\beta\gamma\sin A'}{2}$ and thus

$$\cos A = \cos A' - \frac{\Theta\sin A'}{3r^2},$$

whence it follows that, up to quantities of order $\frac{1}{r^4}$,

$$A = A' + \frac{\Theta}{3r^2}.$$

The last formula is obtained since $\cos(A'+\frac{\Theta}{3r^2}) = \cos A'\cos\frac{\Theta}{3r^2} - \sin A'\sin\frac{\Theta}{3r^2}$ and as $\frac{1}{r}\to 0$ we have $\cos\frac{\Theta}{3r^2}\to 1$ and $\sin\frac{\Theta}{3r^2}\to\frac{\Theta}{3r^2}$.

Similarly,

$$B = B' + \frac{\Theta}{3r^2},\quad C = C' + \frac{\Theta}{3r^2}.$$

Therefore

$$A+B+C = A'+B'+C' + \frac{\Theta}{r^2} = \pi + \frac{\Theta}{r^2}$$

since in each rectilinear triangle the sum of the angles is π. Consequently, $area(ABC) = \frac{\Theta}{r^2}$ and this result is exact up to the quantities of order $\frac{1}{r^4}$. Now, since

$r^2 area(ABC) = area(A'B'C')$, it follows that the areas of spherical triangles on a sphere of large radius are approximated well by the areas of rectilinear triangles whose sides have the same length.

3.7 The Spherical Pythagorean Theorem as a Milestone of Spherical Trigonometry

In this section, we will prove the spherical Pythagorean theorem using Euler's method of "maxima and minima" as explained in his memoir [6]. From this theorem we will derive the cosine law and from the latter we will derive the duality/polarity theorem. Then, by using Lagrange's method, as it is explained in Sects. 3.4 and 3.5, one may obtain any spherical trigonometric formula. In doing this, solid geometry will be not used at all. Consequently, we may consider the Pythagorean theorem as the fundamental equation from which all formulae of spherical trigonometry can be derived.

A Direct Proof of the Spherical Pythagorean Theorem with the Method of Calculus of Variations

Euler obtained many formulae of spherical trigonometry without using solid geometry in his proofs. He used intrinsic methods of the sphere and the duality Theorem 3.3.1. More concretely, Euler obtained the spherical trigonometric formulae, by the method of "maxima and minima" [6], which was the name for what is called today the "calculus of variations". As we pointed out, Menelaus of Alexandria was the first who proved results of spherical geometry intrinsically but his work was not known to Euler.

Euler's method is based on the fact that the edge of a triangle is the shortest path between its extremities. In this way, the intrinsic principles for the study of the differential geometry of surfaces are established, principles which are applicable to spaces which are much more general than the sphere or the Euclidean plane and which will lead to the birth of Riemannian geometry.

In the following, by means of one specific example, namely the proof of the Pythagorean theorem, we will explain the *Euler–Lagrange equation* which is the name given today to the fundamental equation of the calculus of variations, invented by Euler and Lagrange.

Basically, the calculus of variations is the study of certain extrema (maxima and minima) of "functionals" defined on some space of functions. The functionals are usually defined by integrals. In our case below, this space of functions is a space of paths joining two given points on the sphere and we assume that the paths are differentiable, or piecewise differentiable. Applying this method in our special case, one can perceive the method and its strength. For a short description of the method of the "calculus of variations" in relation with Euler, see [12], while in [9] one may find an introduction to this area with many examples.

Our proof follows closely the proof of Theorem 1 in [8], where formulae for hyperbolic right triangles are proved.

On the sphere S^2, we consider the geographical coordinates (t, u), where $-\pi < t < \pi$ and $-\frac{\pi}{2} < u < \frac{\pi}{2}$ are the latitude and longitude respectively. Considering the equator and fixing a meridian, we have a pair of axes whose intersection defines the origin $O = (0, 0)$. Let ABC be a right triangle such that the angle $C = \frac{\pi}{2}$. We take A at the origin O and we may assume that AC is contained in the equator. We also assume that if $B = (t, u)$ then $t > 0$ and $u > 0$, see Fig. 3.5. The line AB in the coordinates (t, u) can be seen as the graph of a function $g : [0, b] \to R$, where b is the length of AC. Let also a be the length of BC.

Thus, if $\gamma(t) = (t, g(t)), t \in [0, b]$ is a parametrization of AB and if M is a point of AB with coordinates $(t, g(t))$, we define the function $r(t) = \text{length}(AM)$. Thus, $r'(t) = |\gamma'(t)|$. Now, with respect to the spherical metric (3.1.1) we have

$$\text{length}(AB) = \int_0^b |\gamma'(t)|dt = \int_0^b r'(t)dt = \int_0^b \sqrt{\cos^2 g(t) + (g'(t))^2}dt.$$

Among all the curves connecting A and B which can be represented in the coordinates (t, u) as graphs of differentiable functions, the line AB has the minimal length. Therefore, if we consider the space $C^1([0, b], R)$ of all real differentiable

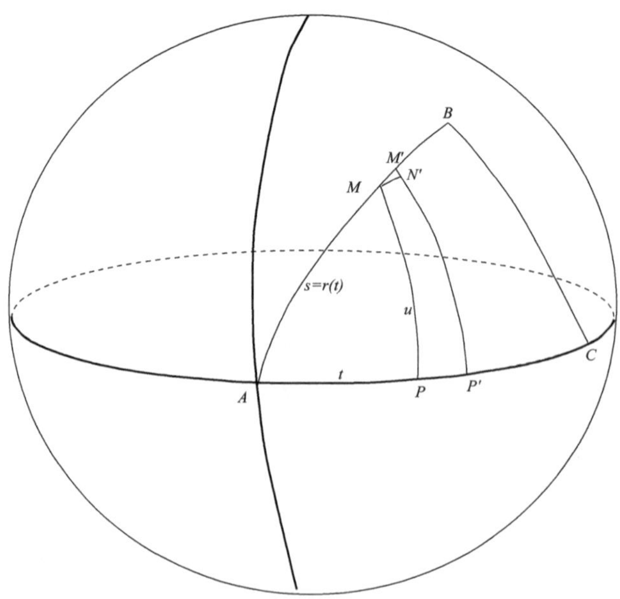

Fig. 3.5 Proof of the Pythagorean theorem

functions $f : [0, b] \to R$ and the functional $L : C^1([0, b], R) \to R$ defined by

$$L(f) = \int_0^b \sqrt{\cos^2 f(t) + (f'(t))^2 dt}$$

then L attains a minimum for $f = g$.

Introducing, for the sake of brevity, the function $F : R^2 \to R$ with $F(x_1, x_2) = \sqrt{\cos^2 x_1 + x_2^2}$, the functional L takes the form

$$L(f) = \int_0^b F(f(t), f'(t))dt.$$

Since L attains a minimum for $f = g$, for every differentiable function $h : [0, b] \to R$ and each real number ε with $g + \varepsilon h \in C^1([0, b], R)$, we get

$$\frac{d}{d\varepsilon}|_{\varepsilon=0} L(g + \varepsilon h) = \lim_{\varepsilon \to 0} \frac{L(g + \varepsilon h) - L(g)}{\varepsilon} = 0.$$

Therefore,

$$\begin{aligned} &\frac{d}{d\varepsilon}|_{\varepsilon=0} \int_0^b (F(g + \varepsilon h, g' + \varepsilon h')dt \\ &= \int_0^b \frac{d}{d\varepsilon}|_{\varepsilon=0} (F(g + \varepsilon h, g' + \varepsilon h')dt = \\ &= \int_0^b [\frac{\partial F}{\partial x_1}(g(t), g'(t))h(t) + \frac{\partial F}{\partial x_2}(g(t), g'(t))h'(t)]dt = 0. \end{aligned}$$

Integration by parts gives

$$\int_0^b [\frac{\partial F}{\partial x_1}(g(t), g'(t)) - \frac{d}{dt}(\frac{\partial F}{\partial x_2}(g(t), g'(t)))]h(t)dt = 0, \ \forall h,$$

which implies that

$$\frac{\partial F}{\partial x_1}(g(t), g'(t)) - \frac{d}{dt}(\frac{\partial F}{\partial x_2}(g(t), g'(t)) = 0 \text{ for all } t \in [0, b]. \tag{3.7.1}$$

Relation (3.7.1) is the so-called Euler–Lagrange equation and if for $i = 1, 2$, we set $G_i(t) = \frac{\partial F}{\partial x_i}(g(t), g'(t))$, it takes the form

$$G_1 = G_2'. \tag{3.7.2}$$

On the other hand, since $r'(t) = F(g(t), g'(t))$, we have

$$r''(t) = \frac{\partial F}{\partial x_1}(g(t), g'(t))g'(x) + \frac{\partial F}{\partial x_2}(g(t), g'(t))g''(t)$$

and by (3.7.2) we have

$$r'' = G_1 g' + G_2 g'' = G_2' g' + G_2 g'' = (G_2 g')'.$$

Thus we deduce that

$$r' = G_2 g' + C, \tag{3.7.3}$$

where C is a constant. Using the previous form of r' and of G_2, (3.7.2) becomes

$$\sqrt{(g')^2 + \cos^2 g} = \frac{(g')^2}{\sqrt{(g')^2 + \cos^2 g}} + C.$$

Resolving this equation in g', we have

$$g'(t) = \frac{\cos(g(t))\sqrt{\cos^2(g(t)) - C^2}}{C}. \tag{3.7.4}$$

Thus, $r'(t) = \sqrt{\cos^2 g(t) + (g'(t))^2}$ becomes

$$r'(t) = \frac{\cos^2 g(t)}{C}. \tag{3.7.5}$$

Now, let $M = (t, u)$, $M' = (t', u')$ be two points on AB and denote by P, P' the projections of M, M' on AC. We consider also a line perpendicular to MP at M and let N' be the intersection point of this line with $M'P'$, see Fig. 3.5. Then the length of $MM' = \Delta s$, the length of $M'N' = \Delta u$ and the length of $PP' = \Delta t$.

Taking $M = A$, the right triangle $AM'P'$ is formed, see Fig. 3.6. In this triangle the following limit Euclidean relation holds:

$$\cos A = \lim_{\Delta s \to 0} \frac{\Delta t}{\Delta s} = \frac{1}{r'(0)}.$$

But from (3.7.5), $\frac{1}{r'(0)} = C$ since $g(0) = 0$. Therefore we have

$$C = \cos A. \tag{3.7.6}$$

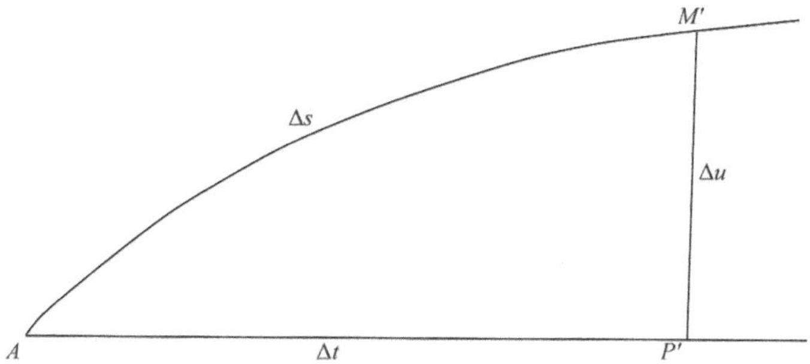

Fig. 3.6 The case $M = A$ and the right triangle $AM'P'$

Now we denote by $w(t)$ $\widehat{AMP}$, see Fig. 3.5. In the triangle $MM'N'$, the angle $\widehat{M'MN'} = \frac{\pi}{2} - w(t)$. Therefore, we have

$$\cos w(t) = \sin(\widehat{M'MN'}) = \lim_{\Delta s \to 0} \frac{\Delta u}{\Delta s} = \frac{g'(t)}{r'(t)} = \frac{\sqrt{\cos^2 g(t) - C^2}}{\cos g(t)}. \tag{3.7.7}$$

From (3.7.7), for $t = b$ we get

$$\cos B = \frac{\sqrt{\cos^2 a - \cos^2 A}}{\cos a}, \tag{3.7.8}$$

since $a = g(b) = length(BC)$. Furthermore, from (3.7.4) and (3.7.5) we get respectively

$$b = \int_0^b dt = \int_0^a \frac{C}{\cos g \sqrt{\cos^2 g - C^2}} dg = \arcsin \frac{C \tan a}{\sqrt{1 - C^2}}$$

and

$$c = r(b) = \int_0^b r'(t)dt = \int_0^a \frac{\cos g}{\sqrt{\cos^2 g - C^2}} dg = \arccos \frac{\cos^2 a - C^2}{1 - C^2}.$$

Therefore, from the previous two relations we get

$$\sin b = \frac{\cos A \sin a}{\sin A \cos a} \tag{3.7.9}$$

and

$$\cos c = \frac{\sqrt{\cos^2 a - \cos^2 A}}{\sin A}. \tag{3.7.10}$$

From the relations established above, we now proceed to derive the spherical Pythagorean theorem.

Corollary 3.7.1 *The following identity holds and is known as the Pythagorean theorem of spherical geometry*

$$\cos c = \cos b \cos a.$$

Proof From (3.7.9) we have

$$\cos b = \frac{\sqrt{\sin^2 A \cos^2 a - \cos^2 A \sin^2 a}}{\sin A \cos a}$$
$$= \frac{\sqrt{(1-\cos^2 A)\cos^2 a - \cos^2 A(1-\cos^2 a)}}{\sin A \cos a}.$$

Therefore,

$$\cos b = \frac{\sqrt{\cos^2 a - \cos^2 A}}{\sin A \cos a} \tag{3.7.11}$$

Obviously, from (3.7.10) and (3.7.11) we have $\cos c = \cos a \cos b$.

The Cosine Law from the Pythagorean Theorem

To prove the spherical law of cosines we transport the method of proof of Theorem 6.9 of [1] from the hyperbolic case to the spherical one.

Let ABC be a spherical triangle whose sides, opposite to A, B, C, have lengths a, b, c respectively. As in [1], we assume that the angle at C is obtuse. The case that this angle is acute can be done with minor modifications.

Let C' be the foot of the altitude from C to AB and B' the foot of the altitude from B to AC. Let v, u, h, k be the lengths of CB', AC', CC' and BB' respectively. From the Pythagorean theorem in the triangles BCC' and ACC' respectively, we have

$$\cos a = \cos h \cos(c-u) = \cos h(\cos c \cos u + \sin c \sin u) \tag{3.7.12}$$

and

$$\cos b = \cos h \cos u. \tag{3.7.13}$$

From (3.7.12) and (3.7.13) we have

$$\cos a = \frac{\cos b}{\cos u}(\cos c \cos u + \sin c \sin u) = \cos b \cos c + \sin c \cos b \tan u$$

hence

$$\cos a = \cos b \cos u + \sin c \sin b \frac{\tan u}{\tan b}. \tag{3.7.14}$$

In the right triangle CBB' we have

$$\cos a = \cos u \cos k, \tag{3.7.15}$$

while in the right triangle ABB' we have

$$\cos c = \cos(b + v) \cos k. \tag{3.7.16}$$

From (3.7.15) and (3.7.16) we have

$$\cos v \cos c = \cos(b + v) \cos a. \tag{3.7.17}$$

Now we need and we will prove a claim.
Claim The following identity is valid

$$\cos b \cos c \cos(b + v) + \sin b \cos c \sin(b + v) = \cos v \cos c. \tag{3.7.18}$$

Proof of the Claim We have

$$\cos b \cos c \cos(b + v) + \sin b \cos c \sin(b + v)$$

$$= \cos b \cos c(\cos b \cos v - \sin b \sin v) + \sin b \cos c(\sin b \cos v + \cos b \sin v)$$

$$= \cos^2 b \cos c \cos v - \cos v \cos c \sin b \sin v + \sin^2 b \cos c \cos v + \sin b \cos c \cos b \sin v$$

$$= \cos v \cos c(\cos^2 b + \sin^2 b) = \cos v \cos c,$$

therefore the claim is proved.

From (3.7.17) and (3.7.18) we get

$$\cos(b + v) \cos a = \cos b \cos c \cos(b + v) + \sin b \cos c \sin(b + v). \tag{3.7.19}$$

Dividing (3.7.19) by $\cos(b + v)$ we get

$$\cos a = \cos b \cos c + \sin b \sin c \frac{\tan(b + v)}{\tan c}. \tag{3.7.20}$$

Therefore, from (3.7.14) and (3.7.20) we have

$$\frac{\tan(b + v)}{\tan c} = \frac{\tan u}{\tan b}. \tag{3.7.21}$$

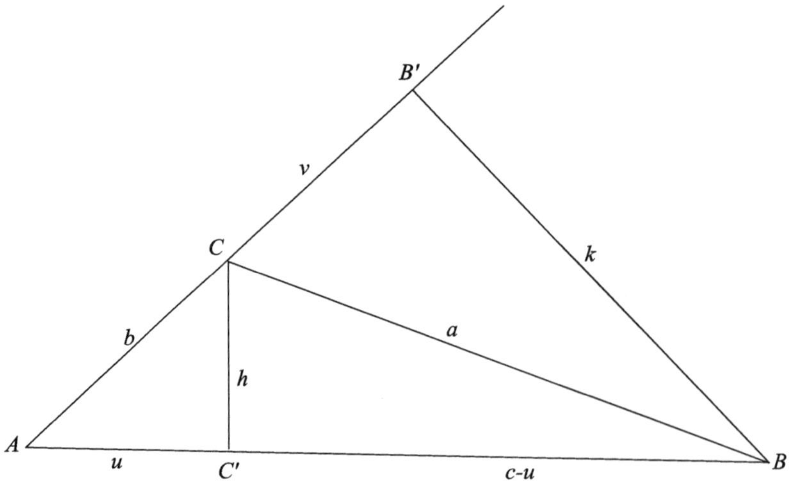

Fig. 3.7 Proof of law of cosines via the Pythagorean theorem

Now from Fig. 3.7, considering the point B fixed, the projection B' is also fixed. Therefore, the ratio $\frac{\tan(b+v)}{\tan c}$ remains fixed independently of the position of the point C. Consequently, from (3.7.21) we have

$$\lim_{b\to 0}\frac{\tan(b+v)}{\tan c} = \lim_{b\to 0}\frac{\tan u}{\tan b}.$$

Using the power expansion of the function $\tan x$ we deduce easily that

$$\lim_{b\to 0}\frac{\tan u}{\tan b} = \lim_{b\to 0}\frac{u}{b} = \cos A. \tag{3.7.22}$$

Finally, from (3.7.20) and (3.7.22) we get the law of cosines, that is,

$$\cos a = \cos b \cos c + \sin b \sin c \cos A.$$

The Duality Theorem: A Second View

In this subsection we will prove the duality theorem as an application of the Pythagorean theorem and the law of cosines.

Let ABC be a spherical triangle with at least two right angles. Then the triangle ABC will be referred to as a *double right triangle*.

Lemma 3.7.2 *If $B = C = \frac{\pi}{2}$ then $b = c = \frac{\pi}{2}$, that is, A is the pole corresponding to BC. Furthermore, $A = a$.*

Proof Since $B = C = \frac{\pi}{2}$, we deduce from the spherical Pythagorean theorem that

$$\cos b = \cos a \cos c \text{ and } \cos c = \cos a \cos b.$$

If $\cos b \cos c \neq 0$ then by multiplying the previous relations we deduce that $\cos^2 a = 1$ which is impossible since $a \neq k\pi$, $k \in \mathbb{Z}$. Therefore $\cos b \cos c = 0$. This implies that at least one of b or c is $\frac{\pi}{2}$. Assuming, without loss of generality, that $b = \frac{\pi}{2}$ we deduce easily from the identity $\cos c = \cos b \cos a + \sin b \sin a \cos C$ that $\cos c = 0$, hence $c = \frac{\pi}{2}$.

Furthermore, in the triangle ABC, the law of cosines gives that $A = a$.

We have the following.

Lemma 3.7.3 *Let ABC be a spherical triangle with $b = c = \frac{\pi}{2}$. Then $B = C = \frac{\pi}{2}$.*

Proof By the law of cosines, we have $\cos b = \cos a \cos c + \sin a \sin b \cos B$. Therefore, $\sin a \cos B = 0$. If $\cos B = 0$ then $B = \frac{\pi}{2}$. If $\cos B \neq 0$ then $\sin a = 0$, hence $\cos a = \pm 1$.

Using the law of cosines, $\cos a = \cos b \cos c + \sin b \sin c \cos A$, we obtain $\cos A = \pm 1$ which is impossible, hence the lemma is proved.

Now we can prove the duality theorem.

Theorem 3.7.4 *The measure of an angle in a spherical triangle and the length of the corresponding side in its dual are supplementary.*

Proof Let $A'B'C'$ be the dual of the spherical triangle ABC. As usual, the lengths of the sides of $A'B'C'$ are denoted by a', b', c'. Then, from Lemma 3.7.3, the triangle $A'CB'$ is a double right triangle, thus, from Lemma 3.7.2, we have $\widehat{A'CB'} = c'$. On the other hand, the triangles $AB'C$, $BA'C$ are double right by construction.

Therefore, we have

$$c' + C = \widehat{A'CB'} + \widehat{ACB} = \widehat{A'CB} + \widehat{ACB'} = \frac{\pi}{2} + \frac{\pi}{2} = \pi$$

which proves the theorem.

References

1. N. A'Campo, A. Papadopoulos, *Notes on Non-Euclidean Geometry, Strasbourg Master Class on Geometry*, ed. by A. Papadopoulos. IRMA Lectures in Mathematics and Theoretical Physics (EMS, Berlin, 2012)
2. S. Alexander, V. Kapovitch, A. Petrunin, Alexandrov geometry: foundations. arXiv:1903.08539v4. https://doi.org/10.48550/arXiv.1903.08539
3. R. Bonola, *La geometria non-euclidea, Esposizione storico-critica del suo sviluppo*, 1st edn. (Ditta Nicola Zanichelli editore, Bologna, 1906). German translation by M. Liebmann in the

collection Wissenschaft und Hypothese, Teubner, Leipzig, 1908. English translation by H. S. Carslaw, Non-Euclidean geometry, A critical and historical study of its development. First edition, Chicago 1912, reprinted by Dover, New York 1955
4. Y. Burago, M. Gromov, G. Perelman, A.D. Alexandrov spaces with curvature bounded below. Russian Math. Surv. **47**(2), 1–58 (1992)
5. J.-P. de Gua de Malves, Trigonométrie sphérique, Mémoire de l'Académie royale des sciences (1873)
6. L. Euler, Principes de la trigonométrie sphérique tirés de la méthode des plus grands et des plus petits. Mémoires de l' Académie des sciences de Berlin, vol. 9 (University of the Pacific, Stockton, 1753–1755), pp. 233–257; Opera Omnia, Series 1, Vol. XXVII, p. 277-308, translated by A. Papadopoulos and A. Zhukova, this volume
7. L. Euler, Trigonometria sphaerica universa, ex primis principiis breviter et dilucide derivata. Acta Acad. Sci. Imper. Petropol. **3**, 72–86 (1782); Opera Omnia Series 1, vol. XXVI, p. 224-236, translated by R. Caddeo, this volume, Chapter 11
8. E. Frenkel, On area and volume in spherical and hyperbolic geometry, PhD thesis, University of Strasbourg (2018). https://tel.archives-ouvertes.fr (HAL Id: tel-01872314)
9. M. Kot, *A First Course in the Calculus of Variations*. Student Mathematical Library, vol. 72 (AMS, Providence, 2014)
10. J.-L. Lagrange, Solution de quelques problèmes relatifs aux triangles sphériques avec une analyse complète de ces triangles. J. de l'École. Polythech. **6**, 331–359 (1799). English translation by V. Alberge and A. Papadopoulos, this volume
11. J.H. Lambert, Theorie der Parallellinien, manuscript (1766). Published in P. Stäckel, F. Engel, Die Theorie der Parallellinien von Euklid bis auf Gauss, eine Urkundensammlung zur Vorgeschichte der nicht-euklidischen Geometrie (B. G. Teubner, Leipzig, 1895). French translation A. Papadopoulos, G. Théret, La théorie des parallèles de Johann Heinrich Lambert, critical edition with French translation and mathematical and historical commentary, Librairie Scientifique et Technique Albert Blanchard, coll. Sciences dans l'Histoire, Paris (2014), 214pp.
12. A. Papadopoulos, On the work of Euler and his followers on spherical geometry, Gaṇita Bhāratī (Indian Mathematics). Bullet. Ind. Soc. History Math. **36**(1–2), 237–292 (2014)
13. A. Papadopoulos, G. Théret, A review of Johann Heinrich Lambert's memoir Theorie der Parallellinien, in *Spherical Geometry in the Eighteenth Century, I: Euler, Lagrange and Lambert*, ed. by R. Caddeo, A. Papadopoulos, Chap. 7 (Springer, Berlin, 2026)
14. R. Rashed (Ed.), Triangle polaire, In: Lexique historique de la langue scientifique arabe, sous la direction de R. Rashed, Hildesheim (2017), pp. 96–100
15. R. Rashed, A. Papadopoulos, On Menelaus' Spherics III.5 in Arabic mathematics, I: Ibn ʻIrāq. Arabic Sci. Philos. **24**(1), 1–68 (2014)
16. R. Rashed, A. Papadopoulos, *Menelaus' Spherics*. Early translation and al-Māhānī/al-Harawī's version. Dual Arabic-English text. Scientia Graeco-Arabica, 21 (De Gruyter, Berlin, 2017). Menelaus' Spherics, De Gruyter, 2017
17. B. Riemann, Ueber die Hypothesen, welche der Geometrie zu Grunde liegen. Habilitationsvortag, Göttingen, 1854; Abhandlungen der Königlichen Gesellschaft der Wissenschaften zu Göttingen XIII (1968), pp. 133–152
18. B.A. Rosenfeld, *History of Non-Euclidean Geometry*. Translated by Abe Shenitzer, Studies in the History of Mathematics and Physical Sciences, vol. 12 (Springer, New York, 1988)

Chapter 4
Notes on Euler's Memoir *A Construction Relative to a Problem of Pappus of Alexandria*

Guillaume Théret

Abstract A number of eminent mathematicians have been interested in the "Pappus problem", a geometrical construction so called by Euler, who gave a solution in his memoir *A construction relative to a problem of Pappus of Alexandria* translated in Chap. 16 of the present volume. In this chapter, we place this result in a more general context, survey its history, and provide detailed versions, and sometimes proofs, due to several mathematicians, ancient and modern, ranging from Greek Antiquity until modern times. We provide the details of Pappus' original proof of this problem, and that of a *generalized Pappus problem*, in the setting of projective geometry, formulated and proved in the modern times. We also give a proof of a related construction of Apollonius of circles tangent to three given objects in the plane.

Keywords Greek mathematics · Pappus problem · Projective geometry · Euclidean geometry · Apollonius problem · Leonhard Euler

AMS Codes 01A50, 01A45, 01A35, 01A20, 51A05, 51N20

4.1 Pappus' Problem and its Solution

In Proposition 117 of Book VII of the *Collection*,[1] Pappus of Alexandria (ca. 290-350 AD) solved the following problem (see Fig. 4.1).

Theorem 4.1.1 (Proposition 117) *Let $\mathcal{C}$ be a circle with center O and let A, B, C be three collinear points. Construct a triangle $A'B'C'$ inscribed in $\mathcal{C}$ such that $A \in B'C'$, $B \in A'C'$ and $C \in A'B'$.*

[1] We are using the numbering in [4].

G. Théret (✉)
Laboratoire Interdisciplinaire Carnot de Bourgogne (ICB), Dijon Cedex, France

R. Caddeo, A. Papadopoulos (eds.), *Spherical Geometry in the Eighteenth Century I*, https://doi.org/10.1007/978-3-032-12466-1_4

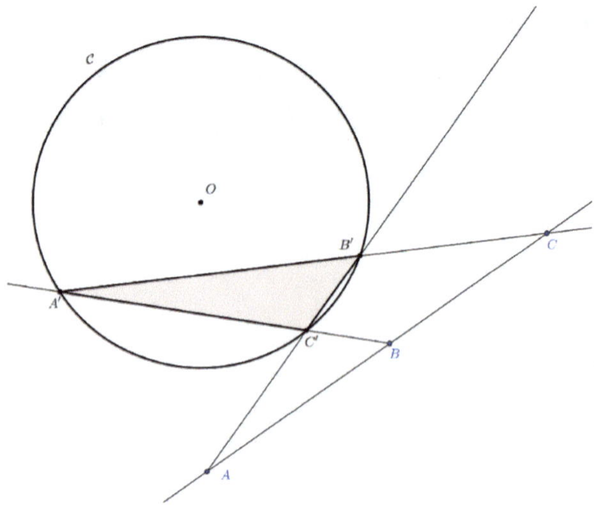

Fig. 4.1 Pappus' theorem

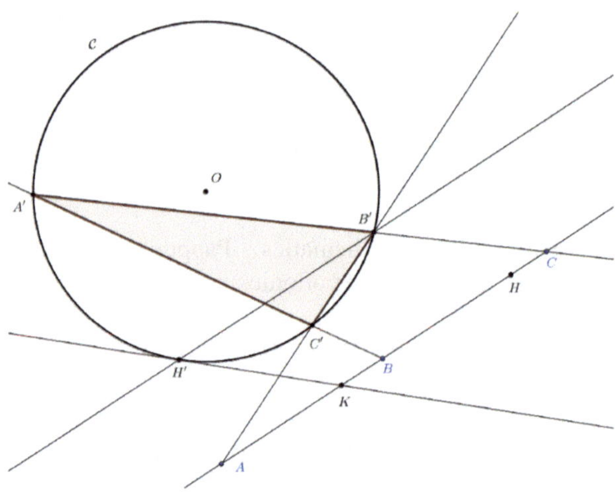

Fig. 4.2 Pappus' construction of the inscribed triangle $A'B'C'$ whose sides pass through the three given points A, B, C

"To construct" means "to construct with unmarked straight edge and compass". This is an easily stated yet difficult problem to solve in Euclidean geometry. Here is a constructive solution to this problem; see Fig. 4.2. (In what follows, $\mathrm{P}_{\mathcal{C}}(M)$ denotes the **power of a point** M with respect to a circle $\mathcal{C}$. This can be defined as the squared distance from the point M and the contact point between $\mathcal{C}$ and the tangent line through M.)

(1) Construct the point H on BC such that $BH \times BC = \mathrm{P}_{\mathcal{C}}(B)$.
(2) Construct the point K on HA such that $HK \times HA = \mathrm{P}_{\mathcal{C}}(H)$.
(3) Construct the contact point H' of the tangent to $\mathcal{C}$ through K.
(4) Let $H'B'$ be the line parallel to HB with B' on $\mathcal{C}$.
(5) Let C' and A' be obtained by extending AB' and BC'.

These steps are not very instructive and it is unclear why they should lead to the solution of the problem. Much clarity is gained by reading is the proof of Pappus himself that we reproduce here, with minor changes, as it appears in his Book VII (Propositions 117 and 105). In particular, it turns out that

- C' is the intersection of HH' with $\mathcal{C}$;
- The points H', A, K, C' are concyclic;
- The points A', C', H, C are concyclic.

4.2 Pappus' Original Proof

Pappus' general method for justifying a construction is to start from the already completed picture and to draw from it a logical thread of consequences which eventually involve points or quantities defined or derived from known data. This is the *analytic* part of his solution. He then reverses the logical sequence of consequences, to obtain the steps of a constructive proof.[2] This is his *synthetic* part of his solution. Usually, he has to check some particular points further to see if the reverse process leads to the proper solution indeed.

Another ingredient entering the proof to follow is his moving the problem to a simpler one which he already solved. Specifically, here, Pappus manages to carry the initial situation to that where one of the three collinear points is at infinity; so he begins with solving this particular case in his Proposition 105 (see Fig. 4.3).[3]

Theorem 4.2.1 (Proposition 105) *Let $\mathcal{C}$ be a circle with center O and let D, E be two points. Construct a triangle $A'B'C'$ inscribed in $\mathcal{C}$ such that $D \in B'C'$, $E \in B'A'$ and $A'C'$ is parallel to DE.*

Pappus' proof of Proposition 105 Suppose that the construction has been completed (see Fig. 4.4). Draw the tangent FA' to $\mathcal{C}$ at point A', where F is the intersection point between this tangent and DE. Since $A'C'$ is parallel to DE, we have $\widehat{A'C'D} = \widehat{C'DE}$. But $\widehat{A'C'D} = \widehat{FA'E}$. Hence $\widehat{C'DE} = \widehat{FA'E}$. Thus the four points A', B', D, F are on a circle. Hence $EA' \times EB' = EF \times ED$. But $EA' \times EB'$ is given, because it is equal to the square of the tangent from E to circle $\mathcal{C}$. Therefore $EF \times ED$ is also given. Since DE is given, EF is given too, which gives F. This, in turns, gives the point A'. Now, the point E being given, this gives points B' and C'.

---

[2] "Demonstrating is reversing the train of thought": a quote from O. Teichmüller, *Extremal quasiconformal mappings and quadratic differentials*, [8].

[3] Pappus' Proposition 107 of the same book is another solution of the same problem as Proposition 105. The difference is that in Proposition 105, the third vertex B' of the inscribed triangle $A'B'C'$ lies between the parallel lines $A'C'$, DE while in Proposition 107, the third vertex is away from both parallel lines. Proposition 108 deals with the case where points D, E lie within the circle $\mathcal{C}$.

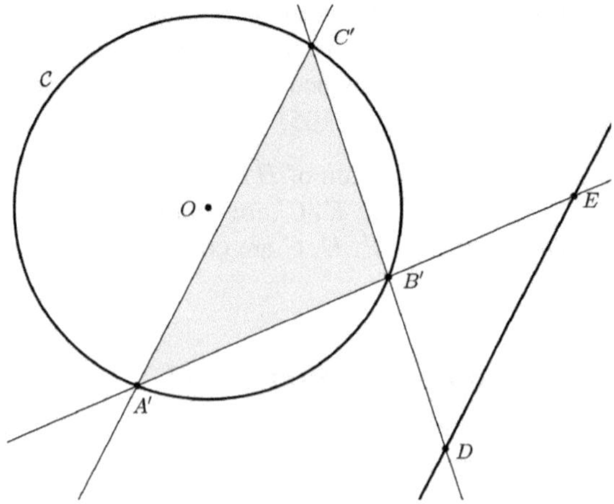

Fig. 4.3 Pappus' Proposition 105. The sides of the inscribed triangle $A'B'C'$ pass through the three given collinear points D, E, F_∞ with F_∞ at infinity

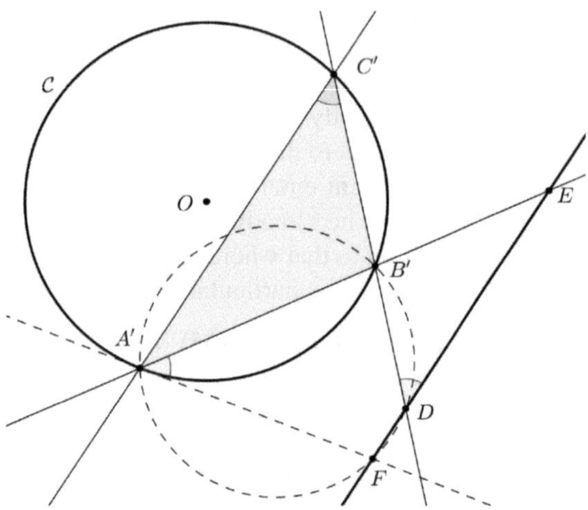

Fig. 4.4 Figure for the proof of Pappus' Proposition 105

Let us proceed now to synthesis. Let circle $\mathcal{C}$ and two points D, E be given. Construct point F on DE such that $DE \times DF$ is equal to the square of the tangent from E. Draw from F the tangent FA' to circle $\mathcal{C}$ at A'. Join $A'E$ which gives point B' on $\mathcal{C}$, and let DB' be joined and extended to C' on $\mathcal{C}$.

I claim that $A'C'$ is parallel to DE.

For, since both $EF \times ED$ and $EA' \times EB'$ are equal to the square of the tangent from E, we have $EF \times ED = EA' \times EB'$. Hence points A', B', D, F are concyclic. Therefore $\widehat{FA'E} = \widehat{B'DE}$. But $\widehat{FA'E} = \widehat{A'C'B'}$. Hence, $\widehat{A'C'B'} = \widehat{B'DE}$ and these are alternate angles, therefore $A'C'$ is parallel to DE.

Let us now proceed with the proof of Pappus' construction for the case of three arbitrary collinear points A, B, C (his Proposition 117).

Proof of Pappus' Proposition 117 Suppose that we have the complete picture in front of our eyes (see Fig. 4.5). Draw from point B' the line $B'H'$ parallel to AC.

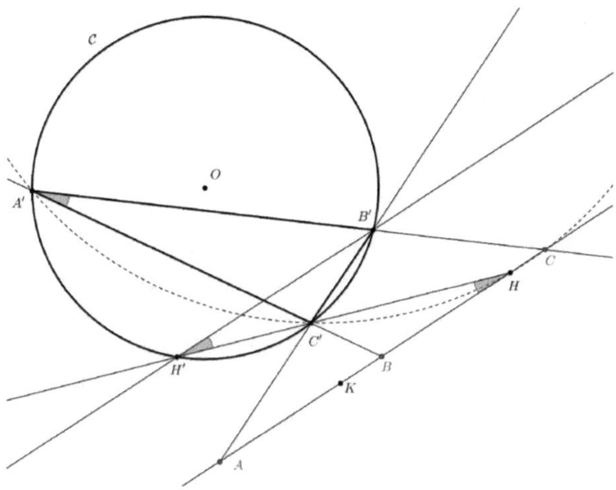

Fig. 4.5 Pappus' construction of the inscribed triangle $A'B'C'$ whose sides pass through the three given points A, B, C

Connect $H'C'$ and extend it to point H on AC. We thus have

$$\widehat{B'H'C'} = \widehat{C'HA}.$$

Now

$$\widehat{B'H'C'} = \widehat{C'A'B'}.$$

angles $\widehat{BA'C}$ and $\widehat{C'HC}$ are thus supplementary, hence points A', C', H, C are concyclic; we therefore have

$$BA' \times BC' = BC \times BH.$$

But $BA' \times BC'$ is equal to the square of the tangent drawn from B to the circle, hence it is a given quantity. Therefore $BC \times BH$ is also a given quantity. Since BC is given, so is BH. Now point B is given, so point H is also given. But point A is given. From Proposition 105, we know how to construct points B', C', H' from the two given points H, A in such a way that the line $B'H'$ is parallel to line HA. Thus, point C' is given and since point B is also given, the line $C'B$ gives A'. Since point C is given, so is the line CA'. This is what we needed.

The inscribed triangle $A'B'C'$ can then be constructed as follows.

Consider a circle $\mathcal{C}$ and three collinear points A, B, C. Construct a point H on BC such that product $BC \times BH$ is equal to the square of the tangent from B to circle $\mathcal{C}$. Given the points A, H, construct point C' on circle $\mathcal{C}$ such that $B'H'$ is parallel to HA, where B' and H' are the intersection points of this parallel line with $\mathcal{C}$ (this construction is carried out in Proposition 105 of the same book of the *Collection*). The line through the points B, C' intersects $\mathcal{C}$ in another point A'. We claim that the three points A', B', C are collinear.

To see this, notice that each of the products $BA' \times BC'$ and $BC \times BH$ is equal to the square of the tangent from B to circle $\mathcal{C}$, so $BA' \times BC' = BC \times BH$. Therefore C, H, C', A' are concyclic. Since the angles $\widehat{B'H'C'}$ and $\widehat{C'HA}$ are equal (because of parallel lines), and since, in the circle, angle $\widehat{B'H'C'}$ is equal to angle $\widehat{B'A'C'}$, then $\widehat{B'A'C'} = \widehat{C'HB}$. But the points A', C', H, C are concyclic. Therefore the line $A'B'$ is the same line as $B'C$.

Pappus makes extensive use of Propositions 35 and 36 of Book III of Euclid's *Elements*, today known as the intersecting chord theorem and tangent-secant theorem. The converse is also used, as well as the inscribed angles theorem (Book III, Proposition 20 of the *Elements*). Equality of alternate angles between parallel lines (Proposition 29 of Book I of the *Elements*) is also an important ingredient. Many of these results are based upon Euclid's parallel postulate.[4]

4.3 Euler's Construction

The story tells that Cramer[5] proposed in 1742 a natural generalization of this problem, where the three given points A, B, C need not be aligned (see Fig. 4.6).[6]

Theorem 4.3.1 *Let $\mathcal{C}$ be a circle with center O and let A, B, C be three points. Construct a triangle $A'B'C'$ inscribed in $\mathcal{C}$ such that $A' \in BC$, $B' \in AC$ and $C' \in AB$.*

The first published solution to Cramer's problem is due to Giovanni Francesco Salvemini da Castiglione, known in short as Castillon, in 1776. Today, this problem is known as Cramer–Castillon problem and, according to Castillon himself, it is "regardé comme fort difficile" (regarded as very difficult). It has become ever since the kind of problem every great mathematician with strong inclination toward

[4] Thus, working in an axiomatic, *model-free* hyperbolic or spherical plane, these proofs cannot be adapted as they are.

[5] This is the Swiss mathematician Gabriel Cramer (1704–1752).

[6] It seems obvious that this problem has ever existed much before its official statement by Cramer. Here I am quoting the Italian-born mathematician Johann Castillon (1708-1791): "Selon mes conjectures, quelque Amateur de la Géométrie des Anciens généralisa le Problème de Pappus, et, ne l'ayant résolu qu'avec beaucoup de peine, ou, peut-être, ne l'ayant point résolu, il le proposa verbalement à quelqu'un qui courait la même carrière. Celui-ci suivait l'exemple de son prédécesseur; et de main en main le problème est parvenu jusqu'à moi. Il semble que le petit nombre de Géomètres qui le connaissaient, le gardaient pour embarrasser les autres dans les occasions." (According to my conjectures, some Amateur of Ancient Geometry generalized Pappus' Problem, and, having solved it only with great difficulty, or, perhaps, having not solved it at all, he verbally proposed it to someone who was pursuing the same career. The latter followed his predecessor's example, and from hand to hand the problem has come down to me. It seems that the small number of Geometers who knew it, used to keep it to embarrass others on occasions) [10].

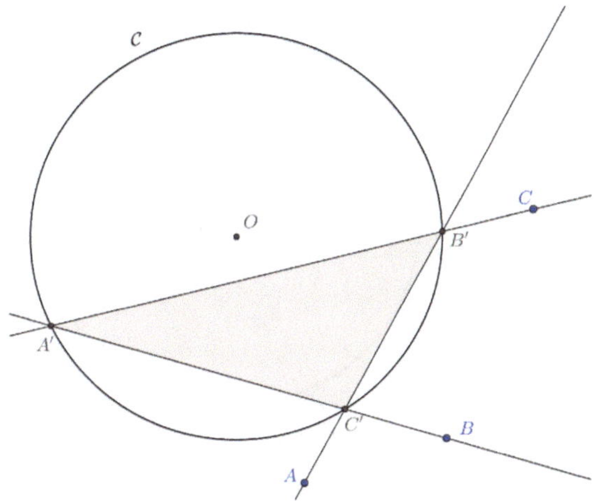

Fig. 4.6 Cramer–Castillon problem

geometry has to struggle with to prove him/herself. Euler, Lagrange and others naturally tackled it, each one with his favourite means.[7]

Euler's solution[8] to the Cramer–Castillon's problem is, according to my opinion, almost impossible to understand without knowing fine details about Pappus' construction; and even in that case, no clues are given that could help to understand why the construction works. Euler's solution then seems almost miraculous. Worse, his paper starts with cumbersome computations that are, as I see them, of no use for understanding his solution to Castillon's problem. Nevertheless, one thing is clear in Euler's intention: to construct a solution only using classical geometry and contradicting Castillon's assertion that Pappus' construction cannot be extended to solve Cramer's problem.[9] And Euler's solution follows similar steps indeed. Here is his solution, almost as it was published (as you will see, it is rather dry; please keep an eye on Fig. 4.7).

(1) Construct a point H on BC such that $BH \times BC = \mathrm{P}_{\mathcal{C}}(B)$.
(2) Construct a point K on HA such that $HK \times HA = \mathrm{P}_{\mathcal{C}}(H)$.
(3) Construct a point H' on $\mathcal{C}$ such that $KO \times \cos \widehat{KOH'} = r \times \cos \widehat{BHA}$ where r is the radius of $\mathcal{C}$.
(4) Bisect the angle $\widehat{BHA}$ with line Ht.

[7] For instance, Lagrange offered an algebraic proof of this problem in his memoir *Solution algébrique d'un problème de géométrie*, Nouveaux Mémoires de l'Académie royale des Sciences et Belles-Lettres de Berlin, 1776, translated in Chap. 17 of the present volume. Note that Euler's proof is prior to Lagrange's, even though the years of publication say the opposite.

[8] As it appeared in *Acta Academiae Scientarum Imperialis Petropolitinae* **4**, 1783, p. 91–96. *Opera Omnia*: Series 1, Volume 26, p. 237–242. See Euler's memoir translated into English in Chap. 14 of the present volume.

[9] As Castillon puts it: "Pappus en donne une solution qui ne peut pas s'appliquer aux autres cas" (Pappus gives a solution that cannot be applied to the other cases).

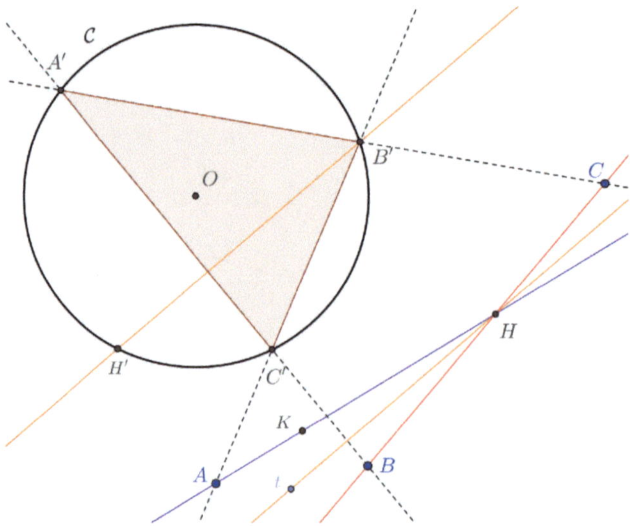

Fig. 4.7 Euler's solution to Cramer's problem

(5) Let $H'B'$ be the line parallel to Ht with B' on $\mathcal{C}$.
(6) Let C' and A' obtained by extending AB' and CB'.

Notice that when A, B, C are aligned, we recover precisely Pappus' solution (since then $\cos \widehat{BHA} = 1$ and hence H' is the contact point of the tangent from K to $\mathcal{C}$).

As Euler notes, his method yields two solutions. He also studies two particular cases where two or all three given points lie at infinite distance from O. Finally, he ends his paper by noticing that his construction can be carried on the sphere, making this problem also a problem in spherical geometry.

In order to at least understand why Euler's construction works, we reproduce below the proof due to Fuss, one of Euler's young collaborators, published in *Acta academiae scientiarum Petropolitanae* **4** I (1780: I), 1783 p. 97–104. Fuss' memoir is very clear and all details are provided for a complete understanding of Euler's construction.

4.4 Fuss' Explanation

Fuss' explanation of Euler's construction (or of his own construction) is similar to Pappus'. He starts with the completed picture and draws from it a list of information which, read the other way around, leads to the construction steps.

Fuss' Explanation Suppose that we have triangle $A'B'C'$ in front of our eyes (refer to Fig. 4.8).

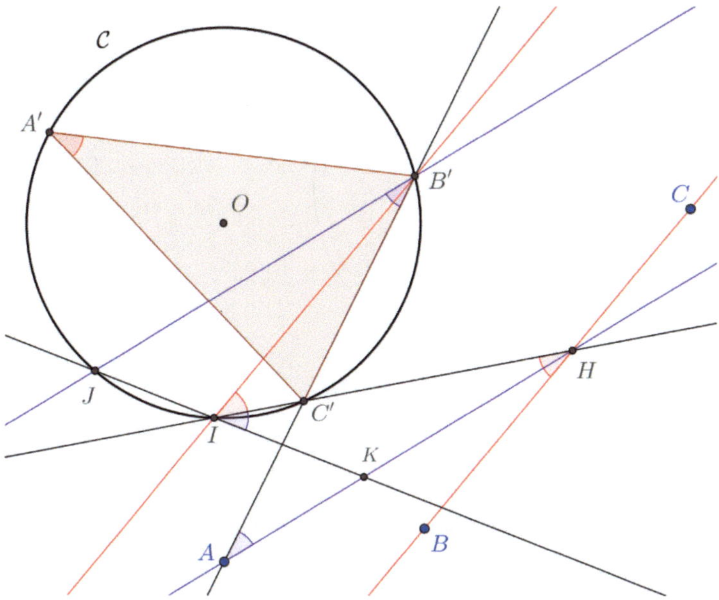

Fig. 4.8 Figure for Fuss' explanation of Euler's solution to Cramer problem

- Draw the line through B' parallel to BC. It cuts $\mathcal{C}$ at point I.
- Draw the line IC' that cuts BC at a point called H.
- Draw AH, then draw the line through B' parallel to AH; it cuts $\mathcal{C}$ at point J.
- Draw the line IJ that cuts AH at a point called K.

We now prove that $BH \times BC = \mathrm{P}_{\mathcal{C}}(B)$ and $HK \times HA = \mathrm{P}_{\mathcal{C}}(H)$. After this is done, the construction amounts to making the synthetis, that is, reversing the logical train of thoughts.

First, $\widehat{BHC'} = \widehat{HIB'}$ (because $B'I$ is parallel to BC). Now $\widehat{B'IC'} = \widehat{B'A'C'}$ (because they sit on the same arc). Hence, $\widehat{BHC'} = \widehat{B'A'C'}$. Therefore triangles $BC'H$ and BCA', having the angles $\widehat{BHC'}$ and $\widehat{BA'C}$ equal and the angle at B in common, are similar. From there it follows that $BC' : BH = BC : BA'$, hence $BC' \times BA' = BC \times BH$.

Now we have $BC' \times BA' = \mathrm{P}_{\mathcal{C}}(B)$, therefore $BC \times BH = \mathrm{P}_{\mathcal{C}}(B)$, which was the first point to be shown.

Now $\widehat{JB'C'} = \widehat{HAC'}$ (because $B'J$ is parallel to AH).

But $\widehat{JB'C'}$ and $\widehat{JIC'}$ are supplementary, hence $\widehat{JB'C'} = \widehat{KIC'}$. Therefore the triangles HAC' and HIK, having the angle at H in common and angles HAC' and HIK equal, are similar. Consequently, $HC' : HA = HK : HI$. Hence $HA \times HK = HC' \times HI = \mathrm{P}_{\mathcal{C}}(H)$. This was the other point.

Here are now the steps of the construction (compare Fig. 4.9):

- Construct H on BC such that $BH \times BC = \mathrm{P}_{\mathcal{C}}(B)$.

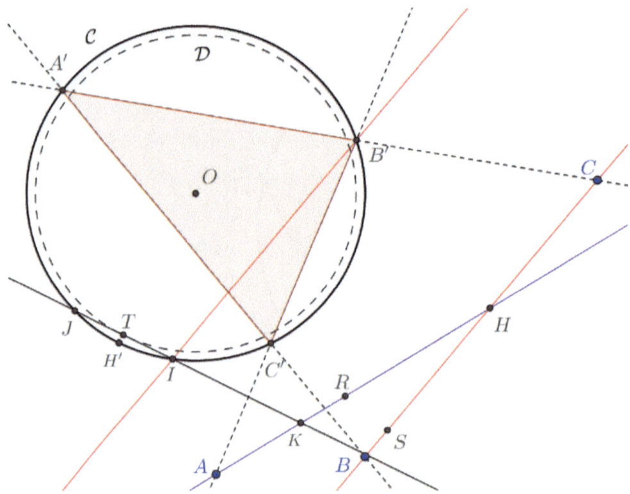

Fig. 4.9 Steps for constructing the solution to Cramer's problem

- Construct the point K on HA such that $HK \times HA = \mathrm{P}_{\mathcal{C}}(H)$.
- Construct the point R on HA such that HR equals the radius of $\mathcal{C}$.
- Drop from R the perpendicular to BC hitting BC at point S (HS then gives $r \times \cos \widehat{BHA}$ in Euler's construction).
- Draw circle $\mathcal{D}$ with center O and radius HS.
- Construct the line through K that is tangent to $\mathcal{D}$ at point T and which intersects $\mathcal{C}$ at points I and J. (Euler's point H' on $\mathcal{C}$ is actually obtained by extending OT until it reaches $\mathcal{C}$. Since points I, J are equidistant from T, the line $OT = OH'$ bisects angle $\widehat{IOJ}$.)
- Draw the line through I and parallel to BC;[10] the other intersection point of this line with $\mathcal{C}$ is B'. (The proof rests on the fact that $\widehat{IB'J} = \widehat{BHA}$, which is equivalent, by the inscribed angle theorem, to Euler's method with the bisector.)
- Complete the picture by drawing the lines $B'C$ and $B'A$ which give the points A', C' on circle $\mathcal{C}$.

The line $A'C'$ passes through B and $A'B'C'$ is the sought-after triangle.

To prove that our construction works, it remains to check that the straight line $B'J$ is necessarily parallel to the straight line HA. This can be proved in the following way. Since in the construction we have supposed that the segment HR is equal to the radius OM of the given circle $\mathcal{C}$, and, RS being drawn, the smallest circle $\mathcal{D}$ has radius $OT = HS$, we have $TJ = RS$. Therefore angles $\widehat{BHR}$, $\widehat{TOJ}$ and $\widehat{IB'J}$ are equal. Consequently, since by construction the straight line $B'I$ is parallel to the straight line BC, the straight line $B'J$ must also be parallel to HA.

[10] There are two possibilities in general depending upon which point is I and J.

Euler's proof of Castillon's problem seems to have been forgotten. One reason may be that the problem was reckoned since then to belong in essence to the realm of projective geometry which is well suited for proving the natural generalization of the problem involving any number of points $A, B, C, D, \ldots$ and replacing the circle by any (non degenerate) conic.

4.5 Generalization and Projective Geometry

It seems unlikely to me that Euler did not think about generalizing Castillon's problem to a situation involving more than 3 given points. But maybe I am wrong and he did not think about it; maybe he thought about it but did something else instead; maybe he thought about it and did not find any answer; maybe he thought about it, found an answer and did not publish it (unlikely).

Generalizing Castillon's problem to more than three points seems to be a very hard task, almost out of reach, until projective geometry comes on the scene. The proof of the general case given below and taken from Berger's *Geometry II* [1], is fantastically elegant and provides a construction using ruler only, whereas the Euclidean construction needs the compass thereupon. Why is it so?

This is because the question asked is all about incidence relationships and, as Daniel Perrin puts it, it belongs to the realm of projective geometry; "le théorème est ici débarrassé de la gangue des notions affines et euclidiennes inutiles" (the theorem is, here, stripped of the gangue of useless affine and Euclidean notions).[11]

However, this requires a good deal of knowledge of projective geometry (axis of homographies and so on). Castillon's problem shows the importance of projective geometry and gives an extra: we then know that this construction works in hyperbolic geometry as well.[12] For those persons who, like me before writing this chapter, have little clue about what the axis of a homography of a conic might be, I straight away add a reminder on this matter.

Let P denote the projective plane and let $P^\star$ denote the dual projective plane. Given a point m in P, we shall denote by $m^\star$ the line in $P^\star$ thought of as the pencil of lines of P through the point m.

Given a (proper) conic $\mathcal{C}$, there is a natural way of associating bijectively points of $\mathcal{C}$ to lines: fix a basepoint $m \in \mathcal{C}$ and associate to any point $n \in \mathcal{C}$ the line nm (which

[11] "On peut dire, en quelque sorte, que chaque théorème possède une (et une seule ?) niche écologique privilégiée, qui correspond au cadre dans lequel il s'énonce avec le plus de généralité et, souvent, où il se démontre avec plus de facilité." (In a way, we could say that each theorem possesses one (and only one?) privileged ecological niche, which corresponds to the framework in which it can be stated with the greatest generality and, often, demonstrated with the greatest ease), D. Perrin [6].

[12] When the conic $\mathcal{C}$ is a circle, one might develop the same arguments in the context of anallagmatic geometry (geometry of inversions in a circle) which is maybe more familiar to some geometers.

is the tangent to $\mathcal{C}$ when $n = m$). Let this map be denoted by π_m. Then it can be shown that, for any two points $m, n \in \mathcal{C}$, the map $\phi_{mn} = \pi_n \circ \pi_m^{-1}$ is a homography from $m^\star$ to $n^\star$, that is, from the pencil of lines through m to the pencil of lines through n. It is then possible to define the **cross-ratio** $[a, b, c, d]$ of four points a, b, c, d on $\mathcal{C}$: choose a point $m \in \mathcal{C}$ and define $[a, b, c, d] := [ma, mb, mc, md]$ using the cross ratio in $m^\star \subset P^\star$. By what precedes, this definition does not depend on the choice of the point m on $\mathcal{C}$.

A bijection f defined over $\mathcal{C}$ is called a **homography** of the conic $\mathcal{C}$ if it preserves the cross-ratio defined above. Homographies of $\mathcal{C}$ form a group. It turns out that any homography f of $\mathcal{C}$ has an **axis**, which is the line Δ such that, for any two distinct points $m, n \in \mathcal{C}$ and $m' = f(m)$, we have $n' = f(n) \Leftrightarrow mn' \cap m'n \in \Delta$. The special case of involutions is of interest. A bijection f of $\mathcal{C}$ is an involution if and only if there is a point $\delta \in P - \mathcal{C}$, called the **Frégier point** of f, such that for any $m \in \mathcal{C}$, $\{m, f(m)\} = \mathcal{C} \cap \delta m$.

We are now ready to deal with the general case of Castillon's problem.

Theorem 4.5.1 *Let $\mathcal{C}$ be conic and let $A_1, A_2, \cdots, A_n$ be n points not on $\mathcal{C}$. Construct a polygon $A_1'A_2' \cdots A_n'$ inscribed in $\mathcal{C}$ such that $A_i \in A_i'A_{i+1}'$ ($i = 1, \ldots, n$).*

Projective Construction of the Inscribed Polygon, Taken from Berger The idea is that the point A_1' should be a fixed point of the homography f of the conic $\mathcal{C}$ defined by $f = \phi_n \circ \cdots \circ \phi_1$, where ϕ_i is the involution of $\mathcal{C}$ having Frégier point A_i.

The explicit construction is carried out by forming $X' = f(X)$, $Y' = f(Y)$ and $Z' = f(Z)$, where X, Y, Z are arbitrary distinct points on $\mathcal{C}$, and then drawing the axis Δ of f as

$$\Delta = < XY' \cap X'Y, YZ' \cap Y'Z >$$

(i.e. Δ is the line through the points $XY' \cap X'Y$ and $YZ' \cap Y'Z$.)
The options for A_1' are the points where Δ and $\mathcal{C}$ intersect. See Fig. 4.10.

Usually, there are two possible constructions, since the axis meets $\mathcal{C}$ in two points. However, there is a special case where there are infinitely many possible triangles $A'B'C'$. This case shows up precisely when $f = id_{\mathcal{C}}$ and has been totally missed out until recently, see [7]. The construction of such a configuration is obtained by drawing a complete inscribed quadrangle $UVWX$ in the conic and by considering the diagonal triangle, whose vertices give the points A, B, C (Fig. 4.11).

There are other proofs of Castillon's problem. Castillon's own proof is worth reading. I also recommend reading Tong Peng's proof, available on the internet, see [5], whose idea is interesting. Tong Peng considers the four point case, which turns out to be much easier to solve, and then carries the three point case to the four point case by adding a special extra point.

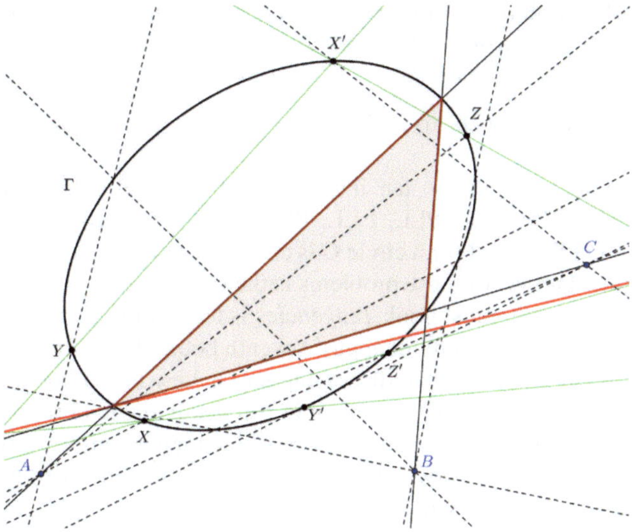

Fig. 4.10 Construction with three points

Fig. 4.11 Construction of the configuration with infinitely many solutions

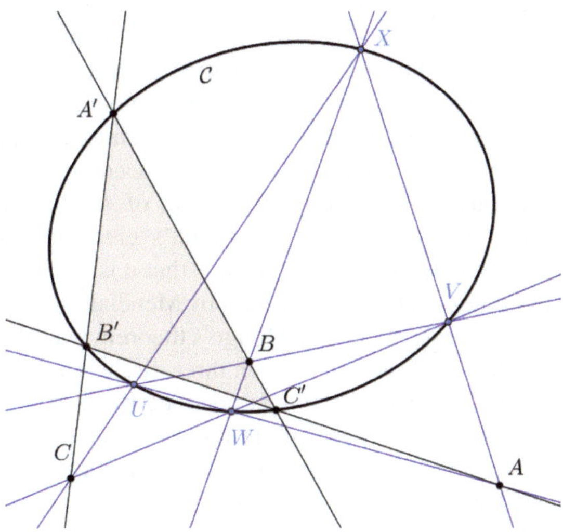

4.6 Apollonius' Famous Set of Tangency Problems

One question that might have struck the reader's mind is "Where does Pappus' problem come from? What led Pappus to this problem?"

In his book *Tangencies*, Apollonius of Perga, the Great Geometer, as he was called (born c. 240 B.C.), stated and solved the following problem regarding circles tangent to three given objects in the plane.

Theorem 4.6.1 (Apollonius Theorem) *Given three objects, which should be chosen among points, lines or circles, construct with straightedge and compass, the circles that are tangent to these three objects simultaneously.*

(Here, tangency with respect to a given point simply means passing through this point.)

This statement gives rise to ten intertwined different problems which can be denoted by PLC, PPP, PPL, PLL, LLL, PPC, PCC, CCC, LLC, LCC according to whether a point P, a line L or a circle C is chosen for one of the three objects. The hardest and most famous of these problems is the CCC problem.

Unfortunately, Apollonius' book *Tangencies* is forever lost. The only account we have of its content is contained in Pappus' seventh book of his *Collection*, precisely where his problem we have been discussing so far lies. In his book, Pappus aims at giving the solution of the CCC problem in full detail and his Proposition 117 we have just been dealing with appears to be one of the major steps toward the solution of that problem. (It is not clear whether this proposition was included in Apollonius' proof or whether Pappus provided it to complete the proof.)

Pappus, however, did not give the proof of the CCC problem. Mathematicians were therefore challenged in reconstructing Apollonius' proof based upon the clues given by Pappus. The first complete solution in the spirit of ancient Greek Geometry was given by François Viète in 1600 [9]. However, his solution does not use Pappus' Proposition 117, which is supposed to be useful in proving the CCC problem. Other solutions were given later with more modern means (René Descartes using algebra and coordinates, Isaac Newton using conics, Joseph-Diaz Gergonne using projective geometry), see Kunkel [3] for a nice account.

Let us now give a possible proof of Apollonius' theorem using Pappus' Proposition 117, as it appears in Heath's great book on Greek Mathematics [2]. One interesting feature of this proof is that it is based upon Monge's theorem which is an easy consequence of a theorem by Menelaus (Menelaus of Alexandria, 70–140 AD). It seems reasonable that Monge's theorem was already known in Apollonius' times. The theorem states that given three circles (none of them lying inside another one), the three intersection points obtained by drawing the external common tangents to each of the three pairs of circles are aligned (see Fig. 4.12). This theorem is one of the most beautiful theorems of classical geometry and it seems impossible that such a truth escaped from great Greek geometers' perspicacity.

It is high time we turn ourselves toward Heath's proof of Apollonius CCC theorem. We reason by analysis/synthetis. Assume that the construction has been completed in front of our eyes (see Fig. 4.13).

Let A, B, C be the centers of the three circles and D, E, F be the collinear intersection points of the tangents as given by Monge's theorem. Denote by P, Q, R the tangent points. It is easy to see that the arcs KP and QL are both homothetic to the arc PQ (by considering the contact points of two circles as the center of a homothety); this is established by Pappus in his book elsewhere. It follows that the line PQ passes through the center of homothety F. Similarly, the lines PR and QR go respectively through D and E. Call M, N the respective points

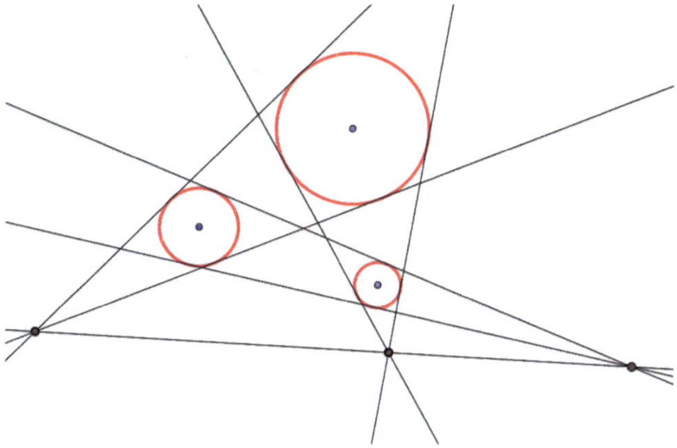

Fig. 4.12 Monge's Theorem

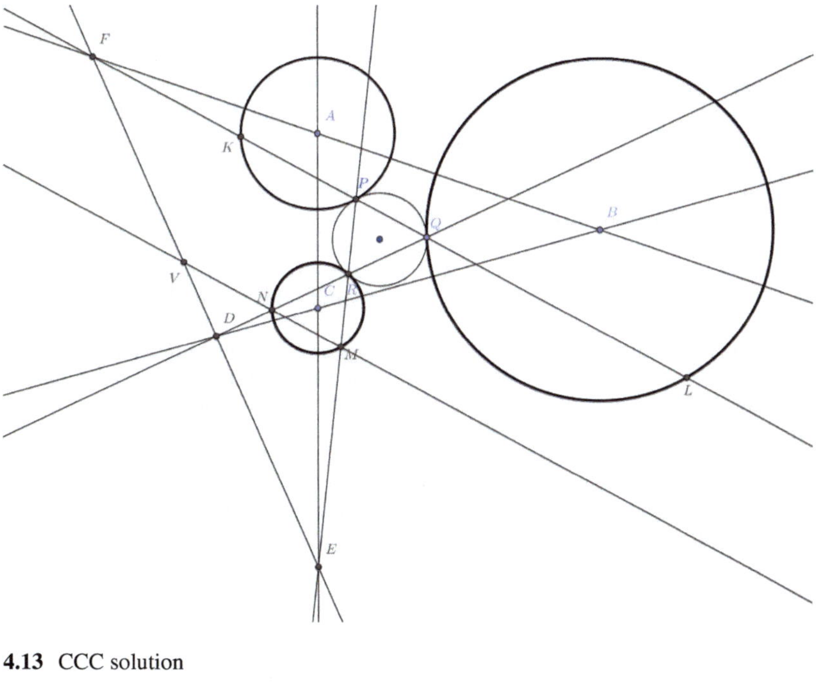

Fig. 4.13 CCC solution

of intersection of PR and QR with the circle of center C. By considering the homothety mapping two contact circles one onto another, we get that the triangles PQR and MNR are similar. In particular, the line MN is parallel to the line PQ. Call V the intersection point between PQ and EF. By Thales' Theorem, $EV/EF = EM/MP = EC/EA$. This proves that V can be constructed from the circles of centres A and C.

Now we synthesize. Given three circles $\mathcal{C}_A, \mathcal{C}_B, \mathcal{C}_C$ with their centers A, B, C, construct the points D, E, F by joining the centers and common tangents. Construct on the line DEF the point V such that $EV/EF = r_C/r_A$. Now, using Pappus' theorem, construct the triangle MNR inscribed in $\mathcal{C}$ such that the lines MN, RM, RN pass through V, E, D respectively. This gives one contact point R, the other two, P, Q being given by the intersections of the lines ER and DR with the other two circles. The solution to the CCC problem is complete.

Now another question pops up. We have been discussing a generalization of Pappus theorem, namely the Castillon theorem. Just as Pappus theorem was a brick for Apollonius' CCC theorem, is Castillon theorem a brick for some generalized Apollonius theorem?

References

1. M. Berger, *Geometry II*. Translated from the French by M. Cole and S. Levy (Springer, New York, 1987)
2. T. Heath, *A History of Greek Mathematics*, vol. II (Oxford University Press, Oxford, 1921)
3. P. Kunkel, The tangency problem of Apollonius: three looks. J. Brit. Soc. History Math. **22**(1), 34–46 (2007)
4. Pappus of Alexandria, *Book 7 of the Collection*, ed. by A. Jones. Sources in the History of Mathematics and Physical Sciences, vol. 8 (Springer, New York, 1986)
5. T. Peng, https://tongpeng.wordpress.com/2015/01/31/a-elementary-solution-to-the-castillons-problem
6. D. Perrin, Géométrie: deux ou trois choses que je sais d'elle, paper available on the web
7. M. Stark, *Castillon's problem, WFNMC-4 Conference*, Melbourne (2002). Available on the web
8. O. Teichmüller, Extremal quasiconformal mappings and quadratic differentials, in *Handbook of Teichmüller Theory, Vol. V*, ed. A. Papadopoulos, IRMA Lectures in Mathematics and Theoretical Physics, vol. 26 (European Mathematical Society, Zürich, 2016), pp. 321–483. Translated from the German by Guillaume Théret
9. F. Viète, Œuvres mathématiques, trad. complète du latin en français, souvent pour la première fois, avec un avertissement et des notes, par Jean Peyroux. Blanchard 1991 Paris
10. J. von Castillon, Sur un problème de géométrie plane qu'on regarde comme fort difficile. Nouveaux Mémoires de l'Académie Royale des Sciences et Belles Lettres de Berlin (1776)

Chapter 5
Notes on Angles and Solid Angles, in Relation with Euler's Memoir *De Mensura Angulorum Solidorum*

Stelios Negrepontis and Athanase Papadopoulos

> *For when I say beauty of form, I am trying to express, not what most people would understand by the words, such as the beauty of animals or of paintings, but I mean, says the dialectics, the straight line and the circle and the plane and solid figures formed from these by turning-lathes and rulers and patterns of angles; perhaps you understand. For I assert that the beauty of these is not relative, like that of other things, but they are always absolutely beautiful by nature and have peculiar pleasures in no way subject to comparison with the pleasures of scratching; and there are colors that possess beauty and pleasures of this character. Do you understand? (Plato,* Philebus *51c-d, In: Plato in Twelve Volumes, Vol. 9, transl. N. Fowler, Harvard University Press, 1925, with modification by the authors)*

> *Je ne parle pas ici, bien entendu, de cette beauté qui frappe les sens, de la beauté des qualités et des apparences ; non que j'en fasse fi, loin de là, mais elle n'a rien à faire avec la science ; je veux parler de cette beauté plus intime qui vient de l'ordre harmonieux des parties, et qu'une intelligence pure peut saisir. (Henri Poincaré,* Science et méthode, *Flammarion, 1908)*

Abstract We provide some historical context to the study of solid angles carried out by Euler in his memoir *De mensura angulorum solidorum* (On the measure of solid angles). We extend our study to the general notion of angle (not only solid). While doing so, we explore some works by Ancient Greek mathematicians and others by Arabs mathematicians of the Middle-Ages as well as some later Western authors

S. Negrepontis
National and Kapodistrian University of Athens, Mathematics Department, Panepistimioupolis, Athens, Greece
e-mail: snegrep@math.uoa.gr

A. Papadopoulos (✉)
Institut de Recherche Mathématique Avancée, Université de Strasbourg et CNRS, Strasbourg, France
e-mail: athanase.papadopoulos@math.unistra.fr

R. Caddeo, A. Papadopoulos (eds.), *Spherical Geometry in the Eighteenth Century I*, https://doi.org/10.1007/978-3-032-12466-1_5

from the Renaissance. In particular, we review the Pythagorean anthyphairetical perspective on angles which establishes the basis of the important relation between the mathematical notion of angle and the philosophical concept of finitization of the Infinite. In doing so, we shall show that questions addressed by Euler lead us to questions raised about 2500 years ago. At the same time, we highlight the fact that mathematics in those times is also today's mathematics. The reader can meditate in this study on the intermingling between mathematics and philosophy.

Keywords Leonhard Euler · Angle · Solid angle · Greek mathematics · Euclid · Proclus · Pythagorean mathematics · Plato · Anthyphairesis · Foundations of mathematics · Finitization of the Infinite · History of angles · History of solid angles · Arabic mathematics · Eighteenth century · Philosophy of mathematics · Eudoxus of Cnidus

AMS Classification 01A20, 01A30, 01A50,00A30, 03A05

5.1 Preamble: On the Mathematical Notion of Angle

In this chapter, we provide a historical context to the study of solid angles carried out by Euler in his memoir *De mensura angulorum solidorum* (on the measure of solid angles). At the same time, we survey some works on the more general notion of angle over a period ranging from Ancient Greece until modern times. While doing so, we discuss the idea of angle in Pythagorean thinking where philosophy was intimately related to mathematics, reporting on an anthyphairetical perspective on angles and the relation the Pythagoreans established between this notion and their idea of finitizing the Infinite.

Where should one start, for a historical discussion on the notion of solid angle? And what about the very notion of angle, say planar angle? No doubt, the notion of angle is part of the heritage of all civilizations: Babylonian, Syriac, Egyptian, Indian, Greek, Chinese, etc., and in fact, there is no question that since man was able to rationalize, there was a need, in one way or another, to compare angles, if not to measure them. For example, we can easily imagine a human mentally evaluating the angle the sun makes above the horizon, in order to get an idea of what time of the day it is.[1] Likewise, at night, by looking at the angle made over the horizon by figures in the sky like the moon, the stars and the constellations, man was able

[1] We note incidentally that the fact of measuring the angle of the sun with the horizon became at some point part of spherical trigonometry. Let us quote the following sentences by Lambert, from the memoir *Anmerkungen und Zusätze zur Trigonometrie* (Notes and comments in trigonometry) whose English translation is contained in the present volume (Chap. 10). From the introduction: "[...] For example, I explained how one could find the height of the sun for every moment from the length of a day by the simple addition of three logarithms." And from §76 of the same memoir: "[...] Suppose that given an angle and its two adjacent sides, we want to find the opposite side. This occurs when calculating tables of solar altitude, and is therefore a frequent problem."

to know (approximately) what time of the night it is. But at what epoch did the mathematical notion of angle arise? This question leads us to another one: what is a mathematical notion? To what extent is a notion that is described by circles, lines and numbers a mathematical notion? The answer, if we refer to Bourbaki, can be found in the opening sentences of his *Éléments de mathématique* (first published in 1939): "Since the Greeks, mathematics has meant proof; some even doubt that outside mathematics there are proofs in the precise and rigorous sense that this word received from the Greeks". A little further, Bourbaki talks about the axiomatic method and formalization. He writes: "The *axiomatic method* is, strictly speaking, nothing else than this art of writing texts whose formalization is easy to conceive." Thus, to answer the question of when did a mathematical notion of angle arise, we shall stick to Bourbaki's view on what is mathematics, putting an accent on formalization, axioms, postulates, theorems and proofs, a point of view that we find reasonable, even if it can be contested.

The earliest known texts in which angles are discussed from the point of view of axioms, postulates, theorems and proofs date back to ancient Greece, and this is why, in this overview, we start there. Philosophy is also involved in this discussion, and it addresses other kinds of questions. Indeed, the Pythagorean's mathematics, and, in their wake, Plato's, were intimately linked to their philosophy, and studying mathematics without philosophy is pointless when it comes to Plato and the Pythagoreans. At the same time, treating such a topic in a few pages is a difficult undertaking. We shall nevertheless try to mention a few essential points of this interweaving between philosophy and mathematics, keeping in mind the notion of angle.

Let us start with a mathematical question about angles. How do we define an angle? And more fundamentally: is it necessary to define an angle, or is this a primary notion, an undefined object? This was a legitimate question in Euclid's time and it was certainly addressed then. And what about the notion of solid angle? How can it be defined, and how can it be computed? If we define a solid angle as an area (as Euler does), then what is area? Do we need measure theory to define area? We read in Heath's translation of the *Elements*[2] [19, Definition 11, Book XI]: "A solid angle is the inclination constituted by more than two lines which meet one another and are not in the same surface, towards all the lines". Is this a clear definition?

All this is meant to say that the mathematical notion of angle is complex. This is the topic that we shall address in the following pages.

5.2 Introduction

Euler's memoir *De mensura angulorum solidorum* [21] is a study of solid angles that culminates in the computation of the measures of those of the five regular convex

[2] For the English translation of the *Elements*, we are using throughout this chapter Heath's edition [19].

polyhedra. At the beginning of his memoir, Euler recalls that the more familiar planar angles are measured by taking a circle of radius 1 centered at the vertex of such an angle and measuring the length of the part of the circle that the angle subtends. Thus, it is natural, by analogy, to define the measure of a solid angle to be the area of the part subtended by this solid angle on a sphere of radius 1 centered at the vertex of the solid angle. Regarding solid angles, Euler starts by considering the simplest example of solid angle, namely, one obtained by gluing three Euclidean triangles pairwise along their edges to form a vertex of a tetrahedron. Taking a sphere centered at this vertex and intersecting it with the three triangles that form this vertex, we obtain a spherical triangle. Indeed, the intersection of the sphere with a plane passing through its origin is a great circle, and a spherical triangle is precisely a region of the sphere enclosed by three segments of great circles joining pairwise three distinct points. Considering this sphere as a unit sphere, the measure of the solid angle is then taken to be the area of this spherical triangle. According to a well-known result attributed to Albert Girard (1595–1632), this area is equal to the angle excess of the spherical triangle, that is, the excess over two right angles of the sum of its three angles. However, it is natural to ask, independently of this formula, what is meant by area for figures that are not spherical triangles. Indeed, there are solid angles that are cones over figures that are much more general than Euclidean triangles or polygons, and whether such a notion of area was available as a mathematical notion before the modern developments of measure theory is a legitimate question.

The Greeks had no formal definition of area, even though there were some extremely important methods, developed particularly by Archimedes, for computing length, area and volume. Let us quote Heath who, in his *History of Greek mathematics* [26, p. 20], recalls that Archimedes "performed in fact what is equivalent to *integration* in finding the areas of a parabolic segment, and of a spiral, the surface and volume of a sphere and a segment of a sphere, and the volumes of any segments of the solids of revolution of the second degree."

Returning to the notion of area in Euler's memoir, let us mention that there is also a questioning about the fact of attributing to Albert Girard the statement that the area of a spherical triangle is equal to its angle excess. Indeed, as we recall in a footnote to the English translation of Euler's memoir which is our main subject here, this formula was known, before Girard, to Thomas Harriot (1560–1621) [3]. Furthermore, Lagrange mentions in his memoir *Solution de quelques problèmes relatifs aux triangles sphériques avec une analyse complète de ces triangles* (Solution of some problems relative to spherical triangles with a complete analysis of these triangles), included in English translation in the present volume, that Girard's proof of this theorem is not satisfying, and that the result should rather be attributed to Bonaventura Cavalieri (1598–1647) [9]. But to simplify, we shall call this result Girard's formula, as Euler does, and as it is often done in the literature on spherical geometry. The point we want to make here is that even if one can *define* the "area" of a spherical triangle to be its angle excess, extending this notion to polygons, and then considering this concept as a special case of a general notion of area leads directly to the question of what is area.

After solid angles that are cones over Euclidean triangles, Euler considers solid angles that are cones over more general Euclidean polygons. But whereas in the first case he deals with triangles with an arbitrary angle at their common vertex, in the more general case, he considers only triangles glued together whose angles at the common vertex are all equal and such that the dihedral angles that any two adjacent triangles make at this vertex are also all equal. In fact, in the application to the computation of the solid angles of the regular polyhedra that Euler carries out in the later sections of his memoir, only this special case is needed. In any case, the measure of the general solid angle that Euler considers is again taken to be the area of its intersection with a sphere of radius one centered at the given vertex.

While doing so, Euler establishes several formulae for the angle excess of a spherical triangle in terms of its side lengths. These formulae involve trigonometric functions of the sides and they are spherical analogues of the well-known Heron formula which gives the area of a Euclidean triangle in terms of its side lengths. Presumably, the formulae given by Euler are the oldest known analogues of Heron's formula in spherical geometry. Lagrange, in his memoir *Solution de quelques problèmes relatifs aux triangles sphériques avec une analyse complète de ces triangles*, published 18 years after Euler's one, establishes again such a formula without mentioning Euler's work on this topic.

As it is often the case in his mathematical memoirs, Euler gives several proofs of these "spherical Heron formulae" and he treats in detail special cases. Of particular interest is the case where the three sides of the triangle are infinitely small, in which case he recovers the familiar euclidean Heron formula, verifying the fact that the sphere with its canonical metric (like any other Riemannian manifold) is locally Euclidean.[3]

As we already said, Euler's memoir concludes with explicit numerical values that are approximations of the solid angles of the five regular polyhedra. Euclid's *Elements*, which culminate in the classification of the five regular polyhedra, do not include such values. (In fact, Euclid's *Elements* do not include explicit computations.)

The aim of the present chapter is to give a brief (perhaps partly subjective) historical account of the notion of angle, from the Greeks to Euler.

The content of the rest of this chapter is the following.

In Sect. 5.3, we start by recalling the importance of the notions of angle and solid angle in the works of the Pythagoreans, Plato and Aristotle, an idea that we shall expand in later sections.

In Sect. 5.4, we consider angles and solid angles in Euclid's *Elements* and in the works of some later commentators of the *Elements*.

Sections 5.5 to 5.7 are concerned with the philosophico-mathematical Pythagorean point of view on angle. We discuss the relation the Pythagoreans made between, on the one hand, the notion of acute and obtuse angle, and, on the other hand, their principle of the Infinite, the latter being interpreted as the infinite anthyphairesis of

---

[3] Here, we are allowing ourselves to use modern concepts.

the diameter to the side of a square.[4] This should not be surprising. As a matter of fact, the Pythagoreans were so impressed by their discovery of incommensurability that they considered that the principles of Finite and Infinite, which constituted the main component of their proof of incommensurability, are at the basis of all their mathematics and philosophy. The principle of the Finite, which they saw as the cause of the right angle, is based on the finitization of the infinite anthyphairesis of the diameter to the side of a square. This is the way we understand the claim made by Proclus (fifth century AD) that the Pythagorean principle of the Finite was for them the cause of the right angle (Proclus 129.17 [49]).[5]

The discussion of angles in Pythagorean thought is an illustration of the fact that their mathematics is inseparable from their philosophy, and also of the fact that they were capable of sweeping generalizations from very limited evidence, which is also the force of mathematicians.

In Sect. 5.8, based on the writings of Proclus, we give our interpretation of Postulate 4 of the *Elements* which says that all right angles are equal.

Since the present essay on angles is a chapter in a book on spherical geometry, it is natural to have a section on angles in non-Euclidean geometry, even if Euler, in his memoir *De mensura angulorum solidorum*, which is the motivation behind this chapter, does not mention this kind of angles. Thus, in Sect. 5.9, we review the relation between the notion of angle and the parallel postulate and we take this opportunity to mention the attempts made by Euler and Lagrange to prove this postulate.

Section 5.10 is a quick overview of the notion of solid angle in the works of Euclid, Apollonius, Heron and Pappus.

In Sect. 5.11, we consider the difficulties that arise in considering angles and solid angles as magnitudes.

Section 5.12 is a short review of the notion of angle and solid angle in the works of some Arab mathematicians of the Middle-Ages, who were the natural heirs of the Greek mathematicians of Antiquity.

Section 5.13 contains a quick overview of the notion of angle in the works of some Renaissance authors.

In Sect. 5.14, we recall the notion of angle in spherical geometry.

We conclude the present chapter, in Sect. 5.15, by pointing out some notions of angle arising in some modern geometrical works.

For the convenience of the reader, we recall, in a first appendix to this paper (Appendix 1), the geometric notion of anthyphairesis. A second appendix, in

[4] We use the term "diameter" of a square to denote what it called today its diagonal. The Greek word (διάμετρον) translates either by diameter or by diagonal. Heath writes in [19, Vol. I, p. 185]: "Diameter was the regular word in Euclid and elsewhere for the diameter of a square, and also of a parallelogram; diagonal (διαγώνιος) was a later term, defined by Heron (Def. 68) as the straight line drawn from an angle to an angle".

[5] See Proclus, Commentary to the First Book of Euclid's *Elements*, usually referred to as *In Euclidem*. The standard reference for the English translation and comments is Glen Morrow's edition [49].

Appendix 2, concerns Eudoxus of Cnidus, one of the central figures of Greek mathematics and whose name appears several times in this chapter. Although this takes us a little away from the main subject of our essay, this excursus on Eudoxus will allow us to talk about one of the main figures of Greek science, and to quote André Weil, Simone Weil and Yuri Manin, three mathematicians, philosophers and historians of mathematics who have integrated the meaning of the Greek heritage into their writings.

5.3 The Pythagoreans, Plato and Aristotle

The Pythagoreans, in the sixth century BC, had some quite precise notions of plane and solid angles, and they used them in their mathematical discoveries, in particular in their work on regular polygons and the classification of regular polyhedra. Less known is the fact that they also established a link between the notion of angle and the concept of Finite and Infinite, via their anthyphairetical proof of the incommensurability of the diameter to the side of a square.[6] The discovery of incommensurability was a brilliant success of the Pythagoreans, and their proof is an example of the fact that Mathematics is very much the art of finitizing the Infinite. About the latter, we can quote a fragment of Philolaos:

---

[6] We quote here a letter sent by André Weil to his sister Simone, dated March 28, 1940. (The letter is sent from Rouen, where André Weil was in prison [62, t. VII, Vol. 1, pp. 554–555]): "*How* do you think they [the Greeks] discovered the immeasurable? I remember that we don't have any sure data on this, and that I saw somewhere an ingenious suggestion, with an infinite descent where the thing could be *seen* on a square and its diagonal (numerically this gives the following: We have $\frac{\sqrt{2}+1}{1} = \frac{1}{\sqrt{2}-1}$; but if $\frac{a}{b} = \frac{c}{d}$ and a, b, c have a common measure, then the same holds for all the lengths obtained by continuing the progression downwards, which is absurd, since these lengths OB', OC', OB'', OC'', etc. become as small as we want)." [*Comment* crois-tu qu'ils aient découvert l'incommensurable? Je me souviens qu'on n'a pas de données certaines là-dessus, et que j'ai vu quelque part une suggestion ingénieuse, avec une descente infinie où la chose se *voyait* sur un carré et sa diagonale (numériquement cela donne ceci : on a $\frac{\sqrt{2}+1}{1} = \frac{1}{\sqrt{2}-1}$; or, si $\frac{a}{b} = \frac{c}{d}$ et a, b, c ont une commune mesure, alors il en est de même de toutes les longueurs obtenues en continuant la progression vers le bas, ce qui est absurde, puisque ces longueurs OB', OC', OB'', OC'', etc. deviennent aussi petites qu'on veut).] The method of infinite descent suggests anthyphairesis. Among the Ancients, Plato gives an anthyphairetic proof of the incommensurability in the *Meno* 80d–86e and 97a–98b provided one interprets this passage properly. Once one realizes that there is such a proof in the *Meno*, then it becomes the most ancient preserved proof of the Pythagorean incommensurability, older by some 35 years of Aristotle's hint of an arithmetical Archytas type proof that is usually thought to be the most ancient. One should note that the *Meno*'s proof is not the original one, because it is extremely streamlined and it also employs the Logos, a concept that was not available to the Pythagoreans. For a discussion of the anthyphairetic proof of incommensurability and its history, see the recent article by Farmaki and Negrepontis [22].

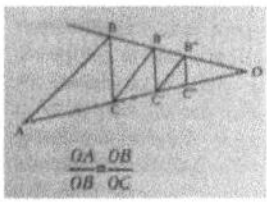

[Fragment 1] The nature (ἁ φύσις) in the [ordered] world (ἐν τῶι κόσμωι) is composed (ἁρμόχθυ) of infinite entities (ἀπείρων) and finitizing entities (περαινόντων), both the world and everything in it.

This idea is developed in detail in the paper [22] by V. Farmaki and the first author of the present chapter. See also the recent paper [45].

It may be useful to recall here that definitions and postulates, and in particular those concerning plane angles, did not start with Euclid. They were established and already used by Greek geometers before Plato. The latter, in the *Republic* 510c2-d3, criticizes the geometers "who base Geometry on arbitrary and unknown hypotheses, definitions and postulates", rather than on intelligible, dialectical principles[7] from which these definitions and postulates result. One specific such criticism concerns precisely the (definition of the) three kinds of angles (γωνιῶν τριττὰ ειδῆ, given in the *Republic* 510c4-5), and in particular what became Postulate 4 in Euclid's *Elements*, the one connected with the right angle. Plato does not explain how the three kinds of angles and Postulate 4 might follow from Intelligible Beings, but Proclus, in a long passage of his *Commentary to the first Book of Euclid's Elements* (*In Euclidem*, 131,9–134, 7) [49], explains that the Pythagoreans, following their epoch-making discovery of the anthyphairesis/continued fraction[8] of the diameter to the side of a square, and of the diameter itself (διαμέτρου αὐτῆς, 510d8), whose sequence of anthyphairetic quotients is [1,2,2,2,...], based the definitions of the three kinds of angles and of Postulate 4 on the closely related to the Pythagorean principle of the Finite and Infinite which essentially describe the two fundamental aspects of this anthyphairesis: the infinite anthyphairetic division, and its finitization by means of periodicity. We shall discuss more thoroughly this question in Sects. 5.5 to 5.7 below. In the rest of the present section, we shall talk about definitions and postulates before Euclid's times.

Plato is a heir of the Pythagoreans. One of the early texts that reached us in which solid angles are studied is his dialogue *Timaeus* [47], written about half a century before Euclid wrote the *Elements*. In this work, Plato exposes his metaphor on the creation of the world which involves the construction of four regular polyhedra. It is generally admitted that Plato learned about the classification of the regular polyhedra from Theaetetus (ca. 417–369 BC), a mathematician who was like him a student of Theodorus of Cyrene (ca. 465–398 BC). Theodorus was certainly influenced by the Pythagoreans. In Plato's mathematical philosophy, the regular polyhedra are neither intelligible nor sensible, but of a third kind, which Plato calls the *receptacle*. The reader interested in Plato's arithmetico-philosophical use of the regular polyhedra and the relation with his notion of the receptacle is referred to the article [43]. Let us now read Plato on regular polyhedra in order to see how solid angles enter in his description:

---

[7] In our interpretation, the word dialectical refers to the Logos criterion of anthyphairesis, as we explain in the sequel.

[8] For the convenience of the reader, we recall, in the appendix to the present paper, the definition of the anthyphairesis of two magnitudes.

> [...] And when four equilateral triangles are combined so that three plane angles meet in a point, they form **one solid angle, which comes next in order to the most obtuse of the plane angles**. And when four such angles are produced, **the first solid figure**[9] is constructed, which divides the whole of the circumscribed sphere into equal and similar parts (*Timaeus* 54e3–55a4).
>
> **And the second solid**[10] is formed from the same triangles, but constructed out of eight equilateral triangles, which produce **one solid angle out of four planes**; and when six such solid angles have been formed, the second body in turn is completed (*Timaeus* 55a4-8).
>
> **And the third solid**[11] is composed of twice sixty of the elemental triangles conjoined, and of twelve solid angles, each contained by five plane equilateral triangles, and it has, by its production, twenty equilateral triangular bases (*Timaeus* 55a8-b3).
>
> Now the first of the elemental triangles ceased acting when it had generated these three solids, the substance of the **fourth kind**[12] being generated by the isosceles triangle. Four of these combined, with their right angles drawn together to the center, produced one equilateral quadrangle; and six such quadrangles, when joined together, formed eight solid angles, each composed of three plane right angles; and the shape of the body thus constructed was cubic, having six plane equilateral quadrangular bases (*Timaeus* 55b3-c4).

Euler considers in his memoir *De mensura angulorum solidorum*—which was the motivation for writing the present chapter—solid angles formed by joining equal triangles. The solid angles of the regular polyhedra which he aims to measure are of this sort. In the passage we recalled above, Plato in the *Timaeus* highlights the fact that the regular polyhedra are constructed using triangles.

Aristotle, who was definitely attentive to almost all the mathematical questions of his times, was naturally interested in solid angles. Heath, in Vol. I, p. 177, of his edition of the *Elements*, reports on a long debate that took place among philosophers, starting in Antiquity and continuing until the early twentieth century Italian commentators, whose object was to know as to what Aristotelian category an angle belongs. Among the questions raised there was whether this notions belongs to the category of *quantity* (*quantum*, ποσόν), *quality* (*quale*, ποιόν), *relation* (πρός τι), etc., whether angles are *magnitudes*, and if yes, whether they are *homogeneous* magnitudes, whether we can compare angles that are within a certain class, and if yes, what is this class and what are the tools used in such a comparison. There is also the question of whether we may apply to angles the known operations on numbers and magnitudes (addition, multiplication, etc.), the theory of proportions and other mathematical operations. All these questions were discussed at length by Aristotle in several treatises, including the *Physics*, the *Metaphysics*, *On the Heavens*, the *Categories*, the *Prior* and the *Second Analytics*, and they remained essential in mathematical thinking during the last two and a half millennia. In his comments on the *Elements*, Heath reviews the points of view of several authors on these questions. Such interrogations were the subject of argumentations between

[9] The tetrahedron (pyramid).

[10] The octahedron.

[11] The icosahedron.

[12] The hexahedron (cube).

mathematicians, and we shall mention later in this chapter the opinions of several of them on this matter.

After Aristotle, it is natural to talk about Euclid, and this is what we do in more detail in the next section. It is likely that the Pythagoreans' interest in the regular polyhedra and Plato's absorption of this theory acted as a motivation for Euclid's complete involvement in this topic and his study of solid angles.

5.4 On Plane Angles in Euclid's *Elements*

In Euclid's *Elements*, angles are introduced in Book I.

Angles first appear at the level of the definitions. Definition 8 of Book I reads:

> A plane angle is the inclination to one another of two lines in a plane which meet one another and do not lie in a straight line.

Definition 9 reads:

> And when the lines containing the angle are straight, the angle is called rectilineal.

Heath makes the following remark, just after these definitions: "It looks as though Euclid really intended to define a rectilineal angle, but on second thoughts, as a concession to the then common recognition of curvilineal angles, altered 'straight lines' into 'lines' and separated the definition into two." The remark is interesting and the modern geometers will be interested to know that Euclid also considered plane angles whose sides are not necessarily straight lines.

In any case, these definitions imply that Euclid considers that there exist more than one species of plane angles. Rectilineal angles form a special class of plane angles, the class where the two curves that contain the angle are straight lines. Other classes include the curvilineal and the mixed. Among the latter are the so-called semi-circle angle, also called contact, horn-like or cornicular angle, the latter name being used by Proclus. This is the angle made by a circle and a straight line tangent to it. We shall say more about this angle now.

Right angles are also mentioned in the section containing the postulates. Postulate 4 reads:

> All right angles are equal.

Euclid considers non-rectilineal angles in Proposition 16 of Book III and in its proof. The proposition reads:

> The straight line drawn at right angles to the diameter of a circle from its extremity will fall outside the circle, and into the space between the straight line and the circumference another straight line cannot be interposed; further the angle of the semicircle is greater, and the remaining angle less, than any acute rectilineal angle.

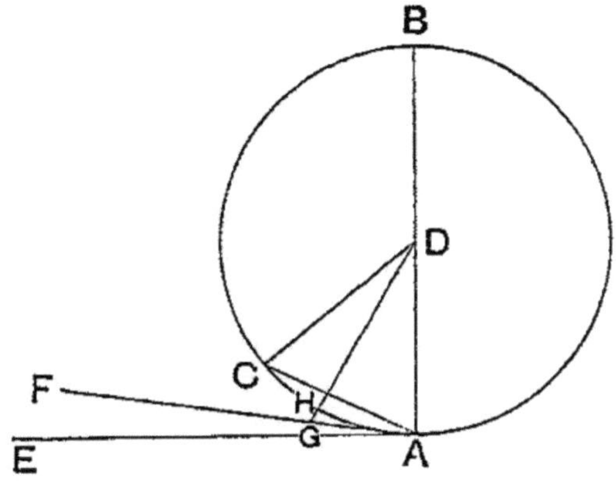

Fig. 5.1 The Figure for Euclid's Prop. III.16

From the proof of the same proposition, we have (Fig. 5.1):

> I say further that the angle of the semicircle contained by the straight line BA and the circumference CHA is greater than any acute rectilineal angle, and the remaining angle contained by the circumference CHA and the straight line AE is less than any acute rectilineal angle.

Giving an exact meaning to this statement and some others in Euclid's *Elements* are among the questions that puzzled ancient mathematicians. Some of these questions found eventually a satisfactory solution with the invention of infinitesimals.

It may be good to recall here that Leibniz, when he introduced infinitesimals (which he also called differentials), postulated them to be greater than zero but smaller than any positive number, and established rules to manipulate them (addition, multiplication, etc.). Less well known is that in the period between Euclid and Leibniz, some Arab mathematicians introduced infinitesimals and treated them in a similar way, in their work on angles. This is a subject highlighted in the relatively recent book *Angles and magnitude* [52] by Roshdi Rashed on which we shall say more in Sect. 5.12.

Heath has a long historical comment on Proposition III.16, and it is interesting to read it. He starts by saying that "this proposition is historically interesting because of the controversies to which the last part of it gave rise from the 13th to the 17th centuries. History was here repeating itself, for it is certain that, in Ancient Greece, both before and after Euclid's times, there had been a great deal of the same sort of contention about the nature of the 'angle of a semicircle' and the 'remaining angle' between the circumference of the semicircle and the tangent at its extremity [...]. Proclus has no hesitation in speaking of the 'angle of a semicircle' and the 'horn-like angle' as true *angles*. Thus he says that 'angles are contained by a straight line and a circumference in two ways; for they are either contained by a straight line and a convex circumference, like that of the semi-circle, or by a straight line and a concave circumference; like the κερατοειδής'[13] (p. 127-11-14). 'There are *mixed* lines, as spirals, and angles, as the angle of the semicircle and the κερατοειδής' (p. 104-16-18). The difficulty which the ancients felt arose from the very fact which Euclid embodies in this proposition. Since an angle can be divided by a line, it would seem to be a magnitude; 'but if it is a magnitude, and all homogeneous magnitudes which are finite have a ratio to one another, then all homogeneous angles, or rather all those on surfaces, will have a ratio to one another, so that the cornicular[14] will also have a ratio to the rectilineal. But things which have a ratio to one another can, if multiplied, exceed one another.[15] Therefore the *cornicular* angle[16] will also sometime exceed the rectilineal; which is impossible, for it is proved that the former is less than any rectilineal angle' (Proclus, p. 121, 24–122, 6)."

Furthermore, dealing with angles, compared to lines, needs additional care, since the value of an angle lies within an interval: in Euclid's *Elements*, this value is always assumed to be between 0 and π. How does one then add two arbitrary angles?

Thus, in many ways, angles do not satisfy the Archimedean axiom. This was already stressed by ancient authors. Furthermore, some operations that can be performed on angles cannot be made on angles that do not belong to the same class (recall that Euclid also considers angles that are not rectilineal, for instance the contact angle between a circle and a tangent, an example of a mixed angle). Hence the need for a careful discussion of angles.

It is interesting to note that even though Euclid introduced, in Definitions I.8 and I.9, a general notion of angle between lines that are not necessarily straight, he uses the general definition only in the case of the space contained by the straight line drawn at right angles to the diameter of a circle and the circle itself (Proposition 16 of Book III, which we already mentioned). Furthermore, even though angles bounded by non-straight lines are considered in this very special case, Euclid, in several propositions, is careful in stressing the fact that the angle he refers to is rectilineal (i.e. bounded by two straight lines); see e.g. Proposition I.10 which we recall next, in which Euclid talks about angles made by a "straight line standing on a straight line"; see also Propositions I.42, I.44, I.45, and there are others.

The next three definitions of Book I are concerned with the attributes right, acute and obtuse for angles. Definition 10 reads:

---

[13] The cornicular, or horn-shaped.

[14] Whereas Euclid considers cornicular angles made by a curve and a line tangent to it. Proclus, in the passage quoted here, considers angles made by a circle and a curve which is not necessarily tangent to it (he talks about an "angle made by a straight line and a concave circumference").

[15] The reference is to the so-called Archimedes axiom which says that given two magnitudes a and b, there exists a natural number n such that $na > b$. The axiom is also called Eudoxus–Archimedes axiom, because Archimedes attributes it to Eudoxus. See Appendix 2 of this chapter.

[16] Here, Proclus is talking about a cornicular angle formed by a circle and a line tangent to it.

> When a straight line standing on a straight line makes the adjacent angles equal to one another, each of the equal angles is right, and the straight line standing on the other is called a perpendicular to that on which it stands.

Definition 11 reads:

> An obtuse angle is an angle greater than a right angle.

Definition 12 reads:

> An acute angle is an angle less than a right angle.

Let us turn back now to the Pythagoreans for a more detailed account of their interpretation of the quality for an angle to be acute, obtuse or right.

5.5 The Pythagorean Principle of the Infinite, Interpreted as the Infinite Anthyphairesis of the Diameter to the Side of the Square, Is the Cause of the Acute and Obtuse Angles

We mentioned in the Introduction that the Pythagoreans established a connection between the principle of the Infinite and the notion of obtuse and acute angles. In this section, we shall explain this in some detail, assuming that the reader is less familiar with this kind of questions than with the other purely mathematical questions discussed in the present chapter.

The key to this connection is the double sequence of the side p_n and diameter q_n of the square, defined by Theon Smyrneus [57], passages 43,5–15, 44,18–45,8, Iamblichus [31], Passages 91,21-92,3, 92,23–93,6] and Proclus [50], passages 2,24,16–25,13, 2,27,1–29,4, by the recursive rule

$$p_1 = q_1 = 1, \; p_{n+1} = p_n + q_n, \; q_{n+1} = 2p_n + q_n \text{ for } n = 1, 2, \ldots.$$

It is easy to prove that this recursive definition is equivalent to the anthyphairetic definition:

$$\text{Anth}(q_n, p_n) = [1, 2, 2, 2, \ldots, 2 \; (n-1 \text{ times})].$$

The line pairs (q_n, p_n) are relatively prime for $n = 1, 2, \ldots$, and their anthyphairesis is a finite initial segment approximating the infinite anthyphairesis of the diameter d to the side s of a square:

$$\text{Anth}(d, s) = [1, 2, 2, 2, \ldots].$$

For each positive integer n, we denote by ω_n the angle formed by the two equal sides in the isosceles triangle with sides p_n, p_n, q_n. Since the double sequence (p_n, q_n) approximates the dyad (s, d) where s is the side and d the diameter of a square, it is intuitively clear that the angle ω_n approximates the right angle, without ever being actually equal to it. It is important to realize that this approximation takes the form alternately of an acute and obtuse angle, exactly because of the fundamental Pell property of the side and diameter numbers, expressed in the next proposition:

Proposition 5.1 *The sequence of the side and diameter numbers satisfies the Pell property* $q_{2n} = 2p_{2n}^2 + (-1)^n$.

The proof is by induction, using Proposition II.10 of the *Elements* stating that for any two line segments a and b, the equality $(a + 2b)^2 + a^2 = 2(a + b)^2 + 2b^2$ holds.[17]

The Pell property combined with the converse of the Pythagorean theorem (Propositions II.12 and 13 of Euclid's *Elements*) has immediate consequences on the kind of angle ω_n, as we shall see now.

Proposition 5.2 *For every* $n = 1, 2, \ldots$, *the angle* ω_{2n-1} *is acute, and the angle* ω_{2n} *is obtuse.*

Proof By the Pell property, we have $q_{2n-1}^2 = 2p_{2n-1}^2 - 1$, hence $q_{2n-1}^2 < 2p_{2n-1}^2$, therefore by Propositions II.4/5,[18] II.12 and II.13, the angle ω_{2n-1} contained by the equal sides p_{2n-1}, p_{2n-1} of the isosceles triangle with sides $p_{2n-1}, p_{2n-1}, q_{2n-1}$ is acute, and by the Pell property, $q_{2n}^2 = 2p_{2n+1}^2$, hence, $q_{2n}^2 > 2p_{2n}^2$. Therefore, again by Propositions II.4/5, II.12 and II.13, the angle ω_{2n} contained by the equal sides p_{2n}, p_{2n} in the isosceles triangle with sides p_{2n}, p_{2n}, q_{2n} is obtuse.

Combining the fact that the side and diameter numbers converge to the dyad (diameter, side) of a square and that this convergence is alternatively from below and from above, it follows that the acute angle ω_{2n-1} becomes, for sufficiently large n, as close to the right angle as desired, and thus greater than any acute angle, while the obtuse angle ω_{2n} becomes, for sufficiently large n, as close to the right angle as desired, and thus smaller than any obtuse angle, thus finally obtaining the following:

---

[17] The use of the Pell property of the side and diameter numbers provides a strong independent argument that the Pythagoreans had indeed a proof of the Pell property of the side and diameter numbers, since now the Pythagoreans actually use this property in order to define the two kinds of angles from the higher principle of the Infinite. It is safe to assume that the Pythagoreans would not use the Pell property if they had not proved it rigorously. We take this opportunity to note that the use of the Pell property in this Proclus passage shows that, despite the claims of Mueller [39], Unguru [60] and Fowler [23], the Pythagoreans not only had knowledge of induction, but in fact employed it in order to prove the Pell property of the side and diameter numbers.

[18] The numbering II.4/5 indicates the position of the statement and proof of the Pythagorean theorem in Book II of the *Elements*, right after II.4 and before II.5, as restored in the paper [42].

Pythagorean Definition of the Acute and the Obtuse Angle in Terms of the Principle of the Infinite

The last sentence of the preceding paragraph justifies the following (Pythagorean) definition of acute and obtuse angle:

> An angle ω is acute if there is an odd number $2n - 1$ such that $\omega < \omega_{2n-1}$, and an angle ω is obtuse if there is an even number $2n$ such that $\omega > \omega_{2n}$.

This Pythagorean definition of obtuse and acute angle is adopted by Plato, as evidenced by the *Republic* passage 510c2-d3, and it becomes a part of his argument that the geometers are mistaken in relying, for the foundation of Geometry, on unknown and arbitrary hypotheses (that is, postulates and definitions), but that they should rather generate Geometry dialectically, namely, from his Intelligible Beings, as we already recalled in Sect. 5.3. For this purpose, Plato repeatedly treats the geometric entity "diameter to the side of a square", as an Intelligible Being, calling it "the diameter itself" in the *Republic* (510d8), and showing, in the *Meno*, 80d-86e and 97a-98b, that its knowledge has the form of "True Opinion plus Logos", precisely as the knowledge of an Intelligible Being. See the article [41] for more details about this complex question.

5.6 Two More Useful Arithmetical Propositions

The next two elementary propositions, based on Proposition I.44 of the *Elements*, will be useful in the next two sections.

Proposition 5.3 *If a, b, c, d are line segments and A, B, C numbers, and if $Aa^2 = Bab + Cb^2$, and $Ac^2 = Bcd + Cd^2$, then $ad = bc$.*

Proof By Proposition I.44 of the *Elements*, there is a line segment e such that $ae = bc$. By assumption, $Aa^2c = Babc + Cb^2c$,
whence $Aa^2c = Baae + Cbae$,
whence $Aac = Bae + Cbe$,
whence $Aacc = Baec + Cbec$,
whence $Ac^2a = Bcea + Ce^2a$,
whence $Ac^2 = Bce + Ce^2$.
Clearly then $d = e$, hence $ad = bc$.

Proposition 5.4 If a, b, c, d are line segments such that $ad = bc$, then $\text{Anth}(a, b) = \text{Anth}(c, d)$.

Proof Let $\text{Anth}(a, b) = [k_0, k_1, ...]$. Thus $a = k_0b + e_1$ with $e_1 < b$. Therefore $cb = ad = k_0bd + e_1d$. By Proposition I.44 of the *Elements*, there is a line segment f_1 such that $e_1d = bf_1$. Since $e_1 < b$, it is clear that $f_1 < d$. Then $cb = k_0bd + e_1d = k_0bd + bf_1$; therefore $c = k_0d + f_1$ with $f_1 < d$.

Thus, we have

$$\text{Anth}(a, b) = [k_0, k_1, \ldots] = [k_0, \text{Anth}(b, e_1)],$$

while

$$\text{Anth}(c, d) = [k_0, \text{Anth}(d, f_1)]$$

and at this point we have established for the tetrad b, e_1, d, f_1 the same condition $bf_1 = de_1$ that we had assumed for the original tetrad a, b, c, d and we are to prove that $\text{Anth}(b, e_1) = \text{Anth}(d, f_1)$. We continue as in the first step, and finish the proof by induction.

5.7 The Pythagorean Principle of the Finite, Interpreted as the Finitization of the Infinite Anthyphairesis of the Diameter to the Side of a Square, Is the Cause of the Right Angle

The Pythagorean principle of the Finite has been interpreted as the Preservation of the Gnomons[19] in the anthyphairesis of the diameter to the side of a square,

$$b^2 = 2bc_1 + c_1^2,$$

$$c_1^2 = 2c_1c_2 + c_2^2.$$

The Preservation of the Gnomons implies, by Proposition 1 of Sect. 5.5, $bc_2 = c_1^2$.

The condition $bc_2 = c_1^2$, by Proposition 5.2 of the *Elements*, implies that $\text{Anth}(b, c_1) = \text{Anth}(c_1, c_2)$, from which we immediately obtain that $\text{Anth}(a, b) = [1, 2, 2, 2, \ldots]$. Thus the Preservation of the Gnomons indeed acts as a Finitizing

[19] The equality $b^2 = 2bc_1 + c_1^2$ can be interpreted as the equality of the small square b^2 with the Gnomon $2bc_1 + c_1^2$ in the greater square $(b + c_1)^2$, hence the (classical) terminology "Preservation of the Gnomons", an expression well understood by the Pythagoreans, see [19, Vol. I p. 47]. It is probably useful to add a few words about the notion of a gnomon. The Pythagoreans had observed that by adding to a square a figure that has the form of a stalk made up of two perpendicular rectangles of the same thickness, the length of each of them being equal to that of the side of the original square and that are joined together by a smaller square, as in the figure,  one obtains then a new square. There are other figures that have similar properties. For instance, parallelograms appear as gnomons in several proofs of propositions in Euclid's *Elements*. Aristotle, in the *Categories* 15a30-31, writes about Gnomons: "[...] But there are things that increase without altering, as the square by surrounding it with a gnomon is increased but is not thereby altered." In the *Physics* 203a10-16, he writes: "Further, the Pythagoreans identify the infinite with the even. For this, they say, when it is cut off and shut in by the odd, provides things with the element of infinity. An indication of this is what happens with numbers. If the gnomons are placed round the one, and without the one, in the one construction the figure that results is always different, in the other it is always the same. But Plato has two infinites, the Great and the Small."

principle, finitizing our complete knowledge of the infinite anthyphairesis of the diameter to the side of a square.

On the other hand, from the Preservation of the Gnomons and its consequence $bc_2 = c_1^2$, since $c_1 = a - b$, $c_2 = b - 2c_1 = b - 2(a - b) = 3b - 2a$, we have, by Proposition II.7,

$$bc_2 = b(3b - 2a) = (a - b)^2 = a^2 + b^2 - 2ab$$

and

$$c_1^2 = 3b^2 - 2ab = a^2 + b^2 - 2ab,$$

resulting in the equality $2b^2 = a^2$.

Thus, the Preservation of the Gnomons gives the equality $a^2 = 2b^2$, which, by the Pythagorean Theorem I.47 of the *Elements* (and II.4/5 in its restored position) and its converse II.12 and 13, is equivalent to the right angle. In this way, Proclus' claim that the Pythagorean principle of the Finite is the cause of the right angle is explained.[20]

5.8 Our Interpretation of Postulate 4 of the *Elements*: All Right Angles Are Equal

Postulate Four of the *Elements* says (see [19, Vol. I, p. 200])

That all right angles are equal to one another.

It is remarkable that Proclus, in *In Euclidem* 188,20–189, 12, produces a proof of this postulate, which is perfectly valid in fact, and agreeing in essence with the proof given in Hilbert's axiomatisation of Euclidean Geometry, concluding with the comment that "this proof has been given by other commentators and required no great study."

In view of the anthyphairetic Pythagorean/Platonic analysis that we pointed out, our interpretation of Postulate 4 of the *Elements* (all right angles are equal) becomes:

Proposition 5.5 *The anthyphairesis of the hypotenuse to the side of every isosceles right triangle is equal to* [1,2,2,2,...].

Proof Let two isosceles right triangles, one with right angle ω_1 and with hypotenuse a and sides b, and the other with right angle ω_2 and with hypotenuse c and sides d. By the Pythagorean theorem I.47 of the *Elements*, $a^2 = 2b^2$, $c^2 = 2d^2$. Then, from

[20] Morrow, in [49, p.104], writes in his Note 93 to Proclus 129.17: "A cryptic reference to the distinction between right angles and obtuse or acute angles. See 131.13–134.7." This remark shows that Morrow has not understood the connection between the Pythagorean treatment of the three kinds of angles and the anthyphairetic proof of incommensurability.

the Pythagorean Principle of the Finite, we obtain, by Proposition 4.1, $ad = bc$, and, by Proposition 4.2, Anth(a, b) = Anth(c, d).

Thus we may interpret the Pythagorean Postulate 4 on the equality of all right angles to mean that in every *isosceles right triangle*, the anthyphairesis of the hypotenuse to the side is equal to [1,2,2,2,. . .]. The postulate is thus seen as a form of Thales' Theorem without ratios of magnitudes. For more details, we refer the reader to the articles [40] and [42].

Proclus, in his comments in *In Euclidem* 191, 5–15, suggests that this content is essentially philosophical, strengthening the dialectic status of the right angle:

> This postulate also shows that the rightness of the angles is akin to the equality, as the acuteness and obtuseness are akin to the inequality. The rightness is in the same column with the equality, for both of them belong under the Finite . . . But the acuteness and obtuseness are akin to the inequality . . . for all of them are the offspring (ἔκγονοι) of the Infinite. (Proclus, *In Euclidem* 191, 5–11.)

Proclus' interpretation of the Pythagorean meaning of Postulate 4 lies in the opposition between the inequality of the acute $q_{2n-1}^2 < 2p_{2n-1}^2$ and obtuse $q_{2n}^2 > 2p_{2n}^2$ angles versus the equality $a^2 - 2b^2$ of the right angle.

Here, p_n, q_n are the n-th side and diameter numbers, respectively, discussed in Section 5.5.

The same philosophical opposition is expressed forcefully in Plato's Symposium 210e1-211b7.

5.9 Angles, the Parallel Postulate, Euler and Lagrange

To conclude our quick review of plane angles, let us recall that this notion is at the heart of the work done on the parallel postulate and the attempts that were made to prove it, an activity that occupied some of the greatest mathematicians over a period which lasted more than two thousand years. Let us first recall the statement of this postulate, in Heath's edition [19, Vol. 1, p. 155]:

> That, if a straight line falling on two straight lines make the interior angles on the same side less than two right angles, the two straight lines, if produced indefinitely, meet on that side on which are the angles less than the two right angles.

It is well known that there are several equivalent forms of this postulate that are formulated in terms of angles. For instance, this postulate can be replaced (keeping all the other postulates unchanged) by one of the following:

(1) The angle sum in (one or in) any triangle is equal to two right angles.
(2) The angle sum in a triangle is independent of the chosen triangle.
(3) There exists non-equal triangles[21] that have equal angles.

[21] That is (using modern terms), non-isometric triangles.

(4) For any point which is interior to an angle whose measure is strictly between 0 and 180°, we can draw a line passing through this point and intersecting the two sides of the angle.

Talking about the parallel postulate and since we are especially interested here in the work of Euler, let us recall that Euler himself attempted proofs of this postulate. Two of his attempts were recorded by Nikolai Ivanovitch (or Nikolaus) Fuss[22] in two manuscripts titled *Euleri Doctrina Parallelismi* and *Euleri Elementa Geometriae ex principio similitudinis deducta*, which were published in the Soviet Union in 1961 and which are analyzed by Pont in [48, p. 281–282].

Euler's first attempt is based on the fact that the set of points in a plane that are a given distance from a line is the union of two lines. The second one is based on the existence of non-congruent homothetic triangles. We know that both hypotheses are equivalent to the axiom of parallels.

Finally, let us also recall that Joseph-Louis Lagrange, who is the second main character in the present book, at some point, believed that he has a proof of the parallel postulate. The following episode is told by Jean-Baptiste Biot[23] in [4, p. 263]:

> One day, Lagrange took out from his pocket a piece of paper which he read to the Academy and which contained a proof of Euclid's famous *postulatum* relative to the theory of parallels. This proof was based on an evident paralogism, which appeared so to everybody, and Lagrange probably realized it while he was reading, because as soon as he finished, he put the piece of paper in his pocket and did not talk about it anymore. A moment of universal silence followed, after which we moved on to other topics.

Lagrange's attempts to prove the parallel postulate are also reviewed by Pont in [48, p. 231–234]; see also Barbarin [2] p. 15 and Bonola [6, p. 52] who quotes de Morgan [15]

Now we pass to solid angles.

---

[22] Nikolai Ivanovitch Fuss (1755–1826) was, first, the secretary and, then, the assistant of Euler, before becoming his dearest friend and the husband of his grand-daughter Albertine Benedikte Philippine Luise, the oldest daughter of his son Johann Albrecht Euler, who was also a mathematician. Fuss was himself a very talented mathematician. He helped Euler in reading and writing his correspondence and in writing his memoirs, during the last twelve years of the latter's life, during which Euler was almost completely blind. After Euler's death, Fuss was appointed professor at the Saint Petersburg Imperial Academy of Sciences. From 1800 and until his death in 1826, he was the permanent secretary of the Academy, and it is at this title that he pronounced a brilliant *Eulogy* of his master, at the latter's death, see [24]. Fuss collected the latter's unpublished works (there were almost 200 such works), he completed those who were not complete, and he prepared them for publication.

[23] Jean-Baptiste Biot (1774–1862) was a mathematician, physicist and astronomer. He was at the same time member of the Académie des sciences and of the Académie française.

5.10 Solid Angles

Solid angles (στερεὰ γωνία) appear in Definition 11 and Propositions 20 to 26 of Book XI of the *Elements*. Later, they are used in the proof of the fact that there are exactly five regular polyhedra (Addendum to Proposition 18 of Book XIII).

Definition 11 of Book XI reads:

> A solid angle is the inclination constituted by more than two lines which meet one another and are not in the same surface, towards all the lines.
>
> Otherwise: A solid angle is that which is contained by more than two plane angles which are not in the same plane and are constructed to one point.

There is a long historical digression by Heath around this definition, in Vol. II, p. 39–42 of his edition of the *Elements* [19], covering a period ranging from ancient commentators like Proclus, until Cardano (1501–1576), Clavius (1538–1612), Viète (1540–1603) and others more recent. Heath comments on this "double definition" (indeed it consists of two statements). He writes that it contains an ambiguity, in particular for the fact that in the first statement Euclid talks about surfaces (ἐπιφάνεια) whereas in the second one, he talks only about solid angles bounded by planes. Heath adds that this is a typical case of a definition which is taken in part from older *Elements*.

After Euclid, one may talk about Apollonius of Perga (3rd-2nd c. BC), who also wrote on solid angles, see Heath's comments in [19] and the article by Paul Tannery [56]. Apollonius defined a solid angle as the "bringing together of a surface of a solid to one point as a broken line or surface" (quoted in [19, vol. III, p. 267]), a definition that Proclus met with scepticism (see *In Euclidem*, 19–125, 3, quoted in [52, Chapter 1]).

Heron of Alexandria (ca. 10–85 AD) considered solid angles defined by joining to a point arbitrary convex surfaces, see [19] and [56]. Definition 24 of his *Book of Definitions* [27, vol. IV, p. 28] says:[24] "A solid angle is in general the bringing together of a surface which has its concavity in one and the same direction."

Pappus of Alexandria (4th c. AD) used solid angles in his work on spherical geometry, namely, in the proof of the spherical triangle inequality, given as Proposition 1 of Book VI of his *Mathematical Collection* [46, p. 37 of t. II of Ver Eecke's French translation]. Like Euler, Pappus considered a solid angle obtained by gluing three triangles and intersecting it with a sphere centered at the vertex of this angle. Then, from the triangle inequality satisfied by the three dihedral angles that form this solid angle (Proposition XI.20 of the *Elements*), he deduced the triangle inequality for spherical triangles. We note incidentally that Menelaus of Alexandria (ca. 70–140) gave in his *Spherics* a proof of the spherical triangle inequality which is, unlike the one of Pappus, intrinsic to the sphere, that is, that does not make use of the ambient space (this is Proposition 5 of Menelaus' *Spherics*, see [53, p. 151]).

[24] Heath's translation of Euclid's *Elements* Vol. III, p. 267.

5.11 Angles and Solid Angles as Magnitudes

We already said that a discussion of the notion of angle and solid angle in the *Elements* must involve that of magnitude. In fact, a question that arises, since Aristotle, is whether solid angles are magnitudes, and if yes, how, since they do not satisfy the usual supposed properties of a magnitude.

Magnitudes are not explicitly defined in the *Elements* but we are said there that lines (or line segments) are examples of magnitudes: they can be compared and the theory of proportions applies to them. They satisfy a certain number of axioms there. For instance, Axiom 1 of Book I concerns magnitudes; it says:

> Things which are equal to the same thing are also equal to one another

and Axiom 5 of the same book says:

> The whole is greater than the part.

But can we compare two magnitudes like a straight line and a curve? If yes, how? If this comparison is done by length, how do we compute lengths without infinitesimal calculus? Aristotle, in the *Physics* VII and in other writings, already addressed the difficulties encountered in comparing an arc of a circle with a straight line. In fact, the theory of proportions developed in the *Elements* applies to magnitudes, but there exist several classes of magnitudes, and only magnitudes of the same class can be compared, added, subtracted, or multiplied. For instance, a surface cannot be compared to a line. The question is then whether there is a class of magnitudes that comprises angles, and what is exactly this class.

Book V of the *Elements*, which contains Eudoxus' theory of proportion, deals with the class of magnitudes in which every two elements have a certain ratio.[25] Definition V.4 says:

> Magnitudes are said to have a ratio to one another which can, when multiplied, exceed one another.

The magnitudes that Euclid considers include lines, areas and solids. Circle segments are examples of non-magnitudes. What about angles? Rectilineal angles are measured by the lengths of the segments they make on a circle of radius 1. This constitutes an indication of the fact that they are *not* magnitudes. Let us investigate this more.

One may first recall that there are no explicit computations of measures of length, area, or angle in the *Elements*, except for saying that some angle is right, or acute, or obtuse, or for statements like *The sum of the three angles in a triangle is equal to two right angles* (Proposition 32 of Book I), or *Two circles are to each other like the square of their radii* (Proposition 2 of Book XII). Quotients, products, etc. of magnitudes are compared, but never computed. The language is geometrical. For

---

[25] We have included in the second appendix to the present chapter (Appendix 2) a short note on Eudoxus.

instance, Euclid talks about the "square on the side", and not the "square of the side". This point of view is in contrast with that of Archimedes, who had a strong inclination for computations. It is well known that the latter computed approximate values for π, of areas under a parabola, of spherical discs, etc.

Difficulties about angles as magnitudes arise for instance when one talks about a cornicular angle, that is, the angle between the circle (or semi-circle) and a tangent line, which is considered in Proposition 16 of Book III of the *Elements* and which we recalled above. Heath, in his edition of the *Elements*, discusses the objections that some Greek mathematicians made on angle being a magnitude. In particular, in his long comment on Proposition III.16, he writes, quoting Proclus and others:

> The difficulty which the ancients felt arose from the very fact which Euclid embodies in this proposition. Since an angle can be divided by a line, it would seem to be a magnitude; "but if it is a magnitude, and all homogeneous magnitudes which are finite have a ratio to one another, then all homogeneous angles, or rather all those on surfaces, will have a ratio to one another, so that the *cornicular* will also have a ratio to the rectilineal. But things which have a ratio to one another can, if multiplied, exceed one another. Therefore the cornicular angle will also sometime exceed the rectilineal; which is impossible, for it is proved that the former is less than any rectilineal angle." (Proclus). The nature of contact between straight lines and circles was also involved in the question, and that this was the subject of controversy before Euclid's time is clear from the title of a work attributed to Democritus (ca. 420-400 BC) [...]

The precise references to Proclus, Democritus and others are given in Heath's passage. Heath then talks about works related to this subject by Cardano, Peletier (Peletarius), Clavius, Viète, Galileo Galilei, Wallis and others.

These and other questions on angles, e.g., whether they can be defined as the "inclination" between two curves, or on how one computes angles that are not bounded by straight lines and how one compare them, lead to infinitesimal mathematics. From a modern viewpoint, topological analysis is also involved in this kind of investigation. Indeed, since one defines an angle as the boundary of a region enclosed by two curves, one needs to make precise the notion of "boundary". Other notions that arise in the context of angles include continuity, convexity, infinite division, etc. Such questions were raised by Aristotle in various treatises, and they became fundamental objects of research at the Renaissance, finding important developments in the works of Galileo Galilei, Wallis, Hobbes and other scientists and culminating in the works of Leibniz and Newton.

Much less known is that, in the period between the ninth and the thirteenth centuries, these questions were thoroughly studied by Arab mathematicians working in cities like Baghdad, Cairo, Damascus, Aleppo and Córdoba, and that their mathematical and philosophical development attained an extremely high level there. This is the subject of the next section.

5.12 Angles and Solid Angles in the Arabic Mathematical Literature

The Arabic tradition of the *Elements* is still a subject about which relatively little is known, since a large number of existing manuscripts of Arabic translations or commentaries of (parts of) the *Elements* have not been examined by mathematicians up to the present day. One may quote here Edward Kennedy, from the introduction to his article *Mathematical geography* [32, p. 185]: "The historian of the Islamic exact sciences is frequently confronted with an *embarras de richesse*—hundreds of manuscript sources which have never been studied in modern times." What Kennedy says about general islamic science is true in particular for the literature on Euclid's *Elements* that still has not been seriously examined for the reconstruction of the "original" version—if one can still talk about an original version. Already at the end of the eighteenth century, the famous French mathematician and historian of mathematics Jean-Étienne Montucla (1725–1799), in his *History of Mathematics*, wrote that "the Arabs provided us with a large number of authors of the same class as Proclus and Theon of Alexandria" [38, Vol. 1, p. 212]. He mentions, on p. 90 of the same volume, that he had learned that the Oxford Library possessed 400 Arabic manuscripts on astronomy. This is only in one university. It is also worth recalling here that many important Greek mathematical works (including writings by the essential Greek mathematicians like Diophantus, Apollonius, Archimedes and Menelaus) have not survived in Greek language, and that it is thanks to the translations, revisions and commentaries made by Arab mathematicians of the Middle-Ages that these texts, or their content in various forms (translations, quotations, paraphrases, commentaries, critiques, corrections, etc.) have come down to us. A large number of still unpublished manuscripts exist. It is worth recalling that Fermat, Descartes, Euler, Huygens and other prominent mathematicians were devoted readers of works of Diophantus, Archimedes and others. Christiaan Huygens' correspondence with Mersenne and others contains episodes that shows us his efforts to obtain an unpublished Arabic manuscript of Apollonius, which he eventually obtained through the Medici and which he tried to decipher himself, see [29, t. 2]. This is also reported on in [8, p. 42–43].

Let us return to angles.

On the notion of (plane and solid) angle in Arabic mathematical manuscripts, a 706-page book was recently published by Roshdi Rashed, titled *Angles et grandeur: d'Euclide à Kamāl al-Dīn al-Fārisī* (Angles and magnitude: From Euclid to Kamāl al-Dīn al-Fārisī) [52]. This book involves mathematics, history and philosophy. It contains critical editions together with analyses of a certain number of manuscripts that deal with the notion of angle written in Arabic, dating from the early period of Arabic scholarship (ninth century) until the end of that period (13th-14th centuries). The manuscripts edited in this book include, among others, translations of parts of Euclid's *Elements* as well as texts originally translated from the Greek on the notion of angle, of which the Greek original is lost. We shall review some of the material contained in this book, especially the part on solid angles. The rest of the present section may be considered as a quick review of Rashed's book and we do this without going into details and explanations, which would have required an entire volume.

According to Rashed, the first known Arab mathematician who wrote on solid angles is Aḥmad ibn 'Abd al-Jalīl al-Sijzī (ca. 945–1020). In a book titled *Introduction to the science of geometry*, al-Sijzī gave a classification of the different kinds of solid angles. He writes [52, p. 328]: "[...] the first [species of solid angle] under a single surface, the second under a surface and a plane, and the third under planes". He gives the following examples of each of these species: "The angle at the apex of a cone of revolution surrounded by the lateral surface of the cone, or that at one of the apexes of the myrobolan-shaped or quince-shaped solid, for the first; the angle at one of the vertices of a half-cone, or a half-solid in the shape of a myrobolan or a quince, or a portion thereof, if they are cut by a plane passing through their vertex, for the second; and finally, the angle between three planar angles—a trihedron—the sum of which is smaller than four right angles, for the third." The reader will notice that we encounter here solid angles of various shapes, that are much more general than those alluded to in Euclid's definition recalled in Sect. 5.10. Al-Sijzī's memoir ends with the words: "These are things concerning angle that leave us puzzled, given that some of its states necessarily imply that it is a magnitude and others that it is not."

Among the other texts presented and discussed in Chapter I of Rashed's book, we mention an anonymous manuscript titled *Treatise on the angle* and referred to as the *Lahore manuscript*, concerned with works of Euclid, Apollonius, the Byzantine scholars Aghānis (fifth century) and Simplicius (sixth century) and containing a wealth of mathematical proofs and technical remarks on the divisibility of various species of angles.

Chapter II of the book is concerned more especially with the research conducted on the notion of magnitude. It contains in particular a discussion on magnitudes that do not satisfy Definition V.4 of the *Elements* or the so-called Eudoxus–Archimedes axiom,[26] which is extensively used in Book XII of the *Elements*. We mention in particular, from that chapter, works by Abū 'Ali al-Ḥasan ibn al-Ḥasan ibn al-Haytham (ca. 965–1040), whose investigations related to the notion of angle are included notably in two essays he wrote on "the explanation and correction of Euclid's *Elements*", in which he considers questions like existence, classification, nature and homogeneity of angles. Rashed also provides a critical edition of a text by the algebraist al-Samaw'al ibn Yaḫyā al-Maghribī (d. 1175) titled *Epistle on the angle of contact*, in which the latter elaborates on the non-homogeneity and the non-comparability of figures, based on the example of the angle of contact. Together with Kamāl al-Dīn al-Fārisī (d. 1319), al-Samaw'al is one of the successors of Ibn al-Haytham who continued the latter's research on angles of contact.

Chapter III of Rashed's book has a more philosophical flavor. It contains an epistle addressed by the famous philosopher and physician Ibn Sīnā (980–1037), known in the Latin World as Avicenna, addressed to another physician and

[26] See Footnote 15.

philosopher, Abū Sahl al-Masīḥī. Notions like quantity, quality, relation, magnitude, figure, limit of angles, and others are considered from the philosophical point of view. Infinite divisibility of angles is also discussed. Ibn Sīnā considers that angles belong to both categories of quality and quantity.

Chapter IV is concerned with solid angles. It contains notably another text by Ibn al-Haytham, who developed the theory of solid angles in the context of his research on isoperimetry and isoepiphany. This theory was motivated by the problem of approximating the volume of the sphere by volumes of convex polyhedra, which he considered in the context of his infinitesimal approach to the study of the sphere. Ibn al-Haytham used the work of Archimedes on the sphere, and also conical projections and spherical geometry. In Ibn al-Haytham's work, solid angles are subject to the usual operations that apply to Archimedean magnitudes including the theory of proportions.

Let us quote now a passage of some Arabic versions of the *Elements* edited in Rashed's book [52] that gives an idea of the questions about solid angles that are dealt with. A version of Euclid's definition of a solid angle by Al-Abharī (d. in 1265) reads [52, p. 308]:

> Solid angle: is any body enclosed by a single surface that terminates at the same point, or more than two plane angles joined at a point, all in the same direction from that point, and such that two of these angles are not in the same plane.

A mathematician will notice that this definition is clearer than the two definitions in Heath's version of the *Elements* that we recalled in Sect. 5.10. Here, [Euclid] talks about "a single surface that terminates at the same point" instead of "two lines which meet one another and are not in the same surface, towards all the lines".

Ibn al-Haytham's work on infinitesimal mathematics, which is a continuation of the works of Archimedes and Apollonius, was already highlighted in another publication by Rashed, *Infinitesimal mathematics from the ninth to the 11th centuries* [51], in 5 volumes. We learn from there that Ibn al-Haytham developed a theory of infinitesimals which he used in the setting of angles, in particular in the comparison between a contact angle and a rectilineal angle. He starts by noting that since these two magnitudes are not Archimedean, a new theory is needed in order to deal with them. Among the arguments he introduced are two sequences, an increasing one and a decreasing one, the second one bounding the first one from above, which he used for the comparison of infinitesimals (see p. 101 of [52]).

In trying to resolve difficulties that appear in the *Elements* and that are inherent to the notion of angle, and in particular of solid angle, Ibn al-Haytham wrote two treatises, the *Explanation of the postulates of the book of Euclid*, of which a fragment survives and is translated and published in [52, p. 286 ff.] and the *Book on the resolution of doubts relative to the book of Euclid on the* Elements *and the explanation of its meanings* (see [52, p. 110 ff.] and [30]) in which he addressed a certain number of difficulties. In these works, he developed a new geometry in which the notion of angle and that of motion[27] of figures are primitive elements

[27] In modern words, this would be called congruence or isometry.

and where the notion of equality of lines or of areas is based on motion. We recall here that motion was avoided in Euclid's *Elements* following a ban by Aristotle who considered that this notion pertains to physics and not to mathematics.[28]

Ibn al-Haytham's work on angles was continued by several Arab mathematicians. We mention in particular a commentary on Archimedes' *Sphere and cylinder* written by Naṣīr al-Dīn al-Ṭūsī (1201–1274) in which the latter addresses the question of the comparability of lines, curves and curvilineal angles. Naṣīr al-Dīn used in particular a notion of "rolling onto each other," while he was comparing lengths of curves (cf. p. 469 ff. of [52]). Incidentally, al-Ṭūsī is also the author of a famous commentary on Euclid's *Elements*. Montucla, in his *History of Mathematics* [38], already wrote in the eighteenth century that the main editor of Euclid in the East was Naṣīr al-Dīn al-Ṭūsī. He speaks of his "learned commentary, printed in Arabic in the magnificent Medici printing house" and adds that "this work, esteemed among those of his nation, would perhaps also be so among us if a more common language had made it accessible to our understanding."

Since Montucla mentions the Medici, let us note that the latter were actively involved in the effort to translate and publish mathematical manuscripts in Arabic. The reader may refer to the recent book by S. Dumas Primbault [18], in particular Chapter 3, in which the author recounts how in 1578 the Patriarch of Antioch, visiting Rome, presented to Cardinal Ferdinando de' Medici (later Grand Duke of Tuscany), as a gift, a number of Arabic manuscripts, including a translation of some books of Apollonius' *Conics* that were thought to be lost. The manuscripts found their way into the Medici library before a Latin translation was published a few years later, an undertaking involving several mathematicians, including Galileo Galilei, Antonio Santini, Vincenzio Viviani, Giovanni Borelli, and others. We already mentioned that Christiaan Huygens, for mathematical reasons (he was highly interested in Apollonius' works), took a keen interest in the arabic manuscripts of the *Conics*; see his correspondence in [29, t. 2] and the report on this subject made in [8, p. 42–43].

We also quote a passage by Ibn al-Haytham, from his epistle *The explanation of postulates*. He writes [52, p. 296] : "Euclid says : 'A solid angle is one that is surrounded by planar angles, more than two angles, that are not in the same plane and that meet at a single point.' This assertion is the definition of a solid angle surrounded by flat surfaces. But there may be solid angles that do not fall under this definition; these are angles surrounded by spherical, cylindrical or conical surfaces, except that the solid angles he has dealt with in this book are those surrounded by plane surfaces only."

After Ibn al-Haytham's study of solid angles, one had to wait for a few centuries for what regards new developments.

---

[28] Bourbaki writes, concerning motion [7, p. 28]: "[...] l'usage des déplacements—notamment dans les 'cas d'égalité des triangles'—longtemps admis comme allant de soi, devait bientôt apparaître à la critique du XIXe siècle comme reposant aussi sur des axiomes non formulés." (The use of motion—especially in "cases of equality of triangles"—long accepted as self-evident, was soon to appear to nineteenth-century critics to also be based on unformulated axioms.)

5.13 Some Notes on Solid Angles from the Renaissance

The history of solid angles starting at the Renaissance involves the works of Johannes Kepler (1571–1630), Girard Desargues (1591–1661), René Descartes (1596–1650), Florimond de Baune[29] (1601–1651), Jean-Paul de Gua[30] (1710–1786) and many others. Let us say a few words on some of these works.

Johannes Kepler, in his *Harmony of the world* [33], considers solid angles of regular and semi-regular polyhedra. In Definition V of Book II, he talks about "the individual angles of several plane figures making up a solid angle". He studies 3-dimensional tilings, a work in which the notion of solid angle plays an important role. He talks there about "regular or semi-regular figures fitted together so as to leave no gap between the sides of the figures". He considers solid figures made of two types of figures fitting together, for instance the rhombic dodecahedron, a figure having two types of angles: eight obtuse trilinear solid angles and six acute quadrilinear solid angles. Kepler writes that "the result is like the way in which bees construct their honeycomb: the cells are contiguous and the end of each is surrounded by three opposing ends while its sides are surrounded by the sides of six more cells. Three more cells could be added at the other end to complete the figure, except that the entrances to the cells must remain open" [33, p. 99–100].

We already mentioned Descartes' *Progymnasmata de solidorum elementis* (Exercises on the elements of solids) [16], a treatise in which the French philosopher-mathematician studies solid angles of convex polyhedra, a study containing a result which is a precursor of Euler's combinatorial formula that gives a relation between the number of vertices, edges and faces of an arbitrary convex polyhedron.

From Mersenne's *Synopsis mathematica* [37], we learn that Desargues wrote a complete treatise on solid angles.

Florimond de Beaune's *Doctrine de l'angle solide* (Doctrine of the solid angle) [12] contains an exposition of various geometric construction problems concerning

[29] Florimond de Beaune (1601–1651) was a jurist and a disciple of Descartes. He is the author of Notes to the latter's *Géométrie* published by van Schooten in Latin translation. De Beaune was an amateur mathematician and he played a certain role in the diffusion of mathematics in the seventeenth century.

[30] Jean-Paul de Gua de Malves (1710–1786), known also as l'Abbé de Gua, was a mathematician, Catholic priest, economist, professor of philosophy at the Collège de France, translator (from English) and initiator of a project that ended up as the *Encyclopédie*. He was an admirer of Descartes and he published a treatise with the significant title *Usage de l'analyse de Descartes pour découvrir, sans le secours du calcul différentiel, les propriétés des lignes géométriques de tous les ordres* (The use of Descartes' analysis to discover, without the aid of differential calculus, the properties of geometric curves of all orders). Lagrange, in the memoir translated in the present volume (Chap. 18), writes: "The late de Gua already had the idea of making all spherical trigonometry dependent on a single general property of spherical triangles; but the Memoir that he gave on this subject in the 1783 volume of the Academy of Sciences contains computations that are so complicated that they seem to be more adapted to show the drawbacks of his method than to make it accepted." [14]

solid angles with three faces, given six of their elements (dihedral angles or face angles).

De Gua's work carries the significant title *Diverses mesures, en partie neuves, des aires sphériques et des angles solides, triangulaires et polygones, dont on est supposé connoître des élémens en nombre suffisant, avec des remarques qu'on croit pouvoir contribuer à simplifier les intégrations de plusieurs équations différentielles à inconnues actuellement séparées* (Various measurements, some of them new, of spherical areas and solid, triangular and polygonal angles, of which we are supposed to know a sufficient number of elements, with remarks that we believe will help simplify the integrations of several differential equations with unknowns that are actually separated) [13].

In all the works mentioned except Euler's, solid angles are studied, but the actual value, or "measure", of a solid angle is not computed. In contrast, this is what Euler does in his memoir [21]. It is also worth noting that Euler used in an essential way the notion of solid angle in his proof of the so-called Euler formula; see his memoir *Elementa doctrinae solidorum* [20].

In 1925, the biologist and mathematician D'Arcy Thompson (1860–1948) published a survey on the semi-regular polyhedra involving the computation of their angles, see [11].

To conclude this section, we note that the notion of solid angle is widely used in optics, thermal radiation, and other fields of physics.

5.14 On the Notion of Angle in Spherical Geometry

A spherical angle is made by the intersection of two lines (great circles) on the sphere.[31] Euclid, in the *Elements*, even though he works with spheres, does not mention spherical angles. A spherical angle is an intrinsic object of the sphere. Euclid does not study intrinsic properties of the sphere. The first geometrical treatise in which the notion of spherical angle is central is the *Spherics* by Menelaus of Alexandria (ca. 70–140 AD). The treatise consists of 91 propositions. Angles appear there generally (but not only) as angles of spherical triangles.[32] A spherical triangle has three sides and three vertices. The angle at each vertex of a spherical triangle is defined as the dihedral angle formed by the planes defining the two sides that

[31] In fact, two great circles make two angles at their intersection point, but in what follows, we shall take the smallest one as the angle made by these two lines.

[32] We recall that on the sphere, considered by all authors of Antiquity as a sphere embedded in three-dimensional Euclidean space, a *great circle* is the intersection of this sphere with a plane passing through its centre, a *segment* on this sphere designates a segment of great circle, and a *spherical triangle* designates the union of three points on this sphere together with three segments joining them. The segments are called the sides of the triangle. For a number of (obvious) reasons, it is generally required that the three points in question and the segments joining them are contained in a half-sphere; we shall assume without further comments that this holds.

contain it. The definition of the dihedral angle formed by two planes is given in Euclid's *Elements* (Book XI, Definition 6).

The first five propositions of Menelaus' *Spherics* are essentially about angles. Proposition 1 concerns the construction of angles, and is used extensively throughout the rest of the treatise. It states that given, on the sphere, a triple consisting of an angle, a line and a point on it, one can construct an angle having this given point as its vertex, a side on the given line and which is equal to the given angle.

Proposition 5 of the *Spherics* expresses the triangle inequality for spherical triangles. We have already encountered this proposition in Sect. 5.10, where we recalled that Pappus used the above definition of angle as a dihedral angle and the fact that the three dihedral angles forming a solid angle satisfy the triangle inequality (Proposition XI.20 of the *Elements*) to prove that the triangle inequality is satisfied by the three sides of a spherical triangle. We also mentioned that this is Proposition 1 of Book VI of Pappus' *Collection*, [46, p. 37 of t. II of Ver Eecke's French translation]. Menelaus, in Proposition 5 of his *Spherics*, gives a proof of this spherical triangle inequality which is, unlike the one of Pappus, intrinsic to the sphere; see [53, p. 151].

Let us finally note that in Menelaus' *Spherics*, as in Euclid's *Elements*, angles are compared with each other but their actual measures are never computed. In Menelaus' treatise, there is a definition of acute, obtuse and right angle, and in fact, this is given right at the beginning (first page) of the treatise (see [53, p. 504]):

> Spherical figures are limited by straight lines, except that their sides are arcs of great circles, each one of them being smaller than a semicircle. One which is bounded by three sides is a triangle and, likewise, those which have four sides. The angles of the triangles are those contained by these sides.
>
> When one of the two planes of the two circles is perpendicular to another one, then their circumferences intersect with right angles. What is smaller than that angle is acute and what exceeds it is obtuse.

5.15 In Guise of a Conclusion: Some Modern Works on the Notion of Angle

The modern period in mathematics is often considered to be the one starting with Euler. Among the "moderns", we start by noting that Euler's contemporary, Jean le Rond d'Alembert, insisted on clear and simple definitions, accessible to everyone. Naturally, he proposed simple definitions of angles. In his *Essai sur les Éléments de philosophie* (An essay on the Elements of philosophy), he gave a definition of angle that is simple enough but he noted that this definition depends on the notion of line and of the uniformity of the circle, which are not simple notions, see [10, p. 319].

David Hilbert, in his *Foundations of geometry* [28] (first version 1898), a treatise considered by many as an updated version of the part of Euclid's *Elements* that is concerned with the axioms of geometry, included the notion of angle in the setting of his congruence axioms (Group IV). Felix Klein, in his *Elementary mathematics from an advanced standpoint* [34] (first ed. 1925) discusses angles at length, making use of the notion of motion, in the third part of this essay, titled *Systematic discussion of geometry and its foundations*.[33] In Birkhoff's axiomatization of geometry [5] (1932), which is minimalistic and assumes the real numbers system, angle belongs to the list of four undefined notions, the other three being point, line and distance.

There is a notion of total angle at a point on an arbitrary surface in the synthetic theory of surfaces developed by A. D. Alexandrov and the school he founded, see [1]. This school is characterized by a return to the techniques of elementary geometry, in the tradition of the Greeks. V. A. Zalgaller, in a tribute to Alexandrov on the occasion of the centenary of his birth [63], describing the Leningrad geometry seminar which the latter led in the period 1945–1952, writes: "When talking about methodology, [Alexandrov] often pronounced the slogan: 'Return to Euclid!', fighting for visual methods." See also the article by Katateladze [35], with the significant title *Aleksandrov of Ancient Hellas*, in which the author writes that "Alexandrov enriched the methods of differential geometry by the tools of functional analysis and measure theory, driving mathematics to its universal status of the epoch of Euclid." One feature of the modern metric theory of surfaces is that it allows angles of arbitrary value (the value could be greater than four right angles).

Let us also mention the notion of orbifold point, and more generally, of cone points on surfaces, see e.g. Chapter 13 of Thurston's *Geometry and topology of 3-manifolds* [58], and his paper *Shapes of polyhedra and triangulations of the sphere* [59]. In the latter paper, a model for a point that has cone angle $\theta \in [0, \infty]$ is the neighborhood of the origin in the metric space obtained by taking the universal cover of the Euclidean plane minus the origin, reinserting the origin and identifying two arbitrary rays issuing from this point by a rotation of angle θ. In the case where $0 \leq \theta \leq 2\pi$, such a point with cone-angle θ is also obtained by taking a sector of the Euclidean plane bounded by two rays emanating from the origin and making an angle θ, and gluing the two boundaries of this sector by a Euclidean rotation of angle θ. A model of the Euclidean cone point of cone angle θ is the neighborhood of the origin in the resulting surface. Thurston's theory, like Alexandrov's, is based on a return to the techniques of elementary geometry.

Finally, let us note that the notion of solid angle is an essential tool in Sullivan's work on the dynamics in hyperbolic space involving measures at infinity, see [54].

Appendix 1: Anthyphairesis

For the convenience of the reader, we recall the notion of anthyphairesis of two magnitudes, used in several sections of this chapter.

[33] This is the part of Klein's treatise that contains a classification of geometries from the point of view of group actions.

Consider two magnitudes a and b, and assume that $a > b$. The *anthyphairesis of the pair* (a, b) is the sequence of operations

$$a = I_0 b + \gamma_1, \text{ with } \gamma_1 < b,$$

$$b = I_1 \gamma_1 + \gamma_2, \text{ with } \gamma_2 < \gamma_1,$$

$$\dots$$

$$\gamma_{n-1} = I_n \gamma_n + \gamma_{n+1}, \text{ with } \gamma_{n+1} < \gamma_n,$$

$$\gamma_n = I_{n+1} \gamma_1 + \gamma_{n+2}, \text{ with } \gamma_{n+2} < \gamma_{n+1},$$

$$\dots$$

The resulting sequence of natural numbers

$$I_0, I_1, I_2, \dots$$

is called the *quotients* of the anthyphairesis, and

$$a > b > \gamma_1 > \gamma_2 > \gamma_3 > \dots$$

is the sequence of *remainders* of the anthyphairesis of a to b.

The anthyphairesis of a to b is *finite* if the last remainder divides the immediately previous one, and in this case, this last remainder is the *greatest common divisor* of a to b. (This is Propositions VII.1 and 2 of Euclid's *Elements* for natural numbers, and Proposition X.3 for magnitudes.)

In the case where no remainder divides the immediately previous one, the anthyphairesis is said to be *infinite*. In this case, a and b are *incommensurable* (Proposition X.2 of Euclid's *Elements*).

Appendix 2: Eudoxus of Cnidus

Eudoxus of Cnidus (408-355 BC) is a mathematician, astronomer, geographer and philosopher to whom we referred in several sections of this chapter. He is an important figure for our subject here because his work has close relations with those the Pythagoreans, of Plato and of Euclid. He was a student of Archytas of Tarentum, the famous Pythagorean, with whom he studied geometry, number theory and music, among others. Eudoxus visited Plato's Academy in Athens at least twice, where he attended Plato's lectures.[34]

Eudoxus is considered as the source of a substantial part of Euclid's *Elements*, both for the axiomatic method and for the theory of proportions, the latter being set out in Books V and VI, and which is at the origin of the modern definition of real numbers. Heath writes in [26, p. 326–327]: "[...] The greatness of the new theory itself needs no further argument when it is remembered that the definition of equal ratios in Eucl. V, Def. 5, corresponds exactly to the modern theory of irrationals due to Dedekind and that it is word for word the same as Weierstrass's definition of equal numbers". We already mentioned the Eudoxus–Archimedes axiom, used in Book V of the *Elements*, which says that given any two magnitudes a and b, there exists a natural number n such that $na > b$ (see Footnote 15.)

As a matter of fact, Scholion X.2 to Euclid's *Elements* is an attack from Platonic circles against Eudoxus' theory of ratios (Book V of the *Elements*) and an unfavorable comparison of his theory to the earlier one by Theaetetus, which was based on equal anthyphairesis, and was defined only for ratios a/b such that the anthyphairesis of a to b is finite or eventually periodic. A reconstruction of the theory and details on Scholion X.2 can be found in [44].

We also owe to Eudoxus the so-called *exhaustion method*, also used by Euclid, a theory that represents the beginning of modern integration theory. Bourbaki, in his *Éléments d'histoire des mathématiques*, writes that it would be more appropriate to call the "Riemann sums" "Archimedes sums", or "Eudoxus sums" [7, p. 248]. In fact, integration started with Book XII of the *Elements*, which is due to Eudoxus. It continued and was greatly extended by Archimedes. Let us quote again André Weil, from the same letter to his sister Simone we quoted in Footnote 6. He writes [62, t. VII, Vol. 1, p. 556]: "It seems to me that Eudoxus should have been the first mathematician in Greek history" [À ce qu'il me semble, Eudoxe a dû être le premier mathématicien dans l'histoire grecque].

On another topic, Eudoxus developed an astronomical theory based on the motions on homocentric (or concentric) spheres that represent the sun, moon and the five planets known at his epoch (Mercury, Venus, Mars, Jupiter, and Saturn). This system agrees with the known observations of the time. Eudoxus' theory had a long development throughout the history of Greek mathematical astronomy. It was taken up later and further developed by Aristotle in the *Metaphysics* and the *De Caelo* and it culminated in the work of Ptolemy about 500 years later. For a recent

[34] Regarding these two visits, Diogenes Laertius, in his *Lives of the eminent philosophers* [17] recounts the following: "[...] Disembarking at the Piraeus, Eudoxus went up to Athens every day, attended the sophists' lectures, and returned to the port", and then:"He then returned to Athens with a great many disciples; some say he did this to annoy Plato, who had originally snubbed him. Some report that at a drinking party given by Plato, Eudoxus, owing to the number of guests, introduced the practice of arranging couches in a semicircle." Thus, Diogenes clearly implies some friction between Eudoxus and Plato in both visits. Some modern scholars believe that Plato set Eudoxus as the director of the Academy during his second visit to Sicily, presumably coinciding with Eudoxus' second visit to the Academy. But there is no ancient source supporting this conjecture.

commentary on Ptolemy's system of homocentric spheres, see Manin's article [36]. For two recent articles on Eudoxus' works on the theory of proportions (Book V of the *Elements*), on the method of exhaustion and on his astronomical model of the universe using homocentric spheres, see the recent articles [40] by Negrepontis and [55] by Sunada. Let us also quote Simone Weil, from her work *Intuitions pré-chrétiennes/Esquisse d'une histoire de la science grecque* [61].

> The same Eudoxus developed an astronomical system to answer Plato's question: "To find the set of uniform circular motions that allow us, concerning the stars, to keep up appearances." It relies on the brilliant idea of composition of motions that is at the basis of our mechanics. Just as we construct the parabolic movement of projectiles with two rectilineal motions, one uniform, the other accelerated, so Eudoxus accounted for all the positions of the stars observed in his time by assuming several uniform circular motions accomplished simultaneously around different axes by a single star. This conception is as daring, as pure as that which defines real numbers and that of the summation of an infinite sum. If Plato wanted only uniform circular motions, it is because this motion alone is divine, and because the stars are, he says, images of the divinity sculpted by the divinity itself. Plato almost certainly has this composition of motions in mind when he talks about the Other, rebellious to unity, harmonized by constraint with the Same. The sun, in its unique motion, is carried along at the same time by the circle of the equator and by that of the ecliptic, which correspond to the Same and the Other; and this makes only one motion. In the following period of Greek science, Ptolemy reproduced, in a much cruder form, Eudoxus' system; Apollonius continued Menechmus[35] discoveries of conics, and Archimedes those of Eudoxus on integration.

The ancient sources for Eudoxus include the *Lives* by Diogenes Laërtius [17], a work we already quoted, and passages from Aristotle's *Metaphysics*, *Nicomachean Ethics* and some other works, and from Archimedes, Callippus, Proclus, Simplicius, and there are others.

References

1. A.D. Alexandrov, *Convex polyhedra*, Russian edition (Gosudarstvennoe Izdatel'stvo Tekhniko-Teoreticheskoi Literatury, Moscow-Leningrad, 1950). English translation by N. S. Dairbekov, S.S. Kutateladze and A. B. Sossinsky, with comments and bibliography by V. A. Zalgaller and appendices by L. A. Shor and Yu. A. Volkov, Springer Monographs in Mathematics, Springer Berlin Heidelberg New York, 2005
2. P. Barbarin, *La Géométrie non-Euclidienne (suivie de Notes Sur la Géométrie non-Euclidienne dans ses Rapports avec la Physique Mathématique par A. Buhl)*, 3rd edn. (Gauthier-Villars, Paris, 1928). (Première éd., 1902)
3. J. Beery, J. Stedall (Eds.), *Thomas Harriot's Doctrine of Triangular Numbers: The "Magisteria Magna,"*. Heritage of European Mathematics Collection (European Mathematical Society, Zürich, 2008)

[35] Menaechmus (c. 380–c. 320 BC) was a geometer and philosopher, disciple of Plato and of Eudoxus. To him is attributed the discovery of conic sections and his solution to the problem of doubling the cube using the parabola and hyperbola.

4. J.-B. Biot, *Mélanges scientifiques et littéraires, t. II, Michel Levy Frères* (Libraires-Editeurs, Paris, 1858)
5. G.D. Birkhoff, A set of postulates for plane geometry (Based on scale and protractors). Ann. Math. **33**(2), 329–345 (1932)
6. R. Bonola, La geometria non-euclidea, in *Esposizione storico-critica del suo sviluppo*, 1st edn. (Ditta Nicola Zanichelli editore, Bologna, 1906). German translation by M. Liebmann in the collection Wissenschaft und Hypothese, Teubner, Leipzig, 1908. English translation by H. S. Carslaw, Non-Euclidean geometry, A critical and historical study of its development. 1st éd., Chicago, 1912, reprint Dover, New York, 1955
7. N. Bourbaki, *Éléments d'histoire des mathématiques*, 1st edn. (Masson, Paris, 1984). New edition, N. Bourbaki and Springer-Verlag, Berlin and Heidelberg, 2007
8. R. Caddeo, X. Hascher, F. Jedrzejewski, A. Papadopoulos, Christiaan Huygens: Écrits sur la musique et le son, œuvres traduites du latin, avec commentaires mathématiques, historiques musicaux, Hermann, coll. Théorie de la musique, Paris (2021), 496pp.
9. B.F. Cavalieri, Directorium generale uranometricum in quo trigonometriæ logarithmicæ fundamenta ac regulæ demonstrantur, astronomicæque supputationes ad solam fere vulgarem additionem reducentur, opus utilisissimum astronomis, mechanicis, geographicis, architectis principue militaribus, mechanicis, geographicis, nec non ipsis philosophis naturalibus; authore Fr. Bonav. Cavalioro, Mediolanensi ordinis Jesuatorum S. Hieronymi, priore titulari ac in almo Bononiensi gymnasio primario mathematicorum professore. Bonaniæ, 1632, in-4
10. d'Alembert (J. Le Rond d'Alembert), *Essai sur les éléments de philosophie*, 1st edn. (1759), New edn. (Fayard, Paris, 1986)
11. D'Arcy Wentworth Thompson, On the thirteen semi-regular solids of Archimedes, and on their development by the transformation of certain plane configurations. Proc. R. Soc. Lond. A **107**(742), 181–188 (1925)
12. F. de Beaune, Doctrine de l'angle solide. Inventaire de sa bibliothèque. Transcription of the ms. by B. Barbiche, Éd. P. Costabel with introduction and commentaries. Paris, J. Vrin (1975)
13. J.-P. de Gua, Diverses mesures, en partie neuves, des aires sphériques et des angles solides, triangulaires et polygones, dont on est supposé connoître des élémens en nombre suffisant, avec des remarques qu'on croit pouvoir contribuer à simplifier les intégrations de plusieurs équations différentielles à inconnues actuellement séparées, Mémoires de l'Académie royale des sciences, année 1783, Paris (1785)
14. J.-P. de Gua, Trigonométrie sphérique, déduite très brièvement et complètement de la seule solution algébrique du plus simple de ses problèmes généraux, au moyen de diverses transformations, dont les rapports des sinus et cosinus, tangentes et cotangentes, sécantes et cosécantes d'un même arc ou d'un même angle-plan, rendent cette solution susceptible, et comprenant quelques formules et observations qu'on croit utiles et neuves, Mémoire de l'Académie royale des sciences, 1783, p. 2291–293
15. A. De Morgan, *A Budget of Paradoxes* (Longmans, Green, London, 1872). Reprint: The Open Court Publishing Co., 1915, and Dover, 1954
16. R. Descartes, Progymnasmata de solidorum elementis, in *Œuvres de Descartes*, ed. by Ch. Adam, P. Tannery, vol. X (Paris, 1908), pp. 265–276; re-edited in René Descartes, Exercices pour les éléments des solides: Essai en complément d'Euclide – Progymnasmata de solidorum elementis, édition critique avec introduction, traduction, notes et commentaires par P. Costabel, Presses universitaires de France, épiméthée, Paris, 1987
17. Diogenes Laërtius, *Lives of eminent philosophers*, ed. by R.D. Hicks (Heinemann, London, 1925)
18. S. Dumas Primbault, Un galiléen d'encre et de papier: Une histoire matérielle des brouillons de Vincenzio Viviani (1622–1703) (Éditions de la Sorbonne, Paris, 2024)
19. Euclid, The thirteen books of Euclid's *Elements*, ed. by T.L. Heath (Cambridge University Press, Cambridge, 1908). Reprint, Dover.
20. L. Euler, Elementa doctrinae solidorum. Novi Comment. Acad. Sci. Petropol. **4**, 109–140 (1758); Opera Omnia: Series 1, Volume 26, p. 71–93

21. L. Euler, De mensura angulorum solidorum. Acta Acad. Sci. Imper. Petropol. **2**, 31–54 (1781); Opera omnia, Series 1, vol. XXVI, p. 204–223. English translation, this volume, Chapter 14
22. V. Farmaki, S. Negrepontis, The paradoxical nature of mathematics, in *Topology and Geometry*. A collection of essays dedicated to Vladimir G. Turaev. ed. by A. Papadopoulos. IRMA Lectures in Mathematics and Theoretical Physics, vol. 33 (European Mathematical Society, Berlin, 2021), pp. 599–642
23. D.H. Fowler, Could the Greeks have used mathematical induction? Did they use it? Physis **31**, 253–265 (1994)
24. N. Fuss, Éloge de Monsieur Léonard Euler, lu à l'Académie Impériale des Sciences, dans son assemblée du 23 Octobre 1783, Saint Petersburg (1783); reprint: Kessinger Publishing's Legacy Reprints, Whitefish
25. A. Girard, *Invention nouvelle en l'algèbre* Amsterdam, Guillaume Iasson Blaeuw, 1629
26. T.L. Heath, *A History of Greek Mathematics* (Clarendon Press, Oxford, 1921). Reprint, Dover
27. J.L. Heiberg, *Heronis Alexandrini opera quae supersunt omnia* Teubner, Leipzig, New edition, 2013
28. D. Hilbert, Grundlagen der Geometrie (Teubner, Leipzig, 1899); English translation, The Foundations of geometry, transl. E. J. Townsend, Reprint, Open Cour Publishing Company, La Salle, Illinois 1950
29. Ch. Huygens, Œuvres complètes, 22 vols (Société hollandaise des sciences, La Haye, 1888–1950)
30. Ibn al-Haytham, Book on the resolution of doubts relative to the book of Euclid on the *Elements* and the explanation of its meanings. Institute for the history of Arabic-Islamic Science, Frankfurt-am-Main (1985)
31. Jamblique, *In Nicomachi arithmeticam*, Edition and translation by N. Vinel, Fabrizio Serra Editore, Pisa (2014)
32. E.S. Kennedy, Mathematical geography, in *Encyclopedia of the History of Arabic Science*, ed. by R. Rashe, vol. 1. Collaboration with R. Morelon (Routledge, London, 1996), pp. 185–201
33. J. Kepler, *Harmonices Mundi* (Johann Planck, Linz, 1619). English translation: The harmony of the world, E. J. Aiton, A. M. Duncan, and J. V. Field, trans., Philadelphia, Pennsylvania: American Philosophical Society, 1997
34. F. Klein, *Elementary Mathematics from an Advanced Standpoint: Arithmetic, Algebra, Analysis*. Translated by E.R. Hedrick, C.A. Noble, 1st edn. (Cosimo, New York, 1924). Reprint, Dover, New York, 1945 and other editions
35. S.S. Kutateladze, Aleksandrov of ancient hellas. Siberian Electron. Math. Rep. **9**, A6–A11 (2012)
36. Y.I. Manin, Time and periodicity from Ptolemy to Schrödinger: paradigm shifts vs. continuity in history of mathematics, in *Geometry in History*, ed by S.G. Dani, A. Papadopoulos (Springer, Cham, 2019), pp. 129–138
37. M. Mersenne, Synopsis mathematica ad reverendiss, in *Patrem Simonem Bachelier* (Robertus Stephanus, Paris, 1626)
38. J.-F. Montucla, *Histoire des mathématiques*, 4 vols (H. Agasse, Paris, 1799–1802)
39. I. Mueller, Philosophy of mathematics and deductive structure in Euclid's Elements (MIT Press, Cambridge, 1981)
40. S. Negrepontis, Plato on geometry and the geometers, in *Geometry in History*, ed. by S.G. Dani, A. Papadopoulos (Springer, Berlin, 2019), pp. 1–88
41. S. Negrepontis, The anthyphairetic interpretation of knowledge as recollection in Plato's Meno 80d-86e and 97a-98b, in *Handbook of the History and Philosophy of Mathematical Practice*, ed. by B. Sriraman (Springer, Cham, 2024), pp. 471–519
42. S. Negrepontis, V. Farmaki, Pythagorean Book II of the Elements restored and Pythagorean incommensurabilities reconstructed, in *Essays on Geometry: Celebrating the 65th Birthday of Athanase Papadopoulos*, ed. by A.M. Uludağ, A. Zeytin, chap. 13 (Springer Cham, 2025), pp. 279–337

43. S. Negrepontis, D. Kalisperi, The mystery of Plato's receptacle in the Timaeus resolved, in *Handbook of the History and Philosophy of Mathematical Practice*, ed. by B. Sriraman (Springer, Cham, 2024), pp. 521–598
44. S. Negrepontis and D. Protopapas, The Reconstruction of Theaetetus' Theory of Ratios of Magnitudes, in: *Essays on Topology Dedicated to Valentin Poénaru*, ed. L. Funar and A. Papadopoulos (Springer, Cham, 2025) pp. 593–652
45. S. Negrepontis, A. Papadopoulos, The mathematics of periodic anthyphairesis as a basis for the full understanding of Plato's philosophy, To appear in: *The intertwined and meaningful world of Mathematics — Forms, concepts and interactions*, ed. L. Boi and Guglielmo Nocera, Birkhäuser.
46. Pappus of Alexandria, *La Collection mathématique de Pappus d'Alexandrie, traduit du grec en français avec une introduction et des notes par P. Ver Eecke* (Desclée de Brouwer, Paris/Bruges, 1933); repr. Paris, A. Blanchard, 1982
47. Plato, *Plato in Twelve Volumes*, vol. 9. Translated by W.R.M. Lamb (Harvard University Press, Cambridge, 1925); London, William Heinemann Ltd. 1925
48. J.-C. Pont, *L'aventure des parallèles. Histoire de la géométrie non-euclidienne: précurseurs et attardés* (Peter Lang, Bern, 1986)
49. Proclus, *Commentary to the First Book of Euclid's Elements*. Translation and comments by Glen Morrow (Princeton University Press, Princeton, 1970)
50. Proclus, *Commentary on Plato's Republic, Volume II, Essays 7–15*. Translated with an introduction and notes by D. Baltzly, J. F. Finamore, G. Miles (Cambridge University Press, Cambridge, 2022)
51. R. Rashed, *Les Mathématiques infinitésimales du IXe au XIe siècle, 5 volumes, French version edited by Al-Furqān Islamic Heritage Foundation, 1996–2006*, English translation, A History of Arabic Sciences and Mathematics (Routledge, London, 2011–2016)
52. R. Rashed, *Angles et grandeur. D'Euclide à Kamāl al-Dīn al-Fārisī*. Scientia Graeco-Arabica, vol. 17 (De Gruyter, Boston, 2015)
53. R. Rashed, A. Papadopoulos, Menelaus' Spherics: Early Translation and al-Māhānī/ al-Harawī's version (Critical edition of Menelaus' *Spherics* from the Arabic manuscripts, with historical and mathematical commentaries). Series: Scientia Graeco-Arabica, vol. 21 (De Gruyter, Boston, 2017), 890pp.
54. D. Sullivan, On the ergodic theory at infinity of an arbitrary discrete group of hyperbolic motions, in *Riemann Surfaces and Related Topics: Proceedings of the 1978 Stony Brook Conference*. Annals of Mathematics Studies, vol. 97 (Princeton University Press, Princeton, 1981), pp. 465–496
55. T. Sunada, From Euclid to Riemann and beyond: how to describe the shape of the universe, in *Geometry in History*, ed. by S.G. Dani, A. Papadopoulos (Springer, Cham, 2019), pp. 213–304
56. P. Tannery, Quelques fragments d'Apollonius de Perge. Bullet. Sci. Math. Astron. **5**(1), 124–136 (1881)
57. Theon of Smyrna, *Mathematics Useful for Understanding Plato*, Translated from the Greek/French edition of J. Dupuis by R. and D. Lawlor (Wizards Bookshelf, San Diego, 1979)
58. W.P. Thurston, *The Geometry and Topology of 3-Manifolds, Mimeographed Lecture Notes* (Princeton University, Princeton, 1977–1978). Published Volume IV of Thurston's Collected Works, AMS, Vol. IV, 2022
59. W.P. Thurston, Shapes of polyhedra and triangulations of the sphere, in *The Epstein Birthday Schrift Dedicated to David Epstein on the Occasion of his 60th Birthday*, ed. by I. Rivin et al., Institute of Mathematics, Geometry & Topology Monographs, vol. 1 (University of Warwick, Warwick, 1998), pp. 511–549
60. S. Unguru, Greek mathematics and mathematical induction. Physis **28**, 273–289 (1991)
61. S. Weil, *Intuitions pré-chrétiennes*. éditions du Vieux Colombier (La Colombe, Paris, 1951). Also published in Tome IV of Vol. 2 of the Complete Works [62]
62. S. Weil, *Complete works*, 16 volumes (Gallimard, Paris, , 1988–2019)
63. V.A. Zalgaller, Memoirs of A. D. Aleksandrov and his Leningrad geometry seminar. Siber. Electron. Math. Rep. **9**, A26–A80 (2012)

Chapter 6
On the Life and Work of Johann Heinrich Lambert

Athanase Papadopoulos

> *The famous Lambert, another Leibniz in the universality and depth of his knowledge, must be placed among the mathematicians who, in a time when the prodigies of analysis occupied all minds, preserved the knowledge and taste of geometry and knew how to make the most skillful applications of it.* (Michel Chasles, *Aperçu historique sur l'origine et le développement des méthodes en géométrie*, Bruxelles, Hayez imprimeur, 1837, p. 186.)

> *We can only admire, along with the authors, the génie, in a way prophetic, of this man who first proved the irrationality of π, and who, at a time when the consideration of complex quantities was still completely foreign to geometers, dared to assert the identity of non-Euclidean geometry with the geometry on a sphere of imaginary radius.* (Jacques Hadamard, Review of Stäckel and Engel's *Die Theorie der Parallellinien von Euklid bis auf Gauss*, Bull. Sc. Math. t. 20, 1896, pp. 279–281.)

Abstract This is a personal and scientific biography of the famous mathematician and polymath Johann Heinrich Lambert (1728–1777). Besides mathematics, we review some of his works on perspective, pyrometry, cosmology, astronomy, philosophy and other topics. We highlight his relations with several thinkers of his times, including some Bernoullis, Lagrange, Kant, and especially Leonhard Euler.

Keywords Johann Heinrich Lambert · Leonhard Euler · Eighteenth century · History of geometry

AMS Codes 01A70, 01A50, 51-03

A. Papadopoulos (✉)
Institut de Recherche Mathématique Avancée, Université de Strasbourg et CNRS, Strasbourg, France
e-mail: athanase.papadopoulos@math.unistra.fr

R. Caddeo, A. Papadopoulos (eds.), *Spherical Geometry in the Eighteenth Century I*, https://doi.org/10.1007/978-3-032-12466-1_6

6.1 Introduction

Of the three main mathematicians whose work is the subject of the present volume, the name of Lambert is certainly the one with whom the mathematical community is less familiar. For this reason, I decided to write this biography and include it in this book.

Johann Heinrich Lambert was born in Mulhouse[1] in 1728, 21 years after Euler and eight years before Lagrange. At age twelve, he was forced to interrupt his schooling in order to work with his father, a tailor of humble means. Within the span of a few years, he managed to acquire a thorough scientific and humanistic culture, essentially by studying every book he could get his hands on. We are told that he used to read at night, after a long day's work, first in his father's workshop, then at home where he was responsible for helping his parents look after his several younger siblings. Lambert practiced a number of different occupations, some of which enabled him to travel and spend some time outside his home country, and these stays abroad contributed in rounding out his culture. Gradually, his scientific production gave him the status of an important scholar and he was eventually acclaimed throughout Western Europe as a versatile scientist: mathematician, philosopher, physicist, astronomer and more. He spent the last 12 years of his life in Berlin, as an ordinary member of the Royal Prussian Academy of Sciences. All this makes Lambert the epitome of the complete autodidact who succeeded in becoming an eminent scholar. He is the author of almost 150 books and memoirs. It is conceivable that the total freedom he enjoyed in his self-education, without parent, tutor or schoolmaster, contributed to his ability to acquire the universal and innovative spirit that characterizes the great men of the Enlightenment.

In this chapter, I will outline some key events in the life and works of Lambert. Regarding his life, I shall be using information from several known sources. Among these sources is the *Éloge* read after his death by Johann Heinrich Samuel Formey, the perpetual secretary of the Berlin Academy of Sciences [24]. Formey was a philosopher and at the same time a Calvinist pastor. He has the advantage over other biographers of Lambert of having known the latter personally and of having been his colleague. I will also make use of the correspondence between Euler and Lambert, which started in 1758. In the letters they exchanged, each of these two scientists used to inform his correspondent about his works and projects. The letters show that the two men had many common interests and that each held the other in high esteem. Among the other sources on Lambert's life, I mention Rudolf Wolf's biography [76], published in 1860, which contains, among other valuable information, excerpts from Lambert's correspondence as well as the correspondence of other scholars and famous people in which he is mentioned. Let me also point out the article on Lambert by Johann III Bernoulli,[2] published in 1777 in the *Nouvelles Littéraires* [4],

---

[1] Now a city in Alsace (France).

[2] Johann III Bernoulli (1744–1807) is one of the important mathematicians of the Bernoulli dynasty (eight outstanding mathematicians and physicists in four generations). He was a grandson of

Johannes Lepsius's biography published in 1881 [59], that of Dieudonné Thiébault, published in 1804 [74], Georges Rémy's 1910 two-part article in the *Revue d'Alsace* [69], Edouard Sitzmann's article in the *Dictionnaire de biographie des hommes célèbres d'Alsace* [72], Christoph Scriba's article in the *Dictionary of Scientific Biography* [71], and the articles by the Mulhouse scholar Roger Jaquel, whose dedicated and meticulous work has greatly contributed to making known the life and work of his illustrious compatriot [31–34]. I also point out the article on Lambert published in the *Nouveau dictionnaire de biographie alsacienne* [65], and Jeremy Gray and Laura Tilling's article on Lambert's work published in 1978 in Historia Mathematica [26]. Finally, let me mention that Lambert kept a diary that came down to us under the name of *Monatsbuch*, which he started in 1752, in which he recorded his ideas, reading notes, travels and publications [54].

At the expense of stating the obvious, let me say that the best way to get to know Lambert is to read his writings, mostly written originally in German, and for some of them, in French. Many of his articles in German were translated into French and English.

I hope that the present volume, together with the volume *Mathematical geography in the eighteenth century: Euler, Lagrange and Lambert* in the same series [11], in which Lambert's work on geography is translated and edited, will contribute to the undertaking of making Lambert's writings better known.

The plan of the rest of this chapter is the following.

In Sect. 6.2, we present some episodes from the childhood, boyhood and youth of Johann Heinrich Lambert that are important in his development. In particular, we report on his encounter with a number of persons who played a crucial role in his development.

Section 6.3 is concerned with his first years as an adult, including his travels in several places in Europe and his first steps as an established scholar.

In Sect. 6.4, several works of Lambert on various topics are presented at the light of his correspondence with Euler.

Section 6.5 is about Lambert's personality and his relations with the scientists he was dealing with.

Section 6.6 is concerned with his work as a member of the Berlin Academy.

In Sect. 6.7, we describe the last moments of Lambert.

In Sect. 6.8 we review some works of Lambert, in particular, in mathematics, cosmology, the theory of perspective, astronomy, cartography, philosophy and logic.

Johann I Bernoulli, the teacher of Leonhard Euler. Johann III Bernoulli will be mentioned several times in this chapter. He was familiar with Lambert's work, first, as a colleague at the Prussian Academy, and second, because at Lambert's death, he was entrusted with the task of sorting out the latter's manuscripts and preparing them for publication. Bernoulli started by publishing a catalog of Lambert's works and he edited a few posthumous manuscripts, including a critical edition of a five-volume *Scientific Correspondence in German*, see [36]. But shortly afterwards, Bernoulli inherited a task that was far more important for him personally, namely, that of handling the Bernoulli family's manuscripts. (He belonged, together with his brother Jacob II Bernoulli, to the last generation of the dynasty.) Thus, Lambert's manuscripts became of secondary interest to him. He sent some of them to journals edited by friends of him to publish them, and he sold others to various libraries. No edition of Lambert's complete works has yet been published.

6.2 Childhood, Boyhood, Youth

Johann Heinrich Lambert was born in Mulhouse on August 26, 1728 to Elisabeth Schmerber and Lukas Lambert, a tailor of modest means. Mulhouse is, today, a town in Upper Alsace (France) which, at the time, had a special status: it was linked to—but not part of—the Helvetic Confederation by a treaty dating back to the early years of the sixteenth century. This treaty guaranteed the city a certain peace from the conflicts raging in neighboring regions, as well as a degree of independence from the Holy Roman Empire. The town thus became a haven for people from various regions who had suffered from the French Wars of Religion. The Lambert family came from the Rhineland Palatinate, a region that had been invaded several times by France. They were Protestants and very devout. One of Johann Heinrich's great-grandfathers, a baker, had left the Palatinate towards the end of the Thirty Years' War (Formey tells us that this great-grandfather "had left France for religious reasons") and had taken refuge in Mulhouse, where he was granted the right of bourgeoisie in 1645.

Several among Johann Heinrich Lambert's biographers tell us that one day, one of his classmates passed him a book on arithmetic, that he was immediately seduced by the subject and read the book in its entirety. A little later, a mason working in his parents' house noticed his aptitude for geometry and lent him a geometry book, which he also read. It was through reading such books—a priori, randomly—that the young Johann Heinrich was introduced to mathematics.

During his school years, as soon as Johann Heinrich got home from school, he had to help his mother with household chores and look after his younger siblings. It seems that reading was the only occupation that gave him pleasure. As soon as a book fell into his hands, he would pounce on it and devour it. This is how he learned a great number of topics on his own. Nobody around him was aware of what the young Johann Heinrich knew.

At elementary school, Johann Heinrich's aptitude for study was noticed by his teacher, who spoke to his father, recommending that he sends his son to be prepared for church ministry. This was certainly the surest way to acquire a social position and a secure future. Formey states in his *Éloge* that, at that time,"the number of learned people in Mulhouse was limited to half a dozen theologians". Lukas Lambert, although he needed his son in his workshop, did not oppose the idea of providing him with a theological training, having realized himself that the boy had no particular talent for sewing. He therefore applied for a scholarship so that the young Johann Heinrich could study theology, but the demand was unsuccessful, and eventually, at age twelve, the young Lambert had to leave school for good and become an apprentice tailor with his father. Indeed, Lukas and Elisabeth Lambert had ten children,[3] and they lacked the means to send their son Johann Heinrich to school beyond the age of twelve. Formey tells us that when Lambert left school,

[3] Some biographers say nine.

"his parents resumed all their severity, and declared to him, albeit by complaining, that it was necessary to work, and that the fingers alone must be the instrument of sustenance. He groaned, but submitted, becoming once again what he had been, tailor and servant" ([24] p. 74).

After leaving school, Lambert continued to read and learn, but was doing so in secret, as his parents preferred to see him handle needle and thread rather than paper and pencil. His curiosity about the motions of the heavens dates from that period. Wolf [76] reports that on clear nights, Johann Heinrich used to spend his time on the roof of his house observing the stars, and that on the evenings when he could not do so, he contented himself with reading, either by moonlight or candlelight. In his *Éloge*, Formey gives us the following details ([24] p. 73):[4]

> To prevent him from studying at night, his mother denied him light. The young Lambert applied himself to calligraphy, which later proved very useful: he used to write and draw very well and he began to produce small drawings which he used to sell to his classmates for a liard or half a sol, depending on whether there were more or fewer faces in them; and with this money he bought candles which he would light when all the other candles in the house were out.

Lambert would later use his drawing skills to illustrate his treatise on perspective *Die freye Perspektive, oder Anweisung jeden perspektivischen Aufriss von freyen Stücken und ohne Grundriss zu verfertigen* (The free perspective, or instructions for making every perspective sketch from free sections and without a ground plane) [40]. This works and others by Lambert are accompanied by beautiful drawings he made. We have reproduced in Fig. 6.1 one of his drawings from the treatise on perspective. Quite conceivably, Lambert's lifelong interest in the theory of perspective began while he was making his little drawings as a child. K. Andersen's book on the history of the mathematical theory of perspective [1] contains a chapter on Lambert's work in this area. We shall say more on Lambert's work on perspective in Sect. 6.8 below.

From an early age, Lambert was interested in all the sciences, and he used to conduct experiments with whatever means presented themselves to him. Johann Bernoulli III writes in [4]:

> It may be said that in this country he probably lacked instruments; but he had a unique resource in this respect in himself: he always employed for his experiments the simplest means, the apparently puniest instruments, most of which he executed alone; he had such a sure tact, such a judicious mind, that he almost always got the same use out of them as others would have done with a large and very expensive apparatus.

Lambert's lively mind and eagerness for reading and for learning were eventually noticed by people around him. Formey writes the following ([24] p. 76):

> In Mulhouse, there were a number of esteemed men, though not theologians, who not only encouraged Lambert, but also gave him free private lessons, for which they felt abundantly rewarded by the pupil's astonishing progress.

[4] In this chapter, the translations from the French are ours.

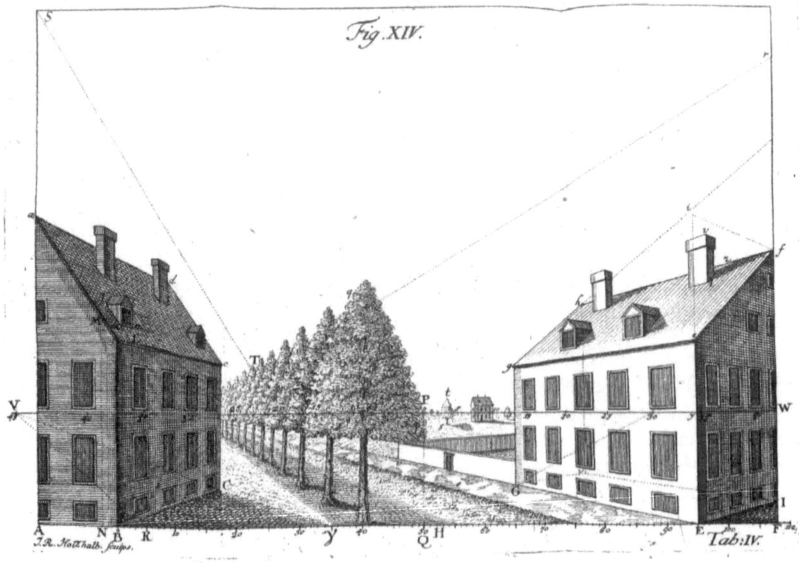

Fig. 6.1 A drawing by Lambert, extracted from his book *Die freye Perspektive*

> In this way, and still in the bosom of his homeland, he further laid the foundations of his philosophical knowledge and even applied himself to oriental languages. At the same time, he perfected his calligraphy skills, which he foresaw would become his first source of income. Indeed, this earned him a position as a copyist at the chancellery, of which Mr. Reber was then the head. At age fifteen, he was keen to learn French;[5] but his parents could not afford to pay a teacher, so he began working as a clerk, or bookkeeper, for a Mr. de Lalande, from Montbéliard, who had a business in the Seppois mines in Upper Alsace. Two years later, believing he knew enough French, Mr. Lambert wished to live in a place where he could satisfy his passion for learning. He was fortunate enough to take up a secretarial position with Mr. Iselin, advisor to the Margrave of Baden-Dourlach, resident in Basel.

Johann Heinrich Reber (1694–1748), mentioned by Formey in the above excerpt, was a well-known Mulhouse town clerk. He hired the young Lambert as a copyist, having found that he had a beautiful handwriting. This freed our future scholar from his domestic duties and his position as apprentice seamstress. It is shortly afterwards, as Formey writes in the above quote, that Reber recommended Lambert to an entrepreneur from Seppois in the Sundgau region, called de Lalande, who is also mentioned by Formey, who took him on as a junior employee, and with whom he stayed from 1744 to 1746. While in Seppois, Lambert worked at a forge. This aroused his interest in pyrometry,[6] an interest he kept for the rest of his life. In 1777, the year of his death, he completed a treatise titled *Pyrometrie oder, Vom Maaße des*

[5] In Mulhouse, in those times, the spoken language was German.

[6] Pyrometry is the science that studies high temperatures and in particular the techniques for measuring them.

Feuers und der Wärme (Pyrometry or measurement of fire and heat) [53], which was published posthumously in 1779, a work in which he studies the dilatation of bodies under the effect of heat. Let us mention that Lambert also used his knowledge in photometry[7] and pyrometry in his work on astronomy, notably on the study of the luminosity of the sun and stars.

During the period 1744–1746, Lambert learned French, as Formey states in the above passage, while continuing to improve his knowledge in various fields of sciences and humanities. It was during the same period that he made observations on the Klinkenberg-Chéseaux comet, which appeared in early 1744. This was a particularly bright comet with six tails, whose orbit he attempted to calculate. The comet attracted the interest of several preeminent astronomers, including Joseph-Nicolas Delisle[8] and Leonhard Euler.[9] It seems that it was Euler and the astronomer

---

[7] Photometry is the science of measure of light, that is, the part of the electromagnetic radiation spectrum that is seen by the human (or animal) eye.

[8] Joseph-Nicolas Delisle (1688–1768) is an important figure of eighteenth-century science, who played an important role in the life of Leonhard Euler. Furthermore, he was a geographer, and Lambert, like Euler, was strongly involved in geography. This is why we shall say a few words about Delisle and his relationship with Euler. Delisle was born in Paris in a family of famous scholars, including several astronomers and geographers. Himself a physicist, astronomer and member of the French Académie royale des sciences, he was among the first scientists that were invited to Saint Petersburg by the Tsar Peter the Great when the latter decided to found the Russian Academy of Sciences. Delisle was appointed director of the astronomical department of the Saint Petersburg Academy. He is the founder of what became known as the first Russian astronomical school, a school that became, a few decades later, one of the most renowned in the world. At the same time, Delisle was in charge there of the department of geography. In fact, the departments of astronomy and geography were both part of the "class" of mathematical Sciences, and Delisle had close relations with mathematicians. During his first years in Saint Petersburg, Delisle collaborated with the mathematician Jakob Hermann, who was, like him, a founding member of the Academy. In 1731, Hermann, who was 29 years older than Euler, left the Russian Academy and returned to his hometown, Basel, which was also Euler's hometown. Delisle started searching for another mathematician who could assist him in his work, and the natural choice was Euler. Delisle's principal task at this department was to collect and establish new precise topographical measurements of all Russia in order to draw maps of that huge Empire. For this, he used the method of triangulations and made extensive use of astronomical data, to measure distances on Earth. Such methods were already known since Greek Antiquity, but Delisle introduced new ingredients, mainly astronomical. In his computations, he considered that the Earth, following a hypothesis emitted by Newton and confirmed later on, is not spherical but has the form of a spheroid flattened at the Poles. From the moment Euler became Delisle's assistant he started working extensively on geography and astronomy. For several years, the two men had very close relations. Delisle's work on geography is introduced and commented on in the volume [11].

[9] The calculation of comet trajectories is one of the subjects addressed, sometimes in great detail, in 23 of the 26 letters exchanged between Euler and Delisle between 1742 and 1745 and published in [6]. Let us also note that Euler published several memoirs and books on comets. We mention in this respect his first book on astronomy, *Theoria motuum planetarum et cometarum* (Theory of the motions of planets and comets) [18], his memoir *Determinatio orbitae cometae qui mense Martio huius anni 1742 potissimum fuit observatus* (Determination of the motion of a comet which was [could be] observed especially in March of this year, 1742) which contains his calculations for the 1742 comet, and the booklet *Beantwortung verschiedener Fragen uber die Beschaffenheit,*

Charles Messier[10] who triggered Lambert's attention on the study of comets; see [64, p. 301] and [29].

Johann Rudolf Iselin, also mentioned by Formey in the passage quoted before, was the publisher of a Basel-based bi-weekly newspaper, the *Basler Zeitung*. He was a member of the Berlin Academy. It was again thanks to a recommendation from Reber that Lambert obtained, in 1746, a job as Iselin's secretary-librarian.

Lambert's observation of the 1744 comet awakened his interest in astronomy, and more generally in cosmology. Later, Johann III Bernoulli wrote in [4]:

> I have spoken of the comet that Mr. Lambert observed in his youth; it seems to have had a strong influence on his subsequent work: it was the first occasion, no doubt, of his ingenious work *Insigniores orbitae cometarum proprietates*[11] and of various good memoirs on comets [contained] in his famous *Beyträge zur angewandten Mathematik*[12] and elsewhere, and that of developing that particular talent he had for geometrical constructions. This talent, which, along with the universality, originality and great clarity of his ideas, is one of the main characteristics of his works, was felt in all his way of thinking and acting; it always had something comparable and resembling to a construction, and was, with a little amour-propre, the source of some singularities and some light spots, very forgivable in such a great man.

Bewegung und Wirkung der Cometen (Answers to various questions about the condition, motion, and effects of comets) [19] that he wrote after the appearance of the 1744 comet.

[10] Charles Messier (1730–1817) was a French astronomer and expert in comets. Member of the Académie Royale des Sciences, he was nicknamed the "ferret of comets". It is considered that he discovered 20 of them and that he thoroughly studied 50.

[11] The translation of the title is: *Notable properties of the orbits of comets*, see [43]. Regarding this book, which Johann III Bernoulli describes as "ingenious", we read the following in Doublet's *Histoire de l'astronomie* [16, p. 538]: "This 24-page pamphlet contains the famous theorem that bears Lambert's name, and which Lagrange considered 'the most beautiful and remarkable discovery in the theory of comets'."

[12] The translation of the title is: *Contributions to applied mathematics*. We know no work by Lambert with the title that Bernoulli gives. It is probable that the latter refers to the *Beiträge zum Gebrauche der Mathematik u. deren Anwendung* (Contributions to the use of mathematics and its application), a collection of works in three parts, each part containing several articles, which appeared on 1765, 1770 and 1772. Article 7 of Part III is titled *Von Beobachtung und Berechnung der Cometen, u. besonders des Cometen von 1769* (On the observation and calculation of comets, and especially of the Comet of 1769). Euler was very much interested in this comet. In this regard, let us quote the following passage from a letter written by Euler's son Johann Albrecht to Daniel Bernoulli, dated December 29 (18), 1769 ([22, p. 838]): "My father has been busy these days calculating the observations made of the comet that appeared last August, September, and October. His research has provided him with a new method for determining with great precision the orbit of any comet, which will be the subject of a work that the Adjunct Mr. Lexell will compose under his direction." (This is Euler's memoir [20]).

6.3 Adulthood and First Published Works

Lambert remained with Iselin for two years, working as his assistant in the newspaper publishing department. During these two years, as soon as his work was done, he could devote himself to reading philosophical and scientific books belonging to his patron's library. Iselin was also a law professor, and he allowed the young Lambert to attend his classes, which Lambert did, even though he generally preferred to learn from books. Bernoulli tells us in [4] that it was by attending these lectures that Lambert acquired a certain number of notions of law. Iselin had a large library, and Lambert could use it as he wished.

Pfeiffer, in the preface to his edition of Lambert's *Freye Perspective* [40], writes that "[Lambert] was inflamed by geometry, physics and astronomy; the scientific books of French authors so enthused him that he decided to publish his future work in French." Indeed, his first published book, *Les propriétés remarquables de la route de la lumière par les airs, et en général par plusieurs milieux réfringents, sphériques et concentriques* (The remarkable properties of the paths of light through the air, and in general through refractive, spherical and concentric media) [39] was written in French. His second book, *Die freye Perspecktive oder Anweisung, jeden perspektivischen Aufriss von freien Stücken und ohne Grundriss zu verfertigen* was written in German and published in 1757, but Lambert translated it himself into French, and published it in that language, in 1759, under the title *La perspective affranchie de l'embarras du plan géométral* (Perspective freed from the embarrassment of the geometrical plane) [40].

Lambert's departure from his hometown in 1746 was definitive, except for a few months' stay in 1759. Speiser writes in [73] that it is conceivable that it was during his stay in Basel that Lambert came into contact with the Bernoulli family, perhaps, first, with Johann I.

Wolf, in [76], quotes a letter dated December 6, 1750, sent to the pastor Rissler of Mulhouse,[13] in which Lambert describes his hopes and intentions during his years in Basel:

> [...] I immediately understood that my efforts had to be directed first and foremost to perfecting my knowledge and making myself happy. But I also understood that naturally depraved intentions cannot improve unless the mind is properly enlightened and freed from acquired prejudices. On this basis, I find these rules, which are very useful for recognizing the mind and its errors and for searching for truth, in Wolff's writings on the power of the human mind, in Malebranche's writings on the search for truth, and in Locke's reflections on the human mind. All this is revealed to us above all in the mathematical sciences, more particularly in algebra and mechanics, which provide clear and thorough examples that have enabled me to confirm the rules I have mentioned and to transform them in some way into my own flesh and blood.

[13] The letter is published in the collection of letters written in German [36].

Lukas Lambert died in 1747, leaving his widow with seven children (five boys and two girls); other children they had died in childhood. Johann Heinrich was nineteen.

In 1748, thanks to Iselin's recommendation, Lambert obtained a position in Chur, a town in the present-day Swiss canton of Graubünden, in the family of Count Peter von Salis, an influential figure and former Swiss ambassador to London. Lambert was responsible for the education of the Count's eleven-year-old grandson, Anton, and of two other boys in the family, Baptista, also eleven, and Johann Ulrich, seven. Peter von Salis was eighty, and he immediately became interested in Lambert's discoveries. For the rest of his life, Lambert remained grateful to the von Salis family, not only for the salary he received from them and the place they offered him in their château, but also for the esteem in which they held him.

The teaching Lambert was to give the von Salis children included languages (French, Italian and Latin), arithmetic, geometry, geography, architecture, history and catechism. The Count and his family were, like Lambert, devout Calvinists. Lambert's stay at the von Salis castle lasted eight years, during which he was able to read and learn at his leisure, as the castle had an extensive library. During this time, he perfected his skills in music, literature, theology, astronomy, natural sciences, philosophy, physics, mathematics, statistics, cartography and poetry. It was during this stay that he invented a calculating machine, a mercury pendulum and an instrument for perspective drawing. In 1750, he began a series of systematic observations of the starry sky, which provided the basis for a theory he later developed concerning the Milky Way, one of the theories for which Lambert is best known in cosmology. In 1753, he became a member of the Helvetic Society of Basel, to which he sent several articles on mathematics and physics that were published from 1755 onwards in the Basel journal *Acta Helvetica*. This brought him into contact with Daniel Bernoulli,[14] who was already aware of Lambert's work and was interested in it. A correspondence developed between the two men; see [5] for a collection of letters, written between February 4,1754 and June 28, 1760.

In 1756, Lambert embarked on an extensive European tour with his two eldest pupils, Anton and Baptista, which took them to several Dutch, French and Italian cities. In Göttingen, Lambert met Abraham Gotthelf Kästner,[15] who was teaching

---

[14] Daniel Bernoulli (1700–1782) was a physicist, medical doctor and mathematician. Born in Gröningen, he preceded Euler to Saint Petersburg and he became friends with him, at the latter's arrival to the Russian capital. Contrary to Euler, Daniel Bernoulli spent only a few years in Saint Petersburg, after which he left for Basel, from where the Bernoulli family originated and where he became a professor, successively of medicine, astronomy, and philosophy.

[15] Abraham Gotthelf Kästner (1719–1800) was a mathematician, poet and historian of mathematics, specialist in the theory of parallels. Kästner was appointed professor at the University of Göttingen in1756, the year Lambert arrived there for a two-year stay. The two men met and kept in touch by letter until Lambert's death. Kästner was very much interested in the question of parallels, and he first tried, like many others before him, to prove the parallel postulate. It is probable that Kästner helped arousing Lambert's interest in this question. He supervised Georg Simon Klügel's doctoral thesis, defended in 1763, in which the latter analyzed twenty-eight attempts by his predecessors, including Saccheri's, to prove the axiom of parallels, exposing the error contained

at the University of Göttingen, with whom he was to remain in correspondence until his death. Lambert's interest in the problem of the independence of the axiom of parallels, of which I will say more in Sect. 6.8 below, may well date from his discussions with Kästner. Besides, the two men shared an interest in perspective theory. Lambert also met the mathematician and astronomer Tobias Mayer,[16] who taught mathematics at the University of Göttingen. His two young students had to follow law courses there, and he was given the opportunity to attend these courses as well. At the same time, he was able to work on the memoirs he was preparing for publication. In the above-mentioned letter, dated December 6, 1750, addressed to Pastor Rissler of Mulhouse and quoted in Wolf [76], Lambert writes:

> I know that in the next three years, I will be traveling to the University of Utrecht, then to England and France, with two young men, to whom I teach languages, catechism, arithmetic, surveying, military construction, geography and history. With Almighty God's help, I have justified hopes for continuing to be happy.

On his arrival in Göttingen, Lambert was elected corresponding member of the Göttingen Royal Society of Sciences. He was already well known, if only for his memoirs already published, on perspective and on the dilatation of bodies due to heat. But he had to leave the city in 1757, when the French occupied it.[17] Lambert went with his two pupils to Utrecht, where he spent most of the following two years. He also stayed in other Dutch cities, in particular The Hague, where he published his first book on light, *Les propriétés remarquables de la route de la lumière par les airs et en général par plusieurs milieux réfringents, sphériques et concentriques* [39].

Lambert's travels with his pupils also took them to Amsterdam, Paris, Marseille, Nice, Turin and other cities. In Paris, Lambert met the astronomer Charles Messier as well as d'Alembert. Wolf reports that Lambert was not appreciated by the latter, see [76]. A correspondence quoted by Wolf between Lambert and the famous Swiss physician and naturalist Albrecht von Haller (1708–1777) indicates that Lambert asked Haller to support his candidacy for a professorship at the University of Göttingen. Despite this support, Lambert did not obtain the position.

We mentioned earlier that Lambert, while in the Netherlands, published his booklet *Les propriétés remarquables de la route de la lumière par les airs, etc.* [39] in The Hague in 1759. During the same trip, he was also working on his *Neues Organon* [44] and *Freye Perspektive* [40].

On his return from his travels in 1759, Lambert stayed in Chur for another year. He passed through Zürich, where he was elected member of the Physical Society.

---

in each of them. Klügel concludes his thesis with the hypothesis that the postulate is unprovable. Other Kästner's doctoral students include Johann Pfaff (who later became the doctoral advisor of Carl Friedrich Gauss), Johann Tobias Mayer, Farkas Bolyai (the father of János Bolyai, one of the three discoverers of non-Euclidean geometry).

[16] Tobias Mayer (1723–1762) was a mathematician, cartographer and astronomer, who, like Lambert, was an autodidact. Without ever having studied at university, he became a professor at the University of Göttingen.

[17] This was the Seven Years' War (1756–1763), a major conflict in Europe that some have called the First World War, pitting France, Prussia and Spain against Great Britain, Austria and Russia, with repercussions on other countries, especially the Colonies, including North America and India.

There, together with the famous physician and naturalist Johannes Gessner,[18] he made astronomical observations, and published his book on perspective, which is also a masterpiece of descriptive geometry. We shall say more about this book in Sect. 6.8.

6.4 On the Correspondence Euler–Lambert

The correspondence between Lambert and Euler started in 1758. Eighteen letters, written in French, exchanged between the two men between the years 1758 and 1772, are published by Bopp in [9]. These letters are concerned with such diverse subjects as the propagation of light, photometry, colors, the law of attraction, the calculation of comet orbits, sound propagation, function theory, and differential equations. They also contain information on Lambert's association with the Academies of St. Petersburg, Berlin and Bavaria.

In a letter written at Utrecht, dated July 1st, 1758, Lambert sent Euler two copies of his work *Les propriétés remarquables de la route de la lumière etc.* [39] which had been published shortly before. He tells his correspondent that one of these copies is a mark of his esteem for him, and that the second one is intended for the Berlin Academy as a sign of esteem for this institution. He explains that the purpose of his memoir was to be able to announce, in the foreword, his work *Photometria sive de mensura et gradibus luminis colorum et umbrae* (Photometry or the measurement and level of light, color and shade) [41], which he intended to publish shortly after. At the same time, Lambert described to Euler the differences between their respective theories on light.

Having received no reply from Euler, Lambert wrote him again 18 months later, in January 1760. In this letter, he described at length some experiments he had carried out on air resistance, namely, by studying and comparing the vibrations of a pendulum in air and in a vacuum.

Euler replied to Lambert on February 2, 1760, saying he had received his first letter, but that he did not know where he could send his reply, as he did not know where Lambert was staying. He praised the work he had received from him, *Les propriétés remarquables de la route de la lumière etc.* In response to questions raised by Lambert about the light emitted by the moon, Euler admitted his lack of knowledge in this subject, but explained to Lambert that their theories did not contradict each other.

On April 4, 1760, Lambert sent to Euler his *Photometria etc.* [41], with a brief explanation of what it contains. In this work, Lambert laid the foundations of what is

[18] Johannes Gessner (1709–1790) was a Swiss medical doctor, naturalist and mathematician, member of several European learned societies and academies. He studied mathematics under Johann I Bernoulli.

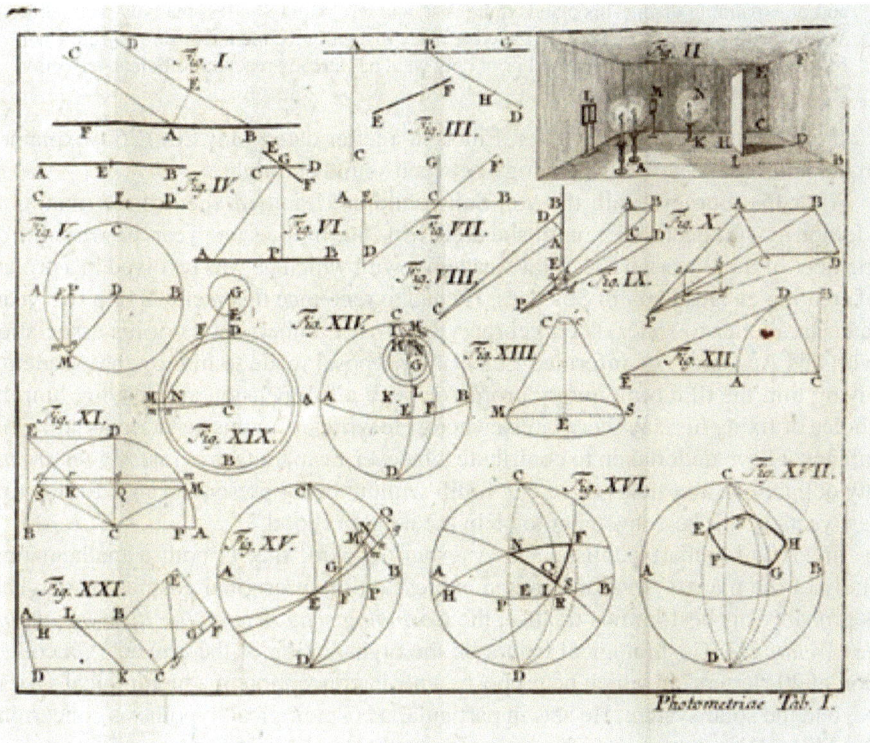

Fig. 6.2 A plate extracted from Lambert's *Photometria sive de mensura et gradibus lumini etc*

known as molecular absorption photometry.[19] Since the present book is concerned with spherical geometry, we have reproduced, in Fig. 6.2, a plate from Lambert's *Photometria*, where the existence of spherical geometry arguments can be seen.

In the same letter, Lambert told Euler about his ideas on the problem of finding the orbit of a comet, as well as his theory of the path of light. He also informed him of his plans:

> It is true, Sir, that for the last three years my stay has not been a fixed one. I spent the first two accompanying Mss. de Salis from the Grisons and taking them out into the world, and since I found that it was time for me to appear in public in my own right, I have formed the plan of successively printing my works, which I had been content to keep in manuscript,

[19] A well-known law in photometry is attributed to Lambert (*Lambert law*, or *Beer–Lambert law*). This is an empirical law for the reduction of the intensity of a light beam in terms of the properties of the medium it passes through and to the thickness crossed. A first (partial) such law was formulated by the French physicist Pierre Bouguer in 1729, before Lambert discovered his law in 1760 [10, pp. 16–22]. An enhanced such law was discovered by the German physicist August Beer, taking into account the concentrations of the physicochemical constituents of the environment crossed [3]. The final law is known law as the Beer–Lambert–Bouguer law.

> and of expanding them. This plan is quite vast, and will require no less than my whole life. While waiting for peace in some academic place that can give me the leisure I need, I will have enough to entertain myself if I print only what has already reached sufficient perfection to be presented to the public.

In response to Lambert's letter, Euler, in a letter dated May 20, 1760, explained to the latter his ideas on the analogy between sound and light.

After his journeys with the von Salis children, Lambert spent three months in Mulhouse with his mother, until she died, in 1760. In the same year, he was elected member of the Bavarian Electoral Academy, with which he was involved in a project of creating an Academy of Sciences. He had to renounce this membership two years later. In a letter to Euler dated February 6, 1761, Lambert spoke of his association with this Academy: he informed Euler of a proposal made to him by the Academy, giving him the title of honorary professor, with a full pension and leaving him the choice of fixing his stay there or elsewhere. He writes: "As this academy is still in its infancy, I have undertaken to contribute whatever I can to its progress, by restricting my obligations to articles that I can fulfill without being present, even if for the first few years it will be convenient to be in the neighborhood."

In 1761, Lambert published in Augsburg a work that is both a mathematical study of the planets' motions and an expression of an original cosmic vision. This is probably his best-known treatise, the *Cosmologische Briefe über die Einrichtung des Weltbaues* (Cosmological letters on the organization of the universe), a collection of 20 "letters" in which he proposes a novel conception of a hierarchical system beyond the solar system. He sets in particular a cosmological hypothesis concerning the Milky Way, based on information about the motions of Saturn and Jupiter and implying the existence of forces that govern the universe that is next to our planetary system; this is his first-order system. Next comes a second-order system, and so on. A large number of Milky Ways forms what is referred to as a fourth-order system. See [42] and [35], for the later editions and re-editions of this work.[20] I will say more about Lambert's *Cosmological letters* in Sect. 6.8.

On April 25, 1761, Euler wrote to Lambert: "Your excellent work on Photometry is becoming more and more widely known and is attracting you the praise that you deserve. What you think of my ideas of the light of colors flatters me greatly, and I have no doubt that your investigations on this subject may well bring it to

[20] The treatise became Lambert's best-known work by the end of the twentieth century, that is, long after his death, especially after the English edition of this work appeared. A French abridged version of this treatise was published in 1770, under the title *Système du monde de M. Lambert*, by Jean-Bernard Merian (1723–1807), who had been a professor at the University of Basel and who had just been called to the Berlin Academy. Lambert himself had translated most of the letters in this French abridged version. This summary was a great success, and was translated into Italian, Russian and English. A complete French translation of Lambert's work, by the Toulouse astronomer Antoine Darquier de Pellepoix (1718–1802), with notes by Jakobus Maurits Karel van Utenhove (1773–1836) of Utrecht, completed in 1785, was published in Amsterdam in 1801 and reproduced for the bicentenary of Lambert's death [57] with a preface by the philosopher of sciences Jacques Merleau-Ponty. (Jacques Merleau-Ponty is the cousin of the philosopher Maurice Merleau-Ponty.)

a higher degree of perfection."[21] In the same letter, Euler tells Lambert that with great satisfaction he learned that the latter was established now in the Bavarian Electoral Academy, and that he hopes that this will give him the occasion to enrich his excellent productions. He informs him that he has proposed his name to the Russian Academy, to fill in the Astronomy position there, which was vacant,[22] but that he did not receive yet any response for this request. He also tells him that at a meeting of the Academy regarding the reception of Foreign members, he had included his name on the list of candidates, and that this was unanimously approved.

In the same letter, Euler tells Lambert that he received his "excellent work on the motion of comets" and that he was quite struck by the beauty of the discoveries it contains, that he read with great pleasure "the beautiful proof of the area of the parabolic sector" and that he was very much surprised to learn about its application to elliptical sectors, which Lambert has used in his study of the fall of the comet towards the sun. Euler tells Lambert that this will lead him to a number of research works before he can understand "all these admirable properties which will contribute to the determination of the comets orbits". He admits that the methods he had personally proposed in the past can be considerably improved.

On August 17, 1761, Euler informs Lambert that the astronomer's position at the Petersburg Academy is still vacant, due to the fact that there is still no nomination of a president of the Academy and that there is still hope that he can obtain for Lambert a nomination as a foreign member with salary. In the same letter, he discusses with him several topics, including tides, the moon's motion, elasticity and sound propagation, and he tells Lambert about his ideas on a machine that could produce sounds that imitate the vowels. Thoughts about this type of machine come up in several later letters.

Between 1761 and 1764, Lambert made several trips between Switzerland and Germany. Some of them are described by Rémy in [69]. In a letter dated July 12, 1762, he informs Euler about some disputes at the Bavarian Academy, in which, he says, he did not take sides. These disputes, as well as other similar problems, bothered Lambert. One year later, on July 3, 1763, after telling his correspondent about the difficulties he is experiencing at this Academy, he writes: "It costs me no trouble to abandon this Academy to its own lights". Shortly after, Lambert resigned.

After Lambert's correspondence with Euler, one may review his correspondence with other scientists; all his letters contain interesting information about him and about science. We limit ourselves here to a few words on his correspondence with Daniel Bernoulli published in [12], on topics different from those addressed in the correspondence between Lambert and Euler.

In several letters exchanged in 1754 and published in [12], Lambert and Daniel Bernoulli discuss the problem of meteorological measurements, particularly the relationship between barometric variation, altitude, and temperature. In letters

[21] According to Nevskaja [64, p. 306], it was Joseph-Nicolas Delisle who encouraged Euler and Lambert to study photometry.

[22] This was the position left vacant by Joseph-Nicolas Delisle, who had returned to France.

exchanged in 1759, Bernoulli explains to Lambert the difficulties inherent in optical experiments. In particular, he questions the law saying that intensity decreases with distance. He congratulates Lambert on his work on the influence of the moon on the barometer. (Lambert believed that the moon had no influence.) Lambert and Bernoulli consider the issue of the subjectivity of optical measurements. They discuss the relationship between luminous intensity and the number of candles, the relationship with distance, and the various experiments used to measure the decrease in the intensity of light with distance.

Before continuing with the chronology of Lambert's travels, let us take a quick look at his personality and character.

6.5 Personality and Character

Lambert's biographers agree that he was a hard worker, his only distraction being a walk, which he used to take only in fine weather. His behavior was somewhat unsociable, and he had little experience in social practices. Rémy tells us that Lambert used to dress extravagantly, in bright colors, so that on the street he was taken for a madman [69]. According to Jaquel [32]: "Lambert had a very peculiar physiognomy, which struck very much his contemporary Lavater,[23] the apostle of physiognomy; but he did not want his portrait done, and there is none done in his lifetime".[24] Unwilling to conform to custom, Lambert talked and laughed loudly, and the food he liked was reportedly rather unrefined. He was also reputed to jostle people as they passed by.

Formey, in his *Éloge* ([24] p. 83), describes Lambert in the following words:

> Preceded by his reputation, escorted, so to speak, by his knowledge, Mr. Lambert was nonetheless an individual to whom the eye and the ear found it hard to become accustomed. Dressed shabbily and singularly, presenting himself in an awkward manner, knowing almost none of the accepted customs or unwilling to conform to them, he seemed occupied only with himself; always thinking. He always became conversational with whoever happened to be facing him, and the flow of his philosophical mouth never dried up until he was alone. I have even seen him having started a conversation with someone who left him, continue and finish it, as if he were in a position to be listened to. With this, flaws of self-esteem, traits of the most advantageous opinion of oneself, returned so often that the consequence was contrary to the premises. One could see, if one wanted to be careful, that it was not pride that made him speak; this passion is more skilful, and does not tend to its ends by

[23] Johann Kaspar Lavater (1741–1801) was a Swiss pastor and thinker, best known for his work on physiognomy, a theory that reached its apogee in the nineteenth century, according to which the observation of a person's physique, particularly the face, provides insight into his personality, feelings and emotions. In 1775 and 1776, he published, in two volumes, his major work *Physiognomische Fragmente, zur Beförderung der Menschenkenntniß und Menschenliebe* (Physiognomic fragments, for the promotion of knowledge of human nature and love of mankind).

[24] Portrait of him were drawn in several decades after his death; see e.g. the portraits in Figs. 6.3 and 6.4.

Fig. 6.3 Portrait of J. H. Lambert; pencil drawing, unknown author and unknown date, after the portrait by Godefroy Engelmann (1829); property of A. Papadopoulos

such crude means: it was a pure and simple intuition of what he was worth, an intimate conviction of his enlightenment and its price, and above all a personal satisfaction that he based on the way in which he had acquired these treasures, by himself, by the strength of his genius, and by the assiduity of his work. Unconcerned about what others might think of him, caring neither to please nor to displease, he showed himself naked; and by dint of

Fig. 6.4 Lithography, D. Chodowiecki and D. Berger, Berlin, 1812

showing himself in this way, he overcame prejudice and forced the admiration of others to identify with his own.

Formey goes on to describe Lambert as permanently driven by profound reflections ([24] p. 86):

> His points of view were limited to the starry vault, to the straight line in front of him, and to the interior of his brain, where he was always confined, even when you thought you were with him, and fixed or at least shared his attention. No digression either to the right or to the left; always in the region of abstractions, the objects we call concrete only grazed him.

Rémy writes in [69] that Lambert "had a tic, which was to never let himself be seen from the front, only from the side. Because of this, he had an accident (falling from the top of a staircase) that put him within an inch of death. The reason is said to be that he did not want to bother his interlocutor with his foul-smelling breath". In his *Éloge*, Formey recounts the following version of the same accident ([24] p. 73):

> M. Lambert gave his treatise *On the path of light*[25] to a bookseller in The Hague. But while he was measuring this road, he found himself in the situation of the astrologer who fell into a well. A most disastrous accident brought him to the brink of death, and shook his constitution to an extent that I suspect he has always felt. As a result of his habit, as strange as it was invariable, of never appearing except from the side, of changing his position accordingly depending on where one stood with respect to him, and of retreating as he was approached, he took a few steps backwards, without thinking of a staircase behind him, and fell backwards from top to bottom. The fall was dreadful: he lost all consciousness, and only recovered after 24 hours. When he reopened his eyes black with blood extracted from his veins, he absolutely refused to believe the doctor who certified the duration of his asphyxia. [...] It took Mr. Lambert a considerable time to recover under the care of the same doctor, Mr. Hahn, a famous professor from Utrecht, who advised him to abstain from strong meditations for a couple of years; but of all the regimes there was none to which he could submit less.

Besides the fact that his social behavior was atypical, Lambert was awkward and anti-conformist, and at the same time he had a very high opinion of himself. He also had difficulty communicating his ideas. Sitzmann's biography reads [72]:

> Lambert had every gift except that of elegant diction. He could speak and write several languages, but he couldn't leave the tone of a dissertation. What he wrote needed to be rewritten; his colleagues took on this task with as much pleasure as success: Merian, from Basel, translated and embellished the *Cosmological Letters*, Tremblay shortened and clarified the *Architectonic* [. . .].

From a religious point of view, Lambert was, like his compatriot Leonhard Euler, a Calvinist Protestant. Both men were devout Christians and notable exceptions to the "spirit of the Enlightenment". It is said, however, that during the last six years of his life, Lambert began to examine his faith through the lens of logic, and by the end of his life, like most Enlightenment philosophers, he had become a deist.

Let me end this section by a few words on Lambert's portrait in Fig. 6.3.

There is no known portrait of Lambert made during his lifetime. The first portrait of him was made in 1828, on the occasion of the centenary of his birth, that is, half a century after the date of his death. This portrait is attributed to Godefroy Engelmann (1788–1839), designer and typographer from Mulhouse, and it is reproduced in many places. I found the drawing reproduced in Fig. 6.3 here by chance, in a bookstore in Mulhouse, a few years ago. I thought it was a nice print representing Lambert, on whom I was working, with Guillaume Théret, on the French edition of his *Theorie der Parallellinien* (1766), which was published by Blanchard in 2014 [66]. I realized, after returning home, that it was a pencil drawing.

[25] This is Lambert's first printed book, *Les Propriétés remarquables de la route de la lumière* [39], which we already mentioned.

6.6 Berlin

Lambert's academic career lasted only a dozen years, and took place in Berlin, where he was appointed ordinary member of the Prussian Academy of Sciences, a position coveted by most scientists of his time.

Lambert arrived in Berlin in 1764, with the intention of obtaining an appointment to the Royal Prussian Academy of Sciences, and, should he fail to do so, continuing on to St. Petersburg where he was quite sure he would obtain a position at the Russian Academy of Sciences, after the recommendation of his friend and compatriot Leonhard Euler. Berlin was almost on the way to St. Petersburg. In Fig. 6.5, the three stars on the map indicate the cities of Mulhouse—Lambert's home city, now in France and at the border with Switzerland —, Berlin and Saint Petersburg; the three stars are almost aligned.

The director of the philosophical section of the Berlin Academy was the Swiss philosopher Johann Georg Sulzer. He had been a student of Johannes Gesture, whom Lambert had met in Zurich. Sulzer considered Lambert a compatriot and he decided to introduce him to Frederick II's entourage, in the hope of getting him recruited to the Prussian Academy.

Fig. 6.5 A map of Europe, reduced by Miss Delisle from a map drawn by Guillaume Delisle (1675–1726), a half-brother of Joseph-Nicolas Delisle (1688–1768). Source: Gallica (Bibliothèque Nationale de France). The three stars, added by Pierre Jehel, indicate the cities of Mulhouse, Berlin and Saint Petersburg. They explain Lambert's stop in Berlin on his way to Saint Petersburg

Lambert talked with the Berlin academicians and made a great impression on a number of them. Especially, he had two strong supporters, his two compatriots, Johann Georg Sulzer, who was the head of the philosophy class, and with whom he had already established a bond of friendship, and Leonhard Euler, with whom he had been in epistolary contact since 1758. Lambert had already been elected to the Berlin Academy in 1761 as a foreign member.

Dieudonné Thiébault,[26] who was member of the Berlin Academy, writes in [74] that above all, it was Lambert's philosophical work, the *Neues organon*, published in Leipzig in 1764, that had a very great effect, because of the strength of the new ideas it contained. This book assured him a great reputation. Thiébault adds that it was jealousy and rivalry that prompted Lambert to leave Leipzig for St. Petersburg, where he initially planned to settle. He confirms that it was the journey to St. Petersburg that took Lambert to Berlin, where Lambert stopped off.

About Lambert's arrival to Berlin in early 1764, Thiébault reports the following [74]:

> [The Academicians] warned the king not to see him, saying: "Sire, we must tell Your Majesty everything; this man, who has such a rare merit, is far from having an exterior that announces it, especially as, having been born very poor, he has not had that first education that corrects forms." The king told them: "You will bring me this Mr. Lambert when it is dark; we will remove the candles: I will not see him, and I will hear him."
>
> "Good evening, Sir" he said; "please tell me which science you have studied in particular.—All of them, Sire.—So you are also a learned mathematician?—Yes, Sire.—And who was the teacher who taught you mathematics?—Myself, Sire.—So you are a second Pascal?—Yes, Sire." At the last word, the King turned his back and returned to his study, finding it infinitely difficult to keep from laughing.
>
> He said to his guests when he was at supper: "Imagine, gentlemen, that this evening my friends wanted me to appoint to my Academy the greatest imbecile I have ever seen." He was told that this imbecile was no less a man of genius. But it took time, perseverance and courage to convince him of this.
>
> Mr. Lambert's friends feared above all that he would escape them and leave for Russia. At last, the King gave in to the unanimous requests of all those present.

Johann Bernoulli, in [4], writes about Lambert's arrival in Berlin:

> He appeared there, if I dare say so, like a man who had fallen from the Moon; so singular and unkempt was his exterior, both in manners and in the way he dressed; and it is not surprising that many people judged the state of his brain unfavorably; but when after a few months he had been given by our August Protector to the Academy, people began to reconsider their assessment of him, and he himself gradually adopted a less striking way of presenting himself, to his disadvantage for those who did not know his merit. He was honored, loved and esteemed by everyone; the King especially gave him proof at various times of his esteem: by appointing him one of the Commissaires for his finances at the Academy; by giving him a position of Senior Advisor to the Department of Buildings, and by considerably increasing the pension he had as an Academician.

[26] Dieudonné Thiébault (1733–1807), whom we quote several times in this essay, is a French scholar, specialist of French grammar. On d'Alembert's recommendation, he was called to Berlin in 1765 by King Frederick II, who gave him a position at the Academy, where he became his reader for everything the King had published, and for all the correspondence he transmitted to the Berlin Academy of Sciences (the King always wrote in French). Thiébault stayed with the King for twenty years, after which he returned to France.

Eventually, the King decided to appoint Lambert after a year's hesitation, and at the express request of a group of academicians who had feared he would leave for the Academy to St. Petersburg, where he had already been offered a position.

In 1765, Lambert became officially member of Frederick II's Academy, in the class of physics—also called the class of experimental philosophy. This class was one of the Academy's four classes.[27] The official appointment took place on January 10, 1765. On the 24th of the same month, Lambert was officially received at the Academy, where he delivered an entrance address on knowledge in physics, the role of experimentation, mathematics, philosophy, metaphysics, history, languages and letters. The title of this address was *On the connection of the knowledge that is the object of each of the Academy's four classes*. One of Lambert's aims in this lecture was to show that the four classes of the Academy to which he was newly appointed were closely interrelated: Mathematics/Experimental Physics, Experimental Physics/Philosophy, Experimental physics/Belles-lettres, see [58] and the comments in [14]. In this connection, let us quote the following comment by Pfeiffer in his preface to the 1977 edition of Lambert's *Freye Perspective*: "Just as Lambert always defended himself against philosophical speculation without algebraic or scientific foundation, he also rebelled against the reign of artistic arbitrariness acting without knowledge of art." Let us also mention Lambert's sentence from his unpublished work *De Pulchritudine* (On beauty),[28] that "Ars sine scientia nihil", i.e., that art without scientific knowledge is worthless.[29]

In the beginning, the King, despite the recommendation of his most prestigious academicians, was not satisfied with this appointment. In 1765, he wrote to d'Alembert (Letter without date [25]):

> I was, so to speak, almost forced to take the most sullen creature in the universe and put him in our Academy. His name is Lambert, and although I can attest that he has no common sense, it is said that he is one of the greatest geometers in Europe. But as this man is ignorant of mortal languages, and speaks only equations and algebra, I do not intend to have the honor of talking with him any time soon. On the other hand, I am very happy with M. Toussaint,[30] whom I have acquired. His science is more human than the other. Toussaint is an inhabitant of Athens, and Lambert a Caraïbe, or some savage from the shores of Cafrerie. However, right down to Mr. Euler, the whole Academy is on its knees before him, and this

[27] The other three classes of the Academy were Mathematics, Speculative philosophy and Belles-lettres.

[28] Lambert's *De Pulchritudine* is still unpublished. This treatise is inspired by Wolff's *Psychologia* [77, 78]. Lambert, in his treatise, develops a phenomenology of beauty, considering the semiotic problem of its translation into the language of truth, with logic serving as a guiding thread. This treatise already announces Lambert's *Neues Organon* (1764) [44], his most important work on logic.

[29] It should be noted that by Lambert's time, such an association between science and art had already taken shape, and for several centuries already, in music, which remains today the privileged domain of this alliance.

[30] In 1764, Frederick II appointed François-Vincent Toussaint (1715–1772) as a member of his Academy. Toussaint had been a lawyer, journalist and translator in Paris, and a contributor to the *Encyclopédie*. In 1748, he published a work titled *Les Mœurs* (which means Moral customs, or "behavior", as a way of life). The book caused a scandal and was banned. Parts of this work were re-used in articles in the *Encyclopédie*.

animal, all crusty from the mire of the crassest pedantry, receives these tributes as Caligula collected those of the Roman people, among whom he wanted to pass himself for a god. Please don't let these little anecdotes from our Academy get out of hand.

It should also be remembered that Frederick II did not like Euler either, even though the latter was the most eminent member of his Academy for a period of twenty-five years (1741–1766). He used to criticize Euler's lack of sensitivity to poetry and the arts in general.

Emmanuel Kant expressed a profound admiration for Lambert. The correspondence between the two men, translated into French, is published in [37]. Let me quote part of a letter that Kant addressed to Lambert, dated December 31, 1765, after the latter's nomination to the Berlin Academy:

No letter could be more pleasing and more desirable to me than the one you have honored me with, since, without saying anything that is not the sincere expression of my opinion, I hold you to be the first genius of Germany, capable of greatly and for a long time perfecting the knowledge with which I am concerned above all else.

It is no mean satisfaction for me to learn from you of the perfect agreement of our methods, an agreement which I had already noticed more than once in your writings, and which contributed to attaching me to them as to a logical test which shows that these thoughts bear the test of the touchstone of universal human reason. Your invitation to share our sketches with one another is very precious to me, and I shall not fail to take advantage of it, although I know myself well enough to know that I can hardly estimate the little science I believe I have acquired after long efforts, since, on the other hand, Sir, your acknowledged talent for combining the highest views with extraordinary penetration into detail, makes me hope that the union of your strengths with my meagre means will result in considerable instruction for me, and perhaps for everyone.

A year after Lambert's appointment to the Berlin Academy, Euler left Berlin to return to St. Petersburg. From there, he maintained with Lambert an important scientific correspondence which we already quoted.

Lambert remained twelve years at the Berlin Academy, from 1765 until his death in 1777. During this period, he published fifty memoirs on a wide variety of subjects, in the theoretical sciences: mathematics, physics and astronomy, as well as in the experimental ones: hygrometry (the study of humidity in the air), topography, energy production (produced by water and wind mills) and others. He wrote essays with the following characteristic titles: *Observations on ink and paper*, *On flutes*, *On the influence of the moon on the weight of the atmosphere*, *On achromatic lorgnettes*, *On the examination of a kind of superstition reduced to the calculation of probabilities*, *Memoir on the photometric part of the whole art of painting*, *On the density of air*, and many others, see [56]. Lepsius, in his 1881 biography *Johann Heinrich Lambert, Eine Darstellung seiner Kosmologischen und philosophischen Leistungen* (Johann Heinrich Lambert, A presentation of his cosmological and philosophical achievements) [59], writes about Lambert's introduction speech to the Berlin Academy:

In this speech, he develops the idea of the unity of method in all the sciences, combining the fruitfulness of experience with the certainty of calculation. [...] His speech was a kind of program for his scientific activities in Berlin. Not only did he believe that the unity of science lies in the unity of method, but he also proved this through action. He was the only member of the Academy to make intensive use of his authorization to read papers in every class.[31]

6.7 Last Moments

Lambert died on September 25, 1777, at age 49, of a disease caused by an ill-treated cold, which he tried to cure on his own, having retained his purely self-taught character right to the end. Formey describes the last years of his life in the following words ([24, p. 88]:

> Mr. Lambert's constitution had been weak in the first years of his life [...] He was not sufficiently attentive to certain precautions which make organs last longer when they are not too tired: but all this was far from heralding any decadence, much less such an imminent end. For a few years, we saw a blossoming stoutness, which was a current sign of good health, rather than of vigor and lasting consistency. A formal evil was needed to shake it. [...] That evil was a heavy cold in the winter of 1775. [...] The illness was long, but its progress was obvious: he alone was unaware of its danger. He consulted doctors only at a very late stage, and as if by way of acquittal, always behaving according to his own principles and supposed remedies. We saw him melt like wax in a fire, until all that remained was a dry, yellow skin stuck to his bones. [...] We saw him again at the assembly of September 18 [1777], more dead than alive; and he even had a convulsive symptom that frightened those who noticed it. He wrote to me on Monday 22nd, sending me a memoir by M. de Segner to present, because he foresaw that he would not have the strength to come to the next assembly. The day of this assembly, September 25, was indeed the day of his death, which was nevertheless unexpected for him.

Let me quote here a letter from Lagrange to d'Alembert, dated October 3, 1777:

> I am deeply saddened by the death of my colleague Mr. Lambert; it is an irreparable loss for our Academy and for Germany in general; he possessed the rare talent of applying calculation to experiments and observations, and of extracting from them, so to speak, everything that was regular. His *Photometry*, a work little known in France and even in Germany, is a true model in this kind of research; moreover, he was well versed in calculus, and ignored none of the various branches of Analysis and Mechanics.

Rémy [69, p. 467] quotes Kant, who wrote, after Lambert's death:

> I had some ideas about perfecting metaphysics that I wanted to mature before submitting them to the judgment and elaboration of my profound friend Lambert. The death of this extraordinary genius has dashed all the hopes I had for such an important collaboration. I deplore this loss all the more because, believing I had found what I was looking for, Lambert was the man whose lucid, inventive mind, unprejudiced by philosophical speculation, made him very capable of appreciating and grasping the whole system of my "Critique of Pure Reason", to point out any mistakes I might have made, and to help me complete and perfect a work which I don't believe impossible now, but which will cost me more time and trouble than if I hadn't been deprived of the help of such a strong head.

[31] That is, in every section of the Academy.

In the rest of this biography, I would like to make a short review of some of Lambert's scientific work.

6.8 On Lambert's Works

Lambert's writings comprise almost 150 items, some of them in several volumes. They are written in German, Latin and French. These writings are remarkable not only for their content, but also for the very broad spectrum of the subjects covered, in both theoretical and applied sciences. Lambert's notes in his *Monatsbuch* indicate that he worked on several subjects at the same time.

Lambert's favorite fields were philosophy and mathematics but, as we already noted, he was involved in other fields, such as astronomy, geography, cosmology, optics and acoustics. He also did important works in more specialized fields such as hygrometry, photometry, pyrometry and magnetism, and there are many others. In what follows, I have chosen to say a few words on his works on the following subjects: mathematics, cosmology, perspective, astronomy, cartography, philosophy and logic.

I will start with mathematics.

Lambert's mathematical work concerns Euclidean, hyperbolic and spherical geometry, arithmetic, the theory of series, probability, integration and other topics. It should also be remembered that his memoirs in several other fields (notably perspective, astronomy, cartography and optics) also contain mathematical results. Thus, a survey of Lambert's purely mathematical writings does not give a complete picture of his contribution to the field. In the next two subsections, I will review some of his results on non-Euclidean geometry and on number theory.

Mathematics: Non-Euclidean Geometry

The memoir *Theorie der Parallellinien* (1766)[32] by Lambert is one of the founding texts of hyperbolic geometry, even if its author—victim of the prevailing ideas of his time—conceived it as an attempt to prove the non-existence of such a geometry. By developing the foundations of a geometry obtained by replacing the parallel postulate by its negation while keeping Euclid's other postulates unchanged and hoping in this way to reach a contradiction, Lambert obtained several fundamental results of hyperbolic geometry whose discovery, by Nikolai Ivanovich Lobachevsky, János Bolyai and Carl Friedrich Gauss, was not to take place until the following century.

[32] See the French translation [66]. No English translation exists up to now.

Indeed, in this memoir, Lambert, like many of his predecessors, tried to show the non-existence of a geometry in which all Euclid's postulates and axioms are satisfied, with the exception of the fifth postulate, known as the *parallel postulate*, and where this postulate is replaced by its negation.[33] It was in investigating the consequences of this hypothesis that Lambert drew remarkable conclusions, which turn out to be theorems of hyperbolic geometry, making this text one of the closest—perhaps the closest—to hyperbolic geometry among the texts that precede the discovery of this geometry by Lobachevsky, Bolyai and Gauss.

One should also note here that at the time of Lambert, it was not an easy matter to accept that these results could be part of a well-grounded geometry. Poincaré writes on this matter, in [67] (1891): "The theorems are, of course, very different from those to which we are accustomed, and they do not fail to disconcert a little at first." Let me give now some of the consequences that Lambert obtained from the assumption that the parallel postulate is replaced by its negation. Let me say however that most of the theorems are not disconcerting for someone familiar with spherical geometry.

- The sum of the three angles of any triangle is strictly smaller than 180°.
- In a triangle, the angles determine the sides.
- The area of a triangle is defined as the *angle defect*, that is, what is missing to the sum of the three angles in order to make two right angles.
- There exist in the plane two disjoint straight lines having a common perpendicular and which diverge from each other on each side of this perpendicular.
- given two straight lines d_1 and d_2 having a common perpendicular, if we raise a perpendicular to d_1 at a point which is far enough from the foot of the common perpendicular, then this perpendicular does not meet d_2.
- There exists a canonical measure for lengths, areas and volumes.[34]

[33] There is an impressive number of predecessors of Lambert who believed at some point that they could prove the parallel postulate, i.e., deduce it from Euclid's other postulates; among them—to mention only Lambert's famous contemporaries and successors—are Euler, Lagrange, Legendre and Fourier. It is also known that the three founders of hyperbolic geometry, Lobachevsky, Bolyai and Gauss, first attempted to prove the parallel postulate over a period of several years, and it is certain that it was the difficulty of this undertaking that made them change their objective and ultimately lead to the discovery of a geometry where this postulate does not hold and where its negation holds. See [68] for a very detailed history of the theory of parallels, see also [7, 70], and the commentary accompanying Lambert's memoir translated into French [66], in which this history is reviewed, and Chap. 7 of the present volume which contains a detailed review and commentary on this memoir.

[34] This last item is less well-known than the preceding ones. There are several ways to see the existence of this canonical measure. One may start with the fact that in hyperbolic geometry there is a unique equilateral triangle with an arbitrary given angle $< 60^{\circ}$. This gives a one-to-one correspondence between the set of angles that lie between 0 and $< 60^{\circ}$ and the set of lengths; a length can be an arbitrary positive number. But there is a canonical measure on angles (this can be deduced from the fact that the total angle at any point is equal to four right angles). Thus, there is a canonical measure for lengths. We refer the reader to Lambert's discussion of this subject in §80 of his dissertation, [66], the commentary on this question in the same volume and Chap. 7 of

Several other results of hyperbolic geometry have been noticed by Lambert. In my sense, one important consequence of his analysis is that in a certain precise sense, there are exactly three geometries: the one where angle sum in a given triangle (or, equivalently, in any triangle) is respectively equal to, strictly greater than, or strictly less than two right angles. These are respectively the Euclidean, spherical and hyperbolic geometries. This observation underpins Lambert's analysis of quadrilaterals, commonly referred to as *Lambert quadrilaterals* or *Ibn al-Haytham–Lambert quadrilaterals*: these are quadrilaterals with three right angles, the fourth of which can be acute, straight or obtuse; they are quadrilaterals with a long history, also reported on in the volume [66].

Another important general fact that Lambert underlined and used extensively in his memoir is the fact that there are strong analogies between certain types of statements in the three geometries, which means that these statements can be treated in parallel and in a unified way. More precisely, there are propositions in all three geometries that are formally identical, up to replacing in some systematic way some inequalities by equalities or vice versa, that is, reversing the direction of monotonicity of certain length or angle functions. One well-known example is the fact that in Euclidean (respectively spherical, hyperbolic) geometry, the sum of angles in a triangle is equal to (respectively strictly greater than, strictly less than) two right angles. To give another example, less well-known but also contained in Lambert's memoir: in an equilateral triangle ABC, if F is the midpoint of BC and D is the point of intersection of the medians, we have $DF = \frac{1}{3}AF$ (respectively $DF > \frac{1}{3}AF$, $DF < \frac{1}{3}AF$) in Euclidean (respectively spherical, hyperbolic) geometry. There is a large number of statements of this type, where to pass from one geometry to another it is sufficient to invert certain inequalities, so that Euclidean geometry appears in several ways as the boundary between spherical and hyperbolic geometries.

Another important remark made by Lambert which I would like to point out is that certain properties of hyperbolic geometry can be obtained by replacing, in the corresponding statements of spherical geometry, the distances by these same distances multiplied by the imaginary number $\sqrt{-1}$. It is to this remark that we must relate Lambert's statement that we "should almost conclude that the third hypothesis[35] occurs for an imaginary sphere." It should be remembered in this respect that it was Lambert who introduced hyperbolic functions (hyperbolic sine and cosine).

---

the present volume. It should be noted that there is a similarly canonical measure for length in spherical geometry, and indeed, in this geometry the usual distance is angular distance. In contrast, there is no canonical measure for length in Euclidean geometry.

[35] The "third hypothesis", along the lines of Lambert's dissertation, is that the sum of the angles in a triangle is strictly less than two right angles; this is the case of hyperbolic geometry.

Mathematics: Irrationality of π, Continued Fractions and Series

One of the best-known mathematical results of Lambert is his proof of the irrationality of π. More generally, Lambert proved that for any non-zero rational number q, e^q and $\tan q$ are irrational. In his proofs of irrationality, he made extensive use of continued fractions, at the same time establishing the modern foundations of this theory. Let us recall that the origin of continued fractions lies in Greek Antiquity in the form of anthyphairesis. Lambert's irrationality results are published in his *Mémoire sur quelques propriétés remarquables des quantités transcendantes circulaires et logarithmiques* (Memoir on some remarkable properties of the circular and logarithmic transcendental quantities) [46]. At the end of this memoir, Lambert attempts to prove that π is transcendental, but he fails. Nevertheless, we owe to him the conjecture that π is transcendental. It wasn't until 1882, that is, more than a hundred years after Lambert had put forward it, that Ferdinand Lindemann proved it, using generalizations of methods of Hermite, who had shown that e is transcendental. As a consequence, the problem of squaring the circle, which is equivalent to that of constructing π with a ruler and compass—a problem that has interested mathematicians for around 2000 years—was seen to be impossible.

In analytic number theory, Lambert's name is attached to the *Lambert series*, used in 1932 by Norbert Wiener in his work on the Tauberian theorem and about which there are still many open questions.

From mathematics, we pass to cosmology.

Cosmological Lettres

Although Lambert was well versed in the astronomy of his time, his cosmological theories, presented in his *Cosmologische Briefe*, are based on teleological principles, in the philosophical tradition of Leibniz and Wolff, and they involve unverifiable hypotheses. Lambert's teleological principles are well summarized by the philosopher Merleau-Ponty in the Preface to the 1977 edition. The latter writes:

> The principles of the *Cosmological letters* are those of a Leibnizian philosopher; they are teleological, but by no means in that naive, elementary way of which the literature of the time gives so many examples. The guiding ideas of Lambert's teleology are those of the maximum (the universe must be populated by the greatest possible number of objects), of harmony (these objects must coexist peacefully, which excludes collision, catastrophe and disorder of all kinds); and finally, the idea of the greatest possible diversity in positions and motions is imposed. The result is a rather subtle dialectic of the simple and the complex, the one and the many, interaction and equilibrium, whose astronomical consequences are sometimes very precise: what is complex under a certain point of view, at a certain order of magnitude, must be simple under another point of view, at another order of magnitude; the trajectories of stars, simple in the closest reference system, become more complicated when we refer them to the centers of larger systems (the Moon, with respect to the Earth, describes an almost perfect ellipse, but an epicycloid with respect to the Sun, etc.).

In the *Cosmological lettres*, Lambert considers that the universe is finite—and here, his opinion differs from Leibniz's—but he was convinced that it is much ampler than everything his predecessors could imagine.[36] Among other things, he formulates in these *Letters* his belief in a multiplicity of animated worlds filling the universe and in the existence of inhabited worlds outside our planetary system. Let me note here that such a conviction that inhabited worlds exist outside our atmosphere was expressed by various scholars before Lambert, among them the great seventeenth-century mathematician and astronomer Christiaan Huygens, who formulates this belief in his work *Cosmotheoros* [27] (first edition 1698). In this treatise, Huygens mentions, among the scientists who shared his conviction on the existence of inhabited worlds other than the planet Earth, Nicolas de Cusa, Giordano Bruno, Johannes Kepler and Tycho Brahe. This belief was also shared by the writer, poet, astronomer, mathematician and philosopher Bernard Le Bouyer de Fontenelle, who discusses this matter in his well-known science popularization work *Entretiens sur la pluralité des mondes* published in 1686. Rudolf Wolf, in his biography of Lambert [76], claims that the work *Cosmologische Briefe* can be considered as an attempt to give a coherent vision of the whole universe, comparable to that which Fontenelle tried to give of the solar system.

Merleau-Ponty concludes his Preface by the following sentence: "what gives these *Lettres* their permanent force and their topicality is precisely the clear awareness of an acute contrast between the rational demand for a coherent thought about the universe and the possibilities offered by the strict application of the positive method."

After cosmology, we consider the theory of perspective.

Perspective

Lambert's *Freye Perspective* was considered a turning point in the theory of perspective. Henri Pfeiffer, the editor of the French version of this work, published in 1977, wrote in the Preface that "this work marks the only major advance in the history of perspective for two thousand years". The mathematician and art theorist Pavel Florensky, who has taken a close interest in the history of perspective, in his book *Reverse perspective* [23], after recalling a few landmarks in the history of the subject, mentioning works of Giotto, Giovanni da Milano, Altichieri, Avanzano, Leonardo, Michelangelo, Filippo Brunelleschi, Paolo Uccello, Leon Battista Alberti, Piero della Francesca, Donatello and others on perspective, refers to

[36] Lambert supported the idea that the distance to the nearest star was five hundred thousand times the distance from the Earth to the sun. Without being able to prove this assertion and others, he claimed he was making "probable" assertions. Merleau-Ponty writes that the *Briefe* "are perhaps less the exposition of a system than an experiment of the thought intended to test the thoughts of a method of discovery and of discussion appropriate to that exceptional domain of empirical research and theoretical speculation which the knowledge of the universe is".

Lambert, among the mathematicians of the eighteenth and nineteenth century who worked on perspective and whose works he analyzed; he writes:

> [...] The few facts I have briefly sketched here are not meant to be generally known historical facts as such, but something quite different. More precisely, their purpose was to recall the complexity and duration of this development, which was only completed in the eighteenth century by Lambert, and later as a branch of descriptive geometry in the works of Loria, Aschieri and Enriques in Italy; Chasles and Poncelet in France; Staudt, Fiedler, Wiener, Kupfer, Burmeister in Germany; Wilson in America; and others who have been part of the mainstream of that extremely important and widespread mathematical discipline, *projective geometry*.

Henri Pfeiffer writes in his Preface to that edition that Johann Rodolf, the publisher and law professor with whom Lambert worked in 1746–1747, was also the author of an encyclopedia bearing his name, which includes articles on perspective, such as the book which Pfeiffer quotes as the *Commentario alle regole della perspettiva* by Ignazio Danti and Giacomo Barozzi; see the exact reference in [2]. Pfeiffer tells us that it was Iselin's love of pictorial art that led Lambert to geometrize perspective drawing so as to create a science of art comparable to the science of music. He reports that Lambert began studying philosophical and artistic texts on perspective dating back to antiquity, copying some of them and translating them into German. Pfeiffer also tells us that some of these handwritten texts are preserved in the Basel University Library, which holds a collection of Lambert's manuscripts.

Pfeiffer also presents in his Preface an overview of Lambert's innovative ideas contained in his *Freye Perspective*, not an easy undertaking, since Lambert himself, in the original preface to his work, writes: "I shall not linger on the subjects I have dealt with in this work. In each section, you will find the reasons which led me to write them, and everyone can judge if they believe them worthy of attention. I will only note that I have mixed my discoveries with those of others."

Lambert tells us in §13 of his *Freye Perspektive* that perspective in drawing is one of those rules that give the work unity, ensuring that landscapes and other parts of the picture are not disparate elements but form a coherent whole. In the same work, Lambert developed his theory of construction with a ruler without a compass, and with a given circle.

Let us quote Pfeiffer again: "The influence of Lambert's metaphysical thinking on Friedrich Nietzsche's perspectivism is obvious, but the Lambertian principle of a mathematization of human thought in all fields is only just beginning to transcend."

Astronomy

We have already mentioned some of Lambert's work on comets, and we have recorded the fact that his work has impressed Euler and Johann III Bernoulli. Calculating orbits of comets is part of Lambert's research on astronomy, but it is also closely related to his work on mathematics. Indeed, he developed geometrical methods of calculating comet orbits. Some of his important results on this topic are contained in his memoir *Observations sur l'orbite apparente des comètes* (Observations on the apparent orbit of comets) [49], and his results generalize Euler's results on the same subject. These results have also attracted the attention of Lagrange, who writes in his memoir *Sur le problème de la détermination des orbites des comètes d'après trois observations* (On the problem of the determination of the orbits of comets, after three observations) [38], after he describes Newton's work on this subject, and the consequences that one can draw from it: "This is what Mr. Lambert has since done in his beautiful treatise *De orbitis Cometarum*, in which he arrived at one of the most elegant and useful theorems that have been found so far on this subject, and which at the same time has the advantage of also applying to elliptical orbits." The memoir that Lagrange mentions is the same one that impressed Johann III Bernoulli, [43]. Lagrange, in the same memoir, reviews in detail several details of Lambert on the question of describing comet orbits.

In celestial mechanics, his name is given to a problem, namely, the *Lambert problem*, which concerns the values at the boundary of the solutions of a differential equation that governs the two-body problem.

From astronomy, we pass to cartography.

Cartography

Gaston Darboux, gave a talk at the Rome 1908 ICM titled *Les origines, les méthodes et les problèmes de la géométrie infinitésimale* (The origins, methods and problems of infinitesimal geometry). In this talk, after recalling that the origin of infinitesimal geometry lies in cartography, he points out a few landmarks of the history of the subject. He writes: "It was one of the most penetrating geometers, Lambert, the very estimated colleague of Lagrange at the Berlin Academy, who, pointing out for the first time a property which is common to the Mercator maps also called *reduced* maps and to those which are obtained by the stereographic projection, was the first to conceive the theory of geographical maps from a really general point of view. He proposed, with all its scope, the problem of representing the surface of the Earth on a plane with the similarity of the infinitely small elements. This beautiful question, which gave rise to the researches of Lambert himself, of Euler, and to two very important memoirs of Lagrange, was treated for the first time in all generality by Gauss. [...]"

More practically, while he was member of the Berlin Academy, Lambert was in charge of an atlas project published under the auspices of this Academy. One may recall here that all kings and emperors of the time were keen to have accurate maps of their territories established, and that several prominent mathematicians (Euler and Gauss, among others) also had the status of cartographer. Lambert's name is attached to a method of drawing a geographical map, and it is also known that he invented and refined more than half a dozen maps. His well-known work on cartography, *Anmerkungen und Zusätze zur Entwerfung der Land- und Himmelscharten* (Notes and Comments on the Composition of Terrestrial and

Celestial Maps) [50], is translated with commentaries by Annette A'Campo and published in [11]. See also the report by Jaquel in [33]. To conclude this section on Lambert's geography, let me mention that the so-called Lambert conformal projection has been compared, using present-day tools, to several other projections, and it turned out that it is the best one, in several precise senses, for the drawing of the 18th Russian Empire at the times of Euler, see the sequence of three papers [61–63]. The drawing of the Russian Empire is an important footstep in mathematical cartography because Euler, during his first stay at the Russian Academy of Sciences in Saint Petersburg, was in charge of drawing such a map, and he developed a theory for this task, see the paper [30], with an English translation in the book [11].

Lambert's writings on cartography naturally contain computations in spherical geometry, see Fig. 6.6.

Philosophy and Logic

Lambert's philosophy is an alliance between metaphysics and logic.

Among the recent publications on Lambert's philosophy, let me mention the volume *Johann Heinrich Lambert : Philosophie*, published in 2018 in *Les Cahiers philosophiques de Strasbourg* [13]. This volume includes an article titled *Quatre essais académiques* (Four academic essays) by Frédéric de Buzon [14], in which the author presents four of Lambert's writings published in the *Histoire* or the *Nouveaux mémoires* of the Prussian Academy. These are:

(1) The *Discours* Lambert delivered on January 24, 1765, on the occasion of his reception at this Academy, and in which he describes his approach to science and to the relationship between science and philosophy [45];
(2) Lambert's memoir *Observations sur quelques dimensions du monde intellectuel* (Observations on some dimensions of the intellectual world), published in 1770, in which he defines his approach to the question of the relationship between the physical and intellectual worlds [47];
(3) Lambert's *Essai de taxéométrie, ou sur la mesure de l'ordre* (An essay on taxeometry, or on the measure of order), a topic that Lambert is considered to be the first to name; in this essay, Lambert deals in a personal and original way with the notion of order and its measurement, used in a metaphysical and ontological framework [51];
(4) Lambert's *Second essai de taxéométrie ou sur la mesure de l'ordre* (Second essay on taxeometry, or on the measure of order). Lambert declares in the opening part of the essay that his goal is "to measure the impression made on us by the greater or lesser beauty of some arrangement, the symmetry, the mixture of resemblances and varieties that we perceive there" [52].

I would like to quote a short paragraph from the last work, to show the conversational style of the author, which is common to all his writings. This is §3 of [52] (my translation).

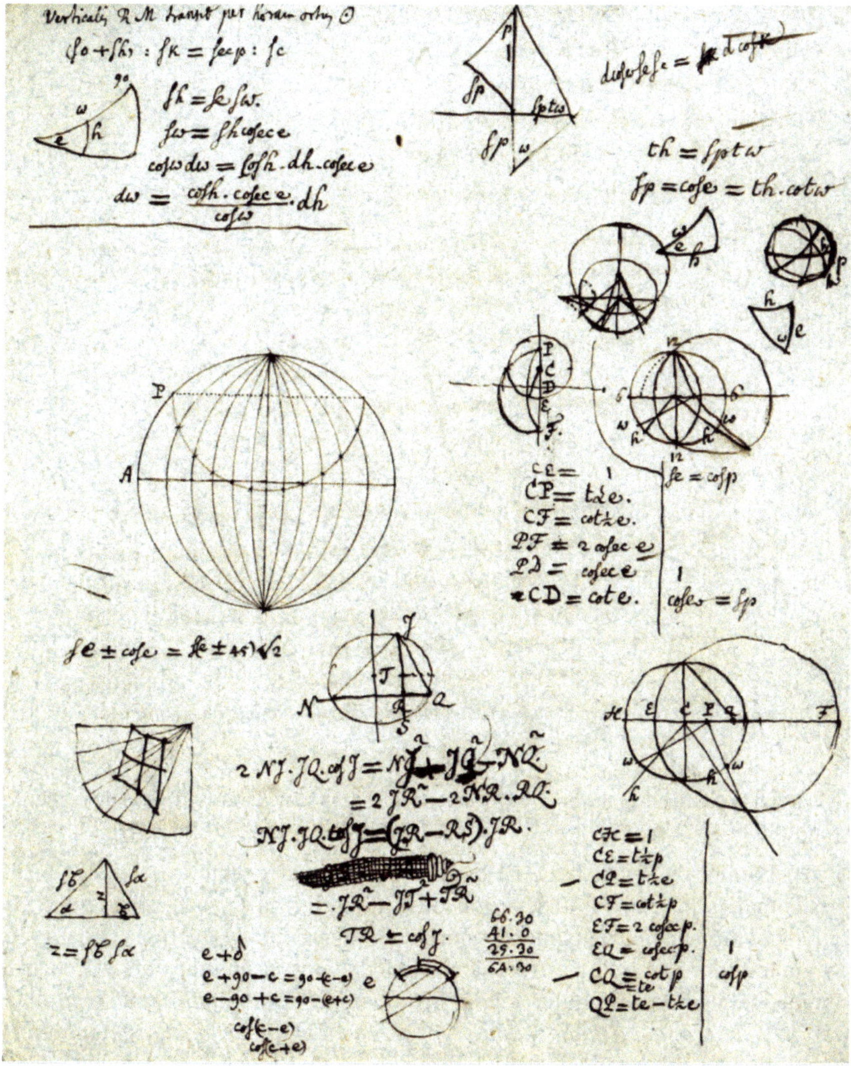

Fig. 6.6 A page from a manuscript by Lambert concerning cartography, with spherical geometry calculations. Archives of Mulhouse, Université de Haute-Alsace

To see clearly in a rather obscure and confusing subject, it is appropriate to go from the simplest to the most complex. I will follow this rule, especially in the choice of the examples on which I am going to report. I proposed them to myself, and I paid attention to the various impressions they made on me, and to the reasons I was able to find; all this with the aim of seeing whether the comparison that I would then make would lead me to some principle that could serve as a basis for computations. I knew in advance that impressions in the intellectual world are very similar to impressions in the physical world, and that the degrees

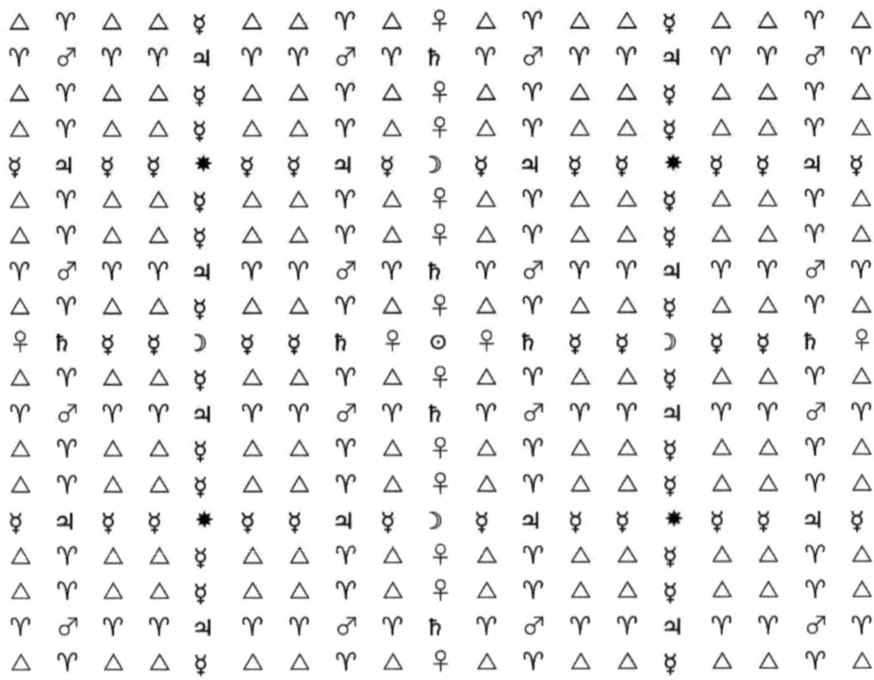

Fig. 6.7 From Lambert's *Second essay on taxeometry, or on the measure of order*

of tension in the fibers of the brain follow more or less the same laws as those in the muscles of the arm, when we compare different weights while holding them in the hand.

From Lambert's *Second essay* I have also extracted a plate, reproduced here in Fig. 6.7, that is part of his discussion of notions like measure, symmetry, order, variety, resemblance, arrangement, interlacing, degree of perfection, linearity, etc. Let us note that some of the astrological signs that appear in this plate were used as mathematical symbols by Euler, e.g. in his memoir *Solutio facilis problematis, quo quaeritur sphaera, quae datas quatuor sphaeras utcunque dispositas contingat* (An easy solution to the problem of finding a sphere that touches four given spheres, arranged in an arbitrary way) [21], mentioned in Chap. 2 of the present volume. Editions of Lambert's four essays are included in the volume [13].

On Metaphysics, theology and morals, let us mention Lambert's essay with the characteristic title *Über die Methode die Metaphysik, Theologie und Moral richtiger zu beweisen* (On the method of proving more correctly metaphysics, theology and morals), which he left in the form of notes and which was eventually published in 1918. [55]

Finally, we approach the field of logic, another area to which Lambert contributed substantially.

Lambert wrote several essays on symbolic logic, a field of which he is considered as a precursor, if not one of the founders. His most important work in this domain is the *Neues Organon*, a treatise whose full title is *Neues Organon oder Gedanken über die Erforschung und Bezeichnung des Wahren und dessen Unterscheidung vom Irrthum und Schein* (The new organon, or thoughts on the search and the designation of the true and its distinction from error and appearance) [44]. This is a 1000-page treatise on logico-mathematical philosophy, published in Leipzig in 1764. In this work, Lambert discusses topics such as the ways in which one may distinguishing between what appears to be objective from what is subjective, making connections with his work on optics. This work found its continuation in another important treatise he wrote on metaphysics, the *Anlage zur Architektonic*,[37] published in Riga in 1771. In the latter work, Lambert sets out a theory of knowledge that he wanted to be both logically and metaphysically valid, and built as a mathematical theory, based on an elementary set of primary notions. In the Preface, he presents this work as "a search for fundamental metaphysical doctrines undertaken in a completely new way".

The title of Lambert's *Neues Organon* refers to Aristotle's *Organon*.[38] and also to Francis Bacon's *Novum Organum*,[39] published in 1620. Lambert's treatise is in four parts, each with a Greek title: "Dianoiology, or theory of the laws of thought", "Alethiology or theory of truth", "Semiotics or theory of the designation of thoughts and things" and "Phenomenology or theory of appearance".

Let us recall that towards the end of the first half of the nineteenth century, Great Britain experienced a renaissance in the field of logic, and in particular an emergence of what is now called formal, or symbolic logic. This emergence was facilitated by a revival of algebra. Several of the authors of this renaissance acknowledged Lambert's influence. Among these authors, we mention John Stuart Mill (1806–1873), the author of *A System of Logic, Ratiocinative and Inductive* [60], published in 1843, in which he praises the research on the four figures of Lambert's syllogism, whose *Neues Organon*, he writes, "contains among other things one of the most

[37] *Anlage zur Architektonic, oder Theorie des Einflachen und Ersten in der philosophischen und mathematischen Erkenntnis* (Plan of an architectonic or theory of what is simple and first in philosophical and mathematical knowledge) [48].

[38] The word *Organon* in Aristotle's work refers to a collection of treatises attributed to him, which are mainly on logic. The word "organon" ὄργανον, in Greek, means "tool", or "instrument". The title *Organon* is not Aristotle's; it is considered that the Peripatetics (that is, those who were part of Aristotle's School), gave the name "organon" to that collection to emphasize that logic is an "instrument" of philosophy. The word "organon" to designate a collection of Aristotle's work is first mentioned by Diogenes Laertius. The precise composition of Aristotle's *Organon* has varied among commentators. Most commonly, it consists of the following six treatises: the *Categories*, *On Interpretation*, the *Prior Analytics*, the *Posterior Analytics*, the *Topics* and *On Sophistical Refutations*, but some commentators also include Aristotle's *Rhetoric* and *Poetics*.

[39] The *Novum organum* is the major work of Francis Bacon (1561–1626), the English philosopher and scientist who supported the idea that scientific knowledge can be based only on inductive reasoning and the careful observation of events in nature. The work was published in 1620 and the exact title is *Novum organum scientiarum*. This expression means "New Organum", that is, "New tool", is a clear reference to Aristotle's *Organon*. Bacon is sometimes considered as the father of modern empiricism.

elaborate and complete expositions of the syllogistic doctrine [and where Lambert] has expressly examined which sort of arguments fall most suitably and naturally into each of the four figures; and his investigation is characterized by great ingenuity and clearness of thought.". [40]

In 1847, the work *The mathematical analysis of logic: Being an Essay Towards a Calculus of Deductive Reasoning* [8] by Georges Boole (1815–1864) was published. This work is sometimes considered as the true beginning of mathematical logic. The same year, the work *Formal logic* [15] by Augustus De Morgan (1806–1871) appeared. In this work, the author writes: "I think it desirable to state all I know of any attempt to deal with the forms of inference otherwise than in the Aristotelian method. Since the time of Wallis, three well-known mathematicians have written on the subject, Euler, Lambert, and Gergonne: there may have been others, but I have not met with them." He then quotes Euler's *Letters to a German princess on various topics in physics and philosophy* [17] and Lambert's *Neues Organon* [44].

Lambert's work is also extensively quoted by John Venn (1834–1923) in his book *Symbolic logic* [75] (1881), where the expression in the title appears for the first time. Venn, in the introduction to his treatise, after mentioning several features of Lambert's work on logic, writes: "[...] These are the main truths of this kind which Lambert had seized. Whatever the defects and limitations in their expression, they represent a very remarkable advance on anything known to have been done before him." In his work *Symbolic logic*, Venn has a long biographical note on Lambert (pp. xxxi–xxxviii). He attributes to him several discoveries in logic. For example, he considers that it was Lambert who first recognized that the operations of addition, subtraction, multiplication and division have their counterparts in logic.

Finally, let us point out that the philosopher and logician Charles Sanders Peirce (1839–1914), in his *Elements of Logic*, compares Lambert to Kant, for whom he also had great admiration. For a rich exposition of Lambert's work as a precursor of mathematical logic, and in particular on the mention of his work by Anglo-Saxon philosopher-logicians of the 19th and 20th centuries, we refer to the article [28] by Roger Jaquel. The article [26] by Gray and Tilling also contains an exposition of Lambert's work in formal logic.

In conclusion, I hope that the reader of the present chapter has got the feeling that Lambert's treatise on spherical geometry which is translated and highlighted in the present volume constitutes only a small portion of the heritage of this great man.

References

1. K. Andersen, *The Geometry of an Art: The History of the Mathematical Theory of Perspective from Alberti to Monge*. Sources and Studies in the History of Mathematics and Physical Sciences (Springer, Berlin, 2007)

[40] Mill [60], Book II, Chapter ii, §1.

2. J. Barozzi, Le due regole della prospettiva pratica di M. Iacomo Barozzi da Vignola. Con i comentarij del R.P.M. Egnatio Danti dell'ordine de Predicatori, Matematico dello Studio di Bologna, Tipografo Vitale Mascardi ad istanza di Filippo de Rossi, Roma (1644)
3. Beer, Bestimmung der Absorption des rothen Lichts in farbigen Flüssigkeiten. Ann. Phys. Chem. **162**(5), 78 (1852)
4. J. Bernoulli (Johann III), Précis de la vie de M. Lambert, in *Nouvelles littéraires de divers pays. Avec des suppléments pour la liste et le nécrologe des astronomes, qui n'ont rien de littéraires*. Troisième cahier, à Berlin chez l'Auteur (Chez Haude & Spener, Berlin, 1777), pp. 73–76
5. D. Bernoulli, *Correspondance entre Daniel Bernoulli et Jean-Henri Lambert*. Annexe aux actes du Colloque J.-H. Lambert. Table ronde Mathématiques et philosophie de l'infini, de 1750 à nos jours, Université Haute Alsace, Ed. (Orphys, Paris, 1979)
6. G. Bigourdan, Lettres de Léonard Euler, en partie inédites. Bullet. Astron. **34**, 258–319 (1917)
7. L. Boi, L. Giacardi, R. Tazzioli (Eds.), La découverte de la géométrie non euclidienne sur la pseudosphère. Les lettres d'Eugenio Beltrami à Jules Hoüel (1868–1881). Introduction, notes and critical commentaries. Preface by Christian Houzel and Eberhard Knobloch. Collection Sciences dans l'histoire. Paris, Librairie Scientifique et Technique Albert Blanchard (1998)
8. G. Boole, *The Mathematical Analysis of Logic, Being an Essay Towards a Calculus of Deductive Reasoning* (Macmillan, Barclay, & Macmillan, Cambridge, 1847)
9. K. Bopp (Ed.), L. Euler und J. H. Lamberts Briefwechsel, aus dem Manuskript hrsg, Abhandlungen der Prussischen Akademie der Wissenschaften. Math.-Phys. Klasse **2**, 1–45 (1924)
10. P. Bouguer, *Essai d'optique sur la gradation de la lumière* (Paris, Claude Jombert, 1729)
11. R. Caddeo, A. Papadopoulos (Eds.), *Mathematical Geography in the Eighteenth Century: Euler, Lagrange and Lambert* (Springer, Cham, 2022)
12. *Correspondance between Daniel Bernoulli et Johann Heinrich Lambert* (in French). Published by P. Radelet-De Grave and V. Scheuber. Association pour la célébration de Jean-Henri Lambert (Mulhouse, Haut-Rhin). Université de Haute-Alsace, Centre de recherches et d'études rhénanes and by the Association pour la célébration de J.-H. Lambert, Table ronde sur Mathématiques et philosophie de l'infini de 1750 à nos jours, 1977, Mulhouse (Ophrys, Paris, 1979)
13. F. de Buzon (Ed.), Johann Heinrich Lambert: philosophie, in *Les Cahiers philosophiques de Strasbourg*, vol. 44 (Presses Universitaires de Strasbourg, 2018)
14. F. de Buzon, Quatre essais académiques. Johann Heinrich Lambert. Les Cahiers Philos. Strasbourg **44**, 127–148 (2018)
15. A. De Morgan, *Formal Logic, or The Calculus of Inference, Necessary and Probable* (1847), New edition (The Open Court, London, 1926)
16. E. Doublet, *Histoire de l'astronomie* (Gaston Doin, Paris, 1922)
17. L. Euler, *Lettres à une princesse d'Allemagne sur divers sujets de physique & de philosophie* (Imperial Academy of Sciences, St. Petersburg, 1768). 2 volumes. Opera Omnia, Series 3, Volume 11 and 12
18. L. Euler, *Theoria Motuum Planetarum et Cometarum* (Ambrosius Haude, Berlin, 1774), pp. 105–251. Opera Omnia, Series 2, Volume 28
19. L. Euler, *Beantwortung verschiedener Fragen uber die Beschaffenheit, Bewegung und Wurckung der Cometen* (Ambrosius Haude, Berlin, 1744), pp. 1–56. Opera Omnia, Series 2, Volume 31, pp. 125–150
20. L. Euler, *Recherches et calculs sur la vraie orbite elliptique de le comète de l'an 1769 et son tems périodique* (Imprimerie de l'Académie Impériale des Sciences, St. Pétersbourg, 1770), 159pp. Opera Omnia, Series 2, Volume 28, pp. 269–314
21. L. Euler, Solutio facilis problematis, quo quaeritur circulus, qui datos tres circulos tangat. Nova Acta Acad. Sci. Imper. Petropol. **6**, 95–101 (1790). Opera Omnia, series 1, volume 26, pp. 270–275
22. L. Euler, *Opera Omnia, Series IV, A: Commercium Epistolicum*, Vol. III, Teil II, ed. by E.A. Fellmann, G.K. Mikhajlov (Birkhäuser, Basel, 2016)

23. P.A. Florensky, Reverse perspective (English Transl.), in *Beyond Vision: Essays on the Perception of Art*, ed. by N. Misler, P. Florensky. Transl. W. Salmond (Reaktion Books, London, 2006). French translation, La perspective inversée, suivi de L'Iconostase et autres écrits sur l'art, transl. Françoise Lhœst, L'Âge d'homme, Lausanne, 1992
24. J.H.S. Formey, Éloge de Lambert. Nouveaux Mémoires de l'Acad. Royale des Sciences et Belles-Lettres de Berlin (1780), pp. 72–90
25. Frédéric II, Œuvres posthumes, t. 1. Correspondence with d'Alembert (1805)
26. J.J. Gray, L. Tilling, Johann Heinrich Lambert, mathematician and scientist, 1728–1777. Historia Math. **5**, 13–41 (1978)
27. Ch. Huygens, *Cosmothéoros*, 1st edn. (1698). Translated from the Latin into French by M. Du Four, under the title La pluralité des mondes, Jean Moreau, Paris, 1702
28. R. Jaquel, Les philosophes anglo-saxons et le philosophe mulhousien Jean-Henri Lambert (1728–1777), part I, Lambert, logicien et précurseur de la logique mathématique, et les philosophes anglo-saxons. Bulletin des professeurs du Lycée d'Etat de Garçons de Mulhouse, No. 1 (1962)
29. R. Jaquel, Jean-Henri Lambert (1728–1777) et l'astronomie cométaire au XVIIIe siècle, Compt. rend. XVIIe Congrès national des Sociétés savantes, Strasbourg et Colmar (1967), sect. scient., t. I, Paris, 1969, pp. 27–56
30. R. Jaquel, Le problème nuancé de la nationalité du "Leibniz alsacien" Jean-Henri Lambert. Bulletin du musée historique de Mulhouse (1971–1973)
31. R. Jaquel (Ed.), Un mulhousien génial du XVIIIe siècle: le savant et philosophe Jean-Henri Lambert (1728–1777), extrait de l'Annuaire statistique de la ville de Mulhouse (1976), pp. 5–12
32. R. Jaquel, Vers les œuvres complètes du savant et philosophe Jean-Henri Lambert (1728–1777): velléités et réalisations depuis deux siècles, in *Le savant et philosophe mulhousien Jean-Henri Lambert, études critiques et documentaires, Université du Haut-Rhin, Centre de Recherches et d'études Rhénanes* (Ophrys, Paris, 1977), pp. 87–97
33. R. Jaquel, Le mulhousien Jean-Henri Lambert inventeur de la cartographie scientifique et ses systèmes de projection, in *Le savant et philosophe mulhousien Jean-Henri Lambert, études critiques et documentaires, Université du Haut-Rhin, Centre de Recherches et d'études Rhénanes* (Ophrys, Paris, 1977), pp. 79–85
34. R. Jaquel, *Le savant et philosophe mulhousien Jean-Henri Lambert (1728–1777). études critiques et documentaires* (Ophrys, Paris, 1977), 170pp.
35. S.L. Jaki, Sur l'édition et la réédition de la traduction française des Cosmologische Briefe de Lambert. Rev. d'histoire Sci. **32**(4), 305–314 (1979)
36. *Johann Heinrich Lamberts Deutscher Gelehler Briefwechsel*, Berlin, 5 volumes, Correspondance en allemand éditée par Johann III Bernoulli (1782–1784)
37. E. Kant, Mélanges de logique, Traduction par Joseph Tissot. Librairie philosophique de Ladrange (1862)
38. J.-L. Lagrange, Sur le problème de la détermination des orbites des comètes d'après trois observations, in *Addition to His Theoretical Results, Lambert Determined Precisely the Orbits of Several Comets (Trois mémoires)* (Nouveaux mémoires de l'Académie royale des sciences et belles-lettres de Berlin, Années 1778–1783), pp. 439–532
39. J.H. Lambert, *Les propriétés remarquables de la route de la lumière par les airs et en général par plusieurs milieux réfringents, sphériques et concentriques, avec la solution des problèmes qui y ont du rapport, comme sont les refractions astronomiques et terrestres, et ce qui en depend*, 1st edn. (The Hague, Nicolaas van Dalen, 1759). Reprint, éditions Alain Brieux, Paris, 1977
40. J.H. Lambert, Die freye Perspektive, oder Anweisung jeden perspektivischen Aufriss von freyen Stücken und ohne Grundriss zu verfertigen, Zürich (1759). French translation: La perspective affranchie de l'embarras du plan géometral, Zurich, chez Heidegguer & Co. 1759, second edition with additions, Zurich, 1774
41. J.H. Lambert, Photometria sive de mensura et gradibus luminis, colorum et umbrae, Augustae Vindelicorum, sumptibus viduae Eberhardi Klett typis Chistophori Petri Detleffsen (1760)

42. J.H. Lambert, *Cosmologische Briefe Cosmologische Briefe Über Die Einrichtung des Weltbaues* (Augsburg, Eberhard Kletts, 1761)
43. J.H. Lambert, *Insigniores orbitae cometarum proprietates* (sumptibus Eberhardi Klett Viduae, Augsburg, 1761)
44. J.H. Lambert, *Neues Organon oder Gedanken über die Erforschung und Bezeichnung des Wahren und dessen Unterscheidung vom Irrthum und Schein*, 2 vols (bey Johann Wendler, Leipzig 1764)
45. J.H. Lambert, *Discours paru dans Histoire de l'Académie royale des sciences et belles-lettres* (Haude & Spener, Berlin, 1765), pp. 506–514
46. J.H. Lambert, *Mémoire sur quelques propriétés remarquables des quantités transcendantes circulaires et logarithmiques*. Histoire de l'Académie Royale des Sciences et Belles-Lettres, Année 1761 vol. XVII, Berlin (1768), pp. 265–322
47. J.H. Lambert, *Observations sur quelques dimensions du monde intellectuel, Histoire de l'Académie royale des sciences et belles-lettres*, vol. XIX (Haude et Spener, Berlin, 1770), pp. 421–438
48. J.H. Lambert, Anlage zur Architektonic, oder Theorie des Einflachen und Ersten in der philosophischen und mathematischen Erkenntnis, Riga, bey Johann Friedrich Hartknoch (1771)
49. J.H. Lambert, Observations sur l'orbite apparente des comètes, Nouveaux Mémoires de l'Académie Royale des Sciences de Berlin (1771), pp. 352–364
50. J.H. Lambert, *Anmerkungen und Zusätze zur Entwerfung der Land- und Himmelscharten*. Hrsg. von A. Wangerin, Berlin, (1772). English translation: Notes and Comments on the Composition of Terrestrial and Celestial Maps, translated by W. R. Tobler, University of Michigan Press, reprinted 2010. New translation and commentary by Annette A'Campo in [11]
51. J.H. Lambert, *Essai de taxéométrie ou sur la mesure de l'ordre, Nouveaux mémoires de l'Académie royale des sciences et belles-lettres, année 1770* (Haude et Spener, Berlin, 1772), pp. 327–342
52. J.H. Lambert, *Second essai de taxéométrie ou sur la mesure de l'ordre*. Nouveaux mémoires de l'Académie royale des sciences et belles-lettres, année 1773 (Haude & Spener, Berlin, 1775), pp. 347–368
53. J.H. Lambert, *Pyrometrie oder Vom Maaße des Feuers und der warme* (Bey Haude und Spener, Berlin, 1779)
54. J.H. Lambert, *Johann Heinrich Lamberts Monatsbuch mit den zugehörigen Kommentaren, sowie mit einem Vorwort über den Stand der Lambertforschung, herausgegeben p. p. K. Bopp*. Abhandlungen der Königl.-Bayerischen Akademie der Wissenschaften, Mathematisch-Physikalische Klasse, vol. 27, 6 (Abhandlung, Munich, 1915–1916)
55. J.H. Lambert, *Über die Methode die Metaphysik, Theologie und Moral richtiger zu beweisen, hrsg* (von K. Bopp, Berlin, 1918)
56. J.H. Lambert, Works of Johann Heinrich Lambert. Hydrograph. Rev. **8**(1), 78–82 (1931)
57. J.H. Lambert, *Lettres cosmologiques sur l'organisation de l'univers, Jakobus Maurits Karel van Utenhove*, Reprint, Paris (Alain Brieux, Paris, 1977). With a preface by Jacques Merleau-Ponty
58. J.H. Lambert, Discours de réception, in *Philosophische Schriften*, ed. by H.-W. Arndt, A. Emmel, A. Spree, vol. VIII-1 (Olms, Hildesheim, 2007)
59. J. Lepsius, *Johann Heinrich Lambert, Eine Darstellung Seiner Kosmologischen Und Philosophischen Leistungen*, ed. Achermann, Munich (1881). reprint 2010, Nabu Press
60. J.S. Mill, A System of Logic, *Ratiocinative and Inductive, Being a Connected View of the Principles of Evidence and the Methods of Scientific Investigation*. First edition, John W. Parker, London, 1843, New edition, University of Toronto Press, Toronto, Buffalo, reprinted 1978, 1981, New edition, (Cambridge University Press, Cambridge, 2012)
61. H. Miyachi, K. Ohshika, Lipschitz and quasiconformal mappings in cartography, in *Essays on Geometry Celebrating the 65th Birthday of Athanase Papadopoulos*, ed. by M. Uludağ, A. Zeytin (Springer, Cham, 2025), pp. 111–130

62. H. Miyachi, K. Ohshika, A. Papadopoulos, Marinus–Ptolemy and Delisle–Euler conical Maps, Gaṇita Bhāratī (Indian mathematics). Bullet. Ind. Soc. History Math. **45**, 227–251 (2013)
63. H. Miyachi, K. Ohshika, A. Papadopoulos, S. Yamada, On the Lambert conformal conical projection and the general map of the Russian Empire, in *Handbook of the Mathematics of the Arts and Sciences*, ed. by B. Sriraman (Springer, Cham, 2026)
64. N.I. Nevskaja, Joseph-Nicolas Delisle (1688–1768). Rev. d'histoire Sci. **26**(4), 289–313 (1973)
65. Nouveau dictionnaire de biographie alsacienne. No. 22, Koe à Lag, Fédération des sociétés d'histoire et d'archéologie d'Alsace, Ed. Jean-Pierre Kintz (1994)
66. A. Papadopoulos, G. Théret, *La théorie des lignes parallèles de Johann Heinrich Lambert*. Collection Sciences dans l'Histoire (Librairie Scientifique and Technique Albert Blanchard, Paris, 2014)
67. H. Poincaré, Les géométries non euclidiennes. Revue générale des sciences pures et appliquées 2e année **23**, 769–774 (1891)
68. J.-C. Pont, *L'aventure des parallèles. Histoire de la géométrie non euclidienne: précurseurs et attardés* (Peter Lang, Bern, 1986)
69. G. Rémy, *Jean-Henri Lambert, sa vie et son œuvre*. Revue d'Alsace, 7e série, 11e année, tome LXI de la collection (A. Picard, Paris, rue Bonaparte, 1910). Part I, pp. 393–406, Part II, pp. 452–468
70. B.A. Rosenfeld, *History of Non-Euclidean Geometry*. Translated by A. Shenitzer, Studies in the History of Mathematics and Physical Sciences, vol. 12 (Springer, New York, 1988)
71. C.J. Scriba, Johann Heinrich Lambert, in *Dictionary of Scientific Biography*, vol. VII (Scribner, New York, 1981), pp. 595–600
72. F.E. Sitzmann, *Dictionnaire de biographie des hommes célèbres d'Alsace: depuis les temps les plus reculés jusqu'à nos jours* (Imprimerie Sutter, Rixheim, 1910)
73. D. Speiser, *Preface of J. H. Lambert, Les propriétés remarquables de la route de la lumière par les airs et en general par plusieurs milieux réfringents, sphériques et concentriques, avec la solution des problèmes qui y ont du rapport, comme sont les réfractions astronomiques et terrestres, et ce qui en depend* (Alain Brieux, Paris, 1977)
74. D. Thiébault, *Mes souvenirs de vingt ans de séjour à Berlin, ou Frédéric le Grand, sa famille, sa cour, son gouvernement, son académie, ses écoles et ses amis littérateurs et philosophes* (F. Buisson, Paris, 1804)
75. J. Venn, *Symbolic logic* (Mc Millan, London, 1881)
76. R. Wolf, Johann Heinrich Lambert von Mühlhausen, 1728–1777, in *Biographien zur Kulturgeschichte der Schweiz*, vol. 3 (Cyclus, Zürich, 1860), pp. 317–356
77. C. Wolff, Psychologia rationalis, Francofurti & Lipsiae, prostat in Officina libraria Rengeriana (1734)
78. C. Wolff, Psychologia empirica, Francofurti & Lipsiae, prostat in Officina libraria Rengeriana (1738)

Chapter 7
Johann Heinrich Lambert's Memoir *Theorie der Parallellinien*: A Review with Commentary

Athanase Papadopoulos and Guillaume Théret

Yet there have been and still are geometers and philosophers, and even some of the most distinguished, who doubt whether the whole universe, or to speak more widely, the whole of being, was only created in Euclid's geometry; they even dare to dream that two parallel lines, which according to Euclid can never meet on earth, may meet somewhere in infinity. I have come to the conclusion that, since I can't understand even that, I can't expect to understand about God. I acknowledge humbly that I have no faculty for settling such questions, I have a Euclidean earthly mind, and how could I solve problems that are not of this world? And I advise you never to think about it either, my dear Alyosha, especially about God, whether He exists or not. All such questions are utterly inappropriate for a mind created with an idea of only three dimensions. (Fyodor Dostoyevsky, The Brothers Karamazov, *transl. by Constance Garnett, [Ivan Karamazov speaking].)*

Like the parallel postulate, the postulate that mathematics will survive has been stripped of its "evidence"; but, while the former is no longer necessary, we would not be able to get on without the latter. (André Weil, The Future of Mathematics, *1946.)*

Abstract We review the memoir *Theorie der Parallellinien* by Johann Heinrich Lambert, written in 1766. Lambert, a victim of the prejudices of his time, conceived this memoir as an attempt to prove the so-called parallel postulate of Euclid's *Elements*, and consequently, the non-existence of the geometry that we now call

A. Papadopoulos (✉)
Institut de Recherche Mathématique Avancée, Université de Strasbourg et CNRS, Strasbourg, France
e-mail: athanase.papadopoulos@math.unistra.fr

G. Théret
Laboratoire Interdisciplinaire Carnot de Bourgogne (ICB), Dijon Cedex, France
e-mail: guillaume.theret71@orange.fr

R. Caddeo, A. Papadopoulos (eds.), *Spherical Geometry in the Eighteenth Century I*, https://doi.org/10.1007/978-3-032-12466-1_7

hyperbolic geometry. In fact, by developing the foundations of a geometry obtained by replacing the parallel postulate with its negation while keeping Euclid's other postulates unchanged, Lambert was hoping to arrive at a contradiction. Of course, he failed in his endeavor, but these attempts at proving the parallel postulate implicitly contain, without Lambert having foreseen it, fundamental results of hyperbolic geometry, the discovery of which, by Lobachevsky, Bolyai and Gauss, was not to take place until the following century. Thus, Lambert's memoir (which he did not intend to publish but which was eventually published in 1895) constitutes one of the founding texts of non-Euclidean geometry.

Spherical geometry is one of the three geometries of constant curvature, the other two being Euclidean geometry and hyperbolic geometry. In this sense, along with hyperbolic geometry, spherical geometry constitutes one of the two non-Euclidean geometries. In fact, Lambert, like Lobachevsky and others after him, understood the deep relationships between the three geometries: Euclidean, spherical, and hyperbolic, in particular the formal and the more profound analogies between the trigonometric formulae, the properties of birectangular isosceles quadrilaterals and of trirectangular quadrilaterals, the monotonicity properties (which can be formulated in terms of convexity properties) which hold in opposite senses in spherical and hyperbolic geometry which at some points he calls a sphere of imaginary radius. It is for these reasons that we decided to include in this volume, dedicated to spherical geometry, a chapter on this important memoir by Lambert, trying to highlight its most important ideas.

Keywords Johann Heinrich Lambert · Neutral geometry · Parallel axiom · Spherical geometry · History of geometry · Hyperbolic geometry · Theory of parallel lines · Lambert quadrilateral · Ibn al-Haytham–Lambert quadrilateral · Greek mathematics · Arabic mathematics · Eighteenth century mathematics · Khayyām–Saccheri quadrilateral · Spherical trigonometry · Parallel postulate

AMS Codes 01A20, 01A30, 01A50, 51-03

7.1 Introduction: The Theory of Parallels

The expression "theory of parallel lines" in the title of Lambert's memoir *Theorie der Parallellinien*, published in [42] (see [28] for a French translation), refers to the various attempts that were made, over a period that lasted almost two and a half millennia, to prove the so-called parallel postulate of Euclid's *Elements*. We recall that this postulate is the statement placed in Book 1 of J. L. Heiberg's classic edition of the *Elementa* as postulate 5 [10, p. 8] and which, in Heath's English edition, based on Heiberg's, says the following [16, Vol. I, p. 202]:

> That, if a straight line falling on two straight lines makes the interior angles on the same side less than two right angles, the two straight lines, if produced indefinitely, meet on that side on which are the angles less than the two right angles.

In some other versions of the *Elements*, this statement is included as the eleventh axiom. Lambert uses the term "eleventh axiom.'

Euclid makes a distinction between a postulate and an axiom. This difference is discussed in detail by Proclus[1] in his *Commentaries on Book I of Euclid's Elements* [31]. To put it in simple words, a postulate asserts that a certain geometric construction is possible, while an axiom may concern properties (such as magnitudes) that are not geometric. The distinction between "postulate" and "axiom" has disappeared in modern axiomatics, that is, those of Peano, Veronese, Pasch, Hilbert, etc., and we shall not make it here. Thus, in this chapter, we will use the words "postulate" or "axiom" interchangeably, and therefore also the expressions "parallel axiom", "fifth axiom", "parallel postulate", "eleventh postulate".

The history of this "theory of parallels" covers a period ranging from Greek antiquity (starting before Plato) until the time (around 1868) when the works of Lobachevsky, Bolyai and Gauss on non-Euclidean geometry started to be considered by the mathematical community as valid. These discoveries showed that Euclid's parallel postulate cannot be deduced from the other postulates, and it gradually put an end to the very numerous attempts to prove this postulate.

Mathematicians addressed the theory of parallels from several facets. First, there are geometers who tried to deduce the fifth postulate from Euclid's other postulates without introducing a new one. We know a posteriori that these attempts were doomed to failure. Other authors modified the definition of parallelism,[2] believing that this would avoid the difficulty, and these also, in general, achieved nothing. But there are others who, finding the postulate complicated and less intuitive than the other postulates, tried to remedy this by replacing it with a simpler and more intuitive one: equidistance of parallel lines, existence of translations, existence of homothetic triangles, etc. Some of them succeeded in their endeavor and obtained valuable results. In Sect. 7.6 of the present chapter, we review several works that have been done on this topic by mathematicians over a period from Greek Antiquity to the nineteenth century, passing through the works of the Arab mathematicians of the Middle Ages.

Lambert holds a special place in this story. His work, although it predates the discovery of non-Euclidean geometry, implicitly contains results from that theory, and this is what we shall try to explain in the following pages.

It is conceivable that the question of parallels is, among all the questions that have been addressed to date by prominent mathematicians, the one about which the volume of existing writings is the largest. Thousands of articles and hundreds of books, of various styles and scopes and spread over a period of more than two thousand years, are devoted to this theory. We may quote here Henri Poincaré who,

[1] Proclus (ca. 412–485 AD) was a Neoplatonist philosopher who studied in Alexandria. His work is an invaluable source of knowledge for historians of Greek mathematics. We have included some information on him in Sect. 7.6 of the present chapter.

[2] In Euclid's *Elements*, this definition is given in Book I (Definition 25): *Parallel straight lines are straight lines which, being in the same plane and being produced indefinitely in both directions, do not meet one another in either direction*. [16, Vol. I, p. 154].

in an exposition of the question published in 1891 [29], writes: "The amount of effort that has been expended on this chimerical hope is truly unthinkable." An article by the French mathematician Georges Brunel (1854–1900) published in 1888 in the *Mémoires de la Société de Sciences Physiques et Naturelles de Bordeaux* [5] mentions one hundred and thirty-four attempts to prove this postulate. The book by Paul Stäckel and Friedrich Engel titled *Die Theorie der Parallellinien von Euklid bis auf Gauss, eine Urkundensammlung zur Vorgeschichte der nicht-euklidischen Geometrie* (Theory of parallel lines from Euclid to Gauss: A collection of sources for the prehistory of non-Euclidean geometry) [42], published in 1895, contains, besides the text by Lambert which is the subject of the present chapter, excerpts of works by Euclid, Wallis, Saccheri, Gauss, Schweikart and Taurinus on the theory of parallels. The bibliography of this book includes three hundred and twenty items about the theory of parallels, published in the period 1482–1837, a bibliography which remains minimalist compared to all that has been written on this subject. Stäckel and Engel, for example, left aside the very important contribution of the Greeks to the question, as well as the no less important contribution of the Arabs, the latter having been, it is true, very little known at the time—with a few notable exceptions; for instance, the work of Naṣīr al-Dīn al-Ṭūsī (thirteenth century) was known and quoted by Wallis and Saccheri. The book by J.-C. Pont, [30], *L'aventure des parallèles* (The Adventure of Parallels), published in 1986, presents a very good overview of this subject, including a section on the work of the Greeks and another one on that of the Arabs.

The plan of the rest of this chapter is the following: Sect. 7.2 contains an outline of the structure of Lambert's memoir. It is followed by an analysis of the three parts of this memoir, made in Sects. 7.3, 7.4, and 7.5. After this, we have included two excursuses. The first one, in Sect. 7.6, contains some milestones on the history of parallel lines, from Greek antiquity until the modern period, and the second one, in Sect. 7.7, is concerned with two sorts of quadrilaterals that play a central role in the theory of parallels, namely, the trirectangular and birectangular isosceles quadrilaterals.

7.2 The Structure of Lambert's Memoir

Lambert's memoir, *Theorie der Parallellinien*, is divided into three parts:

(1) §1 to 11: A description of the problem of parallels and a discussion of Euclid's fifth postulate and its place among the propositions and other axioms of the *Elements*. Lambert discusses there the difficulties posed by this postulate and recalls commentaries and attempts by several of his predecessors to prove it.
(2) §12 to 26: Some propositions of neutral geometry, that is, of Euclidean geometry deprived of the parallel axiom. Neutral geometry is defined axiomatically by taking the Euclidean system of axioms and postulates with the exception of the

parallel postulate. Lambert hopes that these propositions of neutral geometry may be used to prove the fifth postulate.

(3) §27 to 88: Lambert's attempts to prove the parallel postulate, starting from the principle that a negation of the parallel axiom leads to an absurdity. It is in this third part that we find a good number of propositions that are part of the field that was later called hyperbolic geometry.

We shall survey each part of Lambert's memoir separately.

7.3 Part 1 of Lambert's Memoir

In this part, Lambert undertakes an exposition of what he calls "the general problem of parallel lines", a problem which, he says, constitutes "a difficulty that appeared at the dawn of geometry and which, since the time of Euclid, has struck all those who refused to blindly believe, following others, the theories of this science and who on the contrary wanted to be convinced by reason and never to depart from the rigor that they found in most of the proofs." The difficulty, which, he says, jumps out at every reader of the *Elements*, lies in the eleventh axiom. He states this axiom in the following form (Fig. 7.1):[3]

> When two lines CD, BD are cut by a third one BC and when the two internal angles $\widehat{DCB}$ and $\widehat{DBC}$, taken together, are smaller than two right angles, then the two lines CD, BD intersect on the side of D, that is, on the side where we find these angles.[4]

Lambert declares that when we read this axiom, we get the impression that it is a statement requiring proof, an impression reinforced by the fact that in the *Elements*, the converse of this statement is a theorem (Proposition 27 of Book I)[5] and by the fact that Euclid, having postponed the use of this axiom until the 29th proposition of Book I, tried to develop part of his geometry without it. Similar considerations had already been made and widely commented by Lambert's predecessors. Lambert recalls, in §9 of his memoir, that "Proclus himself was troubled by Euclid's eleventh axiom, and required a proof of it, for the simple reason that *its converse is provable*." He mentions in §5 the conceptual difficulty (also raised many times before him) which arises from the fact that in using the parallel axiom, the lines must be extended to infinity. He considers that this use of infinity obfuscates the meaning of this axiom and makes it much less intuitive than the others. We may recall here that Lambert was also a philosopher and that, as such, the notion of infinity was not a mere word from the mathematical vocabulary but a concept that challenged him, as

[3] In the following sentence, and especially in the figure, the letter D indicates a direction rather than a point.

[4] Compare with the statement in Euclid's *Elements*, Heath's translation [16, Vol. I, p. 202] which we recalled in the Introduction to the present chapter.

[5] Proposition 27 says: *If a straight line falling on two straight lines make the alternate angles equal to one another, the straight lines will be parallel to one another.* [16, p. 307].

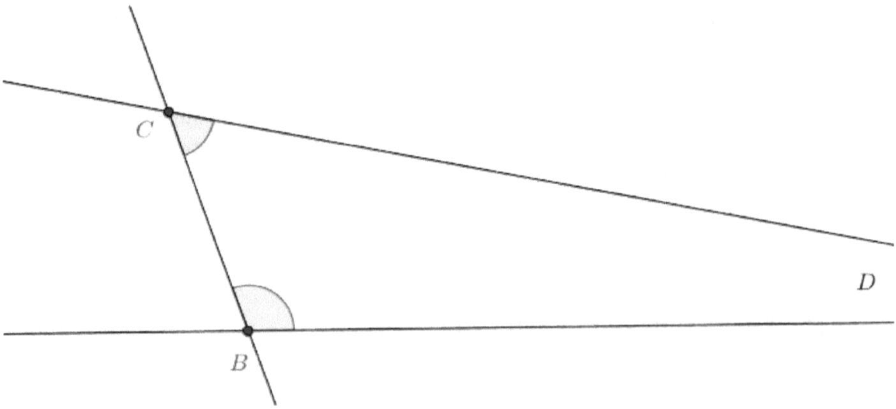

Fig. 7.1 The figure used in Lambert's statement of Euclid's parallel axiom

it challenged other philosophers of mathematics before him (like the neo-Platonic mathematician and philosopher Proclus, whom we have just quoted).

Lambert then discusses the "representability" of the parallel axiom, declaring that if this axiom is not a theorem, then one should accept lines that are not *straight* (Lambert emphasizes this word), but that can be asymptotes to each other. This cannot but remind us of Lobachevsky's definition of parallelism: two lines are parallel if they are asymptotes (in one direction), see [23]. Proclus, in his Commentaries on Book I of Euclid's *Elements*, had already made a similar statement; see e.g. his commentary on Definition 35 in [31, p. 153 & seq.].

Lambert in §8 quotes again Proclus, saying that the latter did not advocate any change in the definition of parallelism but thought that one should be able to prove the parallel axiom without modifying any other axiom (§9). It appears from this passage that Lambert agrees with Proclus, who thought that Euclid had chosen to include the statement on parallels among his axioms because he had not found any proof.

At the beginning of §11, Lambert brings up two possibilities in order to solve the difficulties raised by the parallel axiom: either to deduce it, as a theorem, from the other axioms, or to replace it by one or several axioms that are simpler and clearer. He writes, at the end of this introductive part: "I have no doubt that Euclid had also thought of including his eleventh axiom among the theorems". In the rest of the memoir, Lambert will exploit the first path, of course without any success.

7.4 Part 2 of Lambert's Memoir

Sections 12 to 26 constitute the second part of Lambert's memoir. In this part, Lambert presents a series of propositions that are valid in neutral geometry, that is, the geometry obtained from Euclidean geometry by keeping all the axioms and postulates, except that of parallels, and by adding no others.

Lambert begins by recalling (§12) that Euclid, until proposition 29 of the first book of the *Elements*, did not use the parallel axiom, but that from this proposition onwards, he made extensive use of it.[6] He notes that not only do the proofs of some propositions after the 29th use the parallel axiom, but that conversely, several of these propositions entail this axiom. He mentions as examples the proposition saying that the sum of the angles of a triangle is equal to two right angles (Proposition 32 of Book I) and another one saying that one straight line intersects another if and only if all the parallels to the first intersect the second, which is a proposition of Euclidean geometry that is not in the standard version of the *Elements*.

Lambert then announces that he will prove new propositions that do not use the parallel axiom. In fact, he proves some of the propositions he states, but for others, he does not succeed because, contrary to what he had in mind, they are not valid in neutral geometry: some are valid only under the negation of the parallel axiom, that is to say, in hyperbolic geometry, and others are valid only under the parallel axiom.

The first of these propositions is found in §13 of his memoir. He considers there a triangle ABC which is right at A and a line ED that passes through C and intersects the line AB at D (see Fig. 7.2). He shows that we have the following inequalities:

$$\widehat{ACD} < \widehat{ACB} + \widehat{ABC} < \widehat{ACE}.$$

Lambert proves this proposition and then adds the following in §14: "This theorem now shows very precisely how far we are from Euclid's Proposition 17 concerning the determination of the sum of the three angles of a triangle"[7] and he makes a long comment on his sentence. The interested reader is referred to [28, p. 104 ff.]

In §15, Lambert says he wants to add another proposition, and he proves, by a simple geometric argument, that if the sum of the three angles in any triangle is constant, then this value is necessarily equal to two right angles. He then studies a property that can be stated in terms of regular polygons: Consider a sequence of points $A, B, C, D, E, F \ldots$ such that the consecutive segments $AB, BC, CD, DE, EF, \ldots$ are all equal and the angles at the points

---

[6] In the commonly used versions of *Elements*, the mention of parallel lines begins at proposition 27, not 29, of Book I, but as Lambert says, the use of the parallel postulate only begins at proposition 29. Propositions 27 and 28 state that if two straight lines make equal alternate internal (Proposition 27) or corresponding (proposition 28) angles with a third line that intersects them, then the lines are parallel. These two propositions do not use the parallel postulate, but only the property that the angle sum in a triangle is $\leq 180^\circ$, which is proved using Proposition 16 which does not use the parallel axiom. On the other hand, proposition 29 is the contrapositive of the parallel axiom (and is therefore equivalent to it).

[7] Proposition 17 of Book I says: *In any triangle two angles taken together in any manner are less than two right angles.*

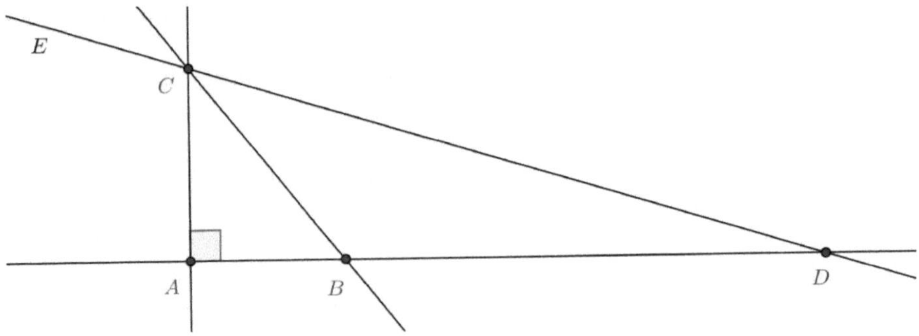

Fig. 7.2 Figure for the Proposition in §13 of Lambert's memoir

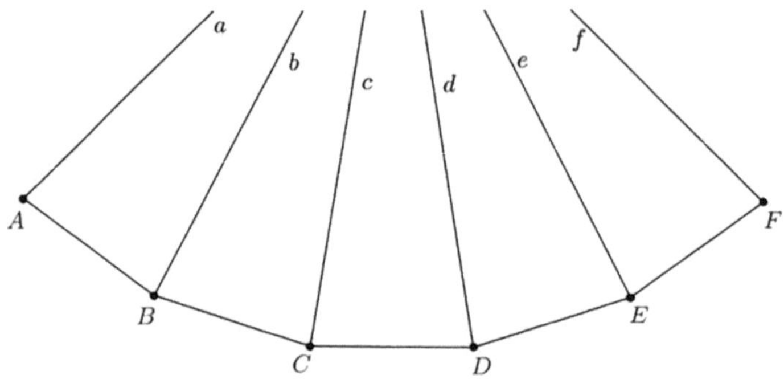

Fig. 7.3 Figure used for an argument in §15 of Lambert's memoir

$B, C, D, E, F, \ldots$ are also all equal and strictly less than two right angles (Fig. 7.3). Consider the angle bisectors Aa, Bb, Cc, Dd, $Ee, \ldots$ of these angles. Then the sequence of points $A, B, C, D, E, F \ldots$ lies on a circle whose center is the common intersection of the bisectors Aa, Bb, Cc, Dd, $Ee, \ldots$

We know that this property does not hold in neutral geometry; in fact, it does not hold in hyperbolic geometry either, because the bisectors in question may not meet (consider consecutive equidistant points on a horocycle and join them by geodesic segments). This property is equivalent to the parallel axiom.

In §16, Lambert states the following property.

> Given a line BD and a perpendicular FG raised at a point F on this line (see Fig. 7.4 extracted from Lambert's memoir). Then from the point G one can draw a line which meets BD at a point A such that the angle $\widehat{AGF}$ differs by as small an amount as one wishes from a right angle.

After a preparation (§17, 18, 19), Lambert outlines three attempts to prove this property (§20, 21, 22) and then acknowledges that he does not succeed. We know that this property does not hold if the parallel axiom is not satisfied. Indeed, in neutral geometry, the angle $\widehat{AGF}$ has an upper limit, called the *Lobachevsky parallelism angle* (Fig. 7.5), whose value depends only on the length of the segment

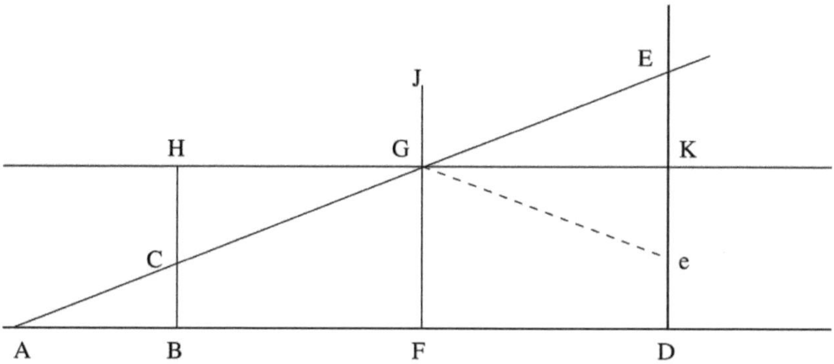

Fig. 7.4 Figure extracted from §16 of Lambert's memoir

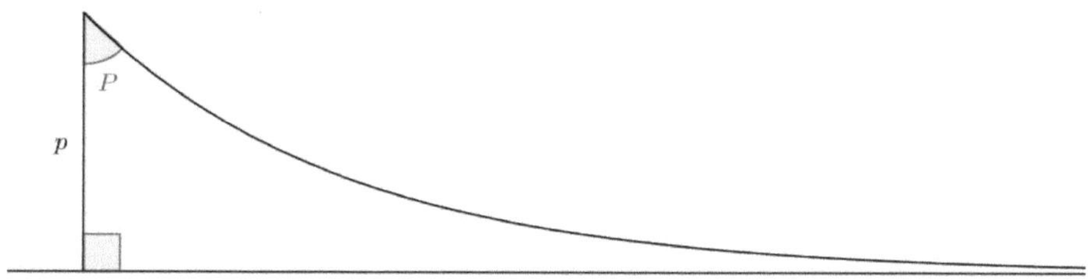

Fig. 7.5 P is the angle of parallelism of the segment p

GF, and which is equal to a right angle if and only if the parallel axiom is satisfied. Under the negation of the parallel axiom, this upper limit is strictly smaller than a right angle. The reader may refer to the English edition of Lobachevsky's *Pangeometry* [23], where Lobachevsky's angle of parallelism plays a paramount role.

In §23, Lambert starts a study of quadrilaterals. He considers a quadrilateral $CBDE$ in which the angles B and D are right angles, and he calls G the intersection point of EC with the perpendicular to BD raised at the point F, the midpoint of this segment (Fig. 7.6). He proves that we have $CB < DE$ if and only if the angle $\widehat{CGF}$ is acute.

In §23, 24 and 25, Lambert studies the monotonicity of the distance function between a point moving on a straight line and a second straight line. We know that this monotonicity requires the parallel axiom, and in fact, it equivalent to it. Lambert eventually concludes (end of §25) that he cannot prove it without this axiom.

In §26, Lambert proves the following fact (without resorting to the parallel axiom). Consider two lines DJ and CL and a point B moving on DJ, from which we draw a perpendicular BA onto CL (Fig. 7.7). If the angle $\widehat{DBA}$ is always acute (regardless of the position of point B) then this angle is constant.

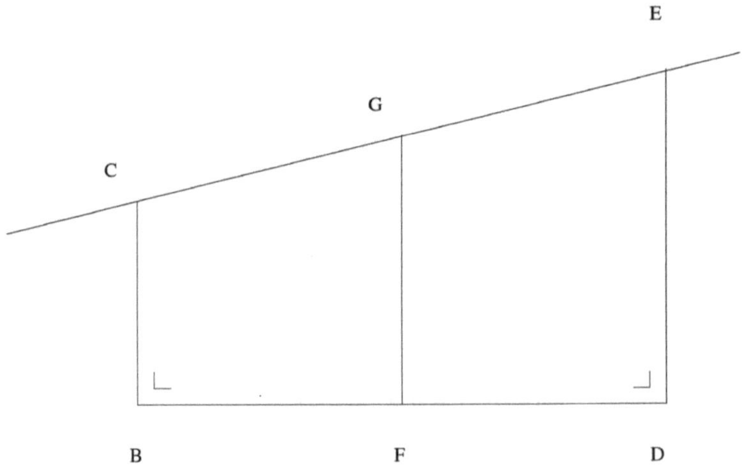

Fig. 7.6 Figure used in §23 of Lambert's memoir

Fig. 7.7 Figure used in §26 of Lambert's memoir

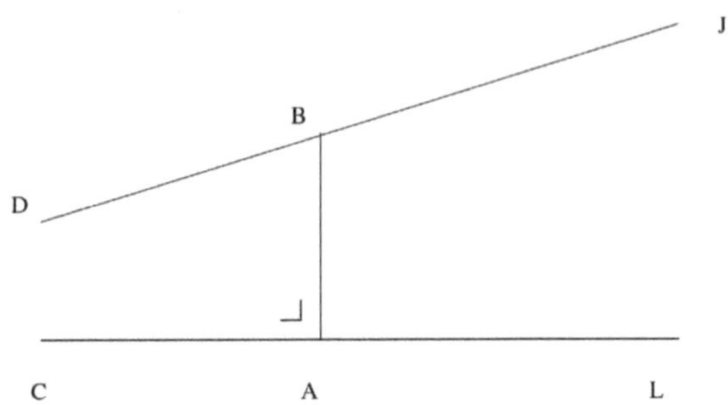

7.5 Part 3 of Lambert's Memoir

The third part of Lambert's memoir (§27 to 88) is the longest. It is titled *Theory of parallel lines*, which is also the title of the whole memoir. At the beginning of this part (§27), Lambert states that his intention is to present a theory that would find its place immediately after Proposition 28 of Book I of Euclid's *Elements*, that is, that would be a continuation of the part of these *Elements* that does not need the parallel axiom. In this way, Proposition 29 and the ones that follow it, instead of resorting to the parallel axiom, would use Lambert's theory.

Lambert states that he considers that certain assertions made by his predecessors, such as the one on the equidistance of parallel lines, are not proved, and that he intends to prove them after having shown the difficulties contained in the statements of these propositions. He first summarizes the approach he will follow, whose central stage consists of successively examining three hypotheses, which will be that the fourth angle of a trirectangular quadrilateral is respectively right, obtuse or acute, and to prove that the second and third hypotheses contradict certain

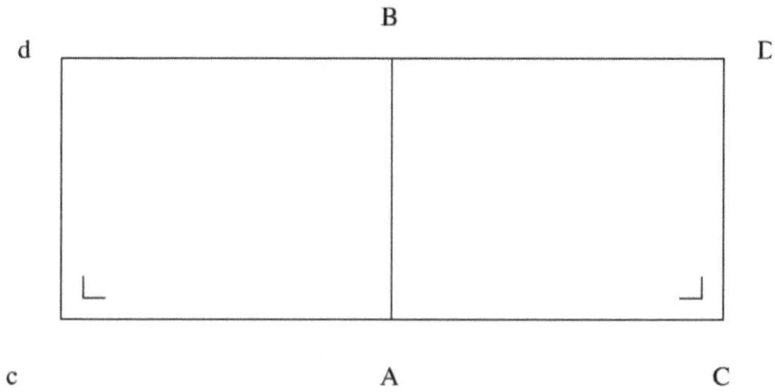

Fig. 7.8 The birectangular isosceles quadrilateral $cCDd$ is obtained by folding the trirectangular quadrilateral $ABCD$ along the side AB; the angles at c and C are right and we have $cd = CD$

axioms of neutral geometry. Trirectangular quadrilaterals were later given the name *Lambert quadrilaterals* or *Ibn al-Haytham–Lambert quadrilaterals*. Indeed, the Arab mathematician Ibn al-Haytham had already, more than seven centuries before Lambert, reduced his approach to the theory of parallels to the analysis of such quadrilaterals. In view of the importance of this topic, we have included, in Sect. 7.7 of this chapter, a detailed discussion of quadrilaterals in the theory of parallels.

In §28 to 38, Lambert recalls some properties of parallel lines that are contained in the *Elements* and some others that easily derive from them. He presents some simple properties of trirectangular quadrilaterals in neutral geometry. In §39, he states again the three hypotheses relating to the trirectangular quadrilateral $ABCD$, represented in Fig. 7.8, in which the three angles A, B and C are right. This figure taken from Lambert's text also represents the isosceles birectangular quadrilateral[8] $cCDd$ in which $cd = CD$, obtained by folding the quadrilateral $ABDC$ along the side AB.[9]

Lambert's three hypotheses are then the following:

(1) $\widehat{BDC} = 90^{\circ}$ (*the right angle hypothesis*) ;
(2) $\widehat{BDC} > 90^{\circ}$ (*the obtuse angle hypothesis*) ;
(3) $\widehat{BDC} < 90^{\circ}$ (*the acute angle hypothesis*).

The first hypothesis is examined in Sections 40 to 51, the second one in Sections 52 to 64, and the third one in Sections 65 to 88.

[8] Birectangular isosceles quadrilaterals are also called *Khayyām–Saccheri quadrilaterals*; see the review we make of quadrilaterals in Sect. 7.7.

[9] Note that in this figure, as in the other figures, Lambert uses the following principle of symmetry in the choice of letters: c and d are the symmetrical points of C and D by the reflection with respect to the line AB. The use of a related notation is already found in the writings of Archimedes. The rule is practical and it indicates that if we exchange all the capital letters with the corresponding lowercase letters, the figure remains valid.

Under the first hypothesis, the quadrilateral $cCDd$ is a rectangle (i.e., a quadrilateral with four right angles; note that this implies that opposite sides are pairwise congruent). Lambert shows that in this case, for any isosceles birectangular quadrilateral (and not only the quadrilateral $cCDd$ on which we made our hypothesis), the perpendiculars drawn from any base to the opposite side are perpendicular to both sides and all have the same length (§41 and 42).

Lambert then proves (§43 to 49) that any straight line L passing through a point on one side of the quadrilateral and making an oblique angle with this side necessarily intersects the opposite side. His proof is based on a monotonicity property we state as follows: by placing side by side a sequence of rectangles congruent to $ABCD$ on a common base, the lengths of the various intersections of the straight line L with the sides crossed in the successive quadrilaterals decrease by a quantity which is bounded from below (stated in §47). The result proved implies that the parallel axiom is valid under the hypothesis of the right angle. The aim of the rest of the memoir is to prove that hypotheses (2) and (3) are not compatible with Euclid's axioms deprived of the parallel axiom.

In Sections 48 to 51, Lambert tries to reverse the arguments and he explains the difficulties encountered in this process. In doing so, he proves a result, which he considers as remarkable, saying that if the first hypothesis holds in one rectangle then it holds in any rectangle. In fact, in doing this, Lambert proves a series of results that the seventeenth–eighteenth century mathematician Giovanni Girolamo Saccheri[10] had obtained in his treatise [39], where he makes three similar hypotheses about triangles. Saccheri proved that if in a particular triangle one of the three hypotheses is valid, then it is also valid in any triangle. The result was later called *Saccheri's theorem* by Bonola in [3].

In Sections 52 to 54, Lambert proves some elementary properties of the trirectangular quadrilateral under the obtuse angle hypothesis. He shows in particular that each of the two sides surrounded by right angles is longer than the opposite side. He then considers (Sections 55 to 57) perpendiculars drawn onto the line containing a side adjacent to two right angles from a sequence of points located on the line containing the opposite side and which are outside the rectangle, on the side of the acute angle and extending further than this acute angle (Fig. 7.9).

He proves that the sequence of lengths of these perpendicular segments is monotonically decreasing, and that the difference between one segment and the previous one is smaller the further apart the points are. It may be recalled here that

[10] Giovanni Gerolamo Saccheri (1667–1733) was a grammar teacher before becoming interested in mathematics, probably under the influence of his teacher and friend the Jesuit priest Tommaso Ceva, brother of the famous mathematician Giovanni Ceva. Saccheri entered the Jesuit order in 1685 and was ordained in 1694. He was appointed professor of philosophy at the University of Turin that same year, and three years later at the University of Pavia, where he taught philosophy, theology and mathematics. He took a close interest in Euclid's *Elements*, following the advice of his first teacher, Ceva. He held this position until his death. Like other great mathematicians, he tackled the question of parallels, and he attempted—a victim, like all his predecessors, of the prejudices of his time—to prove that the parallel postulate is a theorem.

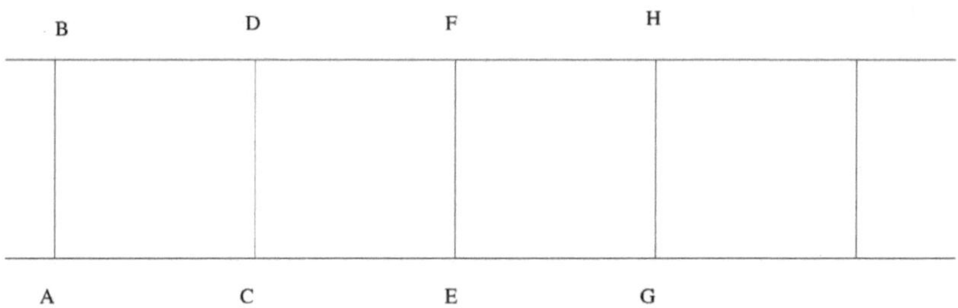

Fig. 7.9 At the points A and B and at points C, E, G, ... the angles are right angles. Under the second hypothesis, the segments CD, EF, GH, ... decrease more rapidly than linearly with respect to the distances from points C, E, G, ... to point A. Similarly, the angles at D, F, H, ... become increasingly obtuse

such monotonicity properties for lengths in spherical geometry are well known, and can be found in the *Spherics* of Theodosius of Tripoli (c. 60-90 BC) and in those of Menelaus of Alexandria (c. 70–140 AD).[11]

After this non-linear monotonicity property for segment lengths, Lambert proves (§58 to 60) a monotonicity property for the angles that the perpendiculars make with the line on which the point from which they are issued is located. These angles become more and more obtuse (Fig. 7.9 again). Monotonicity properties of angles are also known in spherical geometry.[12]

Lambert deduces from this that the lengths of the perpendicular segments cannot become asymptotically shorter (§62) but that these lengths become zero at a certain point, and consequently the line containing the points from which the perpendiculars are drawn must intersect the line containing the opposite side of the quadrilateral. This contradicts the Euclidean axiom which says that two lines cannot intersect at more than one point (§64). The second hypothesis is therefore invalid.

---

[11] The *Spherics* of Theodosius and those of Menelaus are two comprehensive treatises on spherical geometry, that constitute the most important works on this subject that have come down to us from Ancient Greece. It should be noted that both treatises (and especially the second) are very rich from the mathematical point of view. Greek manuscripts of Theodosius' treatise survive, and were the bases of modern editions, but all the Greek versions of Menelaus treatise are lost, and this treatise has come down to us only in the form of Arabic translations. The most faithful edition of Theodosius' *Spherics* remains the one in French, by P. Ver Eecke [43], and it contains valuable notes. Concerning Menelaus' *Spherics*, there is a German translation by M. Krause, based on the Arabic text of Ibn 'Irāq [18], and a recent critical edition with an English translation based on the Arabic translations of al-Māhānī and al-Harawī, by R. Rashed and A. Papadopoulos [33]. The treatises of Theodosius and Menelaus are very different in their approach and in the techniques used. The proofs of the propositions of Theodosius *Spherics* are based on the Euclidean geometry of the ambient space of the sphere considered as embedded in three-dimensional space, whereas those of Menelaus, except for a very small number of propositions, are intrinsic. The monotonicity theorems on lengths to which we refer here are Propositions 6 to 10 of Book III of Theodosius' *Spherics*, and there are a large number of them in Menelaus' *Spherics*, including Propositions 45–61 and 81–91 of al-Māhānī/al-Harawī's version [33].

[12] See Propositions 84–86 of Menelaus' *Spherics* [33].

Lambert was aware of the fact that his second hypothesis is valid on the sphere, but he would only state this in §82, when analyzing his third hypothesis, and when he would state (in comparison) that the latter is realized on a sphere of imaginary radius.

Starting in §65, Lambert examines the third hypothesis, that of the acute angle. He begins by considering a trirectangular quadrilateral (whose fourth angle is therefore acute) and proves (§66 and 67) that each of the two sides adjacent to the acute angle is greater than the side opposite to it. He then shows (§68) that each perpendicular drawn from the extension of one side of the acute angle to the opposite side is shorter than a perpendicular drawn from a point on the same line but further away. Just as under the hypothesis of the obtuse angle, Lambert studies (§69) the monotonicity of the angles formed by such lines. These angles become more and more acute.

Then, as in the previous case, Lambert shows (§70) that the lengths of the perpendiculars not only become larger and larger, but that they can become larger than any value given in advance, whereas under the obtuse angle hypothesis, such strict monotonicity was sufficient to contradict one of the axioms of the *Elements*. From this point on, Lambert rediscovers a property that Saccheri had already emphasized and which is specific to hyperbolic geometry (which, let us recall, was for Lambert, as for Saccheri, only hypothetical): the existence of disjoint lines having a common perpendicular and diverging from each other on each side of this perpendicular. Lambert is struck by this property which makes him imagine lines that do not resemble the usual lines. However, he refuses to be convinced by these representations.

Note that if Lambert had considered, like many of his predecessors, that parallelism implies equidistance between lines, or that equidistance defines parallelism, the work he has undertaken here would already be complete. But Lambert seeks another argument.

In §72, Lambert discusses two properties that could hold under the acute angle hypothesis. They are related to Fig. 7.10, where we have two lines AE and BH with a common perpendicular AB and where the angles at the sequence of points $E, F, G, \ldots$ are right angles. The quadrilaterals $BAEH, BAFJ, BAGK, \ldots$ are trirectangular. The first possibility mentioned by Lambert here is that it is conceivable that when the points $E, F, G, \ldots$ are sufficiently far from the point A,

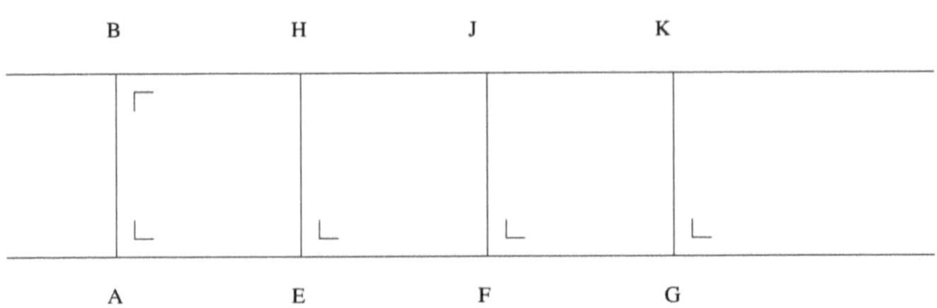

Fig. 7.10 Figure used in §72 and 83 of Lambert's memoir

the perpendicular raised at these points does not meet the line BH. The second possibility is that the angle at $H, J, K, \ldots$ which, as we have already seen, decreases when the point moves away from A, becomes eventually zero, not only at the limit, but for a given point on BH.

In §73 and §74, Lambert proves that under the acute angle hypothesis, the sum of the three angles of each triangle is $< 180^{\circ}$.

In §76 and §77, he shows that in an equilateral triangle ABC, if F is the midpoint of BC and D the intersection point of the medians of this triangle, then we have $DF < \frac{1}{3}AF$. It is easy to see that in neutral geometry, the three medians of an equilateral triangle are concurrent. The case of an arbitrary triangle is more difficult to handle, but Lambert does not discuss it. He notes at the same time that under the obtuse angle hypothesis, we would have $DF > \frac{1}{3}AF$ (and it is well known that $DF = \frac{1}{3}AF$ under the right angle hypothesis).

Lambert notes in §79 that there is a series of results, analogous to the one just stated and to several others that we saw earlier (monotonicity of lengths, angles, etc.), valid under the hypothesis of the acute angle and obtained by reversing the direction of the conclusions of the corresponding results obtained under the hypothesis of the obtuse angle. He notes, however, that, despite this resemblance in the results, the third hypothesis (that of the acute angle) does not lead as easily as the second one (that of the obtuse angle) to a contradiction.

According to Lambert, one of the most important consequences of the third hypothesis is the existence of an absolute measure for lengths, areas, and volumes. He proceeds (§80) as follows. We start from the fact that there is an absolute measure for angles.[13] From this, Lambert notes that there is an absolute measure for the areas of isosceles trirectangular quadrilaterals, i.e., those for which the two sides not adjacent to the acute angle have the same length. Such a quadrilateral is completely determined by the value of the acute angle, and the area of this quadrilateral depends solely on the value of this angle. Thus, the value of this angle is an absolute measure of the area of the quadrilateral. Lambert states that "this consequence has such a force of attraction that it would easily force us to wish that the third hypothesis were true!", but he notes that if this were true, then "trigonometric tables would become interminably long,[14] the similarity and proportionality of the figures would have

[13] Declaring that the total angle at a point is equal to four right angles does not depend on any pre-chosen standard measure. It is in this sense that we have an absolute measure for angles. Using the fact that the lengths of the sides of a triangle are completely determined by their angles, we obtain an absolute measure for lengths. Several years after Lambert, Gauss had also noted this fact. In a letter to Gerling, dated April 11, 1816 (see [11, Vol. VIII, p. 168]), Gauss wrote: "It would have been desirable if Euclidean geometry were not true, because we would have a universal measure a priori. We could use the side of an equilateral triangle with angles 59°59'59",9999.....as a unit of length." [z.B. könnte man als Raumeinheit die Seite desjenigen gleichseitigen Dreiecks annehmen, dessen Winkel = 59°59'59",9999.....]. We recall that in hyperbolic geometry, the common value of the angles in an equilateral triangle is always strictly less than 60°, and this is why Gauss chose such a value.

[14] In non-Euclidean geometry, in a right triangle, the ratio of the length of one side to that of the hypotenuse does not only depend on the angle between these two sides; it also depends on the size

completely disappeared ; no figure could be represented other than in its absolute size."

In §81 and 82, Lambert studies the area function of triangles. He shows that the area of a triangle can be defined as its *angular defect*, that is, the difference between 180° and the sum of the angles. He also points out that under the second hypothesis (that of the obtuse angle), the area of a triangle can be defined as its *angular excess*, that is, the excess over 180° of the sum of the three angles of the triangle.[15] Lambert states that "we should almost conclude that the third hypothesis occurs for an imaginary sphere."

This remark about the imaginary sphere is interesting in several respects, and can be considered prophetic in a certain sense. Indeed, the area of a spherical triangle, on a sphere of radius r, is equal to the angular excess of this triangle, that is, to the quantity $r^2(\alpha + \beta + \gamma - \pi)$ where α, β, γ are the values of the three angles of the triangle. If instead of the radius r we take an *imaginary radius* and multiply this value by the imaginary number $\sqrt{-1}$, we obtain, for the area of a triangle, $r^2(\alpha + \beta+\gamma-\pi) = (\sqrt{-1})^2(\alpha+\beta+\gamma-\pi) = -(\alpha+\beta+\gamma-\pi) = \pi-\alpha-\beta-\gamma$, which is the area of a triangle in the hyperbolic plane whose angle values are α, β, γ. It may be noted here that Lambert, in his *Trigonometric Observations* [19], highlights the analogy between the trigonometric functions sine and cosine and the hyperbolic sine and cosine functions, noting that the hyperbolic cosine of an arc can be considered as the cosine of an imaginary arc, thus developing a kind of trigonometry in which the sides of triangles are considered as imaginary. But he does not seem to have made the connection between this and a geometry where the sum of the angles in a triangle is smaller than 180°, even though he was very close to these ideas.

It was not until Franz Adolph Taurinus (1794–1874) and then Lobachevsky that the statement on the imaginary sphere in relation with the area of non-Euclidean triangles was formulated more explicitly. Taurinus is the author of another *Theorie der Parallellinien* also reproduced in the volume by Stäckel and Engel which contains Lambert's text on the theory of parallels, in which Taurinus also attempts to prove Euclid's parallel axiom from the other axioms. Also seeking a contradiction in the negation of this axiom, Taurinus ends up, like Lambert, developing the elements of a non-Euclidean geometry. We owe to Taurinus one of the first attempts to establish a parallel between the trigonometric formulae of the three geometries: spherical, Euclidean and hyperbolic (the existence of the latter having been, for him, as for Lambert, purely hypothetical). In his memoir, Taurinus obtained fundamental formulae of hyperbolic trigonometry by working

---

of the triangle. It is in this sense that we understand the statement that the trigonometry tables, in hyperbolic geometry, are of infinite length.

[15] This theorem is attributed to Albert Girard (1595–1632), who stated it in his *New invention in algebra* (1629). Lagrange, mentions, in his memoir *Solution of some problems relating to spherical triangles with a complete analysis of these triangles* (1800) translated in Chapter 18 of the present volume, that Girard's proof is not sufficient and he considers that this theorem should rather be attributed to Cavalieri, who gave a complete proof in his *Directorium generale uranometricum* (General guide to celestial measurements), Bologna, 1632.

on a sphere of imaginary radius, calling the geometry obtained in this way a *logarithmic-spherical geometry* (*Logarithmisch-sphärischen Geometrie*). He noted a formal transition between the formulae of hyperbolic geometry and those of spherical geometry which consists in replacing in the spherical formulae certain trigonometric functions (sin and cos) by hyperbolic functions (sinh and cosh). It can also be noted that, in the same way, Lobachevsky, in his *Elements of geometry* (1829–1830) and in his *Geometrical research on the theory of parallels* (1840), highlights a passage between the formulae of spherical trigonometry and those of hyperbolic trigonometry which consists of replacing, in the formulae of spherical trigonometry, the lengths of the sides a, b, c of a triangle by the imaginary quantities $a\sqrt{-1}, b\sqrt{-1}, c\sqrt{-1}$. Note also that the transition from spherical trigonometry formulae to hyperbolic trigonometry formulae involved the replacement of certain trigonometric functions with hyperbolic functions, and that Lambert was one of the first to systematically use hyperbolic functions. Finally, note that Lambert was among the mathematicians who developed the geometric foundations of the theory of complex numbers.

Let us now return to Lambert's *Theory of the parallel lines*.

In §83, Lambert proves a property he had already mentioned in §72, which states (using the notation in Fig. 7.10) that the perpendiculars raised at points $E, F, G, \ldots$ to the line AE do not necessarily meet the line BH, when the point where the perpendicular is raised is far enough from A.

The last paragraphs of the memoir (§84–88) contain sketches of ideas for proving that the acute angle hypothesis leads to a contradiction, but Lambert acknowledges that most of them are unsuccessful. In the last paragraph, he mentions a new attempt, in passing over several details, and the text does not allow us to know whether he himself was convinced of this or not.

7.6 Excursus: On the Theory of Parallel Lines in History

Greek Antiquity

We recalled in Sect. 7.3 that Lambert refers to Proclus regarding the fifth postulate. Let us quote Proclus' passage in question (Proclus 192, [32, p. 150–151]):

> [The fifth postulate] ought to be struck from the postulates altogether. For it is a theorem—one that invites many questions, which Ptolemy proposed to resolve in one of his books—and requires for its proofs a number of definitions as well as theorems. And its converse is proved by Euclid himself as a theorem. But perhaps some persons might mistakenly think that this proposition deserves to be ranked among the postulates on the ground that the angles being less than two right angles makes us at once believe in the convergence and intersection of the straight lines. To them Geminus has given the proper answer when he said that we have learned from the very founders of this science not to pay attention to plausible imaginings in determining what propositions are to be accepted

> in geometry. Aristotle likewise says that to accept probable reasoning[16] from a geometer is like demanding proofs from a rhetorician.[17] And Simmias is made by Plato to say, "I am aware that those who make proofs out of probabilities are impostors."[18] So here, although the statement that the straight lines converge when the right angles decrease is true and necessary, yet the conclusion that because they converge more as they are extended farther they will meet at some time is plausible, but not necessary, in the absence of an argument proving that this is true of straight lines. That there are lines that approach each other indefinitely but never meet seems implausible and paradoxical, yet it is nevertheless true and has been ascertained for other species of lines. May not this, then, be possible for straight lines as for those other lines? Until we have firmly proved that they meet, what is said about other lines strips our imagination of its plausibility. And although the arguments against the intersection of these lines may contain much that surprises us, should we not all the more refuse to admit into our tradition this unreasoned appeal to probability?

Let us say a few more words on Proclus' own contribution. We have repeatedly pointed out that his work contains valuable information on the history of Greek geometry. It is not certain that Proclus obtained original results, but a "proof" of the parallel axiom in his *Commentaries on Book I of Euclid* is attributed to him. Not being satisfied with the proofs of this axiom given by his predecessors, Proclus gave his own proof, based on the hypothesis that if two coplanar lines are disjoint, their distance is bounded, a hypothesis which, as already noted, is equivalent to the parallel axiom. Based on work by Geminus, Proclus conceived the possibility of a geometry involving lines that are not necessarily straight but which can be asymptotes. We have already quoted a passage in which Proclus discusses this question; see also his commentary on Definition 35, p. 153 et seq. of [31]. All this has led some commentators to consider Proclus as one of the precursors of hyperbolic geometry.

We also noted that doubts concerning the parallel axiom had already been raised before Euclid. Aristotle (384–322 BC) discusses the theory of parallels and the difficulties posed by this theory in the *First Analytics* (II, 16, 65a 4;II, 17, 66a 11–15) and the *Second Analytics* (II, 15, 65a 4–7). He declares there that there is a vicious circle in the theory of parallels, that is, something that is impossible to prove, "because of a difficulty in the definition of parallels". Archimedes (ca. 287–212 BC) wrote a treatise, *On parallel lines*, which does not survive but which is quoted by the Arab bibliographer al-Nadīm in his *Book of Bibliography* (Kitāb Fihrist).

We know from Proclus that Heron of Alexandria (10-70 AD) admitted only three axioms; they are quoted in [31, p. 173]. Proclus writes: "However, we should not reduce the number of axioms to a minimum like Heron, who only set three". In the same passage, Proclus. tells us that Pappus of Alexandria (ca. 290–350) added other axioms to those of Heron.

Ptolemy (90–168) wrote a treatise on parallels, titled *On the intersection of straight lines*, in which he gave a proof of the parallel axiom, based on the

---

16 The expression "probable reasoning" here refers to the fact that lines would "eventually" intersect, if they are sufficiently extended.

17 Aristotle, *Nicomachean Ethics*, 1094b26f.

18 Plato, *Phaedo* 92d.

assumption that two parallel lines intersecting a third make equal alternate interior angles with it. This assertion is equivalent to the parallel axiom. He attempted to prove Euclid's proposition I.29 without using the parallel axiom. This work has not reached us, but this proof is reported by Proclus.[19]

Attempts by the Neoplatonic philosopher Simplicius (c. 490–c. 560) to prove the parallel postulate have come down to us in the form of extensive excerpts in the writings of the tenth-century mathematician Al-Nayrīzī. This brings us to the Arab mathematicians.

The Arabs

In the eighth century AD, an Arabic-language science was born. It quickly reached a high degree of maturity and remained alive for about six centuries, with some very talented mathematicians, several of whom involved in the problem of parallels. We shall review some of their works on this subject. To tell the story in a few simple words, the Byzantines, who had ordered the closure of the Philosophical schools of the Empire, had at the same time condemned Hellenistic mathematics, which was taught in the Philosophical schools. Taking advantage of this opportunity, from the eighth century onwards, several Arab missions were sent to the Byzantine court with the aim of acquiring copies of ancient mathematical manuscripts. Because of their significant interest in the Greek's mathematics and philosophy, the Arabs became their natural heirs. They studied Aristotle, for whom they had great respect, and whom they often called "the philosopher" or "the wise". They extensively translated and commented his works along with those of Greek mathematicians, and the problems raised by the question of parallels naturally emerged in these works.

'Abbās Ibn Sa'īd al-Jawharī (ca. 800–860), who worked in Bagdad and Damascus, is one of the first known Arab mathematicians who studied the parallel problem. Naṣīr al-Dīn al-Ṭūsī, in his *Rectification of the book of the Elements*, describes an attempt of a proof of the fifth postulate by al-Jawharī, based on the fact that through any point inside an angle whose value is strictly between 0 and 180°, we can draw a line that intersects both sides. We know that this property is equivalent to the parallel axiom and therefore does not follow from the other axioms. Al-Ṭūsī described new propositions added by Al-Jawharī to Euclid's *Elements*, see Rosenfeld [37] p. 47.

We also mention Thābit Ibn Qurra (826–901), an Arab mathematician whose mother tongue was Syriac and who worked in Mesopotamia (Harrān, then Baghdad). He translated treatises of Apollonius, including several books of the *Conics*, as well as works of Archimedes, Euclid and Ptolemy. He wrote at least two original treatises on parallels, one of them titled *On the Proof of Euclid's Famous Postulate* and another one titled *Book on the fact that if a straight line falls on two straight lines and makes with them and on the same side less than two right angles, the*

[19] See [31] p. 312 et seq.

two straight lines, if extended on that side, meet. Each of these treatises contains an (attempted) proof of the parallel axiom. The first proof is based on the assumption that if two lines diverge on one side, they converge on the other, and the second is based on the existence of equidistant lines. We know that both hypotheses are equivalent to the parallel axiom. The two treatises of Thābit Ibn Qurra have come down to us. The reader is referred to the paper by Rashed and Houzel [35] which contains a French translation of these papers as well as a commentary and references to previous works. The reader may also consult Rosenfeld [37]. According to Rashed and Houzel [35], Ibn Qurra was the first to give a rigorous proof of the symmetry of the equidistance relation (i.e., if all points of a line d are equidistant from a line d', then all points of d' are equidistant from d), once the existence of equidistant lines is admitted (which Thābit Ibn Qurra attempted to prove in the second treatise we just mentioned).

We also mentioned Al-Nayrīzī, the tenth-century mathematician who wrote a *Treatise on the proof of Euclid's famous postulate*. He used the notion of equidistance as a definition of parallelism. Al-Nayrīzī quotes the Arabic translation of a text by Simplicius in which the latter reports on a proof of Aghānis,[20] having also "proved" the parallel axiom using the notion of equidistance and having further assumed that this notion is symmetric.

Other mathematicians of the tenth-century include Yuḥannā Ibn Yūsuf al-Quss (died c. 980), an Arab Christian priest who wrote a book dedicated to the Emir of Aleppo Sayf al-Dawla titled *On the intersection of two straight lines rising from a straight line and making angles with it less than two right angles*, and Abū ʻAlī Ibn Sīnā (981–1037), the famous philosopher, physicist, mathematician, physician, astronomer and music theorist, known in the Western literature under the name Avicenna. Ibn Sīnā's famous treatise, *The book of science*, contains a chapter on geometry, in which parallel lines are defined using the concept of equidistance.

Then one has to mention the great mathematician Abū ʻAlī Ibn al-Haytham (d. after 1040), known in Latin as Alhazen.[21] He is the author of several treatises on geometry, including the *Book explaining the postulates of Euclid's book on the Elements* and the *Book on the resolution of doubts raised by Book I of Euclid's Elements*. In the first treatise, Ibn al-Haytham calls "parallel to a given line" the locus of the end of a segment that moves perpendicularly to the given line while maintaining a constant length. He proves that this locus is a straight line. His "proof" of the parallel axiom is based on the consideration of a quadrilateral with three right angles,[22] see Fig. 7.11.

[20] Aghānis is probably an Arabized name of a fifth-century Greek mathematician, who knew Simplicius.

[21] Born in Basra and working in Cairo, Ibn al-Haytham was an engineer, physicist, and astronomer. He is sometimes called the father of modern optics. His treatise on optics has long been known in the West thanks to Gerard of Cremona's translation. A crater on the moon bears his name.

[22] Pont and Rosenfeld call such a quadrilateral, which Ibn Qurra had already considered, *Ibn al-Haytham–Lambert quadrilateral*, see [30, p. 171] and [37, p. 65]. We use both terminologies.

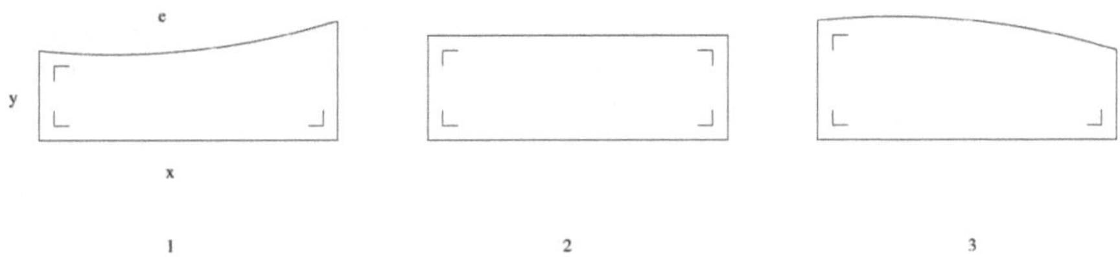

Fig. 7.11 Trirectangular quadrilaterals. In the figure on the left (hyperbolic geometry), the fourth angle is acute; in the middle figure it is right (Euclidean geometry), and in the right figure it is obtuse (spherical geometry)

Ibn al-Haytham analyzes the three cases. Like Lambert in his treatise, he "proves" that only the first hypothesis does not lead to a contradiction. This analysis is reviewed by Youschkevitch in [46, p. 117], Rosenfeld [37, p. 59] and Pont [30, p. 169]. With this work, Ibn al-Haytham was a direct precursor of Lambert, who followed exactly the same approach (except for the definition of parallels). We refer to [34, vol. V] for the translation and commentaries of the mathematical works of Ibn al-Haytham.

Next, we must mention 'Umar al-Khayyām, the mathematician born in Nayshābūr in 1048 and who died in the same city in 1131. He was also a philosopher and was interested in the philosophical foundations of mathematics. He studied the works of Ibn al-Haytham. According to Rashed (Introduction to [36]), Khayyām developed a geometric theory of equations that can be considered as the beginning of algebraic geometry. Khayyām completed his *Commentary on the difficulties of certain postulates in Euclid's work* in 1077. We have already quoted a passage from this work in which the author points out that his predecessors did not correctly address the question of parallels, claiming that they have neglected certain principles that Aristotle had stated.

Khayyam's work consists of three parts. The first, which interests us here, concerns the theory of parallels. The second concerns the theory of proportions, and the third concerns compounded ratios.

In the first part of this book, Khayyām studies (Proposition 3) quadrilaterals with two right angles adjacent to one side, the other two angles being equal. These quadrilaterals are the so-called *birectangular isosceles rectangles* or *Khayyām–Saccheri quadrilaterals*, and we shall discuss them in more detail in Sect. 7.7. The three quadrilaterals in Fig. 7.11 represent the three cases of trirectangular quadrilaterals. In the figure on the left (hyperbolic geometry), the fourth angle is acute; in the middle figure it is right, and in the right figure it is obtuse (spherical geometry).

In the same part, Khayyām makes a detailed analysis of the consequences of the hypothesis on the other two angles, depending on whether they are right, acute, or obtuse. He proves in particular that the hypothesis of the acute angle implies the existence of two straight lines perpendicular to the same straight line and diverging on both sides of this straight line. This is a well-known phenomenon in hyperbolic geometry (but which Khayyam dismisses, considering that it contradicts a first principle). Gerolamo Saccheri, whom we have already mentioned,[23] studied the same quadrilaterals several centuries later, and he obtained the same conclusions.

Finally, let us mention, among the group of Arab mathematicians, Naṣīr al-Dīn al-Ṭūsī, born in Ṭūs (in present-day Iran) in 1201 and died in Baghdad in 1274,[24] of which we have already spoke. Two of his treatises on the theory of parallels are known, the *Treatise which delivers from doubts concerning parallel lines* and his *Redaction of Euclid*, of which a Latin translation and commentary are contained in Vol. 2 of the *Complete Works* of John Wallis, published in Oxford in 1693. Like 'Umar al-Khayyām, Naṣīr al-Dīn al-Ṭūsī made an analysis of the birectangular isosceles quadrilaterals which we already discussed. These quadrilaterals will be mentioned again later in this chapter.

The Renaissance

At the Renaissance, several mathematicians in Western Europe were naturally interested in the problem of parallels. Among them we mention Christophorus Clavius (1538–1612), Giovanni Alfonso Borelli (1608–1679), John Wallis (1616–1703) and Vitale Giordano da Bitonto (1633–1711). All these mathematicians were familiar with the Greek writings on the subject (especially the work of Proclus), and some were also familiar with the writings of the Arabs. The writings of Naṣīr al-Dīn al-Ṭūsī were known in Europe thanks to their Latin translation published by Wallis (see Fig. 7.12). Let us also recall that this translation contains a summary of the work of 'Umar al-Khayyām and Ibn al-Haytham on parallels. It is also known that Borelli collaborated with Abraham Ecchelensis (1605–1664), a Christian Arab (Syro-Lebanese) Catholic philosopher and linguist, to translate from Arabic manuscripts a commentary on books V and VII of Apollonius' *Conics*.

Borelli had access to a large Arabic mathematical literature. The proof he gave of the parallel postulate is essentially that of Ibn Qurra and Ibn al-Haytham, see Rosenfeld, [37, p. 96].

Christophorus Clavius, the German Jesuit mathematician and physicist, translated Euclid into Latin and attempted at least two proofs of the fifth postulate, one of which is based on an argument similar to that of Naṣīr al-Dīn al-Ṭūsī. Clavius gave this proof in 1574 under the assumption that a curve coplanar to a straight line and equidistant from it is also a straight line. Clavius' proof is analyzed by Pont in [30, p. 195–199]. Borelli [4] and Giordano da Bitonto [12] showed that Euclid's axioms, without the one on parallels, imply this equidistance property, after introducing a hypothesis that is equivalent to the parallel axiom, see [3] p. 13 and [30] p. 200–203

[23] See Footnote 10.

[24] Youschkevitch ([46] p. 414) considers that the *Treatise on the complete quadrilateral* of Al-Ṭūsī made its author "the most illustrious scholar of Islam in the field of trigonometry".

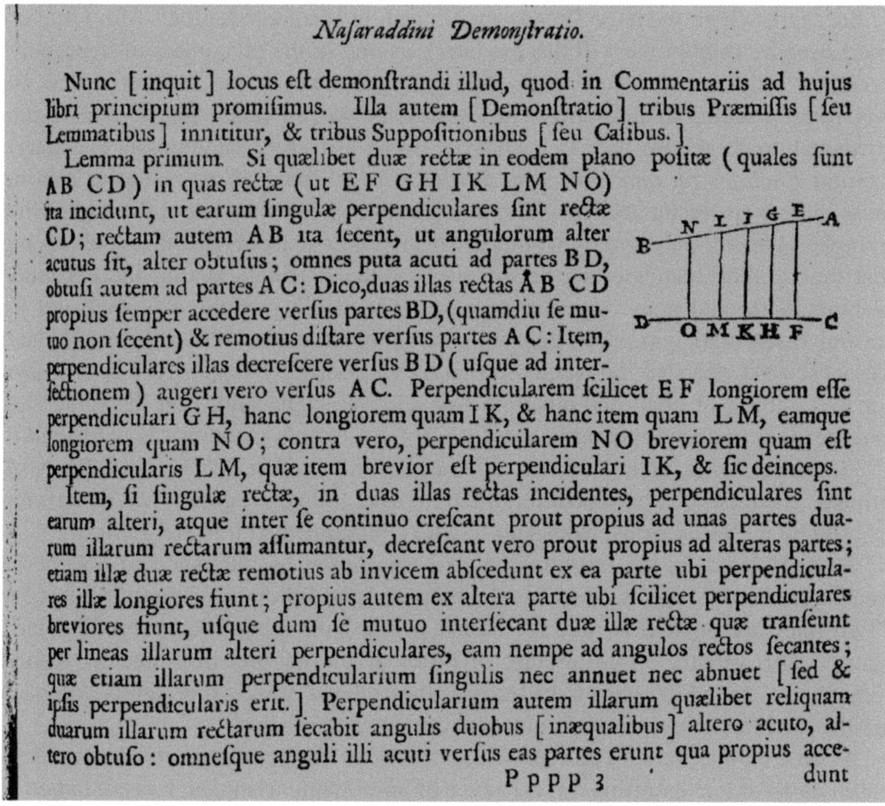

Naſaraddini Demonſtratio.

Nunc [inquit] locus eſt demonſtrandi illud, quod in Commentariis ad hujus libri principium promiſimus. Illa autem [Demonſtratio] tribus Præmiſſis [ſeu Lemmatibus] innititur, & tribus Suppoſitionibus [ſeu Caſibus.]

Lemma primum. Si quælibet duæ rectæ in eodem plano poſitæ (quales ſunt AB CD) in quas rectæ (ut EF GH IK LM NO) ita incidunt, ut earum ſingulæ perpendiculares ſint rectæ CD; rectam autem AB ita ſecent, ut angulorum alter acutus ſit, alter obtuſus; omnes puta acuti ad partes BD, obtuſi autem ad partes AC: Dico, duas illas rectas AB CD propius ſemper accedere verſus partes BD, (quamdiu ſe mutuo non ſecent) & remotius diſtare verſus partes AC: Item, perpendiculares illas decreſcere verſus BD (uſque ad interſectionem) augeri vero verſus AC. Perpendicularem ſcilicet EF longiorem eſſe perpendiculari GH, hanc longiorem quam IK, & hanc item quam LM, eamque longiorem quam NO; contra vero, perpendicularem NO breviorem quam eſt perpendicularis LM, quæ item brevior eſt perpendiculari IK, & ſic deinceps.

Item, ſi ſingulæ rectæ, in duas illas rectas incidentes, perpendiculares ſint earum alteri, atque inter ſe continuo creſcant prout propius ad unas partes duarum illarum rectarum aſſumantur, decreſcant vero prout propius ad alteras partes; etiam illæ duæ rectæ remotius ab invicem abſcedunt ex ea parte ubi perpendiculares illæ longiores fiunt; propius autem ex altera parte ubi ſcilicet perpendiculares breviores fiunt, uſque dum ſe mutuo interſecant duæ illæ rectæ quæ tranſeunt per lineas illarum alteri perpendiculares, eam nempe ad angulos rectos ſecantes; quæ etiam illarum perpendicularium ſingulis nec annuet nec abnuet [ſed & ipſis perpendicularis erit.] Perpendicularium autem illarum quælibet reliquam duarum illarum rectarum ſecabit angulis duobus [inæqualibus] altero acuto, altero obtuſo: omneſque anguli illi acuti verſus eas partes erunt qua propius accedunt

Pppp 3

Fig. 7.12 Page extracted from Wallis' *Complete Works*, a section titled *Nasaraddini Demonstratio* in which the latter sets out the proof of Naṣīr al-Dīn al-Ṭūsī

and 368. Giordano also proved that if three points on a given line are equidistant from another line, then the two lines are equidistant.

Wallis, for his part, showed that the parallel postulate is equivalent to the assertion that there exist, for a given triangle, homothetic triangles that are not congruent to it and that have an arbitrary size [45]. Having considered, like his predecessors, that the parallel postulate deserved a thorough examination, Wallis proposed that, in the teaching of geometry, this postulate be replaced by a postulate stipulating the existence of homothetic figures for arbitrary figures (and not only for triangles).[25] Wallis justified this idea by the fact that one of Euclid's postulates states that there exist homothetic circles of any size.[26]

---

[25] *Opera*, vol. II, Oxford, 1693, p. 667.

[26] Third postulate, Heath's edition of the *Elements* [16, p. 199]: "To describe a circle with any centre and distance." (A more faithful translation would be: "And [let it be postulated] that with

Saccheri, whom we have already mentioned, is considered, along with Lambert and Legendre (whom we will discuss later), as one of the precursors of hyperbolic geometry who came closest to it. Beltrami, in an article on Saccheri titled *Un precursore italiano di Legendre e di Lobatschewski* [2], notes that several results attributed to Legendre and Lobachevsky were already contained in Saccheri's memoir *Euclides ab omni nævo vindicatus* (Euclid freed of every flaw). Among these there is the result affirming that in neutral geometry the sum of the angles in a triangle is smaller than two right angles. This memoir also contains the result saying that the parallel postulate can be replaced by the postulate stating the existence of homothetic and non-congruent triangles. Saccheri's memoir was analyzed by several mathematicians after Beltrami, including Mansion [25], Segre [41] and Veronese [44], see also the recent edition by De Risi [39]. Saccheri is best known for his analysis of quadrilaterals with two adjacent right angles on the same side and the other two angles congruent (Fig. 7.16). Saccheri's memoir was published in 1733, the year of its author's death. It is not known whether Saccheri saw his work printed. The memoir was almost forgotten for over 150 years, until it was discovered in 1889 by the Jesuit priest Manganotti, who communicated it to Beltrami.[27]

G. B. Halsted, in the introduction to a translation he published of Saccheri's treatise (see [39]), reports that Borelli proposed the following definition in 1658: "Two lines are parallel if they are coplanar and have a common perpendicular". Halsted also reports on an attempt of proof of the parallel postulate by Robert Simson in 1756, which starts from the hypothesis that a straight line cannot approach another and then move away from it without intersecting it.[28]

As an example, let us consider Wallis' proof of the parallel postulate, which is based on an axiom asserting the existence of homothetic triangles. It is contained in Wallis' article, and is reproduced by Stäckel and Engel in [42], p. 21–30. It is also reported on by Pont in [30] and by Bonola in [3].

Consider in the neutral plane (i.e. the plane geometry in which all of Euclid's axioms are satisfied except that of parallels, the latter being deleted from the axioms without being replaced by another axiom) two lines l_1 and l_2, meeting a third line l_3 at two points A and B and making with it at these two points, and on the same side, angles a and b satisfying $a + b < \pi$. (These are lines considered in the statement of the parallel axiom; see Fig. 7.13.)

We move continuously the point B on the line l_3 towards the point A, by a motion during which the line l_2 moves at the same time while keeping a constant angle with l_3. When the point B reaches a position B' close enough to A, the line l_2 (in its new position, which we call CP) necessarily meets the line l_1. (This is a result of neutral geometry, which Wallis established before this proof.) Let C be

any centre and any distance a circle can be drawn"; in Greek: "kai panti kentro kai diastémati kuklon gráfestai".

[27] The memoir is cited, however, in Klügel's dissertation (1763) [17], and Lambert was familiar with this thesis.

[28] Such a property of monotonicity is contained in the work of Khayyām (see [36] p. 283).

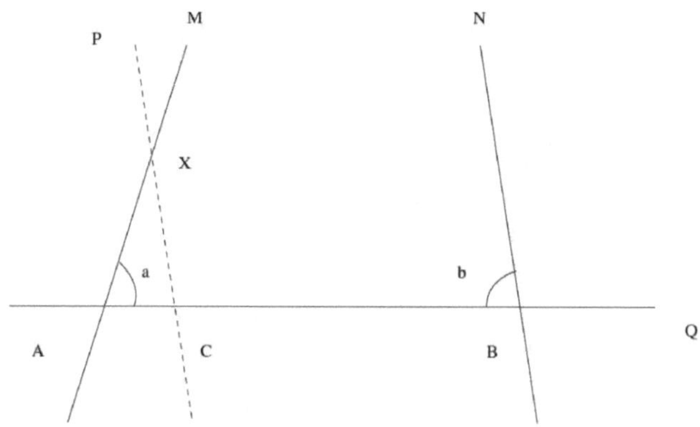

Fig. 7.13 Figure illustrating Wallis' "proof" of the parallel postulate

the intersection point of l_4 with l_1. We consider the triangle $AB'C'$ and we follow the construction that we have just made, in reverse order. That is, now we move the point B' continuously towards B, the line l_4 being carried by the point B' during this motion in such a way that it always makes the same angle with line l_3. The axiom affirming the existence of homothetic triangles says that there exists a triangle ABC homothetic to $AB'C'$ whose side AB' carries the line l_3, and such that the triangle ABC is located on the same side as $AB'C$ with respect to line l_3. This shows that lines l_3 and l_2 intersect (at the point C).

This argument is problematic and in fact, in hyperbolic geometry (which is part of neutral geometry), when we move the point B' and the line l_4 that contains it so that it always maintains the same angle with the line l_3, this line ends up no longer meeting the line l_1.

The Modern Period

The modern period in mathematics can be considered to begin with Leonhard Euler. We have mentioned in [27] two attempts of the latter, as well as one by Lagrange, to prove the parallel postulate. Let us also quote d'Alembert, who, in his works on mathematics, called for simple definitions that are within everyone's reach. He declared that the problem with the question of parallel lines lies in the definitions, writing his famous sentence, that "the definition and properties of the straight line are the pitfall and the scandal of Geometry", see [7, p. 318].

Let us mention now an attempt by Abraham Gotthelf Kästner (1719–1800). Kästner was appointed professor of mathematics at the University of Göttingen in 1756, the year Lambert arrived in that city for a stay that was to last two years. The two men met and remained in contact by letter until Lambert's death. Kästner was interested in the question of parallels, and he himself attempted to prove the fifth postulate. It is likely that Kästner contributed to awakening Lambert's interest in the subject. He had a reputation as a very good teacher and was known for his introductory lectures on various subjects. He supervised the doctoral thesis of Georg Simon Klügel, defended in 1763 [17], in which the latter analyzes twenty-eight attempts to prove the parallel axiom made by his predecessors, including that of Saccheri, exposing the error contained in each of them. In conclusion of his thesis, Klügel puts forward the hypothesis that the postulate is not provable. He writes in particular (translation of Pont [30] p. 463 et seq.):

> Some have tried to exclude this proposition from the list of axioms and have tried to prove it. But these attempts are not free from errors. Other axioms, neither simpler nor more certain, have often been used instead of Euclid's. It thus appears, when we consider all these attempts, that Euclid did the right thing by including among the axioms a proposition that could not be properly established with the help of the other propositions.

Lambert held the same opinion regarding his predecessors; see §3 of his *Theorie der Parallellinien*. He was familiar with Klügel's thesis, which he quotes in his memoir.

Gauss, during his studies in Göttingen, attended Kästner's lectures (but not assiduously; it is said that he found them too elementary). Other students of Kästner included J. M. C. Bartels, who later became a professor at Kazan University where his lectures were attended by Lobachevsky, and Wolfgang Bolyai (1775–1856),[29] who also attended his lectures in Göttingen. Thus, Kästner had a direct or indirect influence on four of the greatest names in the theory of parallels: Lambert and the three founders of non-Euclidean geometry.

The mathematician Johann Friedrich Lorenz (1738–1807) gave a proof of the equivalence between the parallel axiom and the statement saying that a line passing through a point interior to an angle whose value is strictly between 0 and 180° necessarily meets one of the two sides of the angle [24].

The equivalence between the parallel axiom as stated in the *Elements* and the statement that for any line l and for any point A that is not on l, there exists a line l' containing A and disjoint from l, is attributed to the mathematician John Playfair (1748–1819), even though this axiom was already contained in Ibn al-Haytham's *Book on the resolution of doubts raised by Book I of Euclid's Elements* which we already mentioned, cf. Pont [30, p. 171].

The equivalence between the parallel axiom and the fact that there is a quadrilateral with four right angles is contained in the work of Alexis-Claude Clairaut (1713–1765).

[29] Wolfgang Bolyai is the father of János Bolyai, one of the three inventors of hyperbolic geometry. He was himself a talented mathematician, poet, and violinist, and a close friend of Gauss. He was interested in the foundations of geometry and, like many geometers before him, he tried to prove the parallel postulate. He eventually concluded that this was an extremely difficult problem, and when he learned that his son Janos was working on it, his first reaction was to advise him to abandon this research, see [3, 13–15, 26]. He is credited with a proof that the parallel axiom is equivalent to the statement that three non-aligned points always lie on a circle. Rosenfeld points out in [37, p. 109] that Wolfgang Bolyai, who had not realized the importance of his son's discovery, himself published, after this discovery (and therefore after Lobachevsky's), a proof of the parallel postulate.

Pierre-Simon de Laplace (1749–1827), in his *Exposition of the System of the World*, proposed replacing the parallel axiom with an axiom of similarity, much more natural than Euclid's, see [8, p. 386] and [9, Vol. VI, p. 472]. Lazare Carnot (1753–1823) made the same proposal in his *Géométrie de position* ([6] p. 481, in a note).

Adrien-Marie Legendre (1752–1833) is also known for his work on the theory of parallels. In the third edition of his *Elements of Geometry* (published in 1800), he provides a proof of the theorem saying that in neutral geometry the sum of the angles in any triangle is $\leq \pi$, without using the parallel postulate [20, Proposition 19 of Book I]. This theorem, which is already found in Saccheri's memoir, is often still called Legendre's first theorem. The reason this result is attributed to Legendre is that, as already noted, Saccheri's work had been forgotten for more than 150 years, and only reappeared in 1889. Saccheri's proof of this theorem is reproduced in Bonola [3, p. 56 of the English translation]. Gauss also states the Saccheri–Legendre theorem, with a proof analogous to that given by Legendre (see [11], Vol. VIII). The reader is referred to [1] for the place this theorem occupies in the development of neutral geometry and for an exposition of the proof Legendre gave of it.

It should be noted here that in the same edition of his *Elements of Geometry*, Legendre introduced a "proof" of the fact that this sum cannot be strictly less than two right angles, again without using the axiom of parallels. (We know that this proposition is equivalent to the axiom of parallels). Legendre himself acknowledged his mistake later, in an article he published in 1833, the year of his death; see [21, p. 371–372].

Legendre is considered, along with Lambert and Saccheri, to have been one of the closest to the discovery of hyperbolic geometry. It is interesting to read what Lobachevsky himself wrote about Legendre [22, p. 7]:

> In the 1833 *Mémoires de l'Académie de France*, [Legendre] adds the proposition that the sum of the angles is equal to π in all triangles if it has this value in a single one. I also had to prove the same thing in my theory written in 1826. I even find that Legendre sometimes stumbled upon the path I so happily followed; but no doubt a prejudice in favor of the universally accepted proposition led him, at every step, to conclude hastily or to introduce prematurely notions that should no longer be admitted in the new hypothesis.

The proposition about the sum of the angles of triangles that Lobachevsky talks about is already found in Saccheri's memoir.

Another prominent mathematician who attempted to prove the parallel postulate is Joseph Fourier (1768–1830). Between 1820 and 1827, he wrote a 250-page manuscript containing nearly twenty attempts to prove this postulate. Fourier's manuscript is analyzed at length by Pont in [30, p. 531–586]. He used the following definition of parallelism:

> A line l is parallel to another line l' located in the same plane if l separates the lines in the plane that intersect l' from those that do not.

This is interesting because this definition is very close to the one used by Lobachevsky, Bolyai and Gauss in the setting of hyperbolic geometry. Fourier then attempted to prove the parallel postulate by using the fact that a set of points equidistant from a line is the union of two lines (cf. [30] p. 554).

Let us conclude this subsection on the theory of parallels in the modern period by quoting Arthur Schopenhauer (1788–1860) who was of the opinion that the mathematicians' attempts to prove the parallel postulate was pointless. He writes [40, Vol. 2 Chap. 13]:

> The Euclidean method of proof has brought forth from its own womb its most striking parody and caricature in the famous controversy over the theory of parallels, and in the attempts, repeated every year, to prove the eleventh axiom (also known as the fifth postulate). The axiom asserts indeed through the indirect criterion of a third intersecting line, that two lines inclined on each other (for this is the precise meaning of "less than two right angles"), if produced far enough, must meet. Now this truth is supposed to be too complicated to pass as self-evident, and therefore needs a proof; but no such proof can be produced, just because there is nothing more immediate.

7.7 Excursus 5: On Quadrilaterals in the Theory of Parallels

We have already mentioned that attempts to prove the parallel postulate from the notion of equidistance gave rise to the analysis of two remarkable types of quadrilaterals, namely, trirectangular quadrilaterals, also called Ibn al-Haytham–Lambert quadrilaterals, and birectangular isosceles quadrilaterals (or the Khayyām–Saccheri quadrilaterals). Because of their importance, we summarize here the history and some properties of these two types of quadrilaterals. A detailed analysis which shows their place in the development of non-Euclidean geometry is made in [1].

A *trirectangular quadrilateral* in the neutral plane (hyperbolic or Euclidean) or on the sphere, is a quadrilateral with three right angles. We have already encountered such a quadrilateral in the works of Ibn al-Haytham, namely, in his *Book Explaining the Postulates of Euclid's Book on the Elements*, based on an idea contained in the works of Thābit Ibn Qurra. We also saw that it appeared later in Lambert's *Theorie der Parallellinien*. These authors made a thorough analysis of the geometry of these quadrilaterals, distinguishing three cases for the remaining angle: acute, right or obtuse (see Fig. 7.14). Ibn al-Haytham and Lambert both tried to show that the existence of such quadrilaterals where the fourth angle is acute or obtuse contradicts the axioms of neutral geometry. In the case of the obtuse angle, it is known that this indeed contradicts the axioms of this geometry—this is Legendre's theorem

Fig. 7.14 Figures (i), (ii), (iii) represent a trirectangular quadrilateral in hyperbolic, Euclidean and spherical geometry respectively

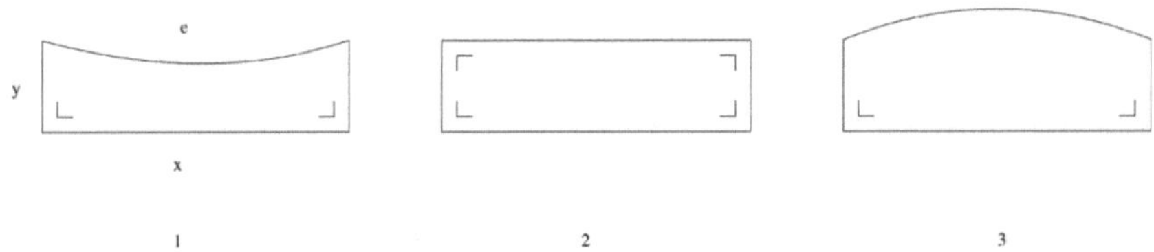

Fig. 7.15 Figures (i), (ii), (iii) represent an birectangular isosceles quadrilateral in hyperbolic, Euclidean and spherical geometry respectively

already mentioned—and the conclusion and arguments of these authors are correct. Trirectangular quadrilaterals exist in spherical geometry where other axioms of Euclid are not satisfied. In the case of the acute angle, Ibn Qurra and Lambert tried to find a property contradicting the axioms of neutral geometry, but they could not succeed, since such quadrilaterals do indeed exist in hyperbolic geometry.

The history of birectangular isosceles quadrilaterals is similar.

A birectangular isosceles quadrilateral (also called a Khayyām–Saccheri quadrilateral), in the neutral plane or on the sphere, is a quadrilateral with two opposite sides congruent and forming a right angle with a third side (Fig. 7.15). In the Euclidean plane, such a quadrilateral is (as is the case for trirectangular quadrilaterals) a rectangle. By cutting a birectangular isosceles quadrilateral along an axis of symmetry, we obtain two trirectangular quadrilaterals.

We have already mentioned that birectangular isosceles quadrilaterals were studied by ʿUmar al-Khayyām in Book I of his *Commentary on the Difficulties of Some Postulates in Euclid's Work*, cf. [36] p. 306 et seq. and [38] p. 467. They were also studied by Naṣīr al-Dīn al-Ṭūsī in his two treatises, *The treatise that delivers from the doubt concerning parallel lines* and the *Redaction of Euclid*, cf. [16] Volume I, p. 209 and [38] p. 468. They were rediscovered by Saccheri, in his *Euclides ab omni nœvo vindicatus* [39] (1733) who, like his predecessors, analyzed them by distinguishing three cases, depending on whether the remaining angles (which are equal to each other) are acute, right or obtuse. As with the case of trirectangular quadrilaterals, all these authors considered that the cases of acute and obtuse angles are excluded from neutral geometry. Lambert, in his *Theorie der Parallellinien*, reasons several times on trirectangular quadrilaterals obtained by cutting them into two isosceles birectangular quadrilaterals, notably when he studies the properties of monotonicity of the lengths of the perpendiculars drawn from the line containing the side adjacent to the angle which is not assumed to be a right angle a priori, and those of the angles which these perpendicular segments make with the line from which they are drawn. We have already discussed Lambert's analysis. Let us say a few words on the analysis of ʿUmar al-Khayyām. It is made in Proposition 3 of Book I of his *Commentary* (p. 324 of the translation in [36]). Khayyām considers the quadrilateral $ABCD$ where the angles at A and B are right. The proposition says that the angles C and D are also right angles. (In Proposition 1 of the same book, Khayyām had already proved that the angles at C and D are equal.) The proof of Proposition 3 begins with a construction that is erroneous.

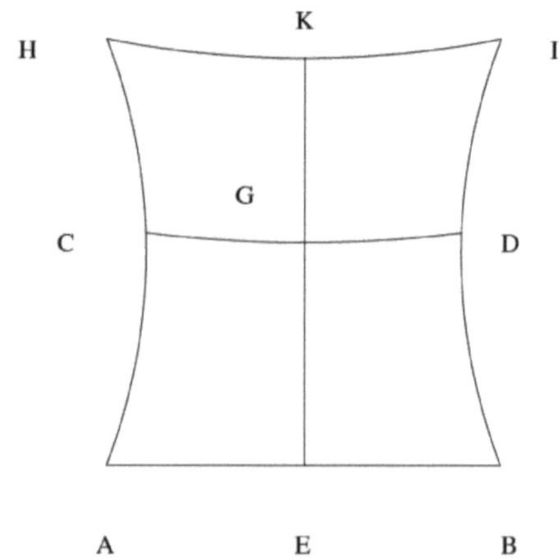

Fig. 7.16 In his analysis of the Khayyām–Saccheri quadrilateral (with right angles at A, B) Khayyām considers that AC and BD necessarily intersect the perpendicular at the point K to EG, which is not true

Indeed, Khayyām takes the perpendicular bisector of the segment AB from E, the midpoint of AB (Fig. 7.16). This perpendicular bisector cuts the segment CD at a point he calls G, and he extends it to a point K satisfyings $GK = EG$. Through the point K, he draws the perpendicular HKI to the line KE. He then extends the lines AC and BD and says that they meet HKI at two points he calls H and I. This is not correct, because in neutral geometry (and in particular in hyperbolic geometry) the lines AC and BD do not necessarily intersect the line HKI.[30]

Saccheri, in the memoir in [3] we already mentioned, analyzes the three possibilities for the angles at C and D (which he knows to be congruent): acute, right, or obtuse. He proves that if one of the three hypotheses holds for a particular birectangular isosceles quadrilateral, then it holds for all birectangular isosceles quadrilaterals (Propositions V, VI, and VII). The obtuse angle hypothesis is ruled out by an argument using the work of Naṣīr al-Dīn al-Ṭūsī, which is quoted by Saccheri.

Saccheri then proves that the right, obtuse, and acute angle hypotheses imply that the sum of the angles in a triangle is equal to π, greater than π, and smaller than π, respectively. In doing so, he establishes a series of results valid in neutral geometry, with the aim of finally proving that the acute angle hypothesis cannot hold. To conclude his argument, and having not reached a proof, he notes that the acute angle hypothesis implies the existence of perpendiculars "located at infinity", that is, the existence of two disjoint lines having no common perpendicular, a property that he excludes *a priori*. We know from the later work of Lobachevsky that this situation arises in hyperbolic geometry.

[30] Khayyām, in his "proof" (which is defective), passes over a number of details, with statements such as: "This is one of the things that have already been treated by the Philosopher" (i.e., Aristotle); "This is one of the things that you will be able to recognize with a minimum of reflection and investigation", or again, "It will be recognized with a minimum of reflection and we will therefore omit it in order to avoid prolixity. Thus, whoever wishes to establish this here according to the mathematical order, let him do so; we will not prevent him!" (quoted by Rashed and Vahabzadeh in [36] p. 326).

References

1. N. A'Campo, A. Papadopoulos, Notes on hyperbolic geometry, in *Strasbourg Master class on Geometry*, ed. by A. Papadopoulos. IRMA Lectures in Mathematics and Theoretical Physics, vol. 18 (European Mathematical Society (EMS), Zürich, 2012), pp. 1–182
2. E. Beltrami, Un precursore italiano di Legendre e di Lobatschewski. Rendiconti della Reale Accademia dei Lincei Roma **5**, 441–448 (1889). Œvres complètes de Beltrami, Vol. IV p. 348
3. R. Bonola, La geometria non-euclidea. Esposizione storico-critica del suo sviluppo. 1st ed., Ditta Nicola Zanichelli editore, Bologna, 1906. German translation by M. Liebmann in the collection Wissenschaft und Hypothese, Teubner, Leipzig, 1908. English translation by H. S. Carslaw, *Non-Euclidean Geometry, A Critical and Historical Study of its Development*. 1st édn., Chicago (1912). Reprint Dover, New York, 1955
4. G.A. Borelli, Euclides restitutus, siue, Prisca geometriae elementa, Pisa (1658)
5. G. Brunel, Notice sur l'influence scientifique de Guillaume-Jules Hoüel. Mémoires de la Société de Sciences Physiques et Naturelles de Bordeaux **4**, 1–78 (1888)
6. L.N.M. Carnot, *Géométrie de position* (J. B. M. Duprat, Paris, 1803)
7. d'Alembert (J. Le Rond d'Alembert), *Essai sur les éléments de philosophie*, 1st edn. 1759, new edition (Fayard, Paris, 1986)
8. P.-S. de Laplace, *Exposition du système du monde*, 1st edn. (Imprimerie du Cercle-Social, Paris, 1796)
9. P.-S. de Laplace, *Œuvres complètes*, 14 volumes, publiées sous les auspices de l'Académie des sciences (Gauthier-Villars, Paris, 1878–1912)
10. Euclides, *Elementa post I. L. Heiberg edidit E. S. Stamatis*. (T. I to V of the Euclidis opéra omnia.) Vol. I: Libri I-IV cum appendicibus, vol. II : Libri V-IX cum appendice (B. G. Teubner, Leipzig, 1883–1888). New printing, 1969–1970, 2 vol. in-8°, XLii-190 p. et vm-239 p.
11. C.F. Gauss, Werke, Königliche Gesellschaft der Wissenschaften, Göttingen (1900)
12. V. Giordano da Bitonto, Euclide restituito overo gli antichi elementi geometrici ristaurati e facilitati, Rome (1680)
13. J. Gray, *Ideas of Space: Euclidean, Non-Euclidean and Relativistic* (Clarendon Press, Oxford, 1979)
14. J. Gray, Worlds out of nothing: a course in the history of geometry in the 19th century. Springer Undergraduate Mathematics Series, 1st edn. (Springer, London, 2007); 2nd ed., 2010
15. M.J. Greenberg, *Euclidean and Non-Euclidean Geometries*, 4th edn. (W. H. Freeman, New York, 2008). (1st ed. 1980)
16. T. Heath, The *Thirteen Books of Euclid's Elements Translated from the text of Heiberg with Introduction and Commentary*, 3 vol., 1st edn. (Cambridge University Press, Cambridge, 1908). Several later editions
17. G.S. Klügel, Conatuum praecipuorum theoriam parallelarum demonstrandi recensio, quam publico examini submittent Abraham Gotthelf Kästner et auctor Georgius Simon Klügel, Göttingen (1763), 30pp.
18. M. Krause, Die Sphärik von Menelaos aus Alexandrien, Abhandlungen der Gesellschaft der Wissenschaften zu Göttingen, phil.-hist. Klasse, 3, 17, Berlin (1936)
19. J.H. Lambert, Observations trigonométriques. Mém. Acad. sc. Berlin **24**, 327–357 (1770)
20. A.-M. Legendre, *Éléments de géométrie*, 1st edn. (Firmin Didot, Paris, 1798); several later editions
21. A.-M. Legendre, Réflexions sur les différentes manières de démontrer la théorie des parallèles ou le théorème sur la somme des trois angles du triangle. Mémoires de l'Académie des sciences de l'Institut de France **12**, 367–410 (1833)
22. N.I. Lobachevsky, New principles of geometry with a complete theory of parallels (in Russian). Uchenye zapiski kazanskogo imperatorskogo universiteta **3**, 3–48 (1835);1836, No. 2, p. 3–98 & No. 3, p. 3–50;1837, No. 1, p. 3–97;1838, No. 1, p. 3–124 & No. 3, p. 3–65. French translation by par F. Mailleux, Mém. Soc. Royale sci. de Liège (3), 2 No. 5, p. 101;3–101 No. 2, p. 32–32 (1899)

23. N.I. Lobachevsky, Pangéométrie ou précis de géométrie fondée sur une théorie générale et rigoureuse des parallèles, in *Recueil d'articles écrits par les professeurs de l'Université de Kazan à l'occasion du cinquantenaire de sa fondation*, vol. I (1856), pp. 277–340. New edition commented and translated into English by A. Papadopoulos, Heritage of European Mathematics, Vol. 4, European Mathematical Society, Zürich, 2010
24. J.F. Lorenz, *Grundriß der reinen und angewandten Mathematik, oder der erste Cursus der gesamten Mathematik* (Helmstädt, Fleckeisen 1791)
25. P. Mansion, Analyse des recherches du P. Saccheri, S. J., sur le postulatum d'Euclide. Annales de la Société scientifique de Bruxelles **14**, 2ème partie, 46–59 (1889–1890)
26. H. Meschkowski, *Non-Euclidean Geometry*. English translation by A. Shenitzer (Academic Press, New York, 1964)
27. S. Negrepontis, A. Papadopoulos, Notes on angles and solid angles, in relation with Euler's memoir De mensura angulorum solidorum, in *Spherical Geometry in the Eighteenth Century, I: Euler, Lagrange and Lambert* (Springer, Berlin, 2026)
28. A. Papadopoulos, G. Théret, *La théorie des parallèles de Johann Heinrich Lambert, critical edition with French translation and mathematical and historical commentary*. Librairie Scientifique et Technique Albert Blanchard, coll. (Sciences dans l'Histoire, Paris, 2014), 214pp.
29. H. Poincaré, Les géométries non euclidiennes. Revue générale des sciences pures et appliquées **23**, 769–774 (1891)
30. J.-C. Pont, L'aventure des parallèles. Histoire de la géométrie non-euclidienne: précurseurs et attardés (Peter Lang, Bern, 1986)
31. Proclus, Commentaires sur le Livre I des éléments d'Euclide, French translation by P. Ver Eecke, Collection des Travaux de l'Académie Internationale des Sciences, vol. 1 (Desclée de Brouwer & Cie, Bruges, 1948)
32. Proclus, *A Commentary on the First Book of Euclid's Elements*. Translated with Introduction and Notes by Glenn R. Morrow (Princeton University Press, Princeton, 1970)
33. R. Rashed, A. Papadopoulos, *Menelaus' Spherics: Early Translation and al-Māhānī/al-Harawī's version: Critical edition of Menelaus' Spherics from Arabic manuscripts, with historical and mathematical commentaries*. De Gruyter, Series: Scientia Graeco-Arabica, vol. 21 (De Gruyter, Berlin, 2017), 890pp.
34. R. Rashed, *Les mathématiques infinitésimales du IXème au XIème siècle*. 5 vols (Al-Furqan Islamic Heritage Foundation Publication, London). Vol. I : Fondateurs et commentateurs : Banū Mūsa, Ibn Qurrā, Ibn Sīnān, al-Khāzin, al-Quhi, Ibn al-Sām, Ibn Hund, 1996 ; Vol. II : Ibn al-Haytham, 1993 ; Vol. III : Ibn al-Haytham. Théorie des coniques, constructions géométriques et géométrie pratique, 2000 ; Vol. IV : Ibn al-Haytham. Méthodes géométriques, transformations ponctuelles et philosophie des mathématiques, 2002 ; Vol. V : Ibn al-Haytham. Astronomie, géométrie sphérique et trigonométrie, 2006
35. R. Rashed, C. Houzel, Thābit Ibn Qurra et la théorie des parallèles. Arabic Sci. Philos. **15**, 9–55 (2005)
36. R. Rashed, B. Vahabzadeh, *Al-Khayyām mathématicien*. Collection Sciences dans l'Histoire (Librairie Scientifique et Technique Albert Blanchard, Paris, 1999)
37. B.A. Rosenfeld, *History of Non-Euclidean Geometry*. Transl. A. Shenitzer, Studies in the History of Mathematics and Physical Sciences, vol. 12 (Springer, New York, 1988)
38. B.A. Rosenfeld, A.P. Youschkevitch, Geometry, in *Encyclopedia of the History of Arabic Science*, ed. by R. Rashed, R. Morelon, vol. 2 (Routledge, London, 1996), pp. 447–494
39. G.G. Saccheri, *Euclides ab omni nœvo vindicatus: sive conatus geometricus quo stabiliuntur prima ipsa universae geometriae principia* (Pauli Antoni Montani, Milan, 1733). German transl. of Book I (l'ouvrage de Saccheri en contient deux) by P. Stäckel dans [42]. English transl.: Euclid freed of every flaw: geometrical essay in which are established the fundamental principles of universal geometry by G. B. Halsted, Chicago and London: the Open court publishing Co., 1920. Italian edition with commentaries by V. De Risi, Pisa, Edizioni della Normale, 2011

40. A. Schopenhauer, *The World as will and Representation*, vol. 1, Ch. Janaway (ed.), Transl. J. Norman, A. Welchman (Cambridge University Press, Cambridge, 2010)
41. C. Segre, Congetture intorno all'influenza di Girolamo Saccheri sulla formazione della geometria non-euclidea. Atti della Reale Accademia delle scienze di Torino **38**, 535–547 (1902–1903)
42. P. Stäckel, F. Engel, *Die Theorie der Parallellinien von Euklid bis auf Gauss, eine Urkundensammlung zur Vorgeschichte der nicht-euklidischen Geometrie* (B. G. Teubner, Leipzig, 1895)
43. Theodosius, *Les Sphériques de Théodose de Tripoli, Œuvres traduites pour la première fois du grec en français avec une introduction et des notes par P. Ver Eecke* (Desclée de Brouwer, Bruges, 1927); repr. Paris, Librairie Albert Blanchard, 1959
44. G. Veronese, *Fondamenti di geometria a più dimensioni e a più specie di unità rettilinee esposti in forma elementare: Lezioni per la scuola di magistero in matematica* (Tipografia del Seminario, Padova, 1891)
45. J. Wallis, *Demonstratio Postulati Quinti Euclidis, in Operum Mathematicorum*, vol. II (Oxford University Press, Oxford, 1693)
46. A.P. Youschkevitch, Les mathématiques arabes *(*VIIIe-XVe siècles*)*. Translated from the Russian by M. Cazenave et K. Jaouiche, Collection L'Histoire des Sciences, Textes et Études (Librairie Philosophique J. Vrin, Paris, 1976)

Chapter 8
The *Pentagramma Mirificum* from Napier to Lambert: Notes on Lambert's Memoir on Spherical Trigonometry

Annette A'Campo-Neuen

Abstract Well-known relations between any three parts of a right-angled spherical triangle can be summarized by Napier's mnemonic rules. In Napier's book appears a drawing of a closing chain of five right-angled triangles enclosing a self-polar pentagon of constant width $\pi/2$ with remarkable properties. In his memoir on spherical trigonometry, Lambert clarified the meaning of this figure and gave a proof of Napier's rules in terms of cyclic transformations of a given triangle. In this commentary, we focus on Lambert's elegant proof of Napier's rule and the properties of the so-called pentagonal star of triangles.

Keywords Pentagramma mirificum · Spherical trigonometry · Napier's rule · Johann Heinrich Lambert

AMS Codes 01A40, 01A45

8.1 Introduction

Lambert's aim, in his memoir (see [1, 2]) on spherical trigonometry, is to determine unknown sides or angles from given parts of a spherical triangle in the most convenient and beautiful way (see §1):

> If one tries to contribute something to this science, the aim is not to increase the number of possible cases since this would be impossible, but the goal is to make the solution of problems more beautiful, shorter and more convenient. In that respect there is still a lot to be thought about.

For right-angled spherical triangles, Napier in [3] found a convenient way to present all 10 relations between any three of the five parts of the triangle (other than

A. A'Campo-Neuen (✉)
Departement Mathematik und Informatik, Universität Basel, Basel, Switzerland
e-mail: annette.acampo@unibas.ch

R. Caddeo, A. Papadopoulos (eds.), *Spherical Geometry in the Eighteenth Century I*, https://doi.org/10.1007/978-3-032-12466-1_8

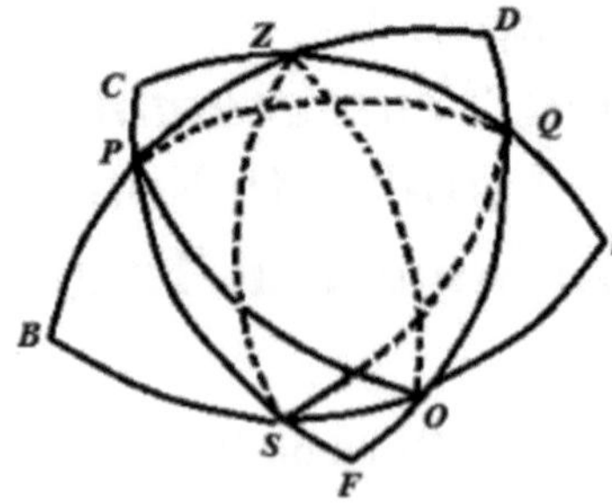

Fig. 8.1 Chain of triangles as depicted by Napier in [4]

the right angle). Namely, he formulated a rule producing all these relations by some simple procedure from two basic ones.

Lambert now gives a geometric interpretation of Napier's rule, possibly inspired by a beautiful configuration of five triangles forming a pentagon that already appeared in Napier's book (see Fig. 8.1). But since Lambert does not refer to Napier's original figure, he may also have found it independently. The idea is the following. From a given right-angled triangle, he constructs a closing chain of 5 right-angled triangles on the sphere such that the data, namely the five angles, are permuted cyclically. And applying the basic identities to all the transformations, all the relations follow (see §12–§19 and §32–§33). Moreover, the configuration of five triangles encloses a self-polar pentagon of constant width $\pi/2$ with remarkable properties (see §20–§31). It was called *pentagramma mirificum*[1] by Gauss (see [5]) who found further properties of this pentagon.

In the second part of his memoir, Lambert considers general spherical triangles. He first looks at triangles where all the angles and sides are $< 90°$. In this case, the six parts are determined by any given three parts, so there are relations between any four parts. Drawing suitable perpendiculars, Lambert uses Napier's relations for right-angled triangles to find relations for the general triangle. Introducing an appropriate notation, he can reduce all the relations to four identities (see §47). He then generalizes his results to the case of triangles with arbitrary angles. And using trigonometric identities, he rewrites his relations in a form most fit for the use of logarithms, namely as products and quotients.

Furthermore, in §64–§68, there is a description of the analogue in planar geometry. The last part of the memoir is devoted to practical applications such as in §80, the calculation of the height of the sun at a given time, from the hour angle, the solar declination and the elevation of the pole.

In this commentary, we will focus on Lambert's elegant proof of Napier's rule and the properties of the pentagonal star of triangles.

[1] Miraculous pentagram.

8.2 Napier's Circular Method

The Scottish mathematician John Napier (1550–1617) discovered that the use of logarithms greatly simplifies the calculations in trigonometry. He managed to produce tables of logarithms of trigonometric functions for angles between 0° and 90° at intervals of 1 minute, an enormous task which took him about 20 years. Finally, he published the tables together with a commentary on how to use them in trigonometry, in the book *Mirifici logarithmorum canonis descriptio ejusque usu in utraque trigonometria* (see [3]). In his introduction, he talks about the idea of a triangulation of a geometric figure which reduces the investigation of the figure to the study of triangles (see [4], p. 31).

> Since geometry is the art of measuring proposed quantities with complete accuracy, and as a figure [..] can be decomposed into the sides of triangles, or by further triangulation into smaller triangles, then a figure is composed of triangles with some of its angles and sides measured, and all the other triangular parts are to be found, and from which the extent of the figure can be found. Therefore it is clear that the arithmetical solution of any geometrical question depends on the principles by which triangles are solved.

Napier discusses both planar and spherical geometry. In a spherical triangle, the sides are circular arcs lying on great circles. Therefore the sides are also measured by angles, and the relations between sides and angles are all expressed in trigonometric functions. The relations are particularly simple for right-angled triangles where they are just proportions. And applying logarithms transforms these proportions into differences. So with the help of a table of logarithms of the trigonometric functions, one can easily determine unknown angles or sides from given ones. That is why his logarithms work particularly well in spherical trigonometry. Napier writes (see [4], p. 38):

> Now spherical triangles follow, the most difficult of all, as commonly set out by others, yet by our logarithms the most easy of all.

Napier considers right-angled spherical triangles where the other two angles and the three sides are strictly smaller than 90°. He calls the vertices S, B, P with a right angle at B (see Fig. 8.2, taken from [4], p. 40). The notation is inspired from his principal celestial example where S is the position of the sun at sunrise, the point B on the horizon indicates the northern direction and P is the celestial pole.

For practical use, Napier proposes a mnemonic rule to produce the relations between any three of the five parts of the triangle which he calls his *circular method*

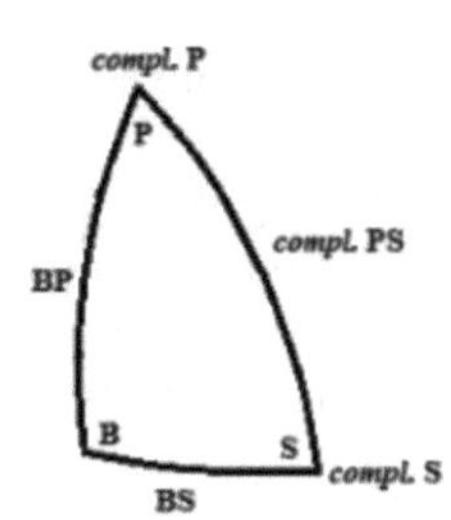

Fig. 8.2 Circular Parts, as depicted by Napier in [4]

(see [4], p. 40):

> For the sake of making the calculation easier, we take the side BP itself unchanged; we take the complement of the angle P; the complement of the side PS; and the complement of the angle S; and the side SB itself unchanged, and we place these five parts to be kept in their natural order, on the margin of a circle, and we call these the circular parts.

Here the complement of an angle α means $\alpha' = 90° - \alpha$. For each of the circular parts, there are two neighboring angles and two opposite ones. Depending on these relative positions, Napier states the following rule (see [4], p. 44):

> The logarithm of the sine of the middle part is equal to the [sum of the] logarithms of the tangents of the [adjacent] extremes, or to the [sum of the] logarithms of the cosines of the opposite extremes.

Napier gives the following first example. If in the above triangle $BS = 70°$ is given and the complement of the angle BPS at P is $16°24'27''$, then these two circular parts are neighbors of the middle part BP, and hence applying the first part of his rule, Napier obtains $BP = 54°$. If we think of the celestial triangle as before, BS is the azimuth of the sun, the complement of the angle at P is the difference of ascension of the sun, i.e. the difference of the sunrise to the sixth hour, and from these data Napier calculates the elevation BP of the pole.

Napier explains that the resulting relations between sides and angles can be verified case-by-case using well-known facts from spherical geometry. But he does not give a detailed proof. And there is no explanation of how he found his rule. He also does not write down formulae but expresses all the statements in words.

However, he notes the following ([4], p. 42, see also Fig. 8.1):

> The uniformity of the circular parts is most apparent with right-angles made on the surface of a sphere from five arcs of great circles, of which the first cuts the second, the second the third, the third the fourth, and finally the fourth cuts the fifth at right angles: thus all the remaining angles are made oblique.

It is this intriguing configuration of five right-angled triangles which will be further investigated by Lambert.

8.3 Lambert's Interpretation of Napier's Rule

Lambert in his memoir [2] considers Napier's rule as well known and does not explicitly describe neither the circular method nor the corresponding relations. Implicitly, he uses a complementary version of Napier's rule where sines and cosines are interchanged as well as tangents and cotangents, and where the logarithms are omitted.

For a given right-angled spherical triangle with vertices A, B, C and a right angle at C, the following parts are read off and written on the margin of a circle in counter-clockwise order: the angle A at the lower endpoint of the hypotenuse, then the arc-length c of the hypotenuse, the angle B at the other endpoint of the hypotenuse and

Fig. 8.3 Napier circle

finally for the two catheti a and b their complements $a' = 90° - a$ and $b' = 90° - b$ respectively (see Fig. 8.3).

For each angle in this circle, there are two neighboring angles and two opposite ones. Depending on these relative positions, Napier's rule says the following:

$$(I) \quad \cos(\text{middle part}) = \sin(\text{right opposite part}) \cdot \sin(\text{left opposite part})$$

$$(II) \quad \cos(\text{middle part}) = \cot(\text{right neighbor}) \cdot \cot(\text{left neighbor})$$

Applying these rules, we obtain the following 10 relations:

$$\begin{array}{lll}
(1)\ \cos(a') = \sin(a) & = \sin(A)\sin(c) & \\
(2)\ \cos(b') = \sin(b) & = \sin(c)\sin(B) & \\
(3)\ \cos(A) & = \sin(B)\sin(a') & = \sin(B)\cos(a) \\
(4)\ \cos(c) & = \sin(a')\sin(b') & = \cos(a)\cos(b) \\
(5)\ \cos(B) & = \sin(b')\sin(A) & = \cos(b)\sin(A) \\
(6)\ \cos(a') = \sin(a) & = \cot(b')\cot(B) & = \tan(b)\cot(B) \\
(7)\ \cos(b') = \sin(b) & = \cot(A)\cot(a') & = \cot(A)\tan(a) \\
(8)\ \cos(A) & = \cot(c)\cot(b') & = \cot(c)\tan(b) \\
(9)\ \cos(c) & = \cot(B)\cot(A) & \\
(10)\ \cos(B) & = \cot(a')\cot(c) & = \tan(a)\cot(c)
\end{array}$$

As Lambert says in his memoir (see [2], §9), each of these 10 relations can be proved with elementary spherical geometry and they were well known before Napier. But Napier's rules make it much easier to memorize the relations.

> One proves the relation for each of these 10 [..] cases separately. This was until Neper to whom we owe such a lot in trigonometry, had the idea to compare the 10 relationships with each other and observed a general principle. Namely, in five cases he found the relationship expressed with sines and in the other five cases with sines and tangents. Introducing a clever notation, he was able to reduce all 10 relations to two general rules that are now included in all introductions to spherical geometry and are used with profit.

Now Lambert asks for a geometric explanation of Napier's rule reducing the proof of all the relations to the proof of the two basic ones (1) and (6). He observes that for such a proof one would need a transformation of the given triangle into a new one with the same circular parts where sides become angles and angles become sides (see [2], §11 and §12).

> Neper's rule without making a difference between angles and sides refers to three chosen parts of the triangle as a middle one and two adjacent or two opposite ones. But in the cases themselves there is a difference between angles and sides, and in order to apply the rule one has to take the complements of the catheti. Thus, a general proof of this rule must be based on the transformation of sides into angles and angles into sides. It is easy to see that one needs five such transformations.

Lambert provides the desired transformations by constructing a chain of triangles from a given first one, similar to Napier's pentagon of triangles (see Fig. 8.4). Note that Napier describes the configuration as given by five great circles intersecting at right angles. But he does not consider a construction from just one triangle to begin with.

Lambert's construction is the following. He starts with a given right-angled spherical triangle with vertices A, B, C such that all angles and arc lengths are $< 90°$ except the right angle at C. The arc b joining A and C lies on a great circle with pole P on the sphere. In order to obtain a conformal drawing in the plane, Lambert uses a stereographic projection from the opposite pole P', so that this great circle appears as a circle with center P in the figure. Moreover, since we have a right angle at C, the great circle through B and C appears as a straight line through the pole P in the projection (see Fig. 8.5). And on the sphere the arc length of $BP = a'$ is complementary to $a = BC$.

The great circle through the hypotenuse c also appears as a circle in the projection, and the straight line through its pole Q and P intersects this circle orthogonally at some point D. In this way we obtain a second right-angled triangle with vertices B, P, D. Let $\tilde{D}$ denote the intersection point of the line PQ and the circle through A and C. Since the point A is at distance 90° both from P and Q on the sphere, it must be the pole of the line joining P and Q. Hence, $DB = c'$, and $\tilde{D}C = b'$ on the sphere. Therefore the angle at P in the second triangle equals b'. Likewise, the angle at A equals $d' = D\tilde{D}$, or in other words $d = PD$ is complementary to the angle at A.

We see that the five angles A, c, B, a', b', occurring in the first triangle as vertex angles, hypotenuse and as complements of catheti, reappear in the second triangle in the following order $B, a', P = b', d' = A, c$ which is just a cyclic permutation.

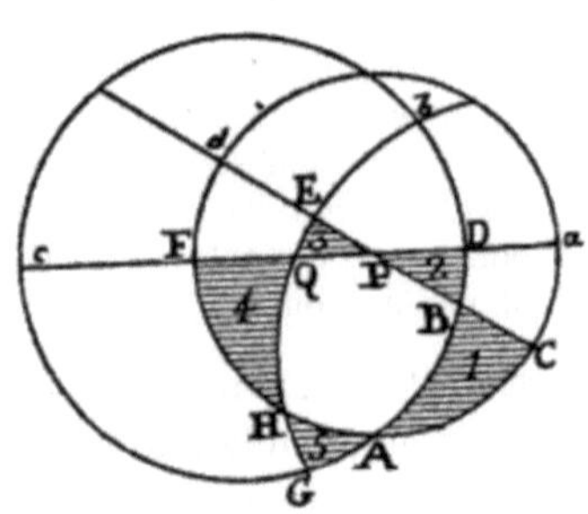

Fig. 8.4 Chain of triangles as depicted by Lambert (Figure 1 in [2])

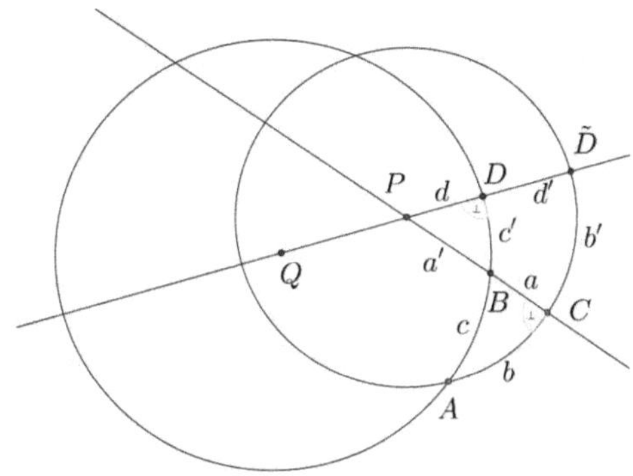

Fig. 8.5 Construction of the first two triangles

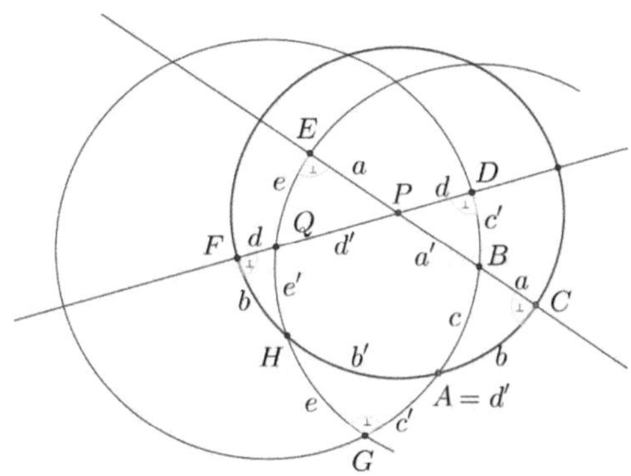

Fig. 8.6 Wreath of triangles

Comparing the data associated with the Napier circle, this amounts to a rotation by $(2/5)$ of the full angle $360°$.

Now let us complete the construction. Let F denote the intersection of the line through P and Q with the circle through A and C. Since P is the pole of the circle, there is a right angle at F. Finally, we draw the circle through Q which is perpendicular to PB and intersects PB at the point E say (see Fig. 8.6). This new circle intersects the circle through A and C at some point H. Moreover, it is perpendicular to the circle through A and B, since Q is its pole. Therefore there is a right angle at the intersection point G. We obtain a closing chain of five right-angled triangles such that their hypotenuses enclose a pentagon and all the triangles lie outside the pentagon.

Arguing as above, we see that the same five parts occur in all five triangles. More precisely, the circular parts are

Triangle 1	A	c	B	a’	b’
Triangle 2	B	a’	P=b’	d’=A	c
Triangle 3	P=b’	d’=A	Q=c	e’=B	a’
Triangle 4	Q=c	e’	H=a’	b’	d’=A
Triangle 5	H=a’	b’	A=d’	c	e’=B

Hence, moving through the chain of triangles, the five circular parts are cyclically permuted as in the transformation from the first to the second triangle, in each step turning the Napier circle by $2/5$ of $360°$.

So this construction provides the desired transformation of angles into sides and sides into angles (see § 18):

> This shows that each side and angle of the first triangle has a special position in each of the other triangles, and that hence these five triangles represent all possible transformations of them. Moreover, remarkably, those parts which according to Neper's rule are called the middle and the adjacent ones, or the middle and the opposite ones respectively, do not change their nature through all five transformations. This is because all five parts occur in all five triangles in the same order if we omit the right angle, since in each triangle the right angle moves between two other places of the five ones and hence goes to all five positions.

So in modern terms, there is an action of a cyclic group of order 5 on the chain of triangles, permuting the five parts cyclically. And this reduces the proof of Napier's rule to checking the two basic relations (1) and (6) as desired.

8.4 Further Properties of the Pentagramma Mirificum

As mentioned above, this configuration of five right-angled spherical triangles enclosing a pentagon, was later called *pentagramma mirificum* and further investigated by Gauss (see [5]) and others. Lambert notes some of its remarkable properties. The enclosed spherical pentagon is self-polar in the following sense: each of its vertices is the pole of the opposite side and hence all the diagonals are of length $90°$ (see §21 and §22). Moreover, at each vertex the angle of the adjacent triangles is equal to the arc length of the opposite hypotenuse.

Furthermore, Lambert proves the following (see §24):

> The area of the five triangles plus twice the area of the pentagon equals $3/8$ of the area of the entire sphere or $3/4$ of the area of the hemisphere.

To see this, Lambert refers to Fig. 8.7 (see Figure 2 in [2]). He uses the fact that the area of a spherical triangle on the unit sphere is given by the excess of the sum of its angles, and similarly the area of a spherical pentagon is given by the sum of its angles minus 3π. Now in the configuration of five triangles, each angle of the pentagon equals π minus the corresponding angle of the adjacent triangle. Denoting the angles of the triangles as in Fig. 8.7 by A, B, C, D, E, Lambert obtains that the area V of the pentagon is the following:

$$V = (\pi - A) + (\pi - B) + (\pi - C) + (\pi - D) + (\pi - E) - 3\pi = 2\pi - (A + B + C + D + E)\,.$$

On the other hand, since each of the angles occurs in two of the enclosing triangles, the sum Δ of the areas of the five triangles amounts to:

$$\Delta = 2(A + B + C + D + E) - 5\pi/2\,.$$

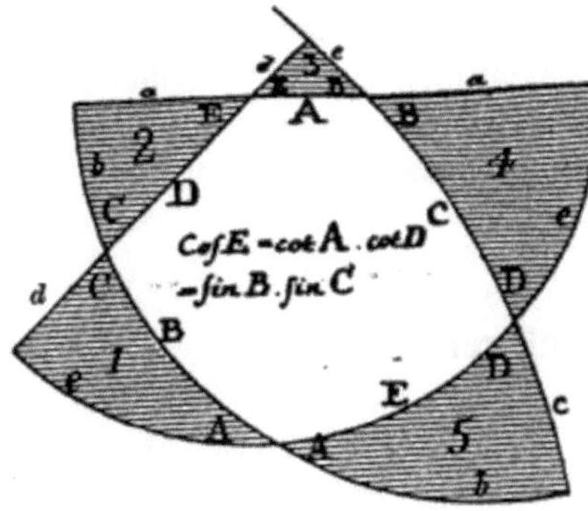

Fig. 8.7 Lambert's Figure 2 from [2]

This immediately implies the claim, namely:

$$\Delta + 2V = 3\pi/2\,.$$

Lambert wonders whether there might be more interesting features. In §31 he writes:

> Though I assume that among so many coincidences there are more general laws hidden, I could not find anything else but the proof of Neper's rule.

In his notes [5], published posthumously, Gauss indeed wrote down both number-theoretic as well as geometric results. He uses complex numbers to describe the vertices of the pentagon and associates the following number to each of the sides of the pentagon in clockwise order (see Fig. 8.7):

$$\alpha = \tan^2(D),\ \beta = \tan^2(A),\ \gamma = \tan^2(C),\ \delta = \tan^2(E),\ \epsilon = \tan^2(B)\,.$$

Between these numbers there are the following relations (see [5], p. 484):

$$1+\delta = \alpha\beta\,, \quad 1+\epsilon = \beta\gamma\,, \quad 1+\alpha = \gamma\delta\,, \quad 1+\beta = \delta\epsilon\,, \quad 1+\gamma = \epsilon\alpha\,.$$

This follows directly from Napier's rule. As inscribed in Fig. 8.7, we have

$$1+\delta = 1+\tan^2(E) = \frac{1}{\cos(E)} = \tan^2(A)\tan^2(D) = \alpha\beta\,,$$

and similarly for the others. Using these relations, Gauss finds the following beautiful identities[2] between those numbers:

$$\alpha\beta\gamma\delta\epsilon = 3+\alpha+\beta+\gamma+\delta+\epsilon = \sqrt{(1+\alpha)(1+\beta)(1+\gamma)(1+\delta)(1+\epsilon)}\,.$$

[2] In his note, Gauss himself calls them "schöne Gleichungen".

Moreover, he considers the projection of the spherical pentagon to a suitable tangent plane from the center of the sphere. The vertices of the resulting planar pentagon lie on an ellipse, and all its heights intersect in the point of contact of the tangent plane with the sphere.

References

1. J.H. Lambert, Anmerkungen und Zusätze zur Trigonometrie, in *Beyträge zum Gebrauche der Mathematik und deren Anwendung*, vol. 1 (Treatise III, Berlin, 1765), pp. 369–424, in: *Johann Heinrich Lambert, Opera Mathematica I*, ed. by A. Speiser (Orell Füssli Verlag Zürich, 1946), pp. 52–84
2. J.H. Lambert, Notes and comments on trigonometry, in *Spherical Geometry in the Eighteenth Century, I: Euler, Lagrange and Lambert*, translation in the present book, Chap. 19 (Springer, Berlin, 2026)
3. J. Napier, Mirifici logarithmorum canonis descriptio, ejusque usus, in utraque trigonometria; ut etiam in omni logistica mathematica, amplissimi, facillimi, & expeditissimi explicatio, Edinburgi, Ex officinâ Andreæ Hart (1614)
4. English translation of Mirifici logarithmorum canonis descriptio etc., Ian Bruce. http://www.17centurymaths.com/contents/napiercontents.html
5. C.F. Gauss, Pentagramma mirificum, in *Collected Works, Volume III: Analysis* (Königliche Gesellschaft der Wissenschaften, Göttingen, 1866), pp. 481–490

Chapter 9
Spherical Trigonometry Before the Modern Era: The Treatise of Naṣīr al-Dīn al-Ṭūsī

Athanase Papadopoulos

Abstract This is an overview of Naṣīr al-Dīn al-Ṭūsī's *Treatise of the quadrilateral*, an invaluable thirteenth century document on spherical geometry which was translated into French in 1891. The title we are using here is the one given by the translator (Alexandre Carathéodory). A title which is closer to the original Arabic is *Disclosing the secrets of the secant figure*. The term "secant figure", to which the title refers, is the so-called "complete (spherical) quadrilateral", that is, the figure that underlies what we call today Menelaus' Theorem. This theorem gives a formula that was extensively used by astronomers in their computations and the establishment of their tables since the first century AD, notably by Ptolemy, in the absence of the spherical trigonometric formulae that were discovered later. Naṣīr al-Dīn al-Ṭūsī's treatise contains much more than Menelaus' theorem, since we find there a complete system of spherical trigonometric formulae, with complete proofs. The treatise includes at the same time invaluable historical information on the discovery of the trigonometric formulae by the Arab mathematicians of the Middle-Ages and the transformation of the field of spherical trigonometry that this discovery led to.

Keywords Naṣīr al-Dīn al-Ṭūsī · Spherical trigonometry · Spherical geometry · Quadrilateral · Compounded ratio · Complete spherical quadrilateral · Sector figure · Menelaus theorem · Discovery of the sine rule

AMS Classification 01A20, 01A30, 01A35

A. Papadopoulos (✉)
Institut de Recherche Mathématique Avancée, Université de Strasbourg et CNRS, Strasbourg, France
e-mail: athanase.papadopoulos@math.unistra.fr

R. Caddeo, A. Papadopoulos (eds.), *Spherical Geometry in the Eighteenth Century I*, https://doi.org/10.1007/978-3-032-12466-1_9

9.1 Introduction

Naṣīr al-Dīn al-Ṭūsī's[1] *Treatise of the quadrilateral*[2] is an invaluable thirteenth century treatise on spherical geometry. It was edited and translated into French by Alexandre Carathéodory in 1891 [17]. The "quadrilateral" to which the French title refers is a complete (Euclidean or spherical) quadrilateral; see Figs. 9.1 and 9.3. It also goes by the names "sector figure", "secant figure" and others. We shall explain these terms below.

Al-Ṭūsī's treatise is a testimony of the transformation of the subject of spherical geometry by the Arab mathematicians that started in the ninth century, when the use of the sine rule and the other formulae of spherical trigonometry replaced that of the sector figure. The exposition of a complete set of spherical trigonometric formulae is carried out in the last part of this work (Book V). The author establishes there the main formulae that are the subject of every modern treatise on spherical trigonometry, including the notion of polar triangle and its use in spherical trigonometry. At the same time, Naṣīr al-Dīn's treatise constitutes a remarkable historical document that makes an overview of the state of the art of the field of spherical geometry as it was developed by the Arabs between the tenth and the thirteenth centuries.

The treatise is divided into five books, with the following titles:

- Book I. On compounded ratios and their rules.

---

[1] Naṣīr al-Dīn al-Ṭūsī was a mathematician, astronomer and philosopher. He belongs to the last period of the Arabic golden era in science. He was born in 1201 in the city of Ṭūs in the province of Khorasān (North-East of Iran). We owe to him commented Arabic editions of several important Greek mathematical texts, including geometrical works of Autolycus of Pitane, Euclid, Apollonius, Archimedes, Theodosius, Menelaus and Ptolemy. These editions played an important role in the transmission of these Greek works to Europe. An exposition of Euclid's *Elements* attributed to him (the *Book of exposition of the* Elements), published in Rome in 1594, contains a commentary that was used and quoted by John Wallis and by Girolamo Saccheri in their works on the parallel problem, see [14–16]. Al-Ṭūsī also wrote a treatise on the theory of parallels, called *Al-risāla al-shāfiya 'an al-shakk fī l-khuṭūṭ* (the letter that heals the doubt raised by the parallel lines). The treatise is analyzed by Rosenfeld in [16, pp. 74–80]. We also mention al Ṭūsī's contribution on the theory of parallels in Chap. 7 of the present volume, written by the author of the present chapter together with Guillaume Théret [10]. Al-Ṭūsī also wrote treatises on logic, astronomy, algebra, number theory and combinatorics, see [2, 8, 9, 11, 12].

[2] We are using the name that the translator, Alexandre Carathéodory (1833–1906), gave to this treatise, in French, *Traité du quadrilatère* [17]. The complete title, in Arabic, is *Kashf al-qināʿ 'an asrār al-shakl al-qaṭṭāʿ* (Disclosing the secrets of the secant figure). The work became known in Europe as *Traité du quadrilatère* (Treatise of the quadrilateral), after Carathéodory's translation. Carathéodory belonged to a distinguished Greek family from Constantinople, who was close to the Sublime Porte. His father, Stephanos Carathéodory, was the personal physician of the Sultan Mahmud II. Alexandre Carathéodory studied law in Paris and obtained there a doctorate, with a dissertation on the philosophical side of theory of error and its applications in the sciences. After his return to Constantinople, he became an Ottoman civil servant and started a diplomatic career. He became ambassador of the Ottoman empire to Rome and the main negotiator at the Preliminary Peace Treaty of San Stefano which ended the Russo-Turkish War (1877–78). He was the great-uncle of the mathematician Constantin Carathéodory.

- Book II. On the plane quadrilateral figure and the ratios that we find therein.
- Book III. On the lemmas for the spherical quadrilateral figure and on what is necessary for using it successfully.
- Book IV. On the spherical quadrilateral figure and the ratios that we find therein.
- Book V. An exposition of the methods that dispense with the theory of the quadrilateral figure, for what concerns the knowledge of the arcs of great circles.

The details of the content of this treatise are given in the rest of the present chapter.

There is a review of a small part of Naṣīr al-Dīn's treatise in Rosenfeld's book on the history of non-Euclidean geometry [16, p. 20ff]. Rosenfeld writes there that this book contains "the most complete exposition of spherical geometry in the East in the Middle Ages, written by the greatest mathematician and astronomer of the thirteenth century." [16, p. 20]

For the general mathematician, the title of Book I of the treatise deserves an explanation, since it is not clear a priori why an object like a compounded ratio (a product of two ratios of numbers or magnitudes, that is, an expression of the form $\frac{A}{B} \cdot \frac{C}{D}$), needs a special chapter in such a book. We shall explain this in Sect. 9.5.

All along this chapter, we shall use modern notation, in particular for what concerns trigonometric functions, as Carathéodory does in his edition.

The first three sections that follow the present introduction are expositions of interesting remarks made by al-Ṭūsī in his treatise, on the intersection of three or four lines in the plane or on the sphere.

In Sect. 9.5, we give some explanations on compounded ratios, which is the subject of the first book of Al-Ṭūsī's treatise.

Section 9.6 is concerned with the Sector Figure, that is, Menelaus' Theorem, which is extensively used and thoroughly discussed in Al-Ṭūsī's treatise.

In Sect. 9.7, we make a quick summary of Book V of the treatise, which is interesting in many aspects. We quote several series of important propositions from this book, and we linger on the importance of the treatise for the historical information it contains.

9.2 On the Figures Formed by the Intersection of Three and Four Lines in the Euclidean Plane

Among the simplest elements of Euclidean plane geometry, there are points, next, pairs of points, and the latter define lines. Then, one may consider a pair of two lines which, if they are generic (that is, if they are not coincident or disjoint), define an angle. Going forward, the next step is to consider a triple of lines, which, again if they are generic, define a triangle. One may continue in the same manner and make a further step, considering the union of four generic lines. The resulting figure contains a subfigure represented in Fig. 9.1, made up of a quadrilateral to which are appended two triangles. This is the so-called *(Euclidean) sector figure*, also called *complete (Euclidean) quadrilateral*.

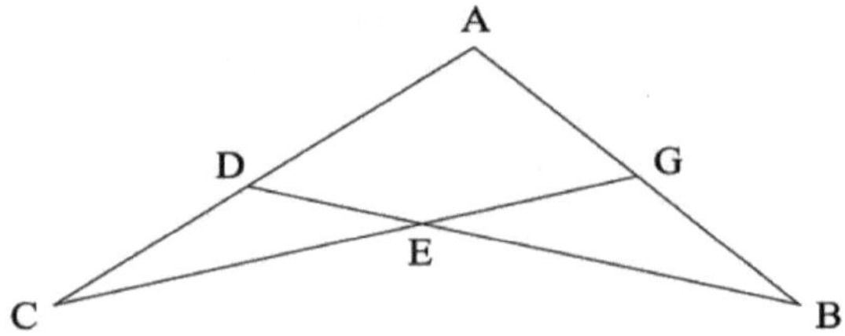

Fig. 9.1 The Euclidean sector figure, or complete Euclidean quadrilateral

The study of the sector figure arises naturally in Euclidean geometry. It comes naturally next to that of triangles. The classical Menelaus Theorem, which we recall in Sect. 9.4, concerns the sector figure.

Now we pass to spherical geometry: first, the intersection of three great circles and then, the intersection of four.

9.3 On the Figure Formed by the Intersection of Three Great Circles

A spherical triangle is determined by three distinct points on the sphere together with the three arcs of great circles that join them pairwise. Before studying spherical triangles, Naṣīr al-Dīn studies the intersection of three great circles (Chap. 2). He notes that a generic triple of great circles divides the sphere into eight spherical triangles (Fig. 9.2). These eight triangles are pairwise equal,[3] since they are pairwise images of each other by the antipodal map of the sphere. For instance, the triangles ACF and DEB of Fig. 9.2 are equal. Thus, we have, in this figure, at most four unequal triangles. Among these four triangles any two have at least one side and one angle in common, and the rest of their elements are supplementary of each other (their sum is equal to two right angles).[4] For example, comparing the triangles ABC and CDF, we see that the two angles at C are equal, $AB = FD$, BC and FC are supplementary, AC and CD are supplementary, the angle ABC is equal to the angle AFC (they are the two angles made by two half great circles with common endpoints) and is supplementary to the angle CFD, and the angle BAC is equal to the angle BDC and is supplementary to the angle CDF.

Now recall that in spherical geometry, triangles are determined by the measures of their angles.[5] Thus, if we know the triangle ABC, then we know the triangle CDF. The same reasoning applies to all the others triangles of the figure, pairwise. Consequently, if one of the eight triangles in this figure is known, then all the other seven triangles are also known.

[3] Following Euclid, we use the term "equal" to means the modern term "isometric".

[4] We recall that the length of a spherical segment is the angle made by the two rays that join the origin of the sphere to the two extremities of this segment.

[5] This is Proposition 19 of Menelaus' *Spherics*, see [13, p. 548].

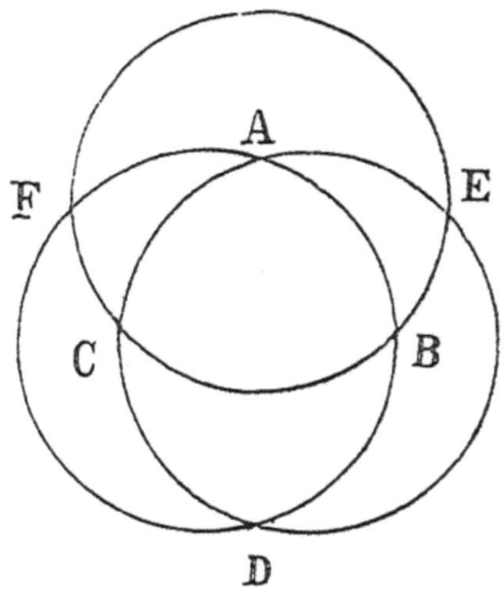

Fig. 9.2 The eight triangles made by three great circles on the sphere. The figure is extracted from Carathéodory's edition of al-Ṭūsī's treatise. Note that there is a triangle with sides EF, FD, DE. It is remarkable that Naṣīr al-Dīn al-Ṭūsī drew an abstract (non-realistic) figure of the intersection pattern of the three great circles of the sphere

9.4 Spherical Complete Quadrilaterals

We have already mentioned complete Euclidean quadrilaterals (see Fig. 9.1). In this section, we examine complete spherical quadrilaterals. Thus, after spherical triangles, we study spherical quadrilaterals, and more especially *complete spherical quadrilaterals*.

A *spherical quadrilateral* is a convex figure on the sphere bordered by four arcs of great circles (the sides of the quadrilateral). A spherical quadrilateral becomes a *complete spherical quadrilateral* if one adds to it two spherical triangles by extending two pairs of opposite sides until in two sides in each such a pair meet. Figure 9.3 is a complete spherical quadrilateral, extracted from Naṣīr al-Dīn al-Ṭūsī's treatise.

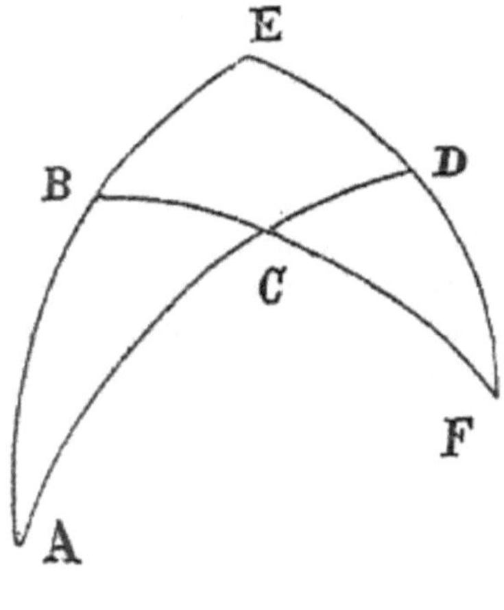

Fig. 9.3 A complete spherical quadrilateral. Figure extracted from Naṣīr al-Dīn al-Ṭūsī's treatise

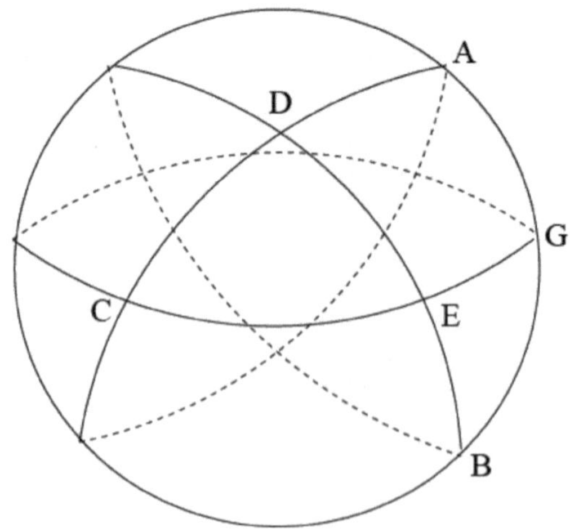

Fig. 9.4 The spherical quadrilateral $ADEG$ to which are appended the triangles DCE and EGB is a complete spherical quadrilateral

In Fig. 9.4 we have drawn four great circles, namely, the three great circles containing the sides of the spherical triangle CDE, and the fourth one, passing by the points A and G. In this figure, also reproduced from a figure in Naṣīr al-Dīn al-Ṭūsī's 's treatise, the quadrilateral $ADEG$ to which are appended the triangles DCE and EGB is a complete spherical quadrilateral. Note that each spherical quadrilateral gives rise to four complete spherical quadrilaterals, depending on the choice of the two adjacent sides and the direction in which we extend them. In Fig. 9.4, made by the intersection of the three circles, we have a total of six quadrilaterals which are equal (by the antipodal map), which makes at most three non-equal quadrilaterals. Since each quadrilateral gives rise to four sector figures, we have in all $6 \times 4 = 24$ sector figures. But we have at most 12 non-equal ones.

The study of complete spherical quadrilaterals was carried out extensively by Menelaus of Alexandria in his *Spherics*. This figure is associated with a proposition of this treatise which had far-reaching applications and which is known in the modern mathematical literature (although stated in a slightly different form) as *Menelaus theorem*, see [13, Proposition 66]. The proposition was used by Ptolemy in Book I, Chap. 11 of the *Almagest*, the famous astronomical treatise written around the middle of the second century AD, as the main lemma for the astronomical calculations that are carried out there, see [5, pp. 50–55]. We have elaborated at length on this proposition and its applications in the commentary contained in our edition of the *Spherics*, see [13, pp. 292–337].

The complete spherical quadrilateral, together with the proposition to which it belongs, also bears the names "quadrilateral figure", "complete quadrilateral","sector figure" or "secant figure", and there are others. The name "secant figure" was used by Ptolemy in the *Almagest*, and it is also the name that was adopted by the Arab mathematicians[6] in their commentaries on Menelaus' and Ptolemy's works and in the treatises on spherical geometry they wrote. The title of Naṣīr al-Dīn's treatise, which is the subject of the present chapter, refers to the complete spherical quadrilateral.

9.5 Compounded Ratios

Book I of Naṣīr al-Dīn's treatise is concerned with compounded ratios. In modern notation, a compounded ratio is an expression of the form $\frac{A}{B} \cdot \frac{C}{D}$. Menelaus' proposition called the Sector Figure, which we shall discuss in the next section and the rest of the treatise, make heavy use of such expressions. Therefore, a skill in the manipulation of compounded ratios is needed in following the arguments.

We remind the reader that expressions involving compounded ratios, such as the equality $\frac{A}{B} \cdot \frac{C}{D} = \frac{A}{D} \cdot \frac{C}{B}$, that are obvious for us, were not so for the Ancients. The reason has to do with the differences in our points of view on such expressions and the operations they involve. Indeed, we are used (consciously or unconsciously) to consider division of quantities and multiplication as formal algebraic operations pertaining to a certain mathematical structure (ring or field), and the identities that they satisfy follow from the axioms of this structure: commutativity, distributivity, etc. For a mathematician of Greek antiquity or of the Arabic period that followed, this was not the way mathematics was conceived. Their point of view goes back to Euclid's *Elements*, where the operations of multiplication and division have a geometric meaning, and the manipulation of expressions involving compounded ratios and equalities between them needed proofs.

As a matter of fact, there is hardly any definition of compounded ratios in Euclid's *Elements*. The first allusion to them is in Definition 5 of Book VI, which is considered to be a later addition; see Heath's comments in [6, Vol. 2, p. 189]. The definition, in Heath's translation, says the following: *A ratio is said to be compounded of ratios when the sizes of the ratios multiplied together make some (? ratio or size).*[7]

Proposition 23 of Book VI of the *Elements* is usually referred to as a typical place where compounded ratios are implicitly used. The proposition is purely geometric and it concerns areas of parallelograms. It says the following [6, Vol. 2, p. 247]: *Equiangular parallelograms have to one another the ratio compounded of the ratios of their sides*. It is interesting to read Heath's notes on this proposition. In his *History of Greek Mathematics*, he writes [7, Vo. 1, p. 393] (see Fig. 9.5 here):

> Proposition 23 is important in itself, and also because it introduces us to the practical use of the method of compounding, i.e. multiplying, ratios which is of such extraordinarily

[6] The name الشكل القطّاع, meaning the "secant figure", or the "cutting figure", was used by Thābit ibn Qurra (826–921) and others.

[7] An uncertainty is expressed by the question mark in Heath's edition.

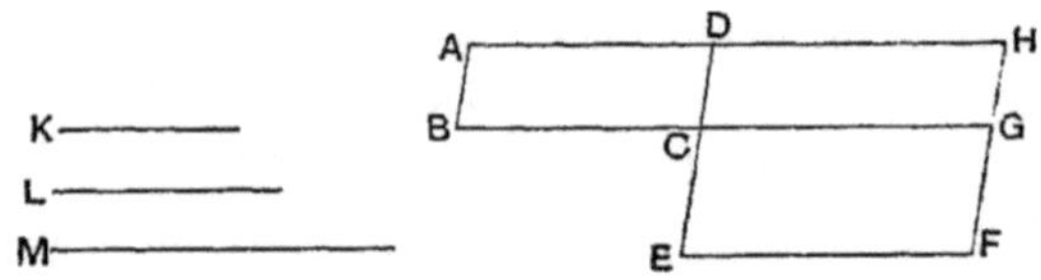

Fig. 9.5 Figure for Euclid's Proposition 23 of Book VI (from Heath's *History of Greek Mathematics*)

wide application in Greek geometry. Euclid has never defined "compound ratio" or the "compounding" of ratios; but the meaning of the terms and the way to compound ratios are made clear in this proposition. The equiangular parallelograms are placed so that two equal angles as BCD, GCE are vertically opposite at C. Complete the parallelogram $DCGH$. Take any straight line K, and find another L, such that

$$BG : CG = K : L,$$

and again another straight line M, such that

$$DC : CE = L : M.$$

Now the ratio compounded of $K : L$ and $L : M$ is $K : M$; therefore $K : M$ is the "ratio compounded of the ratios of the sides". And

$$(ABCD) : (DCGH) = BC : CG = K : L;$$

$$(DCGH) : (CEFG) = DC : CE = L : M,$$

therefore, *ex aequali*,

$$(ABCD) : (CEFG) : K : M.$$

This excursus should explain why Naṣīr al-Dīn makes a systematic study of compounded ratios in his treatise. In Book I, he gives 11 propositions on this subject. As an example, we quote one of them, namely, Proposition 10. It says the following (in modern terms):

> If we have an equality between a ratio of two quantities and a compounded ratio of four quantities expressed as $\frac{A}{B} = \frac{C}{E} \cdot \frac{D}{F}$, then the ratio of any one of the three quantities A, E, F to any one of the three quantities B, C, D is equal to a compounded ratio of the four remaining quantities, whose numerators are taken among the remaining quantities in A, E, F and the denominators among the remaining quantities in B, C, D.

Thus, this proposition says in particular that if $\frac{A}{B} = \frac{C}{E} \cdot \frac{D}{F}$, then $\frac{A}{C} = \frac{B}{E} \cdot \frac{D}{F}$. Naṣīr al-Dīn gives a proof of this proposition which is based on Propositions 33 and 34[8] of Book XI of the *Elements*, which say respectively [6, Vol. 3, p. 342 and

[8] We are using the numbering in Heath's edition. Naṣīr al-Dīn's numbering is different.

345] that *Similar parallepipedal solids are to one another in the triplicate ratio of their corresponding sides* and *In equal parallelepipedal solids the bases are reciprocally proportional to the heights; and those parallelepipedal solids in which the bases are reciprocally proportional to the heights are equal.* Thus, the language for compounded ratios is still geometric.

Now we pass to the rest of Naṣīr al-Dīn's treatise, where compounded ratios are extensively used.

9.6 The Sector Figure Theorem

We start by a few words of explanation on the expression "sector figure". This expression refers to a proposition (or a formula) and the corresponding figure. There is a Euclidean and a spherical sector figure.[9]

We start with the Euclidean proposition. The associated figure is represented in Fig. 9.1, in which we have a convex quadrilateral $AGED$ whose two pairs of opposite sides AG and DE are produced until they intersect. In other words, to the quadrilateral $AGED$ in Fig. 9.1, we adjoin the two triangles CED and BEG.

With this notation, the Euclidean sector figure says that

$$\frac{AG}{BG} = \frac{AC}{CD} \cdot \frac{DE}{EB}.$$

The use and the proof of this proposition also include the cases where one or two opposite edges AD and GE are parallel.

There is a similar construction on the sphere which gives a *spherical sector figure*. Using the same notation as in Fig. 9.1 and assuming now that the quadrilateral is spherical, this proposition says:

$$\frac{\sin AG}{\sin BG} = \frac{\sin AC}{\sin CD} \cdot \frac{\sin DE}{\sin EB}.$$

This is Menelaus' Theorem on the sphere, proved in [13] and used by Ptolemy in the *Almagest*. It is cumbersome, compared to the sine rule. This is why the Arabs, when they discovered the sine rule, called it sometimes "the proposition which exempts" (that is, it exempts from using Menelaus' Theorem). Naṣīr al-Dīn al-Ṭūsī uses this name as well.

[9] In Arabic, the same word is used for Figure and Proposition, "shakl" (شَكل). This is not unrelated to the fact that in Euclid's *Elements*, in Theodosius' *Spherics*, in Menelaus' *Spherics* and in other geometrical Greek treatises, to each proposition corresponds one and only one figure. Thus, the references to some proposition of one of these treatises is at the same time a reference to the corresponding figure.

9.7 A Summary of Book V and Some Propositions

Book V of Naṣīr al-Dīn's treatise contains seven chapters which are concerned with spherical trigonometry. The author studies in particular the following topics:

- (Chapter 2) A complete classification of the figures that are formed by three great circles on the sphere.
- (Chapter 3) A complete classification of spherical sector figures, regarding their side lengths (more precisely, whether they are less, equal to greater than a quadrant) and their angles (acute, right or obtuse) and the consequences of this information on the geometry of these figures.
- (Chapter 4) How to find the elements (sides and angles) of a triangle from some other given elements of this triangle
- (Chapter 5) The Sine Rule with a large number of different proofs, due to several mathematicians.
- (Chapter 6) Other trigonometric formulae, in particular formulae involving the cosine and the tangent functions.
- (Chapter 7) Given the sizes of three elements of a triangle (sides or angles), how to obtain the others using trigonometry and the polar triangle.

An Excerpt from Chap. 2 We now reproduce, from Chap. 2 of Book V a discussion concerning the union of three great circles which is more extensive than the one we had in Sect. 9.3. We reproduce it in full to give an idea of Naṣīr al-Dīn's style and his attention to details. The title of this chapter is *Properties of the triangles formed by the intersections of the great circles on the surface of the sphere and their different species*. The figure is the same, and we reproduce it here (Fig. 9.6). Naṣīr

Fig. 9.6 The eight triangles made by three great circles on the sphere

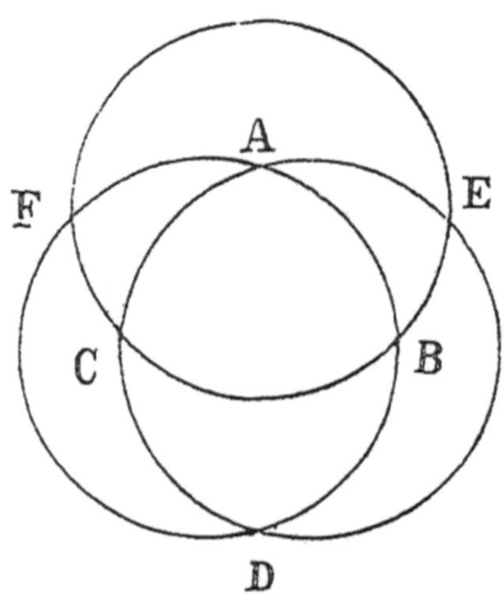

al-Dīn writes:

When three great circles intersect on the surface of the sphere in such a way that they form a triangle, they form at the same time seven other spherical triangles; in this way the surface of the sphere is divided into eight triangles, with six intersection points, twelve arcs of circle and twenty-four angles, and as we already said, among these eight triangles, the four situated in a hemisphere are equal and similar to the four others situated in the other hemisphere, each to each.

For example, the triangle ACF is equal to the triangle DEB because
$AF = \frac{1}{2}\text{circ.} - AB = DB$;
$AC = \frac{1}{2}\text{circ.} - CD = ED$;
$FC = \frac{1}{2}\text{circ.} - BC = EB$;
$\widehat{FAC} = \widehat{EAB}$ (as they are vertically opposite, and because $\widehat{EAB} = \widehat{ABC}$) $= \widehat{EDB}$;
$\widehat{AFC} = \widehat{ABC} = \widehat{EBD}$;
$\widehat{ACF} = \widehat{BCD} = \widehat{BED}$.

The same will hold for the other triangles, in such a way that the four triangles situated in opposite hemispheres will be pairwise equal, for what regards the angles as well as the sides.

For what concerns the four triangles situated in the same hemisphere, comparing any arbitrary two among them, we find that one of their angles and two of their sides are equal, and their other elements are supplementary of each other. Thus, if we compare the triangle ABC with the triangle CDF, we see that:

These two triangles have the angles at C equal, as vertically equal, and the side $AB = FD$ (both being the supplement of the arc AF), whereas $BC =$ suppl. $\widehat{FC}$; $AC =$ suppl. $\widehat{CD}$; the angle $\widehat{ABC} = \widehat{AFC} =$ suppl. $\widehat{CFD}$; and the angle $\widehat{BAC} = \widehat{BDC} =$ suppl. $\widehat{CDF}$.

With this, it suffices to know only one of these eight triangles for all of them to be determined.

Moreover, according to whether each side of a triangle is equal to, greater or smaller than a semicircle, we have the following ten species of triangles:[10]

[10] In the following table and later in the text, the letter D designates a quadrant, that is, a quarter of a great circle. In his translation, Carathéodory writes sometimes q and other times Q. We have opted for the capital Q.

$$A \begin{cases} \text{1st case}: \ 3 \text{ sides} = 3 \text{ Q}; \\ \text{2nd case}: \ 2 \text{ sides} = \text{Q}; \ 1 \text{ side} < \text{Q}; \\ \text{3rd case}: \ 2 \text{ sides} = \text{Q}; \ 1 \text{ side} > \text{Q}; \\ \text{4th case}: \ 1 \text{ side} = \text{Q}; \ 2 \text{ sides} < \text{Q}; \\ \text{5th case}: \ 1 \text{ side} = \ \text{Q}; \ 2 \text{ sides} > \text{Q}; \\ \text{6th case}: \ 1 \text{ side} = \ \text{Q}; \ 1 \text{ side} < \text{Q}; \ 1 \text{ side} > \text{Q}; \\ \text{7th case}: \ 3 \text{ sides} < \text{Q}; \\ \text{8th case}: \ 2 \text{ sides} > \text{Q}; \ 1 \text{ side} < \text{Q}; \\ \text{9th case}: \ 2 \text{ sides} < \text{Q}; \ 1 \text{ side} > \text{Q}; \\ \text{10th case}: \ 3 \text{ sides} > \text{Q}. \end{cases}$$

However since we find two of these species of triangles each time there is an intersection of three circles, it follows that there is a total of five species of intersections. Indeed, let us suppose that one of the eight triangles belongs to the 7th category, in other words, that each of its three sides is smaller than a quadrant. In this case, the three triangles situated in the same hemisphere belong to the 8th category since they must have necessarily two sides greater than a quadrant and one side smaller than a quadrant. Indeed, since the triangle in question should have its three sides equal to one side of each triangle, these will have a side smaller than a quadrant, whereas the two other sides will be the supplements of those of the first. Thus we can conclude from here that if the first triangle belonged to the 8th category, and if it had two sides greater than a quadrant and one side smaller, two of the remaining sides would be of that same category, whereas the third would belong to the 7th category and would have its three sides smaller than a quadrant. This is how two of the ten species of triangles that we have just enumerated (the 7th and the 8th) coexist and derive from the same nature of intersection. The same can be said of the 4th, the 5th and the 6th species, since among the four triangles, one will be of the 4th species, a second one of the 5th, and the two others of the 6th. Likewise for the 9th and for the 10th species (two of the four triangles belonging to each of these two species). Only the first species is not associated with any other species than itself.

Here are now the five intersections that we have just talked about:

$$B \begin{cases} \text{The 1st intersection comprises triangles of the 1st species.} \\ \text{The 2nd intersection comprises triangles of the 2nd and 3rd species.} \\ \text{The 3rd intersection comprises triangles of the 4th, 5th and 6th species.} \\ \text{The 4th intersection comprises triangles of the 7th and 8th species.} \\ \text{The 5th intersection comprises triangles of the 9th and 10th species.} \end{cases}$$

We can also arrange the spherical triangles according to their angles, depending on whether they are right, acute or obtuse, in ten categories as follows:[11]

$$
A' \left\{ \begin{array}{l}
\text{1st case : Each of the three angles} = \text{D.} \\
\text{2nd case : 2 angles} = \text{D, 1 angle} < \text{D.} \\
\text{3rd case : 2 angles} = \text{D, 1 angle} > \text{D.} \\
\text{4th case : 1 angle} = \text{D, 2 angles} < \text{D.} \\
\text{5th case : 1 angle} = \text{D, 2 angles} > \text{D.} \\
\text{6th case : 1 angle} = \text{D; 1 angle} > \text{D; 1 angle} < \text{D.} \\
\text{7th case : 3 angles} < \text{D.} \\
\text{8th case : 1 angle} < \text{D, 2 angles} > \text{D.} \\
\text{9th case : 3 angles} > \text{D.} \\
\text{10th case : 1 angle} > \text{D, 2 angles} < \text{D.}
\end{array} \right.
$$

The number of intersections giving rise to these last ten categories is also five:

$$
B' \left\{ \begin{array}{l}
\text{The 1st intersection comprises triangles of the 1st species} \\
\text{The 2nd intersection comprises triangles of the 2nd and 3rd species} \\
\text{The 3rd intersection comprises triangles of the 4th, 5th and 6th species} \\
\text{The 4th intersection comprises triangles of the 7th, and 8th species} \\
\text{The 5th intersection comprises triangles of the 9th and 10th species}
\end{array} \right.
$$

For what regards the way of finding the association of the various categories of triangles noted under each intersection species, we shall adopt an approach which is similar in every point to the one we used while we treated the first five intersections (B) relative to the ten categories of triangle (A), ordered according to the magnitude of their sides.

An Excerpt from Chap. 3 Without going into the proofs, we quote now a few propositions from Chap. 3 of Book V. These propositions are more involved than those of Chap. 2. This chapter is titled *On the rules relative to the various species of triangles and on the general and particular considerations attached to them.*

Proposition I. The angles of any triangle whose sides are quadrants are necessarily equal to 90°. The intersections of the sides constitute [coincide with] the poles of the sides, each vertex is the pole the opposite side.

[11] In the following table, we use the letter D to mean a right angle, following Carathéodory who uses D for "Droit".

Proposition II. If a spherical triangle has two sides equal to a quadrant, and the third side less than a quadrant, then it will have two angles equal to 90° and an acute angle. The vertex of the latter is the pole of the arc that is opposite to it, and the poles of the other two sides are placed on the arc of the acute angle, outside the triangle.

Proposition III . If two sides of a spherical triangle are equal to a quadrant and the third greater than a quadrant, then this triangle has two right angles and an obtuse angle (the one opposite to the side greater than a quadrant). The vertex of the latter is the pole of the arc that is opposite to it. As for the two other sides, their poles lie on the arc of the obtuse angle in the interior of the triangle.

Proposition IV. If a spherical triangle has one side equal to a quadrant and the two others smaller than a quadrant, then the angle opposite to the quadrant is obtuse, and the two other angles are acute. Furthermore, the three poles will lie inside the triangle.

Proposition V. If a spherical triangle has one side equal to a quadrant and the two others greater than a quadrant, then all its angles are obtuse. The poles will be situated in the interior of the triangle.

Proposition VI. If a spherical triangle has one side equal to a quadrant, one side smaller and one side greater than a quadrant, then the angle opposite to the greatest side is obtuse and the two other angles are acute. The three poles fall outside the triangle.

Proposition VII. If the three sides of a triangle are less than a quadrant, then the triangle has two acute angles. The third angle can be right, obtuse or acute. The three poles of this triangle fall outside the triangle.

Proposition VIII. If a triangle has two sides greater than a quadrant and the third one smaller than a quadrant, then its angles are necessarily of one of the following types:

(1) One right angle and two obtuse angles.
(2) One right angle, one acute angle and one obtuse angle.
(3) One acute angle and two obtuse angles.
(4) One obtuse angle and two acute angles.
(5) Three obtuse angles.

Proposition IX. If a triangle has one side greater than a quadrant and the two others smaller than a quadrant, then the angle opposite to the greatest side is obtuse and the two others are acute. The three poles fall outside the triangle.

Proposition X. If the three sides of a triangle are all greater than a quadrant, then the three angles are obtuse and the three poles fall in the interior of the triangle.

Chapters 4 to 6 of Naṣīr al-Dīn's treatise are dedicated to the proofs of the spherical trigonometric formulae. The proofs use the notion of polar triangle.

An Excerpt from Chap. 5 Chapter 5 is titled "On the so-called supplementary figure and its different species". Naṣīr al-Dīn mentions in this chapter a large number of mathematicians who contributed to the formulae of spherical trigonometry. They include in particular Abu al-Rayḥān al-Bīrūnī, Abū Naṣr Manṣūr Ibn ‘Irāq, Abū al-Wafā’ Muḥammad Ibn-Muḥammad al-Būzjānī, Abū Maḥmūd Ḥamīd ibn al Azhar al-Khujandī, Abul Faḍl al-Nayrīzī, in his commentary on the *Almagest*, Abū Ja‘far al Khāzin in his book titled *Partial researches on the inclination of the inclinations and introduction of the straight sphere*, Kūshyār ibn Labbān al-Jīlī, Thābit ibn Qurra and several others. The history of spherical trigonometry in the medieval Arab world has been investigated relatively recently, and all the authors mentioned by Naṣīr al-Dīn play some role in this history, see [1, 3, 4, 14].

The chapter starts with the following central principle, namely, the sine rule:

Proposition *The ratios of the sines of the sides of the triangles formed on the surface of a sphere by the intersection of arcs of great circles are equal to the ratios of the sines of the angles opposite to these sides.*

Naṣīr al-Dīn writes: “We usually start by establishing this principle for what concerns the right triangle and to this effect we follow various ways that are all exposed in the book of the scholar Abu al-Rayḥān al-Bīrūnī titled *The keys of the knowledge of the figures of the spherical and other surfaces*”.[12]

He then gives proofs of this special case, each time mentioning their author (Abū Naṣr, Abū al-Wafā’, al-Khujandī, Abu al-Rayḥān, etc.) One of these proofs uses the sector figure. He deduces from this several trigonometric formulae.

At the end of Chap. 5, Naṣīr al-Dīn writes the following:

> Abū Maḥmūd al-Khujandī gave to this figure [the supplementary figure] the name “astronomy rule”. Others called it *The figure which exempts from the quadrilateral*. In his book titled *The keys of the knowledge of what happens on the surface of the sphere*, Abu al-Rayḥān declares that it is indeed the Emir Abū Naṣr who was the first to use the supplementary figure instead of the quadrilateral, but that the name which it carries comes from Kūshyār ibn Labbān al-Jīlī. However, this assertion comes up against a difficulty, because the Emir Abū Naṣr, in the second part of the first book of the work titled *The Royal Almagest* at the beginning of Chapter III, where this figure is mentioned, writes verbatim: “The third chapter of what can exempt from the quadrilateral figure”; and then, after he has mentioned the treatise of Thābit ibn Qurra on the different varieties that occur in the quadrilateral figure, he adds: “And Thābit ibn Qurra also wrote a treatise on what exempts from the quadrilateral figure, but he who utilizes it must know the use of the compounded ratios; but I will show right here a procedure which exempts you from the quadrilateral as well as from the compounded ratios.” These words prove indeed that the term “supplementary figure” itself is due to the Emir Abū Naṣr who received it from Thābit ibn Qurra.[13]

We end by quoting a few results from Chap. 6.

[12] See Al-Bīrūnī, *Kitāb Maqālīd ‘Ilm al-hay’a (The Keys to the Science of Astronomy): La trigonométrie sphérique chez les Arabes de l’Est à la fin du Xe siècle*, Marie-Thérèse Debarnot (editor and translator), Damascus, 1985.

[13] Carathéodory writes in a footnote that a note in the margin of the manuscript reads: “I add: That the Emir Abū Naṣr talked about this in his *Commentary on Menelaus*, as I have already reported on, on the occasion of the Supplementary Figure.”

A Few Results from Chap. 6 This chapter is titled "On the figure called shadowed, on its consequences and its accessories".

The shadow is what we call the tangent of an angle. Again, in addition to the mathematics, Naṣīr al-Dīn provides us with historical information. He writes that the priority, for what concerns this figure, is due to Abū al-Wafā' al-Būzjānī, an information provided by Abu al-Rayḥān. He states the following proposition which concerns right spherical triangles:

$$\frac{\text{the sine of one of the sides of the right angle}}{\text{the sine of the right angle}}$$

$$= \frac{\text{the shadow of the other side of the right angle}}{\text{the shadow of the angle opposite to this side}}.$$

From this proposition, Naṣīr al-Dīn deduces several corollaries, among which we state the following:

Corollary *In a right triangle, the cosine of the acute angle that we assume it has is to the sine of the given right angle like the tangent of the complement of the arc opposite to this right angle to the tangent of the complement of the side contained between the right angle and the angle that we assumed to be acute, or else as the tangent of the side contained by the two angles is to the tangent of the side opposite to the given right angle.*

Naṣīr al-Dīn ends this chapter by several remarks on the tangent function, saying that one shortage is the rapid increase of the tangent of the arcs that exceed 45°. He writes: "The tangents increase rapidly beyond the radius, because already the tangent of the eighth part of the circle (45°) is equal to the radius. Thus if you write the tangents in a table in which the arcs increase by equal degrees, the differences between the tangents beyond 45° become very large. This is why we do not have much confidence in taking these tables and modifying what is actually between the lines, as it is done for the other tables. However, if we divide the arcs (in two categories), we will find that the objection that we made on this account to this figure is groundless, since it is not necessary to take tangents in the tables. In any circumstance, we will be able to operate by using the tangents of the eighth part of the circle, because the tangents have properties that we do not find in the sines, and which allow us to use some of them for the others. We have said a few words about this matter at the beginning of each chapter, and we shall soon show how we can use all the tangents by knowing only the tangents of the arcs that are less than the eighth part of the circle."

We have reproduced these excerpts from Book V of Naṣīr al-Dīn al-Ṭūsī's treatise, written before the introduction of the trigonometric functions in Western Europe, to give an idea of its content, both from the mathematical and from the historical point of view. We have also tried to suggest the preciseness of Naṣīr al-

Dīn's writing and his aim to be exhaustive: he always considers all the cases of a construction or a proposition.

References

1. H. Bellosta, Le traité de Thābit ibn Qurra sur la figure secteur. Arabic Sci. Philos. **14**(1), 145–168 (2004)
2. B. Carra de Vaux, Les sphères célestes selon Naṣīr al-Dīn al-Ṭūsī, in *Recherches sur l'histoire de l'astronomie ancienne*, ed. by P. Tannery (Gauthier-Villars, Paris, 1893), Appendice VI, pp. 337–61
3. M.T. Debarnot, *La trigonométrie sphérique chez les Arabes de l'Est à la fin du Xe siècle* (Institut français de Damas, Damascus, 1985)
4. M.T. Debarnot, Trigonometry, in *Encyclopedia of the History of Arabic Science*, ed. by R. Rashed, 3 vols, vol. 2 (Routledge, New York, 1996), pp. 470–513
5. N. Halma, (Ed.), *Composition mathématique de Claude Ptolémée, traduite par Nicolas Halma et suivie des notes de Jean Baptiste Delambre*, 2 vols. (Paris, Henri Grand, 1813). Reprint, 1988, Librairie Blanchard, Paris
6. T.L. Heath, *The Thirteen Books of Euclid's Elements*, 3 vols (Cambridge University Press, Cambridge, 1908). Reprint, Dover
7. T.L. Heath, *A History of Greek Mathematics*, 2 vols (Clarendon Press, Oxford, 1921). Reprint, Dover
8. E.S. Kennedy, Two Persian astronomical treatises by Naṣīr al-Dīn al-Ṭūsī. Centaurus **27**(2), 109–120 (1984)
9. A. Kubesov, The commentaries of Naṣīr al-Dīn al-Ṭūsī on the treatise of Archimedes *On the sphere and cylinder* (in Russian). Voprosy Istor. Estestvoznan. i Tehn. Vyp. **2**, 23–28 (1969)
10. A. Papadopoulos, G. Théret, Johann Heinrich Lambert's memoir *Theorie der Parallellinien*, A review with commentary, in *Spherical Geometry in the Eighteenth Century, I: Euler, Lagrange and Lambert*, chap. 7 (Springer, Berlin, 2026)
11. F.J. Ragep, *Naṣīr al-Dīn al-Ṭūsī's Memoir on Astronomy*, 2 vols (Springer, New York, 1993)
12. R. Rashed (Ed.), *Encyclopedia of the History of Arabic Science*, 3 vol (Routledge, Milton Park, 1996)
13. R. Rashed, A. Papadopoulos, *Menelaus' Spherics: Early Translation and al-Māhānī/al-Harawī's Version* (Critical edition of Menelaus' Spherics from the Arabic manuscripts, with historical and mathematical commentaries). De Gruyter, Series: Scientia Graeco-Arabica, vol. 21 (De Gruyter, Berlin, 2017), 890pp.
14. B.A. Rosenfeld, On the mathematical works of Naṣīr al-Dīn al-Ṭūsī (in Russian). Istor.-Mat. Issled. **4**, 489–512 (1951)
15. B.A. Rosenfeld, A.P. Yushkevich, *Notes on the Treatise of Naṣīr al-Dīn al-Ṭūsī on Parallel Lines* (in Russian). Istor.-Mat. Issled. **13**, 525–532 (1960)
16. B.A. Rosenfeld, *A History of Non-Euclidean Geometry: Evolution of the Concept of a Geometric Space*. Translated by A. Shenitzer (Springer, New York, 1988). First Russian edition, 1976
17. Traité du quadrilatère, attribué à Nassiruddin-el-Toussy, d'après un manuscrit tiré de la bibliothèque de S. E. Edhem Pacha, ancien grand vizir, traduit par Alexandre Pacha Carathéodory, ancien ministre des affaires étrangères, Constantinople, Typographie et Lithographie Osmanié (1891)

Chapter 10
The Polar Triangle, from Ibn ‘Irāq to Euler and Lagrange

Athanase Papadopoulos

Abstract The notion of polar triangle plays a central role in spherical geometry. Euler and Lagrange used this triangle in their respective memoirs *Trigonometria sphaerica universa, ex primis principiis breviter et dilucide derivata* and *Solution de quelques problèmes relatifs aux triangles sphériques, avec une analyse complète de ces triangles*, when they noticed that we can transform some formulae of spherical trigonometry into others by interchanging the letters for sides and angles after replacing the values of the angles and sides by their supplements to two right angles. This chapter is a historical on the notion of polar triangle since its discovery by the Arab mathematician Ibn ‘Irāq (10th-11th c.) until the modern period.

Keywords Polar triangle · Spherical geometry · Spherical trigonometry · Polarity · Duality · Leonhard Euler · Joseph-Louis Lagrange · Ibn ‘Irāq · Naṣīr al-Dīn al-Ṭūsī

AMS Classification 01A30, 01A50, 01A55, 51-03

10.1 Introduction

Leonhard Euler, in §7 of his memoir *Trigonometria sphaerica universa, ex primis principiis breviter et dilucide derivata* (General spherical trigonometry, deduced from fundamental principles in a brief and clear manner) [7] in which he establishes a comprehensive set of formulae of spherical trigonometry, points out a remarkable analogy between some of these formulae. In §7, he writes, with reference to the two formulae

$$\cos A \sin c = \cos a \sin b - \sin a \cos b \cos C$$

A. Papadopoulos (✉)
Institut de Recherche Mathématique Avancée, Université de Strasbourg et CNRS, Strasbourg, France
e-mail: athanase.papadopoulos@math.unistra.fr

R. Caddeo, A. Papadopoulos (eds.), *Spherical Geometry in the Eighteenth Century I*, https://doi.org/10.1007/978-3-032-12466-1_10

and

$$\cos a \sin C = \cos A \sin B + \sin A \cos B \cos c$$

which he had established in §5 and 6 respectively, referring to a spherical triangle of sides a, b, c and angles A, B, C opposite to them respectively:

> "[...] and this [formula] differs from the preceding one only by the fact that the capital letters are exchanged with the small letters and moreover all the cosines are taken with a negative sign.

In §12 of his memoir, he states the following as a theorem:

> Given an arbitrary spherical triangle, with angles A, B, C and sides a, b, c, it is always possible to construct another triangle, whose angles are the complements to two right angles of the sides of the given triangle, whereas the sides are the complements to two right angles of the angles.

Likewise, Lagrange, in §18 of his memoir *Solution de quelques problèmes relatifs aux triangles sphériques avec une analyse complète de ces triangles* (Solution of some problems relative to spherical triangles with a complete analysis of these triangles) [10], whose goal is also to establish a set of formulae of spherical trigonometry, notes, regarding the two formulae

$$\cos a = \cos b \cos c + \sin b \sin c \cos A$$

and

$$\cos A = -\cos B \cos C + \sin B \sin C \cos a,$$

which he had proved in §14 and 17 respectively, that the second one is obtained from the first by replacing the sides a, b, c by the angles A, B, C, keeping the sines unchanged and replacing the cosines by their opposite value. Lagrange is more explicit than Euler about this triangle, and he gives it a name, "supplementary triangle". He writes (§18):

> [...] It follows from here the well-known property of spherical triangles, that every spherical triangle can be changed into another one whose sides and angles are respectively the supplements of the angles and sides of the first one; and we know that this new triangle, which we call *supplementary*, is the one that is formed on the sphere by the three poles of the arcs that constitute the sides of the given triangle, by joining these poles by arcs of great circles.

Lagrange gives no indication on the origin of this idea, whereas for several other notions and proofs in the same memoir he mentions names of various mathematicians: Girard, Cavalieri, Wallis, Euler, de Gua and Legendre.

At any rate, the remarks made by Euler and Lagrange are immediate consequences of the existence of a triangle dual to the triangle ABC, better known as the *polar triangle*, whose angles (respectively sides) are the complements to two right angles of its sides (respectively angles).[1] Lagrange calls this triangle the *supplementary triangle*, because of the property that the sides (respectively angles) of the first one are the supplement of the angles (respectively sides) of the other one. This is the polarity, or duality[2] property expressed by the two mathematicians.

For a long time, the oldest reference to the polar triangle was attributed to François Viète (1540–1603). But some manuscripts, which are now published, show that the theory of the polar triangle was developed by Abū Naṣr Manṣūr ibn Alī Ibn 'Irāq[3] (end of 10th century), and was used by several later mathematicians in the period called the Golden Age of Arabic mathematics. It seems that the first person to report on this work of Ibn 'Irāq was Paul Luckey [11]. See also the reference [5] by Debarnot and the entry "Triangle polaire" in [16].

In the rest of this chapter, we shall make the above points more explicit. We already started by quoting Euler and Lagrange, whose works on the subject are highlighted in the present volume. In the next section, I will review the way some mathematicians, from the period that followed that of Euler and Lagrange, treated the history of the polar triangle. Next, I shall talk about the polar triangle in the work of the thirteenth century mathematician Naṣīr al-Dīn al-Ṭūsī, and then in that of the tenth–eleventh century mathematician Ibn 'Irāq. This is not the chronological order, but it is closer to the order in which the writings of the various authors on this subject became known in the West.

10.2 The Polar Triangle in the Eighteenth and Nineteenth Centuries

The history of the polar triangle was very poorly known for many centuries. Until the nineteenth century, the reference to this notion was François Viète (1540–1603), with sometimes references to Adrian Metius (1571–1635), Giovanni Antonio Magini (1555–1617), Bartholomeo Pitiscus (1561–1613), John Napier (1550–1617), Bonaventura Cavalieri (1598–1647) and Willebrord Snell (1580–1626). Let

[1] We recall that distances on the sphere are also angles: On the sphere of radius 1 embedded in Euclidean 3-space, the distance between two points is the angle made by two rays starting at the origin and joining these points.

[2] For our purposes here, we do not make a difference between polarity and duality; the result, for what concerns the trigonometric formulae, is the same.

[3] Ibn 'Irāq belongs to a group of Arab mathematicians from the 10th-11th centuries who worked on a renewal of spherical geometry. Both Ibn 'Irāq and Naṣīr al-Dīn al-Ṭūsī published editions of Menelaus' *Spherics*, see [13] and [9]. Excerpts from their commentaries from these editions are translated into English in the papers [17], [18] and in the book [19], the latter being the first complete English edition of Menelaus' work (91 propositions), translated from the Arabic version of Abū al-Faḍl Aḥmad ibn Abī Sa'd al-Harawī (d. 1215) based on a first translation from the Greek by Abū 'Abd Allāh al-Māhānī (fl. 853–66). (No Greek version of Menelaus' *Spherics* survives.)

us quote Jean-Baptiste Delambre[4] from his *Histoire de l'astronomie au moyen-âge* (History of astronomy in the middle-ages) (1819) [6, p. 474], where this author seems to suggest that Viète had not completely grasped the concept of the polar triangle:

> *If from the three vertices of an arbitrary triangle taken as poles we draw arcs of great circles, the resulting triangle will be reciprocal to the first one, as much for the angles as for the sides.* Viète does not say more. One is tempted to deduce that he has just defined, under the name of *reciprocal* triangle, the polar or supplementary triangle which all the modern trigonometries deal with. However, in the true polar triangle, the sides are the supplements of the angles of the initial triangle, and the angles the supplements of the sides and conversely. Viète does not pronounce the word *supplement*. One is tempted to believe that the sides simply become the angles, and the angles are transformed into the sides, which is not correct. We shall see the truth later on. In the meanwhile, it suffices to certify that, up to now, nothing proves that Viète was the first to get the idea of the supplementary triangle which the moderns deal with.

Michel Chasles, in his famous treatise *Aperçu historique sur l'origine et le développement des méthodes en géométrie* (Historical preview of the origin and development of methods in geometry), writes that the polar triangle first appeared in Europe in the work of Viète and without criticizing his definition [4, p. 55]:

> We must especially note, in Viète's trigonometry, a new and infinitely fortunate idea, which has a direct relationship with the new doctrines of Geometry; it is the transformation of spherical triangles into others, whose angles and sides correspond, in a certain way, to the sides and angles of the proposed triangles."If, from the three vertices of a spherical triangle," he says, "we describe arcs of great circles as poles, the new triangle that results will be reciprocal to the first triangle, both by the angles and by the sides".

Joseph Serret (1819–1885) uses extensively the polar triangle (which he calls supplementary triangle) in his *Traité de trigonométrie* [20]. In the paper *Triangles polaires réciproques et triangles polaires réciproques supplémentaires* (Reciprocal polar triangles and supplementary polar triangles), published in 1854,[5] the author writes again that the idea of polar triangle is due to Viète, and he provides the reference [22]. But he declares at the same time that the first who described and explicitly used the polar triangle is Willebrord Snellius (1580–1626), and he gives as reference Proposition VIII of Book III of [21].[6] He further writes that "Snellius

---

[4] Jean-Baptiste Delambre (1749–1822) is a famous French astronomer and historian of science. We owe to him an informative biography of Lagrange, to whom he was close. Together with Joseph Serret, Gaston Darboux and Léon Lalanne, Delambre was one of the editor of Lagrange's *Complete Works*.

[5] The paper is published in the *Nouvelles annales de mathématiques* without author attribution, which probably means that it can be attributed to the editor, Joseph Gergonne, [8].

[6] The English translation of Snellius' treatise is: *The four books of the canonical doctrine of [the theory of] triangles, in which is treated in a concise and clear manner the construction of the rule of sines, of tangents and of secants, and the rapid measurement of the triangles, plane as well as spherical.* On the front page of the edition [21], one can read (we are translating again into English): "After the author's death, the edition of the books was directed by Hortensius (Martin van den Hove) from Delft, who added some treatises on geodesics and on spheres, where the use of the main propositions is explained".

does not claim to be the inventor of the theorem [...] He says that he only delivered this theorem *a vinculis*;[7] this probably means that he freed the problem from the unnecessary complications of Viète and others."

Before moving on, let us point out that the idea behind this supplementary or polar triangle is at the basis of general polarity theory in projective geometry. Bourbaki, in his *Éléments d'histoire des mathématiques* [1, p. 83], provides a brief history of polarity, created, he says, in the seventeenth century by Girard Desargues (1591–1661) then brought back into honor towards the end of the eighteenth century by Poncelet [15], Brianchon [2] and Chasles [3]. Bourbaki quotes Chasles for having perceived the kinship of this notion with that of "reciprocal spherical triangle" which, he says, was introduced into spherical trigonometry by Viète and Snellius.

Thus, in the European mathematical literature up until the 19th century, regarding the origin of the polar triangle, several names appear, each offering varying opinions on the precise source of this concept. It was not until the late 19th century that it was first realized that the concept was known to the medieval Arab mathematicians. This is the subject of the remainder of this chapter.

10.3 A Treatise by Naṣīr al-Dīn al-Ṭūsī

In 1891, a French edition of a treatise by Naṣīr al-Dīn al-Ṭūsī (1201–1274), titled *Treatise on the quadrilateral*, was published in Constantinople, translated by Alexandre Carathéodory[8] [12]. In Book V of the treatise, in which the author establishes a complete system of trigonometric formulae, he uses the polar triangle.[9] This triangle appears in the last case discussed in Chapter VII of Book V of the treatise, namely, the case where three angles of a spherical triangle are known and we want to know the sides. Naṣīr al-Dīn considers three sub-cases. The first one is that of a triangle ABC whose sides are all less than a quadrant. He writes, for this case (see Fig. 10.1):

> I. We know the three angles of the triangle ABC.
>
> In the triangle ABC, let us produce AB, AC until AE, AD become quadrants. Likewise, BA, BC in such a way that BF, BH become quadrants., and let us make also CT, CK two quadrants. And let us produce the arcs of great circles DE, FH, TK. The

---

[7] From bonds.

[8] Alexandre Carathéodory was the great-uncle of the mathematician Constantin Carathéodory. Some biographical information of him is contained in Chap. 9 of the present volume.

[9] Book V of al-Ṭūsī's treatise is titled *Explanation of the methods that replace the quadrilateral figure in the research on great circles that intersect on the surface of a sphere*. The *quadrilateral figure* that is referred to in the title is Menelaus' Theorem, which was used in astronomical computations in the absence of the simpler trigonometric formulae. Chapter VII of that book is titled *Complementary developments of the way by which we come to know the unknowns from the knowns in spherical triangles*. A survey of al-Ṭūsī's treatise is contained in Chap. 9 of the present volume [14].

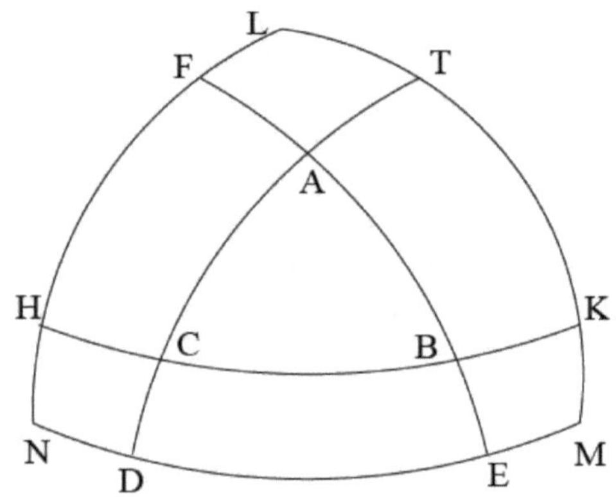

Fig. 10.1 The polar triangle, from al-Tusi's treatise [12]. This is the case where the three sides of the triangle ABC are smaller than a quadrant

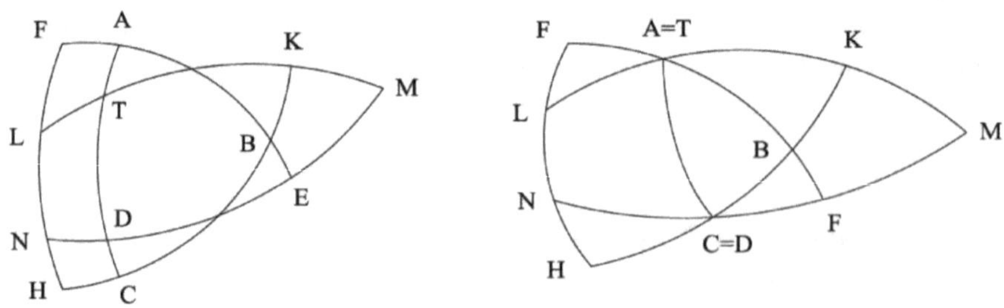

Fig. 10.2 The construction of the polar triangle, from al-Tusi's treatise [12]. The cases where the three sides of the triangle ABC are not all smaller than a quadrant

points where they meet will be L, M, N. In this way we have formed the triangle LMN whose sides are arcs of great circles.

The angles A, B, C being known, we also know DE, TK, FH. And F and H being two right angles, L will be the pole of KH. Likewise, we will have: M pole of TD, and N pole of FE. Now each of TL, KM being the complement of KT, LM, will be known. Likewise, LN and MN. Thus, the three sides of the triangle LMN will be determined, therefore its angles as well, from the preceding case, and therefore the arcs KH, DT, EF [are known]. But each of the arcs KC, BH being equal to a quadrant, the complement of HK will be equal to BC. Thus, BC will be known. We can say the same for AB, AC. Therefore, by the same procedure, we will know the sides of the triangle ABC.

Al-Ṭūsī discusses the cases where one of the sides is equal to or greater than a quadrant or smaller than a quadrant. He gives two figures which we reproduce here (Fig. 10.2), that show how one should construct the polar triangle in these cases.

10.4 Ibn 'Irāq and the Early Version of the Polar Triangle

In this section, we reproduce a translation of the earliest known written text on the polar triangle. The text is an excerpt from a letter sent by Ibn 'Irāq to al-Bīrūnī in

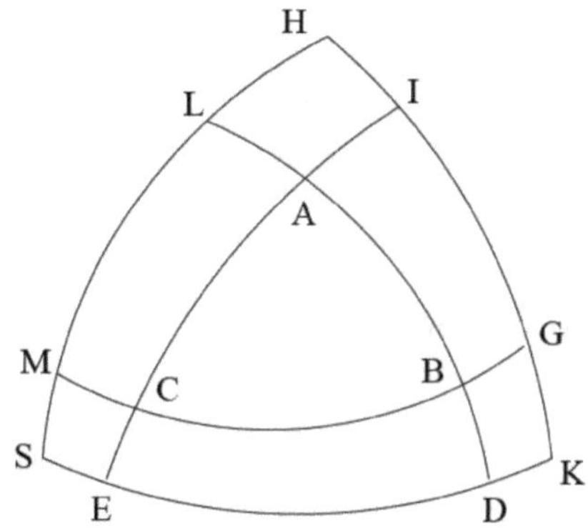

Fig. 10.3 The polar triangle, reproduced from Ibn-Iraq's manuscript

which he corrects a treatise by Abū Ja'far al-Khāzin called *Zīj al-ṣafā'iḥ*.[10] In this letter, Ibn 'Irāq introduces the notion of polar triangle to show that when the sides of a spherical triangle are known, then its angles are known, and conversely (Fig. 10.3).

Abū Naṣr says:

Consider a spherical triangle ABC and suppose that the three angles A, B, C are known and the three sides are unknown. We wish to know the sides.

We complete (the sides of the triangle ABC) to quarter circles and let us draw, using as a pole of each one of the points A, B, C and at a distance equal to a quarter circle, the arcs ED, IG, LM. We then produce these arcs until the three circles meet. As these circles meet at the points K, H, S, we get a triangle KHS made out of great circle arcs. Since the angles A, B, C are known, the arcs DE, IG, LM are known. Since the circle AC passes by the poles of the circles DE, DG, these two circles also pass by the two poles of the circle AC. Therefore the point K is a pole of AC. Since the circle AB passes through the circles DE, LM, then the two circles also pass through the poles of AB, and the point S is the pole of AB. Since the circle BC passes through the poles of the circles IG, LM, these two circles also pass through the poles BC, and the point H is the pole of BC. Since the circle AB passes through the poles of the two circles DE, LM, these two circles also pass through the two poles of AB, and the point S is a pole of AB. Since the circle BC passes through the poles of the two circles GI, LM, these two circles also pass through the two poles B, C. The point H is a pole of BC. Therefore, each of KE, DS, KI, GH, MH, LS is a quarter of a great circle. Since the arcs DE, IG, LM are known, the sides KH, HS, SK are known, because each of them exceeds by a quarter the complement to a quarter of an arc which is known. Therefore the angles K, H, S are known. Therefore, the arcs IE, GM, LD are known, and IE exceeds a quarter by the complement of AC to a quarter, GM exceeds a quarter by the complement of LC to a quarter, and LD exceeds a quarter by the complement of AB to a quarter of a circle. Finally, the remaining arcs AB, AC, BC are known, which is what we wanted[11].

[10] *Risālat Ibn 'Irāq ilā al-Bīrūnī fī taṣḥīḥ mā waqa' li-al-Khāzin min al-sahw fī zīj al-ṣafā'iḥ*, ms. Khudabakhsh, no. 2519, fol. 75r. The English translation of the title is: *The letter of Ibn 'Irāq to al-Bīrūnī on the correction of what happened to al-Khāzin's plane astronomical tables while he was distracted.*

[11] Ms. Khudabakhsh, no. 2519, fol. 75^r, see Footnote 10. See the article *Triangle polaire* in [16], and the complete reference in [19, p. 798], from which this translation is extracted.

Thus, to a triangle ABC, Ibn 'Irāq associates the triangle HKS whose sides are the complements to 180° of the angles of ABC and whose angles are the complements to 180° of the sides of ABC. The construction assumes that the angles of the initial triangle are known; the other cases are left to the reader.

References

1. N. Bourbaki, *Éléments d'histoire des mathématiques*, 1st edn. (Masson, Paris, 1984). New edition, N. Bourbaki and Springer-Verlag, Berlin and Heidelberg, 2007
2. C.-J. Brianchon, *Application de la théorie des transversales, cours d'opérations géométriques sur le terrain, fait a l'école d'artillerie* (Bachelier, Paris, 1818)
3. M. Chasles, *Mémoire de géométrie sur deux principes généraux de la science: la dualité et l'homographie, impr* (Hayez, Bruxelles, 1837)
4. M. Chasles, *Aperçu historique sur l'origine et le développement des méthodes en géométrie, particulièrement de celles qui se rapportent à la géométrie moderne; suivi d'un mémoire de géométrie sur deux principes généraux de la science, la dualité et l'homographie, Mémoires couronnés par l'Académie royale des sciences et belles lettres de Bruxelles*, tome XI (Bruxelles, Hayez, 1837); 1st ed., Paris, Gauthier-Villars, 1875; 3rd ed. 1889
5. M.-Th. Debarnot, Introduction du triangle polaire par Abū Naṣr b. 'Irāq. J. History Arabic Sci. **2**(1), 126–136 (1978)
6. J.-B. Delambre, *Histoire de l'astronomie du moyen-âge* (Courcier, Paris, 1819)
7. L. Euler, Trigonometria sphaerica universa, ex primis principiis breviter et dilucide derivata. Acta Acad. Sci. Imper. Petropol. **3**, 72–86 (1782); *Opera Omnia* Series 1, vol. XXVI, p. 224–236
8. J.D. Gergonne, Triangles polaires réciproques et triangles polaires réciproques supplémentaires. Nouvelles Ann. Math. 1re série **13**, 377–381 (1854)
9. Ibn 'Irāq, Abū Naṣr Manṣūr, Die Sphärik von Menelaos aus Alexandrien in der Verbesserung von Abū Naṣr Manṣūr ben 'Alī Ibn 'Irāq, mit Untersuchungen zur Geschichte des Textes bei den islamischen Mathematiker. Obers. M. Krause, Abhandlungen der Gesellschaft der Wissenschaften zu Göttingen, phil.-hist. Klasse, 3, 17, Berlin (1936)
10. J.L. Lagrange, Solution de quelques problèmes relatifs aux triangles sphériques, avec une analyse complète de ces triangles. J. de l'école Polytech. **2**, 270–297 (1800)
11. P. Luckey, Zur Entstehung der Kugeldreiecksrechnung. Deutsch. Math. **5**, 405–446 (1940)
12. Naṣīr al-Dīn al-Ṭūsī, in Traité du quadrilatère, attribué à Nassiruddin-el-Toussy, d'après un manuscrit tiré de la bibliothèque de S. E. Edhem Pacha, ancien grand vizir, traduit par Alexandre Pacha Carathéodory, ancien ministre des affaires étrangères, Constantinople, Typographie et Lithographie Osmanié (1891)
13. Naṣīr al-Dīn al-Ṭūsī, Iṣlāḥ kitāb Mānālāwus fī al-ashkāl 1 al-kuriyya (Osmania Oriental Publications Bureau, Hyderabad, 1948)
14. A. Papadopoulos, Spherical trigonometry in the medieval era: the treatise of Naṣīr al-Dīn al-Ṭūsī, Chap. 9, in *Spherical Geometry in the Eighteenth Century, I: Euler, Lagrange and Lambert* (Springer, Berlin, 2026)
15. J.-V. Poncelet, *Traité des propriétés projectives des figures: ouvrage utile à ceux qui s'occupent des applications de la géométrie descriptive et d'opérations géométriques sur le terrain*, 2 vols (Gauthier-Villars, Paris, 1865–1866)
16. R. Rashed (Ed.), Lexique historique de la langue scientifique arabe, in *Arabic Science and Culture*, vol. 1 (Hildesheim, Zürich, 2017), pp. 96–100
17. R. Rashed, A. Papadopoulos, *On Menelaus' Spherics III.5 in Arabic Mathematics, I: Ibn 'Irāq, Arabic Science and Philosophy*, vol. 24 (Cambridge University Press, Cambridge, 2014), pp. 1–68

18. R. Rashed, A. Papadopoulos, *On Menelaus' Spherics III.5 in Arabic mathematics, II: Naṣīr al-Dīn al-Ṭūsī and Ibn Abī Jarrāda*. Arabic Science and Philosophy, vol. 25 (Cambridge University Press, Cambridge, 2015), pp. 1–32
19. R. Rashed, A. Papadopoulos, *Menelaus' Spherics: Early Translation and al-Māhānī/al-Harawī's version* (Critical edition of Menelaus' *Spherics* from the Arabic manuscripts, with historical and mathematical commentaries), De Gruyter, Series: Scientia Graeco-Arabica, vol. 21 (De Gruyter, Berlin, 2017), 890pp.
20. J. Serret, *Traité de trigonométrie* (Bachelier, Paris, 1850)
21. W. Snellius, Doctrinae triangulorum canonicae libri quatuor, quibus canonis sinuum, tangentium et secantium constructio, triangulorum tam planorum quam sphaericorum expedita dimensio, breviter ac perspicue traditur; post mortem auctoris in lucem editi a Martino Hortensio Delfensi; qui istis problematum geodeticorum et sphaericorum tractatus singulos adjunxit, quibus praecipuarum utriusque trigonometria propositionum usus declaratur. Lugduni Batavorum (1627)
22. F. Viète, *Opera Mathematica* (Bonaventura and Elzevier, Leyde, 1646)

Part II
Translations of Euler's Memoirs

Portrait of Leonhard Euler Jakob Emanuel Handmann, 1753, Kunstmuseum Basel

Chapter 11
Principles of Spherical Trigonometry Deduced from the Method of Maxima and Minima

Leonhard Euler

Abstract In this memoir, Leonhard Euler establishes a set of formulae of spherical trigonometry using methods of differential geometry and the calculus of variations. The formulae he obtains are, according to his words, very handful for calculations with the aid of logarithms. Using these methods and the formulae he finds, he gives a proof of the fact that the area of a spherical triangle is equal to its angle excess.

Keywords Spherical trigonometry · Leonhard Euler · History of spherical geometry · Calculus of variations · Mathematics in the eighteenth century

AMS Classification 01A50, 51-03, 53A05, 53A35

Since we know that the arcs of great circles, drawn on the surface of a sphere, represent the shortest path from one point to another, a spherical triangle can be defined in such a way that given three points on a spherical surface, if we draw on it the shortest path from one point to another, the space enclosed between these three points will be a spherical triangle. Then, since the sides of a spherical triangle are the shortest lines that we can draw from an angle to the other, the method of maxima and minima[1] can be employed to determine the sides of a spherical triangle; and

[1] This is the name that Euler gave first to the field that is now called "calculus of variations", a topic which he gave rise to and developed together with Lagrange. The name "calculus of variations" was used by Euler in his memoir *Elementa calculi variationum* (Elements of calculus of variations), published in *Novi Commentarii academiae scientiarum Petropolitanae* 10, 1766, p. 51–93, *Opera Omnia* Series 1, vol. XXV, p. 141–176. Before this, Euler had already used this name in his correspondence.

The author Leonhard Euler (1707–1783).
Translated from the French by Athanase Papadopoulos and Alena Zhukova.
Principes de la trigonométrie sphérique tirés de la méthode des plus grands et plus petits, *Mémoires de l'académie des sciences de Berlin*, 9, 1755, p. 223–257. Opera Omnia: Series 1, Volume 27, p.277–308.

R. Caddeo, A. Papadopoulos (eds.), *Spherical Geometry in the Eighteenth Century I*, https://doi.org/10.1007/978-3-032-12466-1_11

therefore we can find the relation that remains between the angles and the sides; and this is what spherical trigonometry consists in. Since the three points where the angles are situated will determine therefore both the three sides and the three angles, and since these six things will always have among themselves such a relation that when any three of them are known, we can determine the other three.

Thus, this is a property that spherical triangles share with planar triangles, which are the object of elementary trigonometry. In fact, as a planar triangle is the space enclosed between three marked points on a plane, when we draw from one to the other the shortest line, which in the plane is a straight line, likewise a spherical triangle is the space enclosed between three marked points on a spherical surface, when we join these three points by the shortest lines that we know how to draw from one to the other on the same surface. But it is clear that a spherical triangle turns into a planar triangle when the radius of the sphere becomes infinitely large, since a plane can be regarded as the surface of an infinitely large sphere.

One might without doubt object that the intention to use the calculus of maxima and minima for the establishment of the foundations of Spherical trigonometry goes against the rules; in addition to the fact that it seems useless to derive them once again from other principles, since those we have used up to now are based on Elementary geometry, whose rigor serves as a rule for all the other parts of Mathematics. But, first, I remark that the method of maxima and minima will acquire from this like a new light, when I will show that it can lead us by itself to the resolution of the spherical triangles; and then, it is always useful to reach the same truths through different paths, since our mind will surely deduce from there new clarifications.

But in addition to this, I can declare that the method of maxima and minima is much more general than the ordinary method. In fact the latter is restricted to triangles formed on surfaces which are either planar or spherical, whereas the former extends equally to arbitrary surfaces. Thus, if we ask the nature of triangles formed on spheroidal or conoidal surfaces, whose sides would be the lines that are the shortest that one can draw from one angle to the other, the ordinary method would not be appropriate for this research: but one should definitely use the method of maxima and minima, without which it is not possible even to know the shortest lines that would form the sides of these triangles.

Therefore we understand that this research can surely become of great importance, since the surface of the Earth, being not spherical, but spheroidal,[2] a triangle formed on the surface of the Earth will belong to the species of which I have just talked about. To see this, it suffices to imagine three points on the surface of the Earth joined by the shortest path that exists from one to the other, or that would be formed by a rope taut from one to the other; indeed, this is how we should imagine the triangles that are formed by the operations that are made for the measure of

[2] A spheroid is a surface obtained by the rotation of an ellipse along one of its axes. It is a special case of an ellipsoid. Considering the trigonometry of the spheroid instead of the sphere was an important point, because at the time of Euler it was already realized that the Earth has a spheroidal rather than a spherical shape, and in principle spheroidal geometry was needed in geography and the art of drawing geographical maps, in which Euler was involved.

the Earth. It is really true that these triangles are usually regarded as planar and rectilinear, and we already obtain a result which is enough accurate if we perform the calculations considering these (planar or rectilinear) triangles as triangles on the sphere. But if we manage to make these triangles much larger and if we want to calculate them with all the possible precision, we would surely be reduced to the search of the true nature of these triangles, which we cannot overcome without resorting to the method of maxima and minima.

Thus, having shown the importance of this method in the subject in question, it will not be out of place to apply this method to the resolution of the spherical triangles, since, on the one hand, this research will serve as a basis and a model for the resolution of the triangles formed on an arbitrary spheroidal surface, and, on the other hand, it will provide us with important clarifications, both on Spherical trigonometry itself and on the method of maxima and minima, of which we shall know more and more the extent and the great use. Because, since we have shown that most of the mechanical and physical problems can be resolved very promptly with the aid of this method, it cannot be but very pleasant to see that the same method brings such a great help for the resolution of the problems of pure Geometry.

To begin this research in a way that it extends equally to the sphere and to an arbitrary spheroid, I start by considering two opposite points on the sphere like its poles; the great circle that is at equal distance from them will represent the equator, and the shortest lines drawn from one pole to each point of the equator will represent meridians, which are perpendicular to the equator. But when it comes to the sphere, we can regard each side of a spherical triangle as being part of the equator, and when the triangle is right, we can always assume that one of the sides of the right angle is a portion of the equator, and the other side as a portion of a meridian, since we can take the two poles at wish; but this will not be the case as soon as the surface is no more spherical, but spheroidal.[3] However, I will talk here only about spherical surfaces, reserving the spheroidal ones for another memoir.[4]

Problem I

1. *Given* (Fig. 11.1) *the arc AP on the equator AB and the arc PM on the meridian OP, to find on the spherical surface the shortest lineAM that can be drawn from point A to point M.*

[3] In the case where the spheroid represents the Earth, the equator is the curve obtained by the rotation of the major axis, since the shape of the Earth is assumed to be a spheroid flattened at the poles.

[4] Euler announces here his memoir *Élémens de la trigonométrie sphéroïdique tirés de la méthode des plus grands et plus petits*, Mémoires de l'Académie royale des Sciences et des Belles-Lettres de Berlin, 1755, 258–293. Opera Omnia: Series 1,Volume 27, p. 309–339. English translation: Elements of spheroidal trigonometry drawn from the method of maxima and minima, In: Mathematical Geography in the eighteenth century: Euler, Lagrange and Lambert (ed. R. Caddeo and A. Papadopoulos), Springer, 2022, p. 205–242.

Fig. 11.1 Figure for Problem I

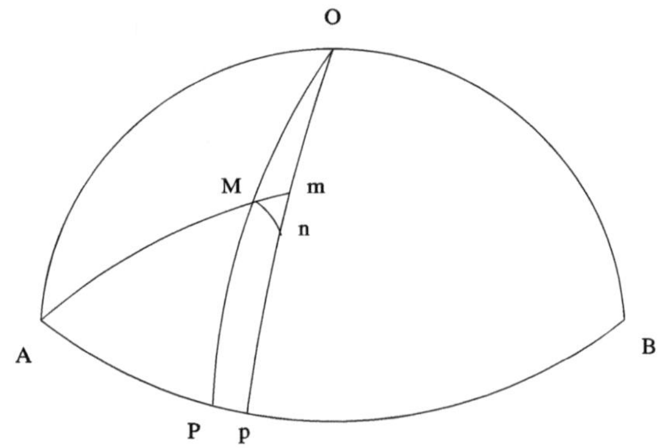

Solution

Setting the semi-diameter of the sphere $= 1$, let the arc of the equator $AP = x$ and the arc of the meridian $PM = y$. Furthermore, let the arc we are looking for $AM = s$, which we extend infinitely slightly to m, in such a way that $Mm = ds$, and let us draw from m the meridian Omp and the element Mn which is perpendicular to it. This being set, we get $Pp = dx$ and $mn = dy$; and since Pp is to Mn as the total sine 1 to the sine of the arc OM,[5] or to the cosine of $PM = y$, we get $Mn = dx \cos y$; and the triangle Mnm which is right at n will give

$$Mm = ds = \sqrt{dy^2 + dx^2 \cos^2 y},$$

therefore

$$AM = s = \int \sqrt{dy^2 + dx^2 \cos^2 y}.$$

Thus, we need to find a relation between x and y such that if we give them definite values such as AP and PM, the value of the integral

$$\int \sqrt{dy^2 + dx^2 \cos^2 y}$$

becomes the smallest possible. For this purpose, let us set $dy = pdx$, in order to reduce the integral to the form

$$\int dx \sqrt{p^2 + \cos^2 y};$$

[5] That is, as the sine of the total meridian $OP = 1$ is to the sine of the arc OM.

and since I have shown that when the integral formula $\int Zdx$, where Z is a function of x, y and p such that $dZ = Mdx + Ndy + Pdp$, must become a maximum or a minimum, this will occur from the equation

$$Ndx - dP = 0.$$

[6]Thus, applying this to our case, we get

$$Z = \sqrt{p^2 + \cos^2 y},$$

hence

$$dZ = -\frac{\sin y \cos y dy}{\sqrt{p^2 + \cos^2 y}} + \frac{pdp}{\sqrt{p^2 + \cos^2 y}},$$

therefore

$$M = 0, \quad N = -\frac{\sin y \cos y}{\sqrt{p^2 + \cos^2 y}} \quad \text{and } P = \frac{p}{\sqrt{p^2 + \cos^2 y}}.$$

But since $M = 0$, and therefore $dZ = Ndy + Pdp$, let us multiply by p the equation $Ndx - dP = 0$, which, since $dy = pdx$, will become $Ndy - pdP = 0$, or $Ndy = pdP$. This value being substituted for Ndy will give

$$dZ = pdP + Pdp,$$

whose integral is $Z = Pp + C$, or else

$$\sqrt{p^2 + \cos^2 y} = \frac{p^2}{\sqrt{p^2 + \cos^2 y}} + C,$$

which reduces to

$$\cos^2 y = C\sqrt{p^2 + \cos^2 y};$$

hence we deduce

$$C^2 p^2 = \cos^2 y(\cos^2 y - C^2)$$

[6] This is the simplest form of Euler's equation that is satisfied by a minimizing curve. This equation is established in Chapter 2 of Euler's *Methodus inveniendi lineas curvas maximi minimive proprietate gaudentes, sive solutio problematis isoperimetrici lattissimo sensu accepti*, Marcum-Michaelem Bousquet, Lausanne & Geneva, 1744, p. 1–322 (Proposition III and Corollary I, Section 21 and 22).

or

$$p = \frac{dy}{dx} = \frac{\cos y\sqrt{\cos^2 y - C^2}}{C}.$$

Thus, the relation between x and y is contained in this separated differential equation:

$$dx = \frac{Cdy}{\cos y\sqrt{\cos^2 y - C^2}},$$

whence we get

$$ds = \sqrt{p^2 + \cos^2 ydx} = \frac{\cos^2 ydx}{C},$$

therefore

$$ds = \frac{\cos ydy}{\sqrt{\cos^2 y - C^2}}$$

and the arc itself [is]

$$s = \int \frac{\cos ydy}{\sqrt{\cos^2 y - C^2}}.$$

Corollary 1
2. Therefore the equation

$$dx = \frac{Cdy}{\cos y\sqrt{\cos^2 y - C^2}}$$

expresses the nature of the line AM which has the property that taking any part of it, it will be the shortest line we can draw between its ends on a spherical surface. But the fact that this line is at the same time a great circle on the sphere, I have already proved elsewhere; and here it is not important for our goal what relation this line has with the sphere, provided we know that it is the shortest between its ends.

Corollary 2
3. Having found

$$dx = \frac{Cdy}{\cos y\sqrt{\cos^2 y - C^2}},$$

we get

$$Mn = \cos y dx = \frac{C dy}{\sqrt{\cos^2 y - C^2}}.$$

But $\frac{Mn}{mn}$ expresses the tangent of the angle AMP, therefore we get

$$\tan AMP = \frac{C}{\sqrt{\cos^2 y - C^2}}.$$

Furthermore, having

$$Mm = ds = \frac{\cos y dy}{\sqrt{\cos^2 y - C^2}},$$

the fraction $\frac{Mn}{Mm}$ expresses the sine of the angle AMP, in such a way that

$$\sin AMP = \frac{C}{\cos y} \text{ and } \cos AMP = \frac{\sqrt{\cos^2 y - C^2}}{\cos y}.$$

Corollary 3
4. Furthermore, setting $y = 0$, the point M will reach A and then the fraction $\frac{dy}{dx}$ will express the tangent of the angle PAM, $\frac{dy}{ds}$ its sine and $\frac{dx}{ds}$ its cosine. But then, having $\cos y = 1$, it will be

$$dx = \frac{C dy}{\sqrt{1 - C^2}} \text{ and } ds = \frac{dy}{\sqrt{1 - C^2}},$$

whence we deduce

$$\tan PAM = \frac{\sqrt{1 - C^2}}{C}, \ \sin PAM = \sqrt{1 - C^2} \text{ and } \cos PAM = C.$$

Corollary 4
5. Thus, if we introduce this angle PAM instead of the constant C, and if we set $PAM = \zeta$, because $C = \cos \zeta$, we get the following two equations:

$$dx = \frac{\cos \zeta dy}{\cos y \sqrt{\cos^2 y - \cos^2 \zeta}} \text{ and } ds = \frac{\cos y dy}{\sqrt{\cos^2 y - \cos^2 \zeta}}.$$

Moreover, if we name the angle $AMP = \theta$, we get:

$$\tan \theta = \frac{\cos \zeta}{\sqrt{\cos^2 y - \cos^2 \zeta}}; \ \sin \theta = \frac{\cos \zeta}{\cos y} \text{ and } \cos \theta = \frac{\sqrt{\cos^2 y - \cos^2 \zeta}}{\cos y}.$$

Corollary 5

6. It also remains to integrate our two differential equations, which express the values of dx and ds. But we will find by integrating:

$$x = \arcsin\left(\frac{C \sin y}{\cos y\sqrt{1-C^2}}\right) \text{ or } \sin x = \frac{C \sin y}{\cos y\sqrt{1-C^2}} = \frac{\cos\zeta \sin y}{\sin\zeta \cos y},$$

and

$$s = \arccos\left(\frac{\sqrt{\cos^2 y - C^2}}{\sqrt{1-C^2}}\right), \text{ or } \cos s = \frac{\sqrt{\cos^2 y - C^2}}{\sqrt{1-C^2}} = \frac{\sqrt{\cos^2 y - \cos^2\zeta}}{\sin\zeta}.$$

Corollary 6

7. Thus, here are, determined from the quantities ζ and y, the other quantities, x, s and θ, in the following manner:

$$\sin x = \frac{\cos\zeta \sin y}{\sin\zeta \cos y}; \ \cos x = \frac{\sqrt{\cos^2 y - \cos^2\zeta}}{\sin\zeta \cos y} \text{ and } \tan x = \frac{\cos\zeta \sin y}{\sqrt{\cos^2 y - \cos^2\zeta}}$$

$$\sin s = \frac{\sin y}{\sin\zeta}; \ \cos s = \frac{\sqrt{\cos^2 y - \cos^2\zeta}}{\sin\zeta} \text{ and } \tan s = \frac{\sin y}{\sqrt{\cos^2 y - \cos^2\zeta}}$$

$$\sin\theta = \frac{\cos\zeta}{\cos y}; \ \cos\theta = \frac{\sqrt{\cos^2 y - \cos^2\zeta}}{\cos y} \text{ and } \tan\theta = \frac{\cos\zeta}{\sqrt{\cos^2 y - \cos^2\zeta}}$$

Corollary 7

8. Since only the following irrational formula $\sqrt{\cos^2 y - \cos^2\zeta}$ is involved in our equations, eliminating it, we obtain:

$$\frac{\cos s}{\cos x} = \cos y; \quad \frac{\cos\theta}{\cos x} = \sin\zeta; \quad \frac{\cos\theta}{\cos s} = \frac{\sin\zeta}{\cos y};$$

$$\frac{\tan x}{\tan s} = \cos\zeta; \quad \frac{\tan x}{\tan\theta} = \sin y; \quad \frac{\tan s}{\tan\theta} = \frac{\sin y}{\cos\zeta};$$

$$\sin x = \frac{\cos\zeta \sin y}{\sin\zeta \cos y}; \quad \cos x \tan s = \frac{\sin y}{\sin\zeta \cos y}; \quad \cos x \tan\theta = \frac{\cos\zeta}{\sin\zeta \cos y};$$

$$\cos s \tan x = \frac{\cos\zeta \sin y}{\sin\zeta}; \quad \sin s = \frac{\sin y}{\sin\zeta}; \quad \cos s \tan\theta = \frac{\cos\zeta}{\sin\zeta};$$

$$\cos\theta \tan x = \frac{\cos\zeta \sin y}{\cos y}; \quad \cos\theta \tan s = \frac{\sin y}{\cos y}; \quad \sin\theta = \frac{\cos\zeta}{\cos y}.$$

Corollary 8

9. Having here five quantities x, y, s, ζ and θ, which belong to the right spherical triangle APM, let us take among the equalities we found those that contain only three [quantities], which will be, reduced to the simplest form:[7]

$$\text{I. } \cos s = \cos x \cos y, \qquad \text{VI. } \sin y = \sin \zeta \sin s,$$

$$\text{II. } \cos \theta = \sin \zeta \cos x, \qquad \text{VII. } \cos s \tan \zeta \tan \theta = 1,$$

$$\text{III. } \tan x = \cos \zeta \tan s, \qquad \text{VIII. } \tan y = \cos \theta \tan s,$$

$$\text{IV. } \tan x = \sin y \tan \theta, \qquad \text{IX. } \cos \zeta = \sin \theta \cos y,$$

$$\text{V. } \tan y = \sin x \tan \zeta,$$

whence, given two arbitrary [among these equations], it is possible to find the three others without need of extracting [square] roots, provided that this tenth [equation] is added to them:

$$\text{X. } \sin x = \sin \theta \sin s,$$

which follows first from the first three formulae to the left part of Paragraph 7.[8]

Problem II

10. *To make an exposition of the rules for the resolution of all the cases of spherical triangles.*

Solution

Let the angles be marked by A, B, C, among which C is the right (Fig. 11.2), and the sides [marked] by the small letters a, b, c, corresponding to the opposite angles, in such a way that c is the hypotenuse and a and b the catheti. Comparing therefore this triangle with the preceding figure, we get:

$$s = c, \; x = b, \; y = a, \; \zeta = A \text{ and } \theta = B.$$

Now everything amounts to the fact that, given two of these five quantities, [we need] to determine from them the other three; but the formulae that have been accounted for will give the following resolutions for all the possible cases.

[7] The first one among the following nine formulae is usually called the *spherical Pythagorean theorem*, because it gives, like the Pythagorean theorem of Euclidean geometry, a relation between the three sides of a right triangle.

[8] In the orignal publication, Euler writes Paragraph 6, but it should be Paragraph 7 or Corollary 6.

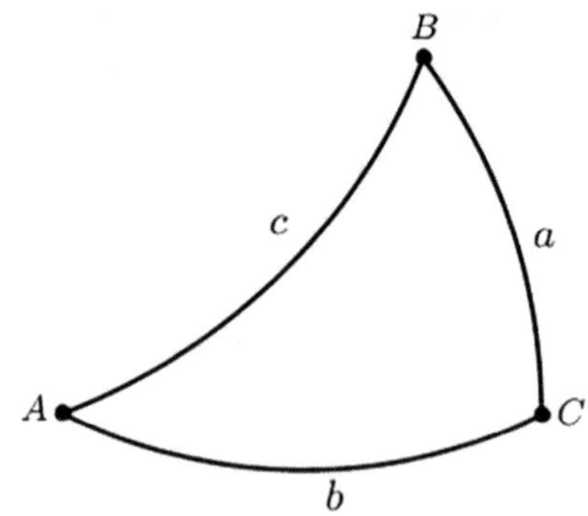

Fig. 11.2 Figure for Problem II

The two given quantities	Determination of the other three
I. a, b	$\cos c = \cos a \cdot \cos b\ ,\ \ \tan A = \frac{\tan a}{\sin b},\ \ \tan B = \frac{\tan b}{\sin a}$
II. a, c	$\cos b = \frac{\cos c}{\cos a}\ ,\ \ \sin A = \frac{\sin a}{\sin c},\ \ \cos B = \frac{\tan a}{\tan c}$
III. b, c	$\cos a = \frac{\cos c}{\cos b}\ ,\ \ \cos A = \frac{\tan b}{\tan c},\ \ \sin B = \frac{\sin b}{\sin c}$
IV. $a,\ A$	$\sin b = \frac{\tan a}{\tan A}\ ,\ \ \sin c = \frac{\sin a}{\sin A},\ \ \sin B = \frac{\cos A}{\cos a}$
V. $a,\ B$	$\tan b = \sin a \tan B\ ,\ \ \tan c = \frac{\tan a}{\cos B},\ \ \cos A = \cos a \sin B$
VI. $b,\ A$	$\tan a = \sin b \tan A\ ,\ \ \tan c = \frac{\tan b}{\cos A},\ \ \cos B = \cos b \sin A$
VII. $b,\ B$	$\sin a = \frac{\tan b}{\tan B}\ ,\ \ \sin c = \frac{\sin b}{\sin B},\ \ \sin A = \frac{\cos B}{\cos b}$
VIII. $c,\ A$	$\sin a = \sin c \sin A\ ,\ \ \tan b = \tan c \cos A,\ \ \tan B = \frac{1}{\cos c \tan A}$
IX. $c,\ B$	$\sin b = \sin c \sin B\ ,\ \ \tan a = \tan c \cos B,\ \ \tan A = \frac{1}{\cos c \tan B}$
X. $A,\ B$	$\cos a = \frac{\cos A}{\sin B}\ ,\ \ \cos b = \frac{\cos B}{\sin A},\ \ \cos c = \frac{1}{\tan A \tan B}$

Corollary 1

11. From here it is evident that the side a with its opposing angle A is involved in these formulae in much the same way as the other side b with its opposing angle B, in such a way that it makes no difference which one of the two sides a and b we are willing to take as a basis, exactly as the nature of the subject requires.

Corollary 2

12. The large number of formulae, which express the relation between various parts of a right triangle, is reduced to the following formulae, whose number is smaller, and which it suffices to know by heart.

$$\text{I. } \sin c = \frac{\sin a}{\sin A} \quad \text{or} \quad \sin c = \frac{\sin b}{\sin B}$$

$$\text{II. } \cos c = \cos a \cos b$$

$$\text{III. } \cos c = \cot A \cot B$$

$$\text{IV. } \cos A = \frac{\tan b}{\tan c} \quad \text{or} \quad \cos B = \frac{\tan a}{\tan c}$$

$$\text{V. } \sin A = \frac{\cos B}{\cos b} \quad \text{or} \quad \sin B = \frac{\cos A}{\cos a}$$

$$\text{VI. } \sin a = \frac{\tan b}{\tan B} \quad \text{or} \quad \sin b = \frac{\tan a}{\tan A}.$$

Corollary 3

13. Thus, we only have to note these six formulae, which contain as much properties of the right spherical triangles, and we will be in a position to resolve all the cases of triangles that can be imagined.

Problem III

14. *To find the area of a right spherical triangle.*

Solution

In the right triangle APM (Fig. 11.1) let the base $AP = x$ and the side $PM = y$, and having drawn the infinitely close meridian Omp, we get $Pp = dx$ and $mn = dy$. Moreover, having $Mn = \cos ydx$, the area element $PMmp$ will be $= \cos ydxdy$, taking dx as constant. Therefore the area of $PMmp$ will be equal to $\sin ydx$, which, being the differential of the area of the triangle APM, will make the latter equal to

$$\int \sin ydx.$$

But we have found

$$dx = \frac{\cos \zeta dy}{\cos y\sqrt{\cos^2 y - \cos^2 \zeta}},$$

where ζ marks the angle PAM, and therefore the area of the triangle will be

$$= \int \frac{\sin y \cos \zeta dy}{\cos y \sqrt{\cos^2 y - \cos^2 \zeta}}.$$

Instead of y, let us introduce the angle $AMP = \theta$; since

$$\sin \theta = \frac{\cos \zeta}{\cos y} \text{ and } \cos \theta = \frac{\sqrt{\cos^2 y - \cos^2 \zeta}}{\cos y},$$

we get

$$\cos \theta d\theta = \frac{\cos \zeta \sin y dy}{\cos^2 y};$$

therefore

$$d\theta = \frac{\cos \zeta \sin y dy}{\cos y \sqrt{\cos^2 y - \cos^2 \zeta}};$$

in this way the area of the desired triangle becomes

$$= \int d\theta = \theta + \text{Const.}$$

To assign the correct value to this constant, we must consider that the area must vanish when the point M falls at A, in which case the angle θ becomes $= 90° - \zeta$; therefore it should be $90° - \zeta + \text{Const.} = 0$, and therefore $\text{Const.} = \zeta - 90°$. Consequently, the area of the triangle APM that we are looking for will be equal to

$$\zeta + \theta - 90°,$$

or, the excess of the sum of the two angles PAM and AMP over a right angle will express the area of the triangle APM.

Corollary 1

15. Therefore the sum of the two angles PAM and AMP is always greater than a right angle, and the excess is all the bigger as the area of the triangle will be bigger. And an arc of a great circle which is a measure of this excess, multiplied by the radius of the sphere will give the area of the spherical triangle.

Corollary 2

16. From here, we can easily deduce the area of an arbitrary spherical triangle: indeed, since such a triangle can be cut into two right triangles, we shall find its area

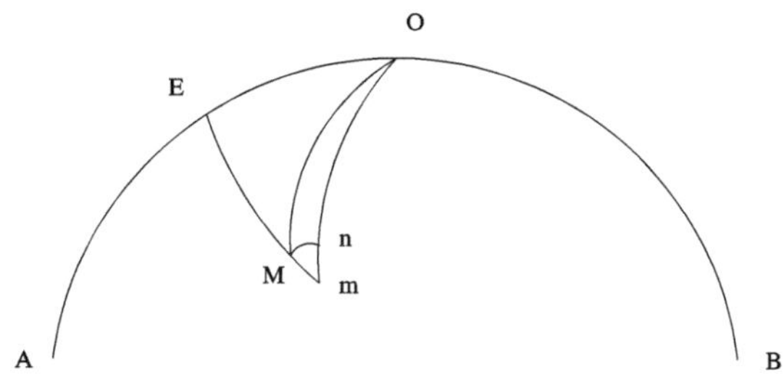

Fig. 11.3 Figure for Problem III

when we multiply the excess over 180° of the sum of its three angles by the radius of the sphere.[9]

Problem IV

17. *Given two arbitrary points E and M on the surface of a sphere (Fig. 11.3), to find the shortest line between these two points.*

Solution

Let us draw from one of the poles O to these points the meridians OE and OM, of which the latter is being considered as a variable. Let us name

$$\text{the meridian } OE = a,\ \ OM = x \text{ and the angle } EOM = y.$$

Furthermore, let the sought for quantities [be called]

$$\text{the arc } \ EM = s, \text{ the angle } OEM = \alpha, \ \ \text{and the angle } \ OME = \varphi,$$

which will vary with the quantities x, y and s, whereas a and α remain invariable. Let us draw the meridian Om infinitely close [to OM], to which one draws from M the perpendicular Mn, and we get

$$mn = dx,\ \text{ the angle } MOm = dy,\ \text{ and } Mn = \sin x dy,$$

taking the unity to mark the radius of the sphere. From this we get

$$\tan\varphi = \frac{Mn}{mn} = \frac{\sin x dy}{dx},$$

or

$$\sin\varphi = \frac{\sin x dy}{ds} \text{ and } \cos\varphi = \frac{dx}{ds}.$$

[9] This is the so-called Girard theorem, which gives the area of a spherical triangle in terms of the angle excess. Euler quotes this theorem and proves it in his other memoirs on spherical geometry as well; see the memoir *Various speculations on the area of spherical triangles*, Chap. 13 of the present volume, and the memoir *On the measure of solid angles*, Chap. 14 of the present volume.

But having $ds = \sqrt{dx^2 + \sin^2 x dy^2}$, the following formula

$$\int \sqrt{dx^2 + \sin^2 x dy^2}$$

must be a *minimum*. For this purpose, let us set $dy = pdx$, in order to make the following formula

$$\int \sqrt{1 + p^2 \sin^2 x} dx = \int Z dx$$

smallest, in such a way that

$$Z = \sqrt{1 + p^2 \sin^2 x}.$$

But in general, if we have $dZ = Mdx + Ndy + Pdp$, the equation for the *minimum* is $Ndx - dP = 0$; therefore, making the application to our case, we have

$$N = 0 \text{ and } P = \frac{p \sin^2 x}{\sqrt{1 + p^2 \sin^2 x}}.$$

But since $N = 0$, our equation will be

$$dP = 0 \text{ and therefore } P = C,$$

where C is a constant. Consequently, we get

$$\frac{p \sin^2 x}{\sqrt{1 + p^2 \sin^2 x}} = C \text{ or } \frac{\sin^2 x dy}{\sqrt{dx^2 + \sin^2 x dy^2}} = C,$$

that is,

$$\frac{\sin^2 x dy}{ds} = \sin x \sin \varphi = C.$$

To determine this constant, one must consider that while the angle $EOM = y$ vanishes, we get $x = a$, and $\varphi = 180° - \alpha$, or $\sin \varphi = \sin \alpha$; therefore in this case we have $\sin a \cdot \sin \alpha = C$. Consequently, the nature of the *minimum* provides the equation:

$$\frac{\sin^2 x dy}{\sqrt{dx^2 + \sin^2 x dy^2}} = \sin a \sin \alpha.$$

But it still remains to integrate the differential equation, which transforms into the following one

$$dy = \frac{C dx}{\sin x \sqrt{\sin^2 x - C^2}}.$$

Substituting C for $\sin a \sin \alpha$, from this, since $ds = \frac{\sin^2 x dy}{C}$, we also have

$$ds = \frac{\sin x dx}{\sqrt{\sin^2 x - C^2}}.$$

But we find by the integration rules:

$$y = -\arcsin \frac{C \cos x}{\sin x \sqrt{1 - C^2}} + \arcsin \frac{C \cos a}{\sin a \sqrt{1 - C^2}}$$

$$= -\arccos \frac{\sqrt{\sin^2 x - C^2}}{\sin x \sqrt{1 - C^2}} + \arccos \frac{\sqrt{\sin^2 a - C^2}}{\sin a \sqrt{1 - C^2}}.$$

$$s = -\arccos \frac{\sqrt{\sin^2 x - C^2}}{\sqrt{1 - C^2}} + \arccos \frac{\sqrt{\sin^2 a - C^2}}{\sqrt{1 - C^2}}$$

$$= -\arcsin \frac{\cos x}{\sqrt{1 - C^2}} + \arcsin \frac{\cos a}{\sqrt{1 - C^2}},$$

where the added constants are such that, making $y = 0$ and $s = 0$, we obtain $x = a$. But the two arcs of circles being reduced to one will give

$$y = \arcsin \frac{C \cos a \sqrt{\sin^2 x - C^2} - C \cos x \sqrt{\sin^2 a - C^2}}{(1 - C^2) \sin a \sin x},$$

$$s = \arcsin \frac{\cos a \sqrt{\sin^2 x - C^2} - \cos x \sqrt{\sin^2 a - C^2}}{1 - C^2},$$

whence we deduce the following two equations:

$$(1 - C^2) \sin a \sin x \sin y = C \cos a \sqrt{\sin^2 x - C^2} - C \cos x \sqrt{\sin^2 a - c^2}$$

$$(1 - C^2) \sin s = \cos a \sqrt{\sin^2 x - C^2} - \cos x \sqrt{\sin^2 a - C^2}.$$

But, by taking the cosines of the angles y and s, we get:

$$(1-C^2)\sin a\sin x\cos y=\sqrt{(\sin^2 a-C^2)(\sin^2 x-C^2)}+C^2\cos a\cos x$$

$$(1-C^2)\cos s=\sqrt{(\sin^2 a-C^2)(\sin^2 x-C^2)}+\cos a\cos x.$$

And putting again for C its value $\sin a\sin\alpha$, since

$$\sqrt{\sin^2 a-C^2}=-\sin a\cos\alpha$$

(because we consider here the angle α as obtuse, so that the angle φ is acute from the point E; because, setting $y=0$, the angle φ becomes $180^\circ-\alpha$, whose cosine is $-\cos\alpha$), we get:

$$(1-\sin^2 a\sin^2\alpha)\sin x\sin y$$

$$=\sin\alpha\cos a\sqrt{\sin^2 x-\sin^2 a\sin^2\alpha}+\sin\alpha\cos\alpha\sin a\cos x;$$

$$(1-\sin^2 a\sin^2\alpha)\sin x\cos y$$

$$=-\cos\alpha\sqrt{\sin^2 x-\sin^2 a\sin^2\alpha}+\sin a\cos a\sin^2\alpha\cos x;$$

$$(1-\sin^2 a\sin^2\alpha)\sin s=\cos\alpha\sqrt{\sin^2 x-\sin^2 a\sin^2\alpha}+\sin a\cos\alpha\cos x;$$

$$(1-\sin^2 a\sin^2\alpha)\cos s=-\sin a\cos\alpha\sqrt{\sin^2 x-\sin^2 a\sin^2\alpha}+\cos a\cos x,$$

to which we must add

$$\sin x\sin\varphi=\sin a\sin\alpha.$$

Corollary 1

18. Since $\sin a\sin\alpha=\sin x\sin\varphi$, we get:

$$\sqrt{\sin^2 x-\sin^2 a\sin^2\alpha}=\sin x\cos\varphi.$$

Therefore our four formulae will be:

$$\text{I. }(1-C^2)\sin y=\sin\alpha\cos a\cos\varphi+\cos\alpha\cos x\sin\varphi$$

$$\text{II. }(1-C^2)\cos y=-\cos\alpha\cos\varphi+\sin\alpha\cos a\cos x\sin\varphi$$

$$\text{III. }(1-C^2)\sin s=\cos a\sin x\cos\varphi+\sin a\cos\alpha\cos x$$

$$\text{IV. }(1-C^2)\cos s=-\sin a\cos\alpha\sin x\cos\varphi+\cos a\cos x,$$

setting for short C^2 instead of $\sin^2 a \sin^2 \alpha$, or [instead] of $\sin^2 x \sin^2 \varphi$.

Corollary 2

19. These four formulae may be combined in several different manners, whence we can deduce simpler formulae. First, let us take

$$I. \cos\alpha + II. \sin\alpha \cos a,$$

we get:

$$(1 - C^2)(\cos\alpha \sin y + \sin\alpha \cos a \cos y) = (\cos^2\alpha + \sin^2\alpha \cos^2 a) \cos x \sin\varphi,$$

but

$$\cos^2\alpha + \sin^2\alpha \cos^2 a = 1 - \sin^2\alpha \sin^2 a = 1 - C^2,$$

whence we deduce:

$$\cos\alpha \sin y + \sin\alpha \cos a \cos y = \cos x \sin\varphi = \frac{\sin a \sin\alpha}{\tan x},$$

or

$$\tan x \sin y + \tan\alpha \tan x \cos a \cos y = \tan\alpha \sin a.$$

Corollary 3

20. Let us make the combination

$$I. \sin\alpha \cos a - II. \cos\alpha,$$

so that we have

$$(1 - C^2)(\sin\alpha \cos a \sin y - \cos\alpha \cos y) = (\sin^2\alpha \cos^2 a + \cos^2 x\alpha) \cos\varphi$$

$$= (1 - C^2) \cos\varphi,$$

whence we deduce, by dividing by $1 - C^2$,

$$\sin\alpha \cos a \sin y - \cos\alpha \cos y = \cos\varphi.$$

Corollary 4

21. The combination

$$I. \sin x - III. \sin\alpha$$

gives $(1 - C^2)(\sin x \sin y - \sin \alpha \sin s) = 0$, or

$$\sin x \sin y = \sin \alpha \sin s,$$

whence, because of $\sin x \sin \varphi = \sin a \sin \alpha$, we deduce

$$\sin a \sin y = \sin \varphi \sin s,$$

or this proportion:

$$\sin a : \sin \varphi = \sin x : \sin \alpha = \sin s : \sin y.$$

Corollary 5
22. The combination

$$I. \sin a \cos \alpha \sin x + IV. \sin \alpha \cos a$$

gives

$$(1 - C^2)(\sin a \cos \alpha \sin x \sin y + \sin \alpha \cos a \cos s)$$
$$= \cos x (\sin a \cos^2 \alpha \sin x \sin \varphi + \sin \alpha \cos^2 a).$$

But since $\sin x \sin \varphi = \sin a \sin \alpha$, the value of this formula will be

$$\sin \alpha \cos x (\sin^2 a \cos^2 \alpha + \cos^2 a)$$
$$= (1 - \sin^2 a \sin^2 \alpha) \sin \alpha \cos x = (1 - C^2) \sin \alpha \cos x,$$

whence, dividing by $(1 - C^2)$, we obtain

$$\sin a \cos \alpha \sin x \sin y + \sin \alpha \cos a \cos s = \sin \alpha \cos x,$$

which, since

$$\sin y = \frac{\sin \alpha \sin s}{\sin x},$$

is transformed into

$$\sin a \cos \alpha \sin s + \cos a \cos s = \cos x.$$

Corollary 6

23. But this combination

$$I.\cos a - IV.\cos\alpha\sin\varphi$$

gives

$$(1 - C^2)(\cos a\sin y - \cos\alpha\sin\varphi\cos s)$$

$$= \cos\varphi(\sin\alpha\cos^2 a + \sin a\cos^2\alpha\sin x\sin\varphi),$$

whose value, since $\sin x\sin\varphi = \sin a\sin\alpha$, will be:

$$\sin\alpha\cos\varphi(\cos^2 a + sin^2 a\cos^2\alpha) = (1 - C^2)\sin\alpha\cos\varphi.$$

Therefore, dividing by $1 - C^2$, we get

$$\cos a\sin y - \cos\alpha\sin\varphi\cos s = \sin\alpha\cos\varphi,$$

which, because $\sin y = \frac{\sin\varphi\sin s}{\sin a}$, transforms into the following one:

$$\cos a\sin\varphi\sin s - \sin a\cos\alpha\sin\varphi\cos s = \sin a\sin\alpha\cos\varphi,$$

or

$$\tan\varphi\sin s - \cos\alpha\tan a\tan\varphi\cos s = \sin\alpha\tan a.$$

Corollary 7

24. Let us consider the combination

$$II.\cos a\sin x + III.\cos\alpha,$$

which gives

$$(1 - C^2)(\cos a\sin x\cos y + \cos\alpha\sin s)$$

$$= \cos x(\sin\alpha\cos^2 a\sin x\sin\varphi + \sin a\cos^2\alpha),$$

whose value, because $\sin x\sin\varphi = \sin a\sin\alpha$, will be

$$\sin a\cos x(\sin^2\alpha\cos^2 a + \cos^2\alpha) = (1 - C^2)\sin a\cos x.$$

Therefore, dividing by $1 - C^2$ we get

$$\cos a\sin x\cos y + \cos\alpha\sin s = \sin a\cos x,$$

and since $\sin s = \frac{\sin x \sin y}{\sin \alpha}$, we obtain

$$\sin\alpha\cos a\sin x\cos y + \cos\alpha\sin x\sin y = \sin\alpha\sin a\cos x,$$

or

$$\tan\alpha\cos a\tan x\cos y + \tan x\sin y = \tan\alpha\sin a,$$

just like in Paragraph 19.

Corollary 8
25. The combination

$$-II.\sin a\cos\alpha + III.\cos a\sin\alpha\sin\varphi$$

gives

$$(1-C^2)(\cos a\sin\alpha\sin s\sin\varphi - \sin a\cos\alpha\cos y)$$
$$= \cos\varphi(\sin a\cos^2\alpha + \cos^2 a\sin\alpha\sin x\sin\varphi),$$

whose value, since $\sin x\sin\varphi = \sin a\sin\alpha$, is

$$\sin a\cos\varphi(\cos^2\alpha + \cos^2 a\sin^2\alpha) = (1-C^2)\sin a\cos\varphi.$$

Therefore, dividing by $1-C^2$ we get:

$$\cos a\sin\alpha\sin s\sin\varphi - \sin a\cos\alpha\cos y = \sin a\cos\varphi.$$

But having $\sin s = \frac{\sin a\sin y}{\sin\varphi}$, we obtain

$$\cos a\sin\alpha\sin y - \cos\alpha\cos y = \cos\varphi,$$

just as in Paragraph 20.

Corollary 9
26. But the combination

$$II.\ \sin a\sin x - IV.1$$

gives

$$(1-C^2)(\sin a\sin x\cos y - \cos s) = \cos a\cos x(\sin a\sin\alpha\sin x\sin\varphi - 1),$$

whose value, since $\sin x \sin \varphi = \sin a \sin \alpha$, is

$$\cos a \cos x(\sin^2 a \sin^2 \alpha - 1) = -(1 - C^2) \cos a \cos x.$$

Therefore, dividing by $-(1 - C^2)$ we get

$$\cos s - \sin a \sin x \cos y = \cos a \cos x.$$

Corollary 10
27. The combination

$$II.1 - IV.\sin \alpha \sin \varphi$$

gives

$$(1 - C^2)(\cos y - \sin \alpha \sin \varphi \cos s) = \cos \alpha \cos \varphi(\sin a \sin \alpha \sin x \sin \varphi - 1),$$

therefore

$$\sin \alpha \sin \varphi \cos s - \cos y = \cos \alpha \cos \varphi.$$

Corollary 11
28. The combination

$$III.\sin a \cos \alpha + IV.\cos a$$

gives

$$(1 - C^2)(\sin a \cos \alpha \sin s + \cos a \cos s) = \cos x(\sin^2 a \cos^2 \alpha + \cos^2 a),$$

therefore:

$$\sin a \cos \alpha \sin s + \cos a \cos s = \cos x,$$

just like in Paragraph 22.

Corollary 12
29. Finally, the combination

$$III.\cos a - IV.\sin a \cos \alpha$$

gives

$$(1 - C^2)(\cos a \sin s - \sin a \cos \alpha \cos s) = \sin x \cos \varphi(\cos^2 a + \sin^2 a \cos^2 \alpha),$$

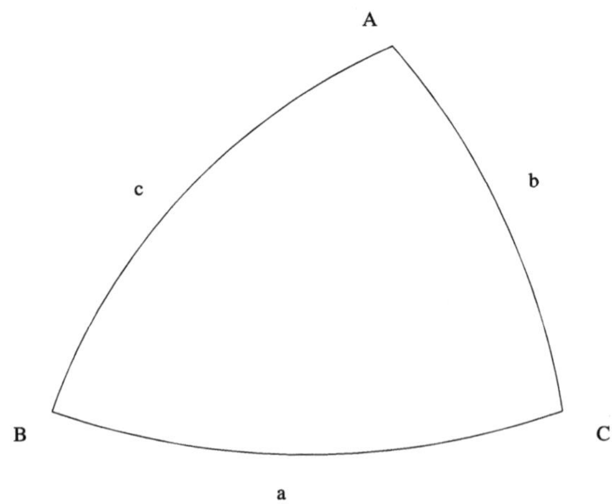

Fig. 11.4 Figure for Problem V

therefore

$$\cos a \sin s - \sin a \cos \alpha \cos s = \sin x \cos \varphi = \frac{\sin a \sin \alpha \cos \varphi}{\sin \varphi},$$

or

$$\tan \varphi \sin s - \tan a \tan \varphi \cos \alpha \cos s = \tan a \sin \alpha,$$

just as in Paragraph 23

Problem V

30. *To find the properties between the sides and the angles of an arbitrary spherical triangle* (Fig. 11.4).

Solution

For any proposed spherical triangle ABC, we can consider one of its angles A as the pole of the sphere; then the sides AB and AC will be two meridians, and the third side BC the shortest line that can be drawn on the surface of the sphere from the point B to the point C, in such a way that this triangle ABC may be compared with the figure ECM that we have just considered in the preceding problem. Therefore, if we use the letters A, B, C to denote the angles of the same name, and if for the sides we set $AB = c$, $AC = b$ and $BC = a$, the preceding denominations reduce to the present ones in the following manner:

$$\text{Previous denominations} \ldots a,\ x,\ s,\ y,\ \alpha,\ \varphi$$

$$\text{Present denominations} \ldots c,\ b,\ a,\ A,\ B,\ C$$

Now the formulae found in the corollaries of the previous problem will provide us with the following properties for the spherical triangle ABC:

$$I.\ \sin a : \sin A = \sin b : \sin B = \sin c : \sin C \text{ by Paragraph 21}$$

$$II. \begin{cases} \cos C = \cos c \sin A \sin B - \cos A \cos B & \text{by Paragraph 20} \\ \cos B = \cos b \sin A \sin C - \cos A \cos C & \text{by analogy} \\ \cos A = \cos a \sin B \sin C - \cos B \cos C & \text{by Paragraph 27} \end{cases}$$

$$III. \begin{cases} \cos c = \cos C \sin a \sin b + \cos a \cos b & \text{by analogy} \\ \cos b = \cos B \sin a \sin c + \cos a \cos c & \text{by Paragraph 22} \\ \cos a = \cos A \sin b \sin c + \cos b \cos c & \text{by Paragraph 26} \end{cases}$$

$$IV. \begin{cases} \sin a \tan C - \sin B \tan c = \cos a \cos B \tan C \tan c & \text{by Paragraph 23} \\ \sin b \tan A - \sin C \tan a = \cos b \cos C \tan A \tan a & \text{by analogy} \\ \sin c \tan B - \sin A \tan b = \cos c \cos A \tan B \tan b & \text{by Paragraph 19} \end{cases}$$

And it is to these four properties that all the formulae that we have found in the previous problem are reduced.

Corollary 1

31. The first property contains the very well-known quality of all the spherical triangles, from which we know that the sines of the sides have among themselves the same ratio as the sines of the angles that are opposite to them.

Corollary 2

32. Therefore, if in a spherical triangle we know one side with the angle opposite to it, and besides that, another angle, or side, we shall first find the side or the angle that is opposite to it.

Corollary 3

33. Each of the formulae that we have just found encloses only four quantities that belong to the triangle, and therefore if we know three of them, we will be able to determine the fourth one.

Corollary 4

34. Thus it is from here that we will be able to deduce the rules for the resolution of all spherical triangles. As there are six things in every triangle, namely, the three sides and the three angles, if we know three of them, we will be able to find the three others, as we shall see in the following problems.

Problem VI

35. *In a spherical triangle, given three sides, to find the angles (Fig. 11.4).*

Solution

In the spherical triangle ABC let the three sides $AB = c$, $AC = b$ and $BC = a$ be given. Let us search for the three angles. this will be done by means of the third property, which provides us with:

$$\cos A = \frac{\cos a - \cos b \cos c}{\sin b \sin c},$$

$$\cos B = \frac{\cos b - \cos a \cos c}{\sin a \sin c},$$

$$\cos C = \frac{\cos c - \cos a \cos b}{\sin a \sin b}.$$

Corollary 1

36. Thus, we get

$$1 - \cos A = \frac{\sin b \sin c + \cos b \cos c - \cos a}{\sin b \sin c},$$

or

$$1 - \cos A = \frac{\cos(b - c) - \cos a}{\sin b \sin c},$$

since

$$\cos(b - c) = \cos b \cos c + \sin b \sin c$$

Corollary 2

37. But since in general we have

$$\cos p - \cos q = 2 \sin \frac{1}{2}(q - p) \sin \frac{1}{2}(q + p),$$

our formula will be changed into the following:

$$1 - \cos A = \frac{2 \sin \frac{1}{2}(a - b + c) \sin \frac{1}{2}(a + b - c)}{\sin b \sin c}.$$

Therefore, since

$$1 - \cos A = 2 \sin^2 \frac{1}{2} A,$$

we get:

$$\sin \frac{1}{2} A = \sqrt{\frac{\sin \frac{1}{2}(a - b + c) \sin \frac{1}{2}(a + b - c)}{\sin b \sin c}}$$

and similarly

$$\sin\frac{1}{2}B = \sqrt{\frac{\sin\frac{1}{2}(b-a+c)\sin\frac{1}{2}(b+a-c)}{\sin a \sin c}}$$

$$\sin\frac{1}{2}C = \sqrt{\frac{\sin\frac{1}{2}(c-a+b)\sin\frac{1}{2}(c+a-b)}{\sin a \sin b}}$$

Corollary 3

38. Adding a unit to the found cosines we get

$$1+\cos A = \frac{\cos a - \cos b\cos c + \sin b \sin c}{\sin b \sin c} = \frac{\cos a - \cos(b+c)}{\sin b \sin c}.$$

Therefore, since $1+\cos A = 2\cos^2\frac{1}{2}A$, the same conversion will give

$$\cos\frac{1}{2}A = \sqrt{\frac{\sin\frac{1}{2}(b+c-a)\sin\frac{1}{2}(b+c+a)}{\sin b \sin c}}$$

$$\cos\frac{1}{2}B = \sqrt{\frac{\sin\frac{1}{2}(a+c-b)\sin\frac{1}{2}(a+c+b)}{\sin a \sin c}}$$

$$\cos\frac{1}{2}C = \sqrt{\frac{\sin\frac{1}{2}(a+b-c)\sin\frac{1}{2}(a+b+c)}{\sin a \sin b}}.$$

Corollary 4

39. From here we deduce the tangents of the half of each of the angles A, B, C:

$$\tan\frac{1}{2}A = \sqrt{\frac{\sin\frac{1}{2}(a-b+c)\sin\frac{1}{2}(a+b-c)}{\sin\frac{1}{2}(b+c-a)\sin\frac{1}{2}(b+c+a)}}$$

$$\tan\frac{1}{2}B = \sqrt{\frac{\sin\frac{1}{2}(b-a+c)\sin\frac{1}{2}(b+a-c)}{\sin\frac{1}{2}(a+c-b)\sin\frac{1}{2}(a+c+b)}}$$

$$\tan\frac{1}{2}C = \sqrt{\frac{\sin\frac{1}{2}(c-a+b)\sin\frac{1}{2}(c+a-b))}{\sin\frac{1}{2}(a+b-c)\sin\frac{1}{2}(a+b+c)}}.$$

Corollary 5

40. These formulae are very handful for making the calculation with the aid of logarithms.[10] Thus, having found one of the angles, like A, we will find the two others more easily; from the first property we get:

$$\sin B = \frac{\sin b \sin A}{\sin a} \text{ and } \sin C = \frac{\sin c \sin A}{\sin a},$$

provided we know whether these angles are greater or smaller than a right angle; but by using the formulae we found, this ambiguity vanishes, since we find the halves of the angles, which are always smaller that a right angle.

Corollary 6

41. The tangents of the half-angles provide more remarkable formulae, since multiplying two [of them] together, we get:

$$\tan \frac{1}{2}A \tan \frac{1}{2}B = \frac{\sin \frac{1}{2}(a+b-c)}{\sin \frac{1}{2}(a+b+c)},$$

and since

$$\sin p + \sin q = 2 \sin \frac{1}{2}(p+q) \cos \frac{1}{2}(p-q)$$

$$\sin p - \sin q = 2 \sin \frac{1}{2}(p-q) \cos \frac{1}{2}(p+q),$$

we obtain:

$$1 + \tan \frac{1}{2}A \tan \frac{1}{2}B = \frac{2 \sin \frac{1}{2}(a+b) \cos \frac{1}{2}c}{\sin \frac{1}{2}(a+b+c)}$$

and

$$1 - \tan \frac{1}{2}A \tan \frac{1}{2}B = \frac{2 \sin \frac{1}{2}c \cos \frac{1}{2}(a+b)}{\sin \frac{1}{2}(a+b+c)}.$$

Corollary 7

42. Likewise, adding or subtracting two of these tangents, we get:

---

[10] Euler is thinking here about applications in astronomy, where the fact of being able to compute exact values is essential. One may recall that in those times where calculations were made by hand, logarithm tables were useful because they transform multiplication into addition. Lambert, in §63 of his memoir *Notes and Comments on Trigonometry* translated into English in Chap. 19 of the present volume, explains how certain forms of the spherical trigonometric formulae are convenient for computations using logarithms.

$$\tan\frac{1}{2}A \pm \tan\frac{1}{2}B = \frac{(\sin\frac{1}{2}(a+c-b) \pm \sin\frac{1}{2}(b+c-a))\sqrt{\sin\frac{1}{2}(a+b-c)}}{\sqrt{\sin\frac{1}{2}(b+c-a)\sin\frac{1}{2}(a+c-b)\sin\frac{1}{2}(a+b+c)}}$$

or

$$\tan\frac{1}{2}A \pm \tan\frac{1}{2}B = \frac{\sin\frac{1}{2}(a+c-b) \pm \sin\frac{1}{2}(b+c-a)}{\tan\frac{1}{2}C\sin\frac{1}{2}(a+b+c)},$$

if we introduce the value of the tangent of $\frac{1}{2}C$. Therefore, employing the reduction taught before, we get the following two equations:

$$\tan\frac{1}{2}A + \tan\frac{1}{2}B = \frac{2\sin\frac{1}{2}c\cos\frac{1}{2}(a-b)}{\tan\frac{1}{2}C\sin\frac{1}{2}(a+b+c)}$$

$$\tan\frac{1}{2}A - \tan\frac{1}{2}B = \frac{2\sin\frac{1}{2}(a-b)\cos\frac{1}{2}c}{\tan\frac{1}{2}C\sin\frac{1}{2}(a+b+c)}.$$

Corollary 8
43. But since

$$\tan\frac{1}{2}(A+B) = \frac{\tan\frac{1}{2}A + \tan\frac{1}{2}B}{1 - \tan\frac{1}{2}A\tan\frac{1}{2}B},$$

we will find by the formulae of the two preceding corollaries:

$$\tan\frac{1}{2}(A+B) = \frac{\cos\frac{1}{2}(a-b)}{\tan\frac{1}{2}C\cos\frac{1}{2}(a+b)}$$

and likewise

$$\tan\frac{1}{2}(A+C) = \frac{\cos\frac{1}{2}(a-c)}{\tan\frac{1}{2}B\cos\frac{1}{2}(a+c)}$$

$$\tan\frac{1}{2}(B+C) = \frac{\cos\frac{1}{2}(b-c)}{\tan\frac{1}{2}A\cos\frac{1}{2}(b+c)}.$$

Corollary 9
Similarly, since

$$\tan\frac{1}{2}(A-B) = \frac{\tan\frac{1}{2}A - \tan\frac{1}{2}B}{1 + \tan\frac{1}{2}A\tan\frac{1}{2}B},$$

we get

$$\tan\frac{1}{2}(A-B)=\frac{\sin\frac{1}{2}(a-b)}{\tan\frac{1}{2}C\sin\frac{1}{2}(a+b)}$$

and in a similar way

$$\tan\frac{1}{2}(A-C)=\frac{\sin\frac{1}{2}(a-c)}{\tan\frac{1}{2}B\sin\frac{1}{2}(a+c)}$$

$$\tan\frac{1}{2}(B-C)=\frac{\sin\frac{1}{2}(b-c)}{\tan\frac{1}{2}A\sin\frac{1}{2}(b+c)}.$$

Problem VII
45. *In a spherical triangle* (Fig. 11.4), *given the three angles* A, B, C, *to find the three sides.*

Solution
Let ABC be the spherical triangle, of which are given the three angles A, B, C and of which we must find the three sides

$$AB=c,\ \ AC=b \text{ and } BC=a.$$

But Property II of Paragraph 30 will give us the cosine of these three sides expressed in the following manner:

$$\cos a=\frac{\cos A+\cos B\cos C}{\sin B\sin C}$$

$$\cos b=\frac{\cos B+\cos A\cos C}{\sin A\sin C}$$

$$\cos c=\frac{\cos C+\cos A\cos B}{\sin A\sin B}.$$

Corollary 1
46. Therefore we first deduce the following formulae:

$$1-\cos a=\frac{\cos A+\cos(B+C)}{\sin B\sin C},$$

$$1+\cos a=\frac{\cos A+\cos(B-C)}{\sin B\sin C}.$$

But since in general we have

$$\cos p + \cos q = 2\cos\frac{1}{2}(p+q)\cos\frac{1}{2}(p-q),$$

we get:

$$1-\cos a = -\frac{2\cos\frac{1}{2}(A+B+C)\cos\frac{1}{2}(B+C-A)}{\sin B\sin C},$$

$$1+\cos a = \frac{2\cos\frac{1}{2}(A+B-C)\cos\frac{1}{2}(A-B+C)}{\sin B\sin C}.$$

Corollary 2
47. But since $1-\cos a = 2\sin^2\frac{1}{2}a$ and $1+\cos a = 2\cos^2\frac{1}{2}a$, we get the following formulae:

$$\sin\frac{1}{2}a = \sqrt{-\frac{\cos\frac{1}{2}(A+B+C)\cos\frac{1}{2}(B+C-A)}{\sin B\sin C}}$$

$$\sin\frac{1}{2}b = \sqrt{-\frac{\cos\frac{1}{2}(A+B+C)\cos\frac{1}{2}(A+C-B)}{\sin A\sin C}}$$

$$\sin\frac{1}{2}c = \sqrt{-\frac{\cos\frac{1}{2}(A+B+C)\cos\frac{1}{2}(A+B-C)}{\sin A\sin B}}$$

where one must remark, since the angle sum $A+B+C$ is always greater than two right angles, [that] the half-sum is greater than a right angle and therefore its cosine is negative.

Corollary 3
48. Regarding the cosine of the half-edges, we get

$$\cos\frac{1}{2}a = \sqrt{\frac{\cos\frac{1}{2}(A+B-C)\cos\frac{1}{2}(A-B+C)}{\sin B\sin C}}$$

$$\cos\frac{1}{2}b = \sqrt{\frac{\cos\frac{1}{2}(B+A-C)\cos\frac{1}{2}(B-A+C)}{\sin A\sin C}}$$

$$\cos\frac{1}{2}c = \sqrt{\frac{\cos\frac{1}{2}(C+A-B)\cos\frac{1}{2}(C-A+B)}{\sin A\sin B}}$$

and these formulae facilitate the usage of logarithms.

Corollary 4

49. From the sines and cosines of the shalf-sides, we can easily deduce their tangents, which will be

$$\tan\frac{1}{2}a=\sqrt{-\frac{\cos\frac{1}{2}(A+B+C)\cos\frac{1}{2}(B+C-A)}{\cos\frac{1}{2}(A+B-C)\cos\frac{1}{2}(A-B+C)}}$$

$$\tan\frac{1}{2}b=\sqrt{-\frac{\cos\frac{1}{2}(A+B+C)\cos\frac{1}{2}(A+C-B)}{\cos\frac{1}{2}(B+A-C)\cos\frac{1}{2}(B-A+C)}}$$

$$\tan\frac{1}{2}c=\sqrt{-\frac{\cos\frac{1}{2}(A+B+C)\cos\frac{1}{2}(A+B-C)}{\cos\frac{1}{2}(C+A-B)\cos\frac{1}{2}(C-A+B)}}$$

where we can also easily use the calculus of logarithms.

Corollary 5

50. Multiplying together two of these tangents, we get:

$$\tan\frac{1}{2}a\tan\frac{1}{2}b=-\frac{\cos\frac{1}{2}(A+B+C)}{\cos\frac{1}{2}(A+B-C)}.$$

But from here we will derive the following two formulae:

$$1-\tan\frac{1}{2}a\tan\frac{1}{2}b=\frac{2\cos\frac{1}{2}(A+B)\cos\frac{1}{2}C}{\cos\frac{1}{2}(A+B-C)}$$

$$1+\tan\frac{1}{2}a\tan\frac{1}{2}b=\frac{2\sin\frac{1}{2}C\sin\frac{1}{2}(A+B)}{\cos\frac{1}{2}(A+B-C)}$$

Corollary 6

51. And if we add, or subtract, two of these formulae, we get:

$$\tan\frac{1}{2}a\pm\tan\frac{1}{2}b$$
$$=\frac{(\cos\frac{1}{2}(B+C-A)\pm\cos\frac{1}{2}(A+C-B))\sqrt{-\cos\frac{1}{2}(A+B+C)}}{\sqrt{\cos\frac{1}{2}(A+B-C)\cos\frac{1}{2}(A+C-B)\cos\frac{1}{2}(B+C-A)}}.$$

But since

$$\tan\frac{1}{2}c = \sqrt{-\frac{\cos\frac{1}{2}(A+B+C)\cos\frac{1}{2}(A+B-C)}{\cos\frac{1}{2}(C+A-B)\cos\frac{1}{2}(C-A+B)}},$$

we get

$$\tan\frac{1}{2}a \pm \tan\frac{1}{2}b = \frac{(\cos\frac{1}{2}(B+C-A) \pm \cos\frac{1}{2}(A+C-B))\tan\frac{1}{2}c}{\cos\frac{1}{2}(A+B-C)}.$$

Corollary 7

52. From here we find by the reductions taught:

$$\tan\frac{1}{2}a + \tan\frac{1}{2}b = \frac{2\cos\frac{1}{2}C\cos\frac{1}{2}(A-B)\tan\frac{1}{2}c}{\cos\frac{1}{2}(A+B-C)}$$

and

$$\tan\frac{1}{2}a - \tan\frac{1}{2}b = \frac{2\sin\frac{1}{2}(A-B)\sin\frac{1}{2}C\tan\frac{1}{2}c}{\cos\frac{1}{2}(A+B-C)}.$$

Corollary 8

53. We will find then, as above:

$$\tan\frac{1}{2}(a+b) = \frac{\cos\frac{1}{2}(A-B)}{\cos\frac{1}{2}(A+B)}\tan\frac{1}{2}c$$

$$\tan\frac{1}{2}(a+c) = \frac{\cos\frac{1}{2}(A-C)}{\cos\frac{1}{2}(A+C)}\tan\frac{1}{2}b$$

$$\tan\frac{1}{2}(b+c) = \frac{\cos\frac{1}{2}(B-C)}{\cos\frac{1}{2}(B+C)}\tan\frac{1}{2}a.$$

Corollary 9

54. Likewise, the tangents of the semi-differences of the sides will be:

$$\tan\frac{1}{2}(a-b) = \frac{\sin\frac{1}{2}(A-B)}{\sin\frac{1}{2}(A+B)}\tan\frac{1}{2}c$$

$$\tan\frac{1}{2}(a-c) = \frac{\sin\frac{1}{2}(A-C)}{\sin\frac{1}{2}(A+C)}\tan\frac{1}{2}b$$

$$\tan\frac{1}{2}(b-c)=\frac{\sin\frac{1}{2}(B-C)}{\sin\frac{1}{2}(B+C)}\tan\frac{1}{2}a.$$

The usage of these formulae will be of great importance in the following problems.

Problem VIII

55. *In a spherical triangle* (Fig. 11.4), *given two sides with the angle contained between them, to find the third side and the two other angles.*

Solution

Let ABC be the triangle (Fig. 11.4) of which are given the two sides $AB = c$, $AC = b$ with the angle A contained between them, and let the sides $BC = a$ and the angles B and C be searched for.

The third formula from the third property gives first

$$\cos a=\cos A\sin b\sin c+\cos b\cos c,$$

and the third formula from the fourth property provides the angle B,

$$\tan B=\frac{\sin A\tan b}{\sin c-\tan b\cos c\cos A}$$

whence we deduce, by analogy:

$$\tan C=\frac{\sin A\tan c}{\sin b-\tan c\cos b\cos A}.$$

But the expressions for the cotangents will be more convenient, in such a way that we get, for the solution, the following formulae:

$$\cos a=\cos A\sin b\sin c+\cos b\cos c$$

$$\cot B=\frac{\sin c\cot b-\cos c\cos A}{\sin A}$$

$$\cot C=\frac{\sin b\cot c-\cos b\cos A}{\sin A}.$$

Corollary 1

56. Since

$$\cos b\cos c=\frac{1}{2}\cos(b-c)+\frac{1}{2}\cos(b+c)$$

and

$$\sin b \sin c = \frac{1}{2}\cos(b-c) - \frac{1}{2}\cos(b+c),$$

the cosine of the side a may be expressed by the addition and subtraction of simple cosines in the following manner:

$$\cos a = \frac{1}{4}\cos(A-b+c) + \frac{1}{4}\cos(A+b-c) - \frac{1}{4}\cos(A-b-c)$$
$$- \frac{1}{4}\cos(A+b+c) + \frac{1}{2}\cos(b-c) + \frac{1}{2}\cos(b+c).$$

Corollary 2
57. But if we want to use logarithms, this formula is less convenient. However we can apply logarithms introducing in it a new angle u, setting

$$\tan u = \frac{\cos A \sin b}{\cos b},$$

or else $\tan u = \cos A \tan b$, and having found this angle u we get:

$$\cos a = \tan u \cos b \sin c + \cos b \cos c = \frac{\cos b \cos(c-u)}{\cos u},$$

where we can easily find the side a using logarithms.

Corollary 3
58. The same introduction of the angle u, in such a way that

$$\tan u = \cos A \tan b,$$

makes the other formulae fit as well to applying to them logarithms, since we get:

$$\tan B = \frac{\sin A \tan b}{\sin c - \tan u \cos c} = \frac{\sin A \tan b \cos u}{\sin(c-u)} = \frac{\tan A \sin u}{\sin(c-u)}.$$

Regarding the angle C, we find it by the rule

$$\sin C = \frac{\sin A \sin c}{\sin a}.$$

Corollary 4
59. But the most convenient search for the angles B and C will be deduced from formulae given in Paragraphs 43 and 44, from which we get

$$\tan\frac{1}{2}(B+C)=\frac{\cos\frac{1}{2}(b-c)}{\cos\frac{1}{2}(b+c)}\cot\frac{1}{2}A$$

$$\tan\frac{1}{2}(B-C)=\frac{\sin\frac{1}{2}(b-c)}{\sin\frac{1}{2}(b+c)}\cot\frac{1}{2}A.$$

Now if we have half of their sum with half of their difference, we get each one separately, and therefore we can finally obtain the side a by the rule

$$\sin a=\frac{\sin b}{\sin B}\sin A=\frac{\sin c}{\sin C}\sin A.$$

Problem IX

60. *In a spherical triangle* (Fig. 11.4), *given two angles and the side contained by them, to find the third angle and the two sides.*

Solution

Let ABC be the triangle, in which are given the two angles A and B with the side $AB=c$ contained by them, and let us find the third angle C together with the two other sides $AC=b$ and $BC=a$.

The first formula of the second property (Paragraph 30) gives first

$$\cos C=\cos c\sin A\sin B-\cos A\cos B,$$

and the third formula of the fourth property gives

$$\tan b=\frac{\sin c\tan B}{\sin A+\cos c\cos A\tan B},$$

and by analogy

$$\tan a=\frac{\sin c\tan A}{\sin B+\cos c\cos B\tan A}.$$

Therefore, taking the cotangents, we get the following solution:

$$\cos C=\cos c\sin A\sin B-\cos A\cos B$$

$$\cot a=\frac{\cot A\sin B+\cos c\cos B}{\sin c}$$

$$\cot b=\frac{\cot B\sin A+\cos c\cos A}{\sin c}.$$

Corollary 1

61. The two sides will be more easily found from the formulae of Paragraphs 52 and 53, from where we deduce:

$$\tan\frac{1}{2}(a+b) = \frac{\cos\frac{1}{2}(A-B)}{\cos\frac{1}{2}(A+B)}\tan\frac{1}{2}c$$

$$\tan\frac{1}{2}(a-b) = \frac{\sin\frac{1}{2}(A-B)}{\sin\frac{1}{2}(A+B)}\tan\frac{1}{2}c,$$

where it is easy to use logarithms.

Corollary 2

62. After having found the sides a and b, we will find easily the angle C, since we have

$$\sin C = \frac{\sin A}{\sin a}\sin c = \frac{\sin B}{\sin b}\sin c,$$

or else we can also, if we want, express $\cos C$ by simple cosines in the following way:

$$\cos C = \frac{1}{4}\cos(c+A-B) + \frac{1}{4}\cos(c-A+B) - \frac{1}{4}\cos(c-A-B)$$
$$-\frac{1}{4}\cos(c+A+B) - \frac{1}{2}\cos(A-B) - \frac{1}{2}\cos(A+B).$$

Problem X

63. *In a spherical triangle, given two sides with an angle not contained by them* (Fig. 11.4), *or, which amounts to the same, given two angles with a side not contained by them, to find the other quantities belonging to the triangle.*

Solution

Let ABC be the triangle in which we are given, for the first case, the two sides $BC = a$ and $AC = b$ with the angle A. We shall know first the angle B because

$$\sin B = \frac{\sin A}{\sin a}\sin b.$$

For the other case, let A and B be two given angles, with the side $BC = a$. Then we first have the side b since

$$\sin b = \frac{\sin a}{\sin A}\sin B.$$

From here, in either case, we can regard as given the two sides $BC = a$ and $AC = b$ as well as the two angles A and B which are opposite to them. Thus, we must find its side $AB = c$ and its angle C.

The first formula of the fourth property provides:

$$\sin a \tan C - \sin B \tan c = \cos a \cos B \tan C \tan c,$$

whence, by transposing the sides a and b with the angles A and B we get

$$\sin b \tan C - \sin A \tan c = \cos b \cos A \tan C \tan c.$$

From these two equations, eliminating in one case $\tan C$ and in the other case $\tan c$, we will find

$$\tan c = \frac{\sin A \sin a - \sin B \sin b}{\sin A \cos B \cos a - \cos A \sin B \cos b}$$

$$\tan C = \frac{\sin A \sin a - \sin B \sin b}{\cos B \cos a \sin b - \cos A \sin a \cos b},$$

to which we must add the following equation

$$\sin A \sin b = \sin B \sin a.$$

Corollary 1

64. Since we have $\sin A : \sin B = \sin a : \sin b$, we will also have

$$\tan c = \frac{\sin^2 a - \sin^2 b}{\cos B \sin a \cos a - \cos A \sin b \cos b}$$

and

$$\tan C = \frac{\sin^2 A - \sin^2 B}{\sin B \cos B \cos a - \sin A \cos A \cos b}.$$

Corollary 2

65. But Paragraphs 43, 44, 53 and 54 provide us also with the following more convenient solutions:

$$\tan \frac{1}{2} c = \frac{\cos \frac{1}{2}(A+B)}{\cos \frac{1}{2}(A-B)} \tan \frac{1}{2}(a+b) = \frac{\sin \frac{1}{2}(A+B)}{\sin \frac{1}{2}(A-B)} \tan \frac{1}{2}(a-b)$$

and

$$\tan \frac{1}{2} C = \frac{\cos \frac{1}{2}(a-b)}{\cos \frac{1}{2}(a+b)} \cot \frac{1}{2}(A+B) = \frac{\sin \frac{1}{2}(a-b)}{\sin \frac{1}{2}(a+b)} \cot \frac{1}{2}(A-B),$$

to which one can easily apply the usage of logarithms.

Problem XI

66. *To find (*Fig. 11.3) *the area of an arbitrary spherical triangle.*

Solution

Let EOM be the proposed spherical triangle (Fig. 11.3) and let us put, as above (Paragraph 17) the side $OE = a$, the angle $OEM = \alpha$, the angle $EOM = y$, the side $OM = x$ and the angle $OME = \varphi$. This being set, the trilinear figure MOm will represent the differential of the area we are looking for, and since $mn = dx$ and $Mn = dy \sin x$, the product $dydx \sin x$ will express the differential of MOm, hence

$$MOm = dy \int dx \sin x = dy(1 - \cos x)$$

therefore the area:

$$EOM = y - \int dy \cos x.$$

But we have found

$$dy = \frac{Cdx}{\sin x \sqrt{\sin^2 x - C^2}},$$

in such a way that

$$\text{the area } EOM = y - \int \frac{Cdx \cos x}{\sin x \sqrt{\sin^2 x - C^2}}.$$

Then, having found $\sin \varphi = \frac{C}{\sin x}$, because of $C = \sin a \sin \alpha$ and since

$$\cos \varphi = \frac{\sqrt{\sin x^2 - C^2}}{\sin x},$$

we get

$$d\varphi \cos \varphi = -Cdx \frac{\cos x}{\sin^2 x},$$

therefore

$$d\varphi = -\frac{Cdx \cos x}{\sin x \sqrt{\sin^2 x - C^2}} \text{ and } -\int \frac{Cdx \cos x}{\sin x \sqrt{\sin^2 x - C^2}} = \varphi + \text{Const.}$$

Consequently, the area of the triangle EOM will be

$$= y + \varphi + \text{Const.} = \alpha + y + \varphi - \text{Const.}$$

To know this constant, let us suppose that $y = 0$; since we get $\varphi = 180^\circ - \alpha$, the area of this vanishing triangle will be 180° − Const., therefore Const. $= 180^\circ$. Thus, we have:

$$\text{the area of the triangle } EOM = \alpha + y + \varphi - 180^\circ.$$

Corollary 1

67. Therefore, to find the area of an arbitrary spherical triangle, it suffices to take the excess of the sum of its three angles to two right [angles], when the radius of the sphere is expressed by 1.[11] But on an arbitrary sphere we shall take an arc of a great circle as the measure of the said excess, and the product of this arc by the radius of the sphere will give the area of the spherical triangle we are looking for.

Corollary 2

68. Thus, the more a spherical triangle will be large, the more its angle sum will exceed two right angles, and when the area of the triangle occupies one eighth of the surface of the sphere, the value of this excess will be a right angle. Because an arc of great circle of 90° multiplied by the radius gives half of the area of the great circle, and hence the eighth part of the surface of the sphere. Therefore we shall use this rule to find the area of any spherical triangle. We shall say: like 8 right angles, or 720°, are to the excess of the sum of the three angles over two right [angles], so is the surface of the entire sphere to the area of the proposed triangle.

[11] This is the theorem attributed to Albert Girard which Euler already mentions in §15 and 16 of the present memoir (see Footnote 9) and of which he also gives a proof in the memoir *Various speculations on the area of spherical triangles*, translated in Chap. 13 of the present volume. See also Euler's memoir *Principles of spherical trigonometry deduced from the method of maxima and minima*, translated in Chap. 11 of the present volume.

Chapter 12
General Spherical Trigonometry, Deduced from Fundamental Principles in a Brief and Clear Manner

Leonhard Euler

Abstract In this memoir, Leonhard Euler establishes a set of formulae of spherical trigonometry based on the geometry of the 3-dimensional Euclidean space.

Keywords Spherical trigonometry · Leonhard Euler · History of spherical geometry · Mathematics in the eighteenth century

AMS Classification 01A50, 51-03, 53A05, 53A35

§1. (Figure 12.1) Given an arbitrary spherical triangle, we denote its angles by the capital letters A, B, C and its sides by the small letters a, b, c, as in the figure, in such a way that the small letters become opposite to the same capital letters. Now from the center of the sphere, to which we attribute the letter O, for each angle are raised the straight lines[1] OC, Oa, Ob, which at the center O constitute a solid angle whose measure of plane angles are the sides of the triangle, whereas [the measures of] their mutual inclinations are the angles of the triangle.
§2. (Figure 12.2.) With these premises, we take OC, which is equal to the radius of the sphere, equal to 1, and in the two planes COa and COb we construct the straight lines Ca and Cb perpendicular to OC. Then from b to Ca we draw the

[1] Euler sometimes also uses small letters to denote the vertices of the triangle that are opposite to the sides that are called by these letters. Here for instance he uses Oa and Ob instead of OA and OB. On the other hand, when in the following he writes Ca and Cb, he indicates segments of straight lines, while CA and CB are arcs of great circles. This results in a series of confusions in the notations at the beginning of this memoir that we have not tried to unravel (there was no obvious method for doing so). But the notation he uses in the trigonometric formulae he finds is consistent. The confusion is probably due to the typographer.

The author Leonhard Euler (1707–1783).
Translated from the Latin by Renzo Caddeo.
Trigonometria sphaerica universa, ex primis principiis breviter et dilucide derivata, *Acta Academiae Scientiarum Imperialis Petropolitanae, 3, pro anno 1779, pars prior, 1782, p. 72–86. Opera Omnia: Series 1, Volume 26, p. 224–236.*

R. Caddeo, A. Papadopoulos (eds.), *Spherical Geometry in the Eighteenth Century I*, https://doi.org/10.1007/978-3-032-12466-1_12

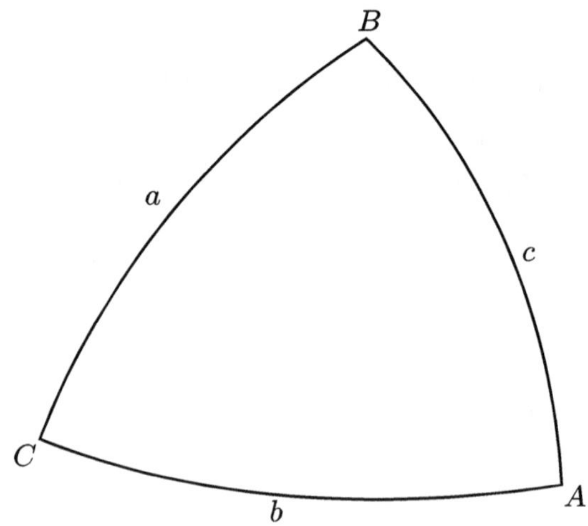

Fig. 12.1 The spherical triangle ABC and its sides a, b, c

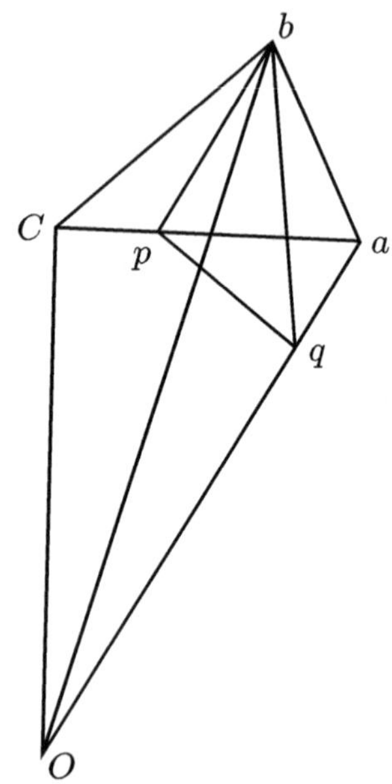

Fig. 12.2 We have reproduced Euler's figure in which abC denotes a planar triangle whose sides are segments of straight lines

perpendicular bp, which will be at the same time perpendicular to the plane COa; and moreover from p we draw pq perpendicular to Oa, and then also bq will be perpendicular to Oa. In this way the whole figure that we need has been constructed.
§3. Now, the angle COa being equal to the side b, we get

$$Ca = \tan b \text{ and } Oa = \sec b = \frac{1}{\cos b}.$$

Similarly, the angle COb being equal to a, we get

$$Cb = \tan a \text{ and } Ob = \sec a = \frac{1}{\cos a}.$$

But on the other hand, since the angle aOb is equal to c and $Ob = \frac{1}{\cos a}$, we get

$$bq = \frac{\sin c}{\cos a} \text{ and } Oq = \frac{\cos c}{\cos a}.$$

From here, to represent the remaining lines of the figure, since the angle aCb is equal to C, we get

$$bp = Cb \sin C = \tan a \sin C$$

and

$$Cp = Cb \cos C = \tan a \cos C,$$

whence moreover we obtain

$$ap = Ca - Cp = \tan b - \tan a \cos C,$$

and as the angle CaO is equal to $90^\circ - b$, we get

$$pq = ap \cos b = \sin b - \tan a \cos b \cos C$$

and

$$aq = ap \sin b = \frac{\sin^2 b}{\cos b} - \tan a \sin b \cos C.$$

Therefore, since $Oq = \frac{\cos c}{\cos a}$, we get

$$Oa = \frac{1}{\cos b} = \frac{\cos c}{\cos a} + \frac{\sin^2 b}{\cos b} - \tan a \sin b \cos C,$$

and in this way we get

$$\frac{\cos c}{\cos a} = \cos b + \tan a \sin b \cos C,$$

that is,

$$\cos c = \cos a \cos b + \sin a \sin b \cos C.$$

§4. Now, since the angle bqp produces the inclination of the plane aOb with respect to aOC, for this angle we get $bqp = A$, and then from the triangle bqp we get first of all

$$\sin A = \frac{bp}{bq} = \frac{\sin a \sin C}{\sin c},$$

that is,

$$\frac{\sin C}{\sin c} = \frac{\sin A}{\sin a};$$

and from here it follows that the sines of the angles of our triangle are proportional to the sines of the opposite sides. Therefore the equation

$$\cos A = \frac{pq}{bq} = \frac{\cos a \sin b - \sin a \cos b \cos C}{\sin c},$$

together with the two preceding ones, embraces all spherical geometry, and this fact requires a larger explanation, in which we want to develop even more each of these three equations.

Development of the First Formula

$$\frac{\sin C}{\sin c} = \frac{\sin A}{\sin a}.$$

§5. Since either the capital letters A, B, C, or the small ones a, b, c can be exchanged among each other, provided the same small letters remain in front of the capital letters, we will also have

$$\frac{\sin C}{\sin c} = \frac{\sin B}{\sin b},$$

and in this way the following triple equation appears:

$$\frac{\sin A}{\sin a} = \frac{\sin B}{\sin b} = \frac{\sin C}{\sin c}.$$

Furthermore, it will also be useful to observe the following equalities:

$$\sin A \sin b = \sin B \sin a,$$

$$\sin A \sin c = \sin C \sin a,$$

$$\sin B \sin c = \sin C \sin b.$$

Development of the Formula

$$\cos A \sin c = \cos a \sin b - \sin a \cos b \cos C.$$

§6. Given that $\sin A \sin c = \sin C \sin a$, we divide the left hand side of this equation by $\sin A \sin c$ and subsequently by $\sin C \sin a$, and we obtain

$$\cot A = \frac{\cos a \sin b - \sin a \cos b \cos C}{\sin a \sin C},$$

and from here, given the two sides a and b with the contained angle C, it is possible to determine the angle A; and in a similar way, from the same data we find the angle B from the formula derived from this one, exchanging the letters A, B, a, b:

$$\cot B = \frac{\cos b \sin a - \cos a \sin b \cos C}{\sin b \sin C}.$$

§7. If moreover, in this same formula that we are considering, we multiply the first term by $\frac{\sin C}{\sin c}$, the second one by $\frac{\sin B}{\sin b}$, and the third one by $\frac{\sin A}{\sin a}$, the following memorable equation will emerge:

$$\cos A \sin C = \cos a \sin B - \cos b \sin A \cos C,$$

or

$$\cos a = \frac{\cos A \sin C + \sin A \cos C \cos b}{\sin B};$$

then, exchanging among themselves the letters B and C, and likewise b and c, we get

$$\cos a = \frac{\cos A \sin B + \sin A \cos B \cos c}{\sin C},$$

that is,

$$\cos a \sin C = \cos A \sin B + \sin A \cos B \cos c,$$

and this one differs from the preceding one[2] only by the fact that the capital letters are exchanged with the small letters and moreover all the cosines are taken with a negative sign.[3]

[2] By "the preceding one", Euler means the formula established at the end of §5.

[3] See the general statement on polarity theory, expressed in the theorem stated in §12 of the present memoir, and the footnote there.

§8 Now if in the last equation we divide the left hand side by $\sin a \sin C$, and subsequently by $\sin A \sin c$, we get the following equation

$$\cot a = \frac{\cos A \sin B + \sin A \cos B \cos c}{\sin A \sin c},$$

which is useful for finding the side a when we are given the two angles A and B and the side c they contain; but then from the same data we also have that the side b is defined by the equation

$$\cot b = \frac{\cos B \sin A + \sin B \cos A \cos c}{\sin B \sin c}.$$

§9. But, in addition, from the same formula presented above, we get the sometimes very difficult case in which given the three angles, the sides are sought. Indeed, having

$$\cos A \sin c = \cos a \sin b - \sin a \cos b \cos C,$$

we will also have, exchanging both the letters A and B, and the letters a and b,

$$\cos B \sin c = \cos b \sin a - \sin b \cos a \cos C.$$

If this is multiplied by $\cos C$ and then added to the first one, the following equation appears

$$\sin c(\cos A + \cos B \cos C) = \cos a \sin b \sin^2 C;$$

but since $\sin b \sin C = \sin B \sin c$, this equation takes the form

$$\cos A + \cos B \cos C = \cos a \sin B \sin C,$$

that is,

$$\cos A = -\cos B \cos C + \sin C \cos a \sin B.$$

Permuting the letters A and C, while B remains unchanged, we get

$$\cos C = -\cos B \cos A + \sin B \sin A \cos c;$$

this equality can be obtained from our third formula,[4] that is,

$$\cos c = \cos a \cos b + \sin a \sin b \cos C,$$

[4] This is formula in the last line of §3.

if the capital letters are exchanged with the small ones, while all the cosines are taken with a negative sign.

Development of the Formula

$$\cos c = \cos a \cos b + \sin a \sin b \cos C.$$

§10. We see immediately here that this formula lends itself to a twofold use: the first, where, given the sides a, b, c, one has to find the angles, and for this we need the following formula:

$$\cos C = \frac{\cos c - \cos a \cos b}{\sin a \sin b};$$

the second use instead, when given two sides a and b and the angle C contained by them, one has to find the third side c, and for this the following formula is once again useful:

$$\cos c = \cos a \cos b + \sin a \sin b \cos C.$$

§11. After that we can now also transfer this use to the angles, since we have recently found the formula

$$\cos C = -\cos A \cos B + \sin A \sin B \cos c.$$

In fact, through this, if the two angles A and B and the side c included between them are given, the third angle C is determined immediately. But then, if all three angles of a spherical triangle are given, an arbitrary side, such as c, is determined by the formula

$$\cos c = \frac{\cos C + \cos A \cos B}{\sin A \sin B}.$$

§12. Since all spherical trigonometry is based on the three equations found above, in general they must be preserved if permutations of the angles and sides are made, provided that all the cosines are taken with the negative sign. In fact, in the first formula

$$\frac{\sin C}{\sin c} = \frac{\sin B}{\sin b} = \frac{\sin A}{\sin a},$$

this permutability results in itself already as evident, as there is no cosine there, and consequently we obtain immediately the same permutability for the other two formulas, and from these considerations the notable theorem follows.

Theorem
Given an arbitrary spherical triangle, with angles A, B, C and sides a, b, c, it is always possible to construct another (analogous) triangle, whose angles are the complements to two right angles of the sides of the given triangle, whereas the sides are the complements to two right angles of the angles.[5]

In fact, in this way, all the sines remain the same, whereas all the cosines become their negative, and then also the tangents and the cotangents. One notices moreover that such triangle is formed by the poles of the three sides of the given triangle.[6]
§13. Therefore, for practical needs, all the rules may be represented in four types, among which there are two pairs whose rules are so much connected that in each pair the rules can be obtained from each other exchanging among themselves the capital letters and the small ones, and taking the cosines with the minus sign, in such a way that it is sufficient to keep in memory only two types. Thus, we report on these four types in an explicit way, with all the variants that can be obtained through transposition of the letters.

First Type
§14. This type includes two cases: in the first of them, from the three sides we obtain a certain angle; in the second, instead, given two sides and the angle contained by them, we determine the third side:

$$\cos A = \frac{\cos a - \cos b \cos c}{\sin b \sin c}$$

$$\cos B = \frac{\cos b - \cos a \cos c}{\sin a \sin c}$$

$$\cos C = \frac{\cos c - \cos a \cos b}{\sin a \sin b}$$

$$\cos a = \cos b \cos c + \sin b \sin c \cos A$$

$$\cos b = \cos a \cos c + \sin a \sin c \cos B$$

$$\cos c = \cos a \cos b + \sin a \sin b \cos C.$$

[5] This property, which is a version of polarity in spherical trigonometry, was already noticed by the Arab mathematicians of the tenth century; see Chap. 10 in the present volume.

[6] Lagrange makes a similar remark in §18 of his memoir *Solution de quelques problèmes relatifs aux triangles sphériques avec une analyse complète de ces triangles* (Solution of some problems relative to spherical triangles with a complete analysis of these triangles), translated in the present volume, Chap. 17. This is a version of polarity theory in spherical geometry, which was discovered by the Arab mathematicians of the 9th-10th centuries, see Chap. 10 of the present volume.

Second Type

§15. This type also contains two cases: in the first one, given the three angles, we want a certain side; in the second one instead, given two angles and the side their contain, we want to know the third angle:

$$\cos a = \frac{\cos A + \cos B \cos C}{\sin B \sin C}$$

$$\cos b = \frac{\cos B + \cos A \cos C}{\sin A \sin C}$$

$$\cos c = \frac{\cos C + \cos A \cos B}{\sin A \sin B}$$

$$\cos A = -\cos B \cos C + \sin B \sin C \cos a$$

$$\cos B = -\cos A \cos C + \sin A \sin C \cos b$$

$$\cos C = -\cos A \cos B + \sin A \sin B \cos c.$$

Third Type

§16. This type contains the case where, given two sides and the angle comprised by them, the other two angles are determined; the formulae with their variations are:

$$\cot A = \frac{\cos a \sin b - \sin a \cos b \cos C}{\sin a \sin C}$$

$$\cot B = \frac{\cos b \sin c - \sin b \cos c \cos A}{\sin b \sin A}$$

$$\cot C = \frac{\cos c \sin a - \sin c \cos a \cos B}{\sin c \sin B},$$

$$\cot B = \frac{\sin a \cos b - \cos a \sin b \cos C}{\sin b \sin C}$$

$$\cot C = \frac{\sin b \cos c - \cos b \sin c \cos A}{\sin c \sin A}$$

$$\cot A = \frac{\sin c \cos a - \cos c \sin a \cos B}{\sin a \sin B}.$$

Fourth Type

§17. This type regards the case where, given two angles and the side between them, we want to determine the other two sides. The corresponding formulae with their variations are:

$$\cot a = \frac{\cos A \sin B + \sin A \cos B \cos c}{\sin A \sin c},$$

$$\cot b = \frac{\cos B \sin C + \sin B \cos C \cos a}{\sin B \sin a},$$

$$\cot c = \frac{\cos C \sin A + \sin C \cos A \cos b}{\sin C \sin b},$$

$$\cot b = \frac{\sin A \cos B + \cos A \sin B \cos c}{\sin B \sin c}.$$

$$\cot c = \frac{\sin B \cos C + \cos B \sin C \cos a}{\sin C \sin a}.$$

$$\cot a = \frac{\sin C \cos A + \cos C \sin A \cos b}{\sin A \sin b}.$$

§18. This simplicity is much more noteworthy because the resolution of the right triangles requires up to six completely different formulas. Because if the angle C is right, and then c is the hypotenuse and both a and b are the cathets, the six required formulas are:

$$\cos c = \cos a \cos b$$

$$\cos c = \cot A \cot B$$

$$\sin a = \sin c \sin A, \quad \text{or} \quad \sin b = \sin c \sin B$$

$$\tan b = \tan c \cos A, \quad \text{or} \quad \tan a = \tan c \cos B$$

$$\tan a = \tan A \sin b, \quad \text{or} \quad \tan b = \tan B \sin a$$

$$\cos A = \cos a \sin B, \quad \text{or} \quad \cos B = \cos b \sin A,$$

and these formulae follow in a natural way from the preceding ones if we set $\cos C = 0$ and $\sin C = 1$.

§19. However, if we want to use logarithms, we need to derive from the previous types other formulae of this kind, constituted by [a product of] factors; this can be obtained through certain transformations with which we are led to the halves either of the angles or of the sides. Now it is possible to introduce briefly these transformations in the following ways.

First Transformation

§20. This transformation is obtained very simply from the following formula of the first type:

$$\cos A = \frac{\cos a - \cos b \cos c}{\sin b \sin c}.$$

From which it follows first of all that

$$1 - \cos A = \frac{\cos(b-c) - \cos a}{\sin b \sin c}$$

$$1 + \cos A = \frac{\cos a - \cos(b+c)}{\sin b \sin c}.$$

From these identities, since we have

$$\frac{1-\cos A}{1+\cos A} = \tan^2 \frac{1}{2} A,$$

we get

$$\tan^2 \frac{1}{2} A = \frac{\cos(b-c) - \cos a}{\cos a - \cos(b+c)};$$

on the other hand, it is clear that we have

$$\cos p - \cos q = 2 \sin \frac{q-p}{2} \sin \frac{p+q}{2},$$

whence we obtain

$$\tan \frac{1}{2} A = \sqrt{\frac{\sin \frac{a-b+c}{2} \sin \frac{a+b-c}{2}}{\sin \frac{b+c-a}{2} \sin \frac{a+b+c}{2}}}.$$

Second Transformation

§21. This is deduced from the first formula of the second type

$$\cos a = \frac{\cos A + \cos B \cos C}{\sin B \sin C}$$

whence we deduce

$$1 - \cos a = \frac{-\cos(B+C) - \cos A}{\sin B \sin C}$$

$$1+\cos a=\frac{\cos A+\cos(B-C)}{\sin B\sin C},$$

and in this way we get

$$\tan^2\frac{1}{2}a=-\frac{\cos(B+C)+\cos A}{\cos(B-C)+\cos A}.$$

Then having

$$\cos p+\cos q=2\cos\frac{p-q}{2}\cos\frac{p+q}{2},$$

we obtain

$$\tan\frac{1}{2}a=\sqrt{-\frac{\cos\frac{B+C-A}{2}\cos\frac{B+C+A}{2}}{\cos\frac{B+A-C}{2}\cos\frac{A+C-B}{2}}}.$$

Third Transformation

§22. This transformation may be also obtained from the first type, combining these two formulae:

$$\cos a-\cos b\cos c=\sin b\sin c\cos A,$$

$$\cos b-\cos a\cos c=\sin a\sin c\cos B;$$

dividing the first by the second, we obtain

$$\frac{\cos a-\cos b\cos c}{\cos b-\cos a\cos c}=\frac{\sin b\cos A}{\sin a\cos B}=\frac{\sin B\cos A}{\sin A\cos B}.$$

Adding 1 to each member, we get

$$(\cos a+\cos b)(1-\cos c)=\frac{\sin(A+B)}{\sin A\cos B},$$

then, subtracting 1 from each member, we get

$$(\cos a-\cos b)(1+\cos c)=\frac{\sin(B-A)}{\sin A\cos B},$$

and if we divide this equation by the preceding, we find

$$\frac{\cos a-\cos b}{\cos a+\cos b}\cot^2\frac{1}{2}c=\frac{\sin(B-A)}{\sin(B+A)}.$$

On the other hand, the following identity is valid:

$$\frac{\cos p - \cos q}{\cos p + \cos q} = \tan\frac{q+p}{2}\tan\frac{p-q}{2},$$

whence we obtain

$$\tan\frac{b-a}{2}\tan\frac{b+a}{2}\cot^2\frac{1}{2}c = \frac{\sin(B-A)}{\sin(B+A)}.$$

§23. Now from the principal property we call the help of the formula

$$\frac{\sin b}{\sin a} = \frac{\sin B}{\sin A},$$

whence we deduce

$$\frac{\sin b - \sin a}{\sin b + \sin a} = \frac{\sin B - \sin A}{\sin B + \sin A}$$

and this equation reduces to the form

$$\tan\frac{b-a}{2}\cot\frac{b+a}{2} = \tan\frac{B-A}{2}\cot\frac{B+A}{2}.$$

Now if we multiply this equation by the one obtained before, we find the equation

$$(\tan\frac{b-a}{2})^2\cot^2\frac{1}{2}c = \frac{(\sin\frac{(B-A)}{2})^2}{(\sin\frac{(B+A)}{2})^2},$$

that is, extracting the root,

$$\tan\frac{b-a}{2}\cot\frac{1}{2}c = \frac{\sin\frac{B-A}{2}}{\sin\frac{B+A}{2}}.$$

If, instead of multiplying these two equations, we divide the former by the second, we find

$$\tan\frac{b+a}{2}\cot\frac{1}{2}c = \frac{\cos\frac{B-A}{2}}{\cos\frac{B+A}{2}}.$$

So with these formulas we solve the case in which two angles A and B and the side c they comprise are given, and we search for the two sides a and b; this can be obtained thanks to the formulae

$$\tan\frac{b-a}{2}=\tan\frac{1}{2}c\frac{\sin\frac{B-A}{2}}{\sin\frac{B+A}{2}}$$

$$\tan\frac{b+a}{2}=\tan\frac{1}{2}c\frac{\cos\frac{B-A}{2}}{\cos\frac{B+A}{2}}.$$

Fourth Transformation

§24. This is deduced in a similar way from the formulae

$$\cos A+\cos B\cos C=\sin B\sin C\cos a$$

$$\cos B+\cos A\cos C=\sin A\sin C\cos b.$$

If we divide the first by the second, we find

$$\frac{\cos A+\cos B\cos C}{\cos B+\cos A\cos C}=\frac{\sin B\cos a}{\sin A\cos b}=\frac{\sin b\cos a}{\sin a\cos b}.$$

Adding 1 to these equalities, and then subtracting 1 from them, we obtain the following new equations:

$$(\cos A+\cos B)(1+\cos C)=\frac{\sin(a+b)}{\sin a\cos b},$$

$$(\cos A-\cos B)(1-\cos C)=\frac{\sin(b-a)}{\sin a\cos b},$$

and dividing the first by the second, we find

$$\frac{\cos A+\cos B}{\cos A-\cos B}\cot^2\frac{1}{2}C=\frac{\sin(a+b)}{\sin(b-a)},$$

that is,

$$\tan\frac{B-A}{2}\tan\frac{B+A}{2}=\cot^2\frac{1}{2}C=\frac{\sin(b-a)}{\sin(b+a)};$$

multiplying and dividing this equation by

$$\tan\frac{B-A}{2}\cot\frac{B+A}{2}=\tan\frac{b-a}{2}\cot\frac{b+a}{2}$$

we obtain the equations

$$\tan\frac{B-A}{2}=\cot\frac{1}{2}C\frac{\sin\frac{b-a}{2}}{\sin\frac{b+a}{2}},$$

$$\tan\frac{B+A}{2}=\cot\frac{1}{2}C\frac{\cos\frac{b-a}{2}}{\cos\frac{b+a}{2}},$$

and these formulae are valid in the case where two sides and the angle they contain are given.

§25. Now that we have reported on all the variations of the preceding types, we also highlight these four cases with all their variations.

$$\tan\frac{1}{2}A=\sqrt{\frac{\sin\frac{a+b-c}{2}\sin\frac{a-b+c}{2}}{\sin\frac{b+c-a}{2}\sin\frac{a+b+c}{2}}}.$$

$$\tan\frac{1}{2}B=\sqrt{\frac{\sin\frac{b+c-a}{2}\sin\frac{a+b-c}{2}}{\sin\frac{a+c-b}{2}\sin\frac{a+b+c}{2}}}.$$

$$\tan\frac{1}{2}C=\sqrt{\frac{\sin\frac{a+c-b}{2}\sin\frac{b+c-a}{2}}{\sin\frac{a+b-c}{2}\sin\frac{a+b+c}{2}}}.$$

$$\tan\frac{1}{2}a=\sqrt{-\frac{\cos\frac{B+C-A}{2}\cos\frac{A+B+C}{2}}{\cos\frac{A+B-C}{2}\cos\frac{A+C-B}{2}}}.$$

$$\tan\frac{1}{2}b=\sqrt{-\frac{\cos\frac{A+C-B}{2}\cos\frac{A+B+C}{2}}{\cos\frac{B+C-A}{2}\cos\frac{A+B-C}{2}}}.$$

$$\tan\frac{1}{2}c=\sqrt{-\frac{\cos\frac{A+B-C}{2}\cos\frac{A+B+C}{2}}{\cos\frac{A+C-B}{2}\cos\frac{B+C-A}{2}}}.$$

$$\tan\frac{b-a}{2}=\tan\frac{1}{2}c\frac{\sin\frac{B-A}{2}}{\sin\frac{B+A}{2}}$$

$$\tan\frac{c-b}{2} = \tan\frac{1}{2}a\frac{\sin\frac{C-B}{2}}{\sin\frac{C+B}{2}}$$

$$\tan\frac{a-c}{2} = \tan\frac{1}{2}b\frac{\sin\frac{A-C}{2}}{\sin\frac{A+C}{2}}$$

$$\tan\frac{b+a}{2} = \tan\frac{1}{2}c\frac{\cos\frac{B-A}{2}}{\cos\frac{B+A}{2}}$$

$$\tan\frac{c+b}{2} = \tan\frac{1}{2}a\frac{\cos\frac{C-B}{2}}{\cos\frac{C+B}{2}}$$

$$\tan\frac{a+c}{2} = \tan\frac{1}{2}b\frac{\cos\frac{A-C}{2}}{\cos\frac{A+C}{2}}$$

$$\tan\frac{B-A}{2} = \cot\frac{1}{2}C\frac{\sin\frac{b-a}{2}}{\sin\frac{b+a}{2}}$$

$$\tan\frac{C-B}{2} = \cot\frac{1}{2}A\frac{\sin\frac{c-b}{2}}{\sin\frac{c+b}{2}}$$

$$\tan\frac{A-C}{2} = \cot\frac{1}{2}B\frac{\sin\frac{a-c}{2}}{\sin\frac{a+c}{2}}$$

$$\tan\frac{B+A}{2} = \cot\frac{1}{2}C\frac{\cos\frac{b-a}{2}}{\cos\frac{b+a}{2}}$$

$$\tan\frac{C+B}{2} = \cot\frac{1}{2}A\frac{\cos\frac{c-b}{2}}{\cos\frac{c+b}{2}}$$

$$\tan\frac{A+C}{2} = \cot\frac{1}{2}B\frac{\cos\frac{a-c}{2}}{\cos\frac{a+c}{2}}$$

§26. From these last formulae we easily solve the case that we have not touched yet, in which we are given two sides with their opposite angles, and we want to find either the third side, or the third angle, which in both cases can occur in two ways. We therefore report below the formulae together with the variations.

$$\tan\frac{1}{2}c = \tan\frac{b-a}{2}\frac{\sin\frac{B+A}{2}}{\sin\frac{B-A}{2}}$$

$$\tan\frac{1}{2}a = \tan\frac{c-b}{2}\frac{\sin\frac{C+B}{2}}{\sin\frac{C-B}{2}}$$

$$\tan\frac{1}{2}b = \tan\frac{a-c}{2}\frac{\sin\frac{A+C}{2}}{\sin\frac{A-C}{2}}$$

$$\tan\frac{1}{2}c = \tan\frac{b+a}{2}\frac{\cos\frac{B+A}{2}}{\cos\frac{B-A}{2}}$$

$$\tan\frac{1}{2}a = \tan\frac{c+b}{2}\frac{\cos\frac{C+B}{2}}{\cos\frac{C-B}{2}}$$

$$\tan\frac{1}{2}b = \tan\frac{a+c}{2}\frac{\cos\frac{A+C}{2}}{\cos\frac{A-C}{2}}$$

$$\cot\frac{1}{2}C = \tan\frac{B-A}{2}\frac{\sin\frac{b+a}{2}}{\sin\frac{b-a}{2}}$$

$$\cot\frac{1}{2}A = \tan\frac{C-B}{2}\frac{\sin\frac{c+b}{2}}{\sin\frac{c-b}{2}}$$

$$\cot\frac{1}{2}B = \tan\frac{A-C}{2}\frac{\sin\frac{a+c}{2}}{\sin\frac{a-c}{2}}$$

$$\cot\frac{1}{2}C = \tan\frac{B+A}{2}\frac{\cos\frac{b+a}{2}}{\cos\frac{b-a}{2}}$$

$$\cot\frac{1}{2}A = \tan\frac{C+B}{2}\frac{\cos\frac{c+b}{2}}{\cos\frac{c-b}{2}}$$

$$\cot\frac{1}{2}B = \tan\frac{A+C}{2}\frac{\cos\frac{a+c}{2}}{\cos\frac{a-c}{2}}$$

Thus, in this way, the present discussion can be seen as the complete system of all spherical trigonometry.

Chapter 13
Various Speculations on the Area of Spherical Triangles

Leonhard Euler

Abstract In this memoir, Euler gives several formulae for the area of a spherical triangle in terms of its side lengths. This is an analogue, in spherical geometry, of the Euclidean Heron formula. He then uses these results to solve the problem of finding the locus of the vertices of spherical triangles having a fixed base and the same area.

Keywords Spherical area · Leonhard Euler · Girard formula · Lexell problem · History of spherical geometry · Mathematics in the eighteenth century

AMS Classification 01A50, 51-03, 53A05, 53A35

§1 The first to teach how to define the area of a spherical triangle, according to the testimony of Wallis, was Albert Girard,[1] who proved that the area of a spherical triangle is always proportional to the excess of the sum of the three angles with respect to two right angles, and that furthermore this area is found by converting this

[1] Albert Girard (1595–1632) is a French mathematician and musician who was settled in the Netherlands, with interests in algebra, military engineering and other fields. Euler refers to John Wallis (1616–1703) who gave a proof of Girard's theorem; see Vol. II of Wallis' Collected Works (*Operum mathematicorum*, Typis Leon, Oxford, 1693).

The author Leonhard Euler (1707–1783).
Translated from the Latin by Renzo Caddeo.
Variae speculationes super area triangulorum sphaericorum, *Nova Acta Academiae Scientiarum Imperialis Petropolitanae*, 10, 1797, p. 47–62. Opera Omnia: Series 1, Volume 29, p. 253–266.

R. Caddeo, A. Papadopoulos (eds.), *Spherical Geometry in the Eighteenth Century I*, https://doi.org/10.1007/978-3-032-12466-1_13

excess into an arc of great circle and multiplying it by the radius of the sphere.[2] But the search for a manner with which the area of a spherical triangle can be determined from its sides requires an investigation of much greater difficulty. But already a while ago I discovered an outstanding theorem[3] through which this determination can be easily constructed, and which establishes the following Theorem 1. *If we denote by the letters a, b, c the sides of a spherical triangle and by Δ the area of this triangle, then we will always have*

$$\cos\frac{1}{2}\Delta = \frac{1+\cos a+\cos b+\cos c}{4\cos\frac{1}{2}a\cos\frac{1}{2}b\cos\frac{1}{2}c}.$$

whose correctness can be proved, not without long and tortuous paths, either by Girard's theorem or directly using integral calculus. Therefore it is worth here to bring to everyone's knowledge both proofs.

Problem 1

If to the two sides AZ and BZ of the spherical triangle AZB constructed on the base AB we add their differentials, we generate a triangle AzB. To investigate the increment that the area of the triangle AZB undergoes in this way.

Solution

§2. Denoting by a the base AB of this triangle (Fig. 13.1), let us call its sides AZ and BZ respectively x and y, whence the sides of the enlarged triangle must be $Az = x + \partial x$ and $Bz = y + \partial y$.

---

[2] Girard stated this theorem in his essay *Invention nouvelle en l'algèbre* (New invention in algebra), Amsterdam, 1629, on p. 9 of an appendix to that essay titled *De la mesure de la superficie des triangles et polygones sphériques, nouvellement inventée* (On the measure and surface of spherical triangles and polygons, recently discovered). The mathematician Eugène Prouhet (1817–1867), in his paper *Note historique sur l'aire du triangle sphérique* (Nouvelles annales de mathématiques, Vol. 14, 152–154), reports that Euler evidently did not read Girard's book, and that Euler attributed to Girard a proof which we owe to Cavalieri. He then summarizes the content of Cavalieri's book, and ends his review with the following biographical note on the author: Born in Milano in 1598, he entered the order of Jesuates and dedicated himself to mathematics. He became a student of Galileo, and died on December 3, 1667. According to Boris Rosenfeld, in his book *A history of non-Euclidean geometry. Evolution of the concept of a geometric space* (Springer-Verlag, 1988, p. 31), a similar rule was discovered in 1603, that is, before Girard, by Thomas Harriot (1560–1621), as indicated by a letter written by Henry Briggs (1561–1630) to Johannes Kepler (1571–1630) written in 1625, reproduced in Kepler's *Gesammelte Werke*, vol. 18, p. 228–229, Munich, 1937–1969. See also *Thomas Harriot's Doctrine of Triangular Numbers: The "Magisteria Magna"*, European Mathematical Society, Heritage of European Mathematics collection, Zürich, 2008, p. 28 ff. Girard's formula is also quoted in Euler's memoirs *Principles of spherical trigonometry deduced from the method of maxima and minima* and *Various speculations on the area of spherical triangles*, and in Lagrange's memoir *Solution of some problems relative to spherical triangles with a complete analysis of these triangles*, all of them edited in English translation in the present volume.

[3] Euler uses the latin expression *Egregium theorema*.

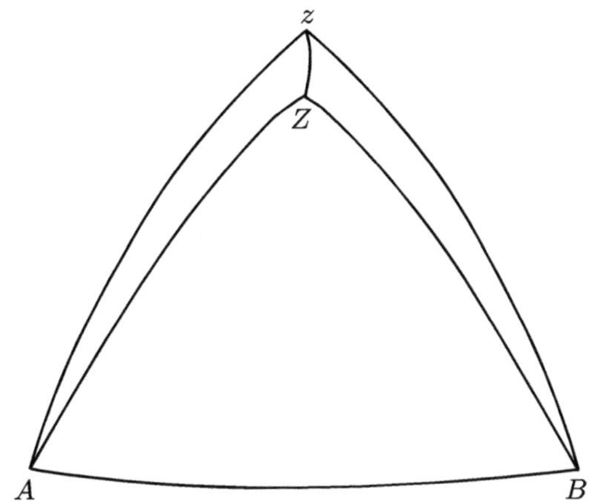

Fig. 13.1 Problem 1

Then, instead, we call ϕ and ψ respectively the angles BAZ and ABZ, in such a way that the elementary angles $ZAz = \partial\phi$ and $ZBz = \partial\psi$ appear, and with this notation it follows that the area of the elementary triangle ZAz is equal to $\partial\phi(1-\cos x)$, whereas that of the elementary triangle ZBz is equal to $\partial\psi(1-\cos y)$.

Since these two elementary triangles express the increase $\partial\Delta$ of the area of the triangle, we get the equation

$$\partial\Delta = \partial\phi(1 - \cos x) + \partial\psi(1 - \cos y).$$

§3. Therefore now we eliminate from the calculation the angles ϕ and ψ, and we introduce instead the sides x and y using the trigonometric rules, which give:

$$\cos\phi = \frac{\cos y - \cos a \cos x}{\sin a \sin x}$$

and

$$\cos\psi = \frac{\cos x - \cos a \cos y}{\sin a \sin y};$$

from here, differentiating, we get

$$-\partial\phi \sin\phi = \frac{\partial x \cos a - \partial x \cos x \cos y - \partial y \sin x \sin y}{\sin a \sin^2 x},$$

and in the same manner we find

$$-\partial\psi \sin\psi = \frac{\partial y \cos a - \partial y \cos x \cos y - \partial x \sin x \sin y}{\sin a \sin^2 y}.$$

But since

$$\cos\phi = \frac{\cos y - \cos a \cos x}{\sin a \sin x},$$

we get

$$\sin\phi = \frac{\sqrt{1 - \cos^2 a - \cos^2 x - \cos^2 y + 2\cos a \cos x \cos y}}{\sin a \sin x},$$

and in a similar manner

$$\sin\psi = \frac{\sqrt{1 - \cos^2 a - \cos^2 x - \cos^2 y + 2\cos a \cos x \cos y}}{\sin a \sin y}.$$

Since the radicands in these two formulas are equal, we set, for brevity,

$$\sqrt{1 - \cos^2 a - \cos^2 x - \cos^2 y + 2\cos a \cos x \cos y} = v,$$

to obtain

$$\sin\phi = \frac{v}{\sin a \sin x}$$

and

$$\sin\psi = \frac{v}{\sin a \sin y}.$$

§4. Once these values have been replaced, we obtain the following differential values:

$$\partial\phi = -\frac{\partial x \cos a + \partial x \cos x \cos y + \partial y \sin x \sin y}{v \sin x}$$

and

$$\partial\psi = -\frac{\partial y \cos a + \partial y \cos x \cos y + \partial x \sin x \sin y}{v \sin y}.$$

From here we obtain that the area increment sought for is

$$\partial\Delta = \frac{\left[\begin{array}{l} -\cos a\big(\partial x \sin y(1-\cos x) + \partial y \sin x(1-\cos y)\big) \\ \quad +\partial x \sin y\big(\cos x \cos y(1-\cos x) + \sin^2 x(1-\cos y)\big) \\ \quad +\partial y \sin x\big(\cos x \cos y(1-\cos y) + \sin^2 y(1-\cos x)\big) \end{array}\right]}{v \sin x \sin y}$$

Expanding the calculations, this formula leads to the expression

$$\upsilon \partial \Delta = +\partial x \sin x(1-\cos y) + \frac{\partial x \cos x \cos y(1-\cos x)}{\sin x} - \frac{\partial x \cos a(1-\cos x)}{\sin x}$$

$$+ \partial y \sin y(1-\cos x) + \frac{\partial y \cos x \cos y(1-\cos y)}{\sin y} - \frac{\partial y \cos a(1-\cos y)}{\sin y}.$$

Now here we observe that

$$\frac{1-\cos x}{\sin x} = \tan \frac{1}{2}x$$

and

$$\frac{1-\cos y}{\sin y} = \tan \frac{1}{2}y;$$

taking this into account, it therefore follows that the term that contains the element ∂x is

$$\partial x \sin x(1-\cos y) + \partial x \cos x \cos y \tan \frac{1}{2}x - \partial x \cos a \tan \frac{1}{2}x.$$

But since we have not only

$$\tan \frac{1}{2}x = \frac{1-\cos x}{\sin x},$$

but also

$$\tan \frac{1}{2}x = \frac{\sin x}{1+\cos x},$$

in the first addend we substitute $\sin x$ by $(1+\cos x)\tan \frac{1}{2}x$, in such a way that ∂x is everywhere multiplied by $\tan \frac{1}{2}x$, and in this way this term [which contains ∂x] is reduced to the form

$$\partial x \tan \frac{1}{2}x(1+\cos x - \cos y - \cos a).$$

In the same manner [one finds that] the other member [that is, the term containing ∂y] is

$$\partial y \tan \frac{1}{2}y(1+\cos y - \cos x - \cos a),$$

and in this way our entire equation will have the expression

$$v\partial\Delta = \partial x \tan\frac{1}{2}x(1+\cos x-\cos y-\cos a)+\partial y \tan\frac{1}{2}y(1+\cos y-\cos x-\cos a).$$

§5. Therefore, if now, to simplify the formulae, we set $\cos a + \cos x + \cos y = s$, we get

$$v\partial\Delta = \partial x \tan\frac{1}{2}x(1-s+2\cos x)+\partial y \tan\frac{1}{2}y(1-s+2\cos y),$$

and this equation can be written in the form:

$$v\partial\Delta = (1-s)(\partial x \tan\frac{1}{2}x+\partial y \tan\frac{1}{2}y)+2\partial x \cos x \tan\frac{1}{2}x+2\partial y \cos y \tan\frac{1}{2}y.$$

Now, since $\tan\frac{1}{2}x = \frac{1-\cos x}{\sin x}$, we get

$$\tan\frac{1}{2}x\cos x = \frac{\cos x-\cos^2 x}{\sin x} = \frac{\cos x - 1 + \sin^2 x}{\sin x} = \sin x - \tan\frac{1}{2}x.$$

In the same manner, we have $\tan\frac{1}{2}y\cos y = \sin y - \tan\frac{1}{2}y$, and once these values have been replaced, the equation becomes

$$v\partial\Delta = -(1+s)(\partial x \tan\frac{1}{2}x+\partial y \tan\frac{1}{2}y)+2\partial x \sin x+2\partial y \sin y.$$

§6. This last form is most noteworthy for the fact that the right hand side is completely integrable when divided by $1+s$. In fact, after this division, our equation leads to the expression

$$\frac{v\partial\Delta}{1+s} = -\partial x \tan\frac{1}{2}x+\partial y \tan\frac{1}{2}y+\frac{2\partial x \sin x+2\partial y \sin y}{1+\cos a+\cos x+\cos y},$$

where we observe that

$$\int \partial x \tan\frac{1}{2}x = -2\ln(\cos\frac{1}{2}x),$$

and analogously

$$\int \partial y \tan\frac{1}{2}y = -2\ln(\cos\frac{1}{2}y);$$

finally,

$$2\int\frac{\partial x \sin x+\partial y \sin y}{1+\cos a+\cos x+\cos y} = -2\ln(1+\cos a+\cos x+\cos y) = -2\ln(1+s);$$

so by integrating we find

$$\int \frac{v\partial\Delta}{1+s} = 2\ln(\cos\frac{1}{2}x) + 2\ln(\cos\frac{1}{2}y) - 2\ln(1+s) = 2\ln(\frac{\cos\frac{1}{2}x\cos\frac{1}{2}y}{1+s}).$$

But in this way the left hand side cannot be integrated; however we can remedy this inconvenience in the following way.

§7. Having already introduced the notation $s = \cos a + \cos x + \cos y$, we set in addition $\cos\frac{1}{2}a\cos\frac{1}{2}x\cos\frac{1}{2}y = q$, and in this way the equation takes the form

$$\int \frac{v\partial\Delta}{1+s} = 2\ln(\frac{q}{(1+s)\cos\frac{1}{2}a});$$

here then we set $\frac{q}{1+s} = p$ and we obtain

$$\int \frac{v\partial\Delta}{1+s} = 2\ln(\frac{p}{\cos\frac{1}{2}a});$$

taking the differential of this equation we have

$$\frac{v\partial\Delta}{1+s} = \frac{2\partial p}{p},$$

whence it follows

$$\partial\Delta = \frac{2\partial p(1+s)}{pv},$$

and therefore this expression can be integrated as long as $\frac{v}{1+s}$ is some function of the same p, which we can assert with certainty, due to the fact that $\partial\Delta$ denotes precisely the differential of the area of the triangle.

§8. To prove this, it is useful to observe that we have

$$\begin{aligned} v^2 + (1+s)^2 &= 2(1 + \cos a + \cos x + \cos y + \cos a\cos x \\ &\quad + \cos a\cos y + \cos x\cos y + \cos a\cos x\cos y) \\ &= 2(1+\cos a)(1+\cos x)(1+\cos y). \end{aligned}$$

On the other hand, it is easy to see that the following identities hold

$$1 + \cos a = 2\cos^2\frac{1}{2}a,$$

$$1 + \cos x = 2\cos^2\frac{1}{2}x$$

and

$$1 + \cos y = 2\cos^2\frac{1}{2}y.$$

Therefore, having set $q = \cos\frac{1}{2}a\cos\frac{1}{2}x\cos\frac{1}{2}y$, we get

$$v^2 + (1+s)^2 = 16q^2$$

and therefore

$$v = \sqrt{16q^2 - (1+s)^2},$$

and from here then

$$\frac{v}{1+s} = \sqrt{\frac{16q^2}{(1+s)^2} - 1}.$$

§9. Therefore since our differential equation was $\frac{v\partial\Delta}{1+s} = \frac{2\partial p}{p}$, taking into account the fact that $p = \frac{q}{1+s}$, it takes the form

$$\partial\Delta\sqrt{16p^2 - 1} = \frac{2\partial p}{p},$$

and therefore

$$\partial\Delta = \frac{2\partial p}{p\sqrt{16p^2 - 1}}.$$

Now we set $p = \frac{1}{r}$, so that we have

$$\partial\Delta = -\frac{2\partial r}{\sqrt{16 - r^2}},$$

and from here, integrating, we get $\Delta = C + 2\arccos\frac{r}{4}$; substituting r by its value, which is

$$r = \frac{1}{p} = \frac{1+s}{q} = \frac{1+\cos a + \cos x + \cos y}{\cos\frac{1}{2}a\cos\frac{1}{2}x\cos\frac{1}{2}y},$$

our integral equation[4] will be

$$\Delta = C + 2\arccos\frac{1+\cos a + \cos x + \cos y}{4\cos\frac{1}{2}a\cos\frac{1}{2}x\cos\frac{1}{2}y}.$$

[4] Euler writes “nostra aequatio integralis”, which means “our integral equation”, but in fact, he means the equation which results from the integration.

§10. Therefore now the whole question is reduced to the study of the value of the constant C introduced in the integration procedure, which of course is appropriate to derive from some known case; on the other hand it is clear that the area of the triangle must vanish when one of the two legs x or y vanishes.[5] Let us suppose then that $y = 0$; then necessarily we have $x = 0$, and therefore in this case our equation becomes

$$0 = C + 2 \arccos \frac{2 + 2\cos a}{4\cos^2 \frac{1}{2}a}.$$

But since $4\cos^2 \frac{1}{2}a^2 = 2 + 2\cos \frac{1}{2}a$ and $\arccos 1 = 0$, it is obvious that we must require $C = 0$, so that we have

$$\Delta = 2 \arccos \frac{1 + \cos a + \cos x + \cos y}{4 \cos \frac{1}{2}a \cos \frac{1}{2}x \cos \frac{1}{2}y},$$

whence in conclusion we obtain

$$\cos \frac{1}{2}\Delta = \frac{1 + \cos a + \cos x + \cos y}{4 \cos \frac{1}{2}a \cos \frac{1}{2}x \cos \frac{1}{2}y},$$

and this expression is in perfect agreement with the aforementioned theorem; it suffices to write the letters b and c instead of x and y.

Another Geometric Proof of the Theorem Reported on at the Beginning

§11. Let then ABC be the spherical triangle considered (Fig. 13.2), let us call a, b, c its sides and α, β, γ the angles opposite to them, whereas we indicate its area, which we want to determine, with the character Δ. Then since by Girard's theorem

[5] [Remarks on Latin, Greek and Arabic terminology for (spherical) triangles] Euler uses instead of *laterum* the term *crurum*, which is less widespread but more precise; this is the plural genitive of *crus* (which usually means *leg*), with reference to the following definition of a triangle as in the Latin: *est figura tribus lineis et tribus angulis contenta. Lineas illas vocant duo latera seu crura, et unam basin*, see Johannes Micraelius (Joh. Micraelii) *Lexicon Philosophicorum Terminorum Philosophis Usitatorum Ordine Alphabetico Sic Digestorum, Ut Inde Facile Liceat Cognosse, Praesertim Si Tam Latinus, Quam Graecus Index Praemissus Non Negligatur, Quid In Singulis Disciplinis Quomodo Sit Distinguendum Et Definiendum*. Stettin, 1662. The first definition of a spherical triangle is due to Menelaus of Alexandria, (see R. Rashed, A. Papadopoulos, Menelaus' Spherics. Early translation and al-Māhānī/al-Harawī's version. Dual Arabic-English text. Scientia Graeco-Arabica, 21. De Gruyter, Berlin, 2017.). The Greek word used by Menelaus is *tripleuron* (τρίπλευρον), that is, [a figure enclosed by] three sides. As a matter of fact, Menelaus' work reached us in Arabic (the Greek original is lost), and the Arabs, in their translation of the *Spherics*, used to call a triangle شَكل ذو ثَلاثة أضلاع ("trilateral figure"), that is, "a figure with three members" (a *member* meaning here either a leg or a hand). Let us also recall that Euclid, in the *Elements* (Definition I.3), defines a trilateral figure as a figure contained by three straight lines. In any case, a triangle is enclosed by three lines and three angles; for visual reasons, in some figures, two of these lines are called *sides* or *legs*, and the third one the *base*.

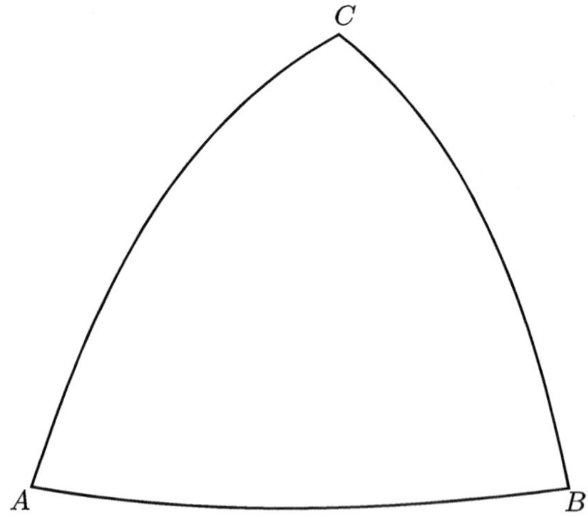

Fig. 13.2 Second Geometrical Proof of Theorem 1

we have

$$\Delta = \alpha + \beta + \gamma - 180^{\circ},$$

we get

$$\cos \Delta = -\cos(\alpha + \beta + \gamma).$$

But now, from the composition of the angles, it follows that

$$\sin(\alpha + \beta) = \sin \alpha \cos \beta + \cos \alpha \sin \beta$$

and

$$\cos(\alpha + \beta) = \cos \alpha \cos \beta - \sin \alpha \sin \beta;$$

taking into account these identities, we obtain

$$\begin{aligned}\cos(\alpha + \beta + \gamma) &= \cos(\alpha + \beta) \cos \gamma - \sin(\alpha + \beta) \sin \gamma \\ &= \cos \alpha \cos \beta \cos \gamma - \cos \alpha \sin \beta \sin \gamma \\ &\quad - \sin \alpha \cos \beta \sin \gamma - \sin \alpha \sin \beta \cos \gamma,\end{aligned}$$

and consequently we have

$$\begin{aligned}\cos \Delta = &\cos \alpha \sin \beta \sin \gamma + \sin \alpha \cos \beta \sin \gamma \\ &+ \sin \alpha \sin \beta \cos \gamma - \cos \alpha \cos \beta \cos \gamma.\end{aligned}$$

§12. On the other hand, we know from spherical trigonometry the identities

$$\cos\alpha = \frac{\cos a - \cos b \cos c}{\sin b \sin c},$$

$$\cos\beta = \frac{\cos b - \cos a \cos c}{\sin a \sin c}$$

and

$$\cos\gamma = \frac{\cos c - \cos a \cos b}{\sin a \sin b},$$

whence we obtain

$$\sin\alpha = \frac{\sqrt{1 - \cos^2 a - \cos^2 b - \cos^2 c + 2\cos a \cos b \cos c}}{\sin b \sin c};$$

$$\sin\beta = \frac{\sqrt{1 - \cos^2 a - \cos^2 b - \cos^2 c + 2\cos a \cos b \cos c}}{\sin a \sin c}$$

and

$$\sin\gamma = \frac{\sqrt{1 - \cos^2 a - \cos^2 b - \cos^2 c + 2\cos a \cos b \cos c}}{\sin a \sin b}.$$

To simplify [the formulae] we set

$$\sqrt{1 - \cos^2 a - \cos^2 b - \cos^2 c + 2\cos a \cos b \cos c} = v,$$

whence we obtain

$$\sin\alpha = \frac{v}{\sin b \sin c};$$

$$\sin\beta = \frac{v}{\sin a \sin c}$$

and

$$\sin\gamma = \frac{v}{\sin a \sin b},$$

and replacing these values, we find

$$\cos\Delta = \frac{v^2(\cos a - \cos b\cos c)}{\sin^2 a\sin^2 b\sin^2 c} + \frac{v^2(\cos b - \cos a\cos c)}{\sin^2 a\sin^2 b\sin^2 c}$$
$$+ \frac{v^2(\cos c - \cos a\cos b)}{\sin^2 a\sin^2 b\sin^2 c}$$
$$- \frac{(\cos a - \cos b\cos c)(\cos b - \cos a\cos c)(\cos c - \cos a\cos b)}{\sin^2 a\sin^2 b\sin^2 c}.$$

Therefore we get

$$\sin^2 a\sin^2 b\sin^2 c\cos\Delta$$
$$= v^2(\cos a + \cos b + \cos c - \cos a\cos b - \cos a\cos c - \cos b\cos c)$$
$$- \cos a\cos b\cos c + \cos^2 a\cos^2 b + \cos^2 a\cos^2 c + \cos^2 b\cos^2 c$$
$$- \cos a\cos b\cos c(\cos^2 a + \cos^2 b + \cos^2 c) + \cos^2 a\cos^2 b\cos^2 c.$$

§13. Now, so that it is possible to deal more easily with these rather complicated formulae, let us set, first of all, for the sake of brevity, $\cos a = A$, $\cos b = B$, $\cos c = C$, in such a way that we have

$$(1 - A^2)(1 - B^2)(1 - C^2)\cos\Delta = v^2(A + B + C - AB - AC - BC)$$
$$- ABC + A^2B^2 + A^2C^2 + B^2C^2$$
$$- ABC(A^2 + B^2 + C^2) + A^2B^2C^2,$$

where

$$v^2 = 1 - A^2 - B^2 - C^2 + 2ABC.$$

§14. Since here the three letters A, B, C enter into the calculation with equal roles [i.e., they are interchangeable with each other], so that they can be seen as roots of a cubic equation, it will contribute significantly to the simplification of the calculations the introduction of the notation

$$A + B + C = P,$$

$$AB + AC + BC = Q,$$

$$ABC = R,$$

and from here we easily obtain that we have

$$A^2 + B^2 + C^2 = P^2 - 2Q$$

and therefore

$$v^2 = 1 - P^2 + 2Q + 2R.$$

Let us then observe that the formula $(1 - A^2)(1 - B^2)(1 - C^2)$ is the product of the two factors $(1+A)(1+B)(1+C) = 1+P+Q+R$ and $(1-A)(1-B^{)}(1-C) = 1 - P + Q - R$, and therefore our equation takes the form

$$(1 + P + Q + R)(1 - P + Q - R)\cos\Delta$$

$$= (1 - P^2 + 2Q + 2R)(P - Q) - R + Q^2 - 2PR - R(P^2 - 2Q) + R^2,$$

whose right hand side expanded becomes

$$P - Q - R - Q^2 + 2PQ + R^2 - P^3 + P^2Q - P^2R.$$

This, divided by $1-P+Q-R$, gives as quotient $P-Q-R+P^2$, and consequently our equation takes the following form:

$$(1 + P + Q + R)\cos\Delta = P - Q - R + P^2.$$

§15. Thus, up to now, we have reached the equation[6]

$$\cos\Delta = \frac{P - Q - R + P^2}{1 + P + Q + R};$$

from this we then derive

$$1 + \cos\Delta = \frac{(1 + P)^2}{1 + P + Q + R} = 2\cos^2\frac{1}{2}\Delta,$$

and consequently we get

$$\cos\frac{1}{2}\Delta = \frac{1 + P}{\sqrt{2(1 + P + Q + R)}}.$$

Therefore, since

$$(1 + P + Q + R) = (1 + A)(1 + B)(1 + C) = (1 + \cos a)(1 + \cos b)(1 + \cos c),$$

introducing the half angles we get[7]

$$1 + P + Q + R = 8\cos^2\frac{1}{2}a\cos^2\frac{1}{2}b\cos^2\frac{1}{2}c.$$

[6] In the latin original, there is a typo in the following formula.

[7] In the latin original, in the right hand side of the next formula, we have cosines instead of the squares of the cosines.

Therefore, since

$$1 + P = 1 + \cos a + \cos b + \cos c,$$

from here we finally get this value:

$$\cos \frac{1}{2}\Delta = \frac{1 + \cos a + \cos b + \cos c}{4 \cos \frac{1}{2}a \cos \frac{1}{2}b \cos \frac{1}{2}c},$$

and thus we have another proof of the theorem mentioned at the beginning.

§16. The occasion for these reflections was given to me by the theorem made known by the illustrious Professor Lexell,[8] on all the spherical triangles built on the same base and having the same area, where with great sharpness he proved that all the vertices of these triangles are always situated on some small circle of the sphere, and this very elegant property can be obtained from our theorem, [albeit] with a very tortuous procedure; however the following consideration opens a very easy way to prove this fact.

On the Remarkable Properties of Two Parallel Circles Equal to Each Other on a Spherical Surface

§17. Let MN and mn be two such parallel circles (see Fig. 13.3); since they are chosen to be equal to each another, they will be equally distant from the equator AB, as well as from the two poles P and p. Here the first property that appears consists in the fact that an arbitrary arc of the great circle Ee, bounded by these two parallels, is equally inclined with respect to both [parallels], and is cut by the equator at O into two equal parts. In fact if from the point O we draw the meridian OPp, which intersects the two parallels in H and in h, since the angles HOE and hOe between them are equal, the two trilines[9] HOE and hOe will be equal and similar to each other, and therefore we get both equality of the arcs $OE = Oe$ and of the angles OEH and Oeh. Here it is also useful to note that if this arc Ee is extended until [it describes] a semicircle, it intersects again the small circle mn, of course in the point that is diametrically opposed to the point E.

[8] Anders Johan Lexell (1740–1784) was a Finnish-Swedish astronomer and geometer who worked at the Saint Petersburg Academy of Sciences at the time when Euler was the leading figure there. Euler refers here to Lexell's memoir *Solutio problematis geometrici ex doctrina sphaericorum* (Solution of a geometric problem in spherical geometry), Acta academiae scientarum imperialis Petropolitinae, 5 (1) (1781), 1784, p. 112–126. Lexell's works on spherical geometry, translated into English, with notes and comments, are contained in a sequel to the present volume carrying the title *Spherical geometry in the 18th century, II: Lexell, Schubert and Fuss*, ed. R. Caddeo, A. Papadopoulos and A. Zhukova, to appear (Springer).

[9] The terms *triline (quadriline)* denote a figure bounded by three (four) sides on the surface of the sphere. The side of this figure (unlike the case for triangles and quadrilaterals) are not necessarily shortest lines. They can be arcs of circles, etc.

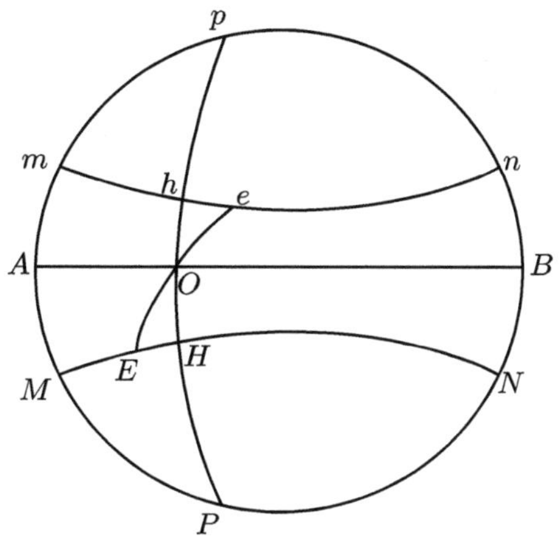

Fig. 13.3 §17. Remarkable Properties of Two Equal Parallels in §17

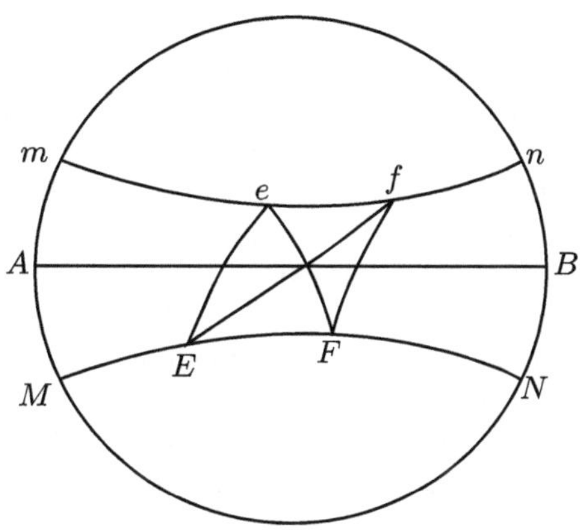

Fig. 13.4 Remarkable Properties of Two Equal Parallels in §18

§18. Moreover, we now draw between these same parallels another arc of great circle Ff (Fig. 13.4), equally inclined with respect to both [circles], like the arc Ee, and it is clear that not only these two edges Ee and Ff are equal to each other, but also the same holds for the arcs of small circles EF and ef. For this reason, since in this quadriline $EFef$ not only the opposite sides but also the opposite angles are equal to each other, the quadriline may well be called a spherical parallelogram, as it has all the properties of [planar] parallelograms. In fact, it is evident that this quadriline is also divided by each of its diagonals Ef and Fe in two equal trilines, and of course the areas of the trilines efF and EsF are both [equal to] half of the area of the parallelogram $EefF$.

§19. Now suppose we elevate on the same arc EF (Fig. 13.5), taken as a base, another spherical parallelogram $EF\zeta\epsilon$ of the same type, and it is easily understandable that the areas of these two parallelograms $EFfe$ and $EF\zeta\epsilon$ are equal to each other. Here in fact, in the same way as in the plane, both trilines $Ee\epsilon$ and $Ff\zeta$ are perfectly equal to each other; and if we cut from them the common triline $oe\zeta$, the remaining quadrilines $Eo\zeta\epsilon$ and $Foef$ will be equal to each other;

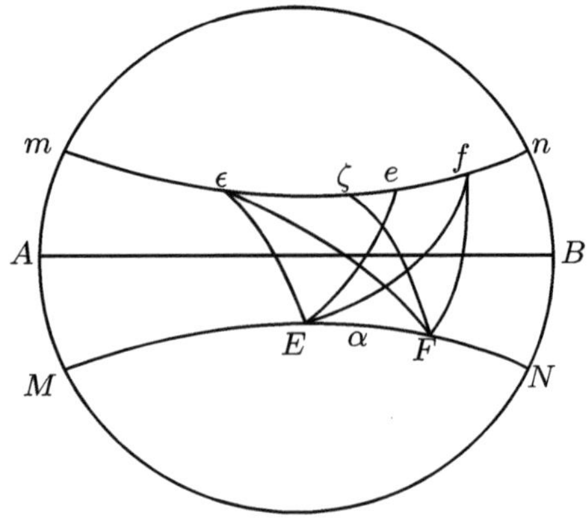

Fig. 13.5 Remarkable Properties of Two Equal Parallels in §19

if we add to these the triline EoF[10] intersect, does not appear in Fig. 13.5, the entire parallelograms will also both be equal, and thus it becomes clear that all the spherical parallelograms between two equal parallel circles, built on the same base EF, are equal to each other.

§20. Thus, given such spherical parallelograms divided by the diagonals into two equal parts, all the trilines built on the same base and having the other vertex on the other parallel mn will have equal areas; evidently in this figure there are four equal trilines: 1. EfF; 2. EeF; 3. $E\zeta F$; 4. $E\epsilon F$.

§21. However, we do not call these trilines triangles, due to the fact that their base EF is not an arc of great circle, which is usually assumed in spherical triangles. But these trilines are transformed into spherical triangles if from E to F we draw the great circle $E\alpha F$, with which the same increment $EF\alpha E$[11] is added to the previous trilines so that now also all the spherical triangles built on the same base $E\alpha F$, having the third vertex on the parallel mn, have equal areas if only the extremities E and F of the base are assumed to be on the other parallel MN and is opposite to them; and in this way it is evident that if an arbitrarily large number of spherical triangles, having equal areas, is constructed on an arbitrary base, all their vertices always lie on a certain small circle of the sphere. This being observed, the problem of the very illustrious Professor Lexell can be very easily solved in the following way.

Problem 2

To construct, on a spherical surface and on a given base EF, all the spherical triangles having an assigned area Δ, where Δ indicates precisely the arc of great circle which, multiplied by the radius of the sphere, produces the prescribed area.

---

[10] The point o in which Ee and $F\zeta$

[11] This is the region bounded by the arc EF and $E\alpha F$ in Fig. 13.5.

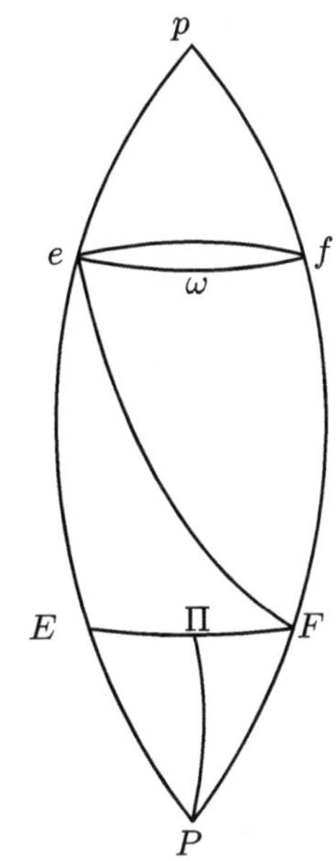

Fig. 13.6 Solution of Problem 2

Solution

§22. Thus, let us assume that EF is the fixed base and that $EF = a$ (see Fig. 13.6); all the question is reduced to the determination of the poles P and p, which satisfy the requirements; in fact, once they have been found, if from the pole p at distance $pe = PE$ we draw the small circle ef, the areas of all the triangles constructed on the base EF and with vertex on the small circle ef will be equal to each other; and it remains only to determine, starting from the fixed surface Δ, the positions of the poles P and p.

§23. Thus, if EF is an arc of great circle equal to a, we set $EP = FP = x$ and we denote by ω the angle EPF; then from spherical trigonometry we know that $\cos\omega = \frac{\cos a - \cos^2 x}{\sin^2 x}$, therefore

$$1 - \cos\omega = \frac{1 - \cos a}{\sin^2 x} = 2\sin^2\frac{1}{2}\omega.$$

Therefore, since $1 - \cos a = 2\sin^2\frac{1}{2}a$, we get $\sin\frac{1}{2}\omega = \frac{\sin\frac{1}{2}a}{\sin x}$ and from here, inverting, $\sin x = \frac{\sin\frac{1}{2}a}{\sin\frac{1}{2}\omega}$. But from the knowledge of the angle ω the total area of the spherical sector between the two semi-circles PEp e PFp will be known, which in fact is equal to 2ω. Of course the arc of great circle equal to 2ω multiplied by the radius of the sphere equal to 1 will give the area of this sector.

§24. Let us now also look for the area of the triangle EPF, and to this end let us call ϕ the angle $PEF = PFE = \phi$, so that the sum of the three angles of this triangle is equal to $\omega + 2\phi$, and thus the area of this triangle will be a equal to $\omega + 2\phi - \pi$.

Therefore if also from e and f we draw the arc of great circle $e\omega f$, then we will also find $\omega + 2\phi - \pi$ as the area of the spherical triangle $pe\omega f$. Thus consequently the area of the spherical quadrilateral $EFfe$ contained by the arcs great circle Ee; Ff; EF and $e\omega f$ will be $2\omega - 2(\omega + 2\phi - \pi) = 2\pi - 4\phi$, whose half clearly represents the area of the spherical triangle EFe.

§25. Therefore since the point e also lies on the small circle ef, the triangle EeF will be one of the infinitely many triangles we have to build on the base EF having area equal to Δ, and thus we have obtained the equation $\Delta = \pi - 2\phi$, whenceget the angle $\phi = \frac{1}{2}\pi - \frac{1}{2}\Delta$. Thus, given the angle Δ, on the assigned base EF we construct at the two endpoints [that is, from the extremities] the two angles $FEP = EFP = 90^{\circ} - \frac{1}{2}\Delta$. In this way the pole P is known, and consequently also its antipode $[p]$; if from there we draw at distance $pe = PE$ a small circle ef, all the triangles constructed on the base EF with vertex on the small circle ef will have their area equal to the area Δ set in advance.

§26. In order to make this construction easier, we draw from the pole P onto the midpoint Π of the base the perpendicular arc $P\Pi$; since for the side $E\Pi$ of the triangle $EP\Pi$ we have $E\Pi = \frac{1}{2}a$, given the angle $PE\Pi = 90^{\circ} - \frac{1}{2}\Delta$, consequently for the side EP we have $EP = x$, whose tangent is $\tan x = \frac{\tan\frac{1}{2}a}{\tan\frac{1}{2}\Delta}$. Therefore now, having found the measures of the arcs EP and FP, their intersection will give the pole P; the small circle drawn at distance $pe = x$ from its antipodal point p constitutes the locus of the vertices of all the triangles which we want to construct on the base EF, and this construction agrees very well with the one discovered by the very illustrious Lexell.

Chapter 14
On the Measure of Solid Angles

Leonhard Euler

Abstract In this memoir, Euler starts by giving a geometrical proof of Girard's formula for the area of spherical triangle. He then gives several formulae for the area of a spherical triangle as a function of its side lengths. He uses these formulae for spherical area to give numerical values of the solid angles of the five regular polyhedra, which is the goal of the memoir.

Keywords Spherical area · Leonhard Euler · Girard formula · Solid angle · Lexell problem · History of spherical geometry · Mathematics in the eighteenth century

AMS Classification 01A50, 51-03, 53A05, 53A35

1. In the same way as planar angles are measured using the circular arcs that they subtend if the vertex of an angle is made to be at the center of a circumference, it seems in accordance with the nature of things to measure the solid angles by means of the portions of the spherical surface that they similarly subtend, if the vertex of an angle is made to be at the center of the sphere. In this way, if a solid angle is made out of three angles,[1] which we denote by a, b, c, and if we draw around the

[1] The solid angles that Euler considers in this memoir are made out of three or more triangles glued pairwise along their sides, forming a vertex of a polyhedron. However, whereas in the case of a solid angle made out of three triangles, Euler considers triangles with arbitrary angles at the vertex forming the solid angle, in the case of a solid angle made out of more than three triangles, he considers only triangles whose angles at the vertex are all equal and where the dihedral angles they make at a vertex are all equal. In the application to the computation of the solid angles of the regular polyhedra that he carries out in the last sections of the memoir, only this special case is needed. For a historical overview of general solid angles, see Chap. 5 of the present volume.

The author Leonhard Euler (1707–1783).
Translated from the Latin by Renzo Caddeo.
De mensura angulorum solidorum *Acta Academiae Scientiarum Imperialis Petropolitanae*, 2, 1781, pp. 31–54. Opera Omnia: Series 1, 26, pp. 204–223 *Leonhard Euler*.

R. Caddeo, A. Papadopoulos (eds.), *Spherical Geometry in the Eighteenth Century I*, https://doi.org/10.1007/978-3-032-12466-1_14

Fig. 14.1 A lune (Lemma 1)

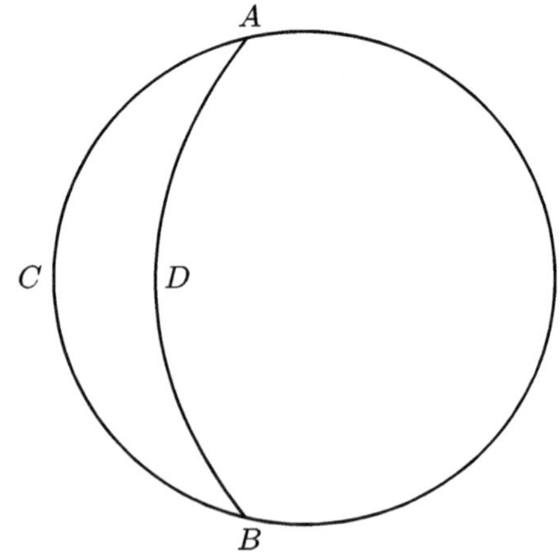

vertex the sphere whose radius is represented by unity, then the measure of this solid angle is usually set to be the area of the spherical triangle having sides equal to $a, b, c,$[2] since these sides are the measures of these planar angles. Likewise, if a solid angle is constituted of four or more planar angles, its measure will be the area of the spherical quadrilateral, or of the polygon with more angles, whose sides are equal to the planar angles that form the solid angle. For this reason, the size of the solid angles then reduces to the determination of the area of the spherical triangle, or of the polygon with more angles, whose sides have been given. Then, since the area of an arbitrary spherical triangle is obtained in a very simple way if its angles are known, as was proved some time ago by the very acute geometer *Albert Girard*,[3] I will report on that proof in what follows here, as it seems that it is not sufficiently well known.

Lemma 1 (Fig. 14.1)
2. *The area of the spherical portion contained by two meridians inclined on each other by an angle* α *is, to the surface of the entire sphere, like the angle* α *to* $360°$.

---

[2] This is the spherical triangle obtained by intersecting the solid angle in Footnote 1 by a sphere centered at its vertex.

[3] This rule attributed by Euler to Albert Girard (1595-1632) is also mentioned and used by Euler in his memoir *Various speculations on the area of spherical triangles* (see Chap. 13 of the present volume). We indicated in a footnote there that this rule should rather be attributed to Thomas Harriot (1560-1621), who found it before Girard. The same formula is proved by Euler, using differential calculus, in his other memoir on spherical geometry, *Principles of spherical trigonometry deduced from the method of maxima and minima* (see Chap. 11 of the present volume). Lagrange, in his memoir *Solution of some problems relative to spherical triangles with a complete analysis of these triangles*, also translated in the present volume (Chap. 18), attributes this formula to Bonaventura Cavalieri.

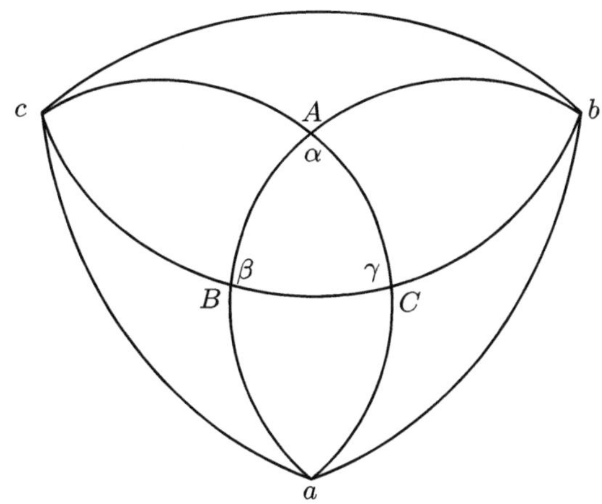

Fig. 14.2 Proof of Theorem 1

Proof
Let A, C, B and A, D, B be two great semicircles on the spherical surface intersecting at the opposite poles A and B and inclined on each other by an angle α; then it is evident that the area of this spherical sector $ACBDA$[4] is contained in the entire sphere as many times the angle α is contained in 360°.

3. Therefore if we set the radius of the sphere to be equal to r, whence the total surface of the sphere measures $4\pi r^2$, where π denotes the perimeter of a circle of diameter equal to 1, then the area of our spherical sector will be equal to $4\pi r^2 \frac{\alpha}{360^o}$, in the case where the angle α is expressed in degrees. If instead α is given in radians,[5] i.e. in parts of the radius, which is always expressed with the unity, being that $360^o = 2\,\pi$, then the area of the spherical sector will be equal to $2\alpha r^2$. As a consequence, if the radius of the sphere is likewise taken of unit length, then this area will be 2α. Therefore, in this way, it is possible to represent the area of this sector simply by the angle, whereas the total surface is 4π.

Theorem 1 (Fig. 14.2) (Albert Girard)
4. *The area of a spherical triangle is always equal to the angle by which the sum of the three angles of the triangle exceeds two right angles.*

Proof
Let ABC be the spherical triangle considered, whose area is to be found, and let us denote its angles by the letters α, β, γ. We first produce on the spherical surface the sides AB and AC until they meet again at the pole a, opposite to the same vertex

[4] This region of the sphere contained by two great semicircles that share the same endpoints is usually called a *lune*.

[5] Euler writes "in partibus radii" which literally means "in parts of the radius". It is clear that here, Euler is measuring angles in radians. F. Cajori, in his History of Mathematics (2nd Ed. New York 1919), p. 147, writes: "the word 'radian' was first used in print by J. Thompson, a brother of Lord Kelvin." This is one century after Euler.

A.[6] Since these arcs ABa and ACa can be seen as two meridians, distant from each other by the angle α, the area of this sector $ACaB$ will be equal to 2α.

Then, in the same way, let us produce the two sides BA and BC until b; this point will be equally a pole, opposite to B, and the area of this sector $BAbC$ will be equal to 2β. Finally, we proceed in the same way with the sides CA and CB until the pole c opposite to C, and the area of the sector $CBcA$ will be equal to 2γ.

In conclusion, if we call S the area of the triangle ABC which is sought-after, it follows from what was said that the areas of the following triangles are known:[7]

1. $aBC = 2\alpha - S$;
2. $bAC = 2\beta - S$;
3. $cAB = 2\gamma - S$.

5.[8] Now, since the points a, b, c of the sphere are diametrically opposite to the points A, B, C, they keep among each other the same distances, even though in the figure they seem much different. Therefore, drawing the arcs ab, bc, ca, one has

$$ab = AB, \ ac = AC, \ bc = BC,$$

therefore the area of this triangle abc, which lies in the rear region of the sphere, will also be equal to S. Thus the whole surface of the sphere contains:

1. the triangles $ABC = S$ and $abc = S$;
2. the triangles $aBC = 2\alpha - S, \ bAC = 2\beta - S, \ cAB = 2\gamma - S$.

But furthermore, the figure contains the triangles abC, acB and bcA, and the areas of the latter may be deduced from those of the previous ones. In fact, regarding the sides of the triangle abC, one has first $ab = AB$, $aC = Ac$ and $bC = Bc$; therefore it is evident that for this triangle one has

$$abC = ABc = 2\gamma - S.$$

In the same way it is understandable that for the triangle acB one must have

$$acB = ACb = 2\beta - S,$$

and finally that

$$bcA = BCa = 2\alpha - S.$$

6. It follows that, since the whole surface of the sphere is cut into eight triangles and given that we have represented here the expression of the area of each of them, the

[6] In the original edition: to the same “angle” A.

[7] With an abuse of notation, the right hand sides represent the areas of the corresponding triangles.

[8] In the original edition, this section is again numbered 4. In this edition, we have called this section 5, and the next sections are shifted accordingly.

sum of these [areas] must be equal to 4π, that is, to the total surface of the sphere. And from this equality one may determine the area S sought for.

Thus, let us summarize these triangles, each one with its area, in the following table:

I. $ABC = S$	III. $aBC = 2\alpha - S$	VI. $Abc = 2\alpha - S$
II $abc = S$	IV. $bAC = 2\beta - S$	VII. $Bac = 2\beta - S$
	V. $cAB = 2\gamma - S$	VIII. $Cab = 2\gamma - S$

Adding the areas of these eight triangles, we find:

$$2S + 2(\alpha + \beta + \gamma) - 3S + 2(\alpha + \beta + \gamma) - 3S$$

or

$$4(\alpha + \beta + \gamma) - 4S.$$

Then, this quantity must be equal to 4π, whereby, dividing by 4, it follows that

$$\alpha + \beta + \gamma - S = \pi,$$

namely,

$$S = \alpha + \beta + \gamma - \pi$$

where $\alpha + \beta + \gamma$ is the sum of all the angles of the given triangle and π is the measure of two right angles, that is, 180°.

Thus, to find the area of the given spherical triangle, one subtracts two right angles from the sum $\alpha + \beta + \gamma$ of all its angles, exactly as the theorem asserts.

7. Therefore, the whole problem of the measure of the solid angles reduces to the following: how to determine the area of a spherical triangle knowing the lengths of its three sides. Therefore, we start by solving the following problem:[9]

[9] Euler establishes a formula for the area of a spherical triangle in terms of its side lengths. Such a formula is also given by Lagrange in §10 of his memoir *Solution de quelques problèmes relatifs aux triangles sphériques avec une analyse complète de ces triangles* (Solution of some problems relative to spherical triangles with a complete analysis of these triangles), see the translation in Chap. 18 of the present volume.

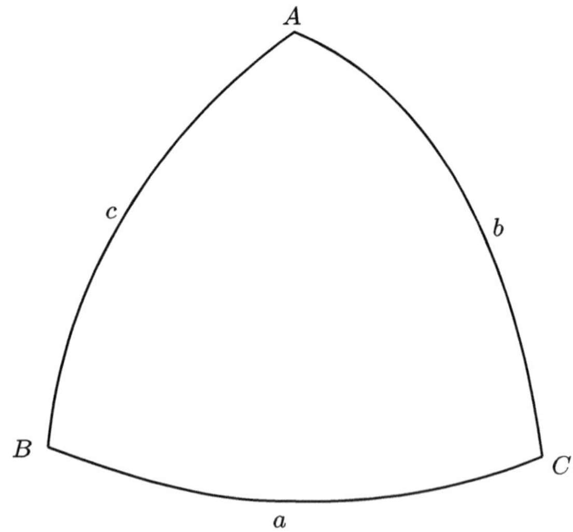

Fig. 14.3 Finding the area of a spherical triangle from its sides

General Problem (Fig. 14.3)
Given a spherical triangle and its three sides $AB = c$, $AC = b$ and $BC = a$, to find the area of this spherical triangle.

Solution
8. We denote by A, B, C the angles of this triangle and by S its area which we want to find, and which, as we have seen, is

$$S = A + B + C - 180^{\circ}.$$

From here it follows that

$$\sin S = -\sin(A + B + C)$$

and

$$\cos S = -\cos(A + B + C),$$

and therefore

$$\tan S = +\tan(A + B + C).$$

Thus, in the computation, one has only to introduce instead of the angles A, B, C the sides a, b, c.

Now, by the rules of spherical trigonometry, the angles are obtained from the sides using the formulae

$$\cos A = \frac{\cos a - \cos b \cos c}{\sin b \sin c},$$

$$\cos B = \frac{\cos b - \cos a \cos c}{\sin a \sin c},$$

$$\cos C = \frac{\cos c - \cos a \cos b}{\sin a \sin b},$$

whence one can deduce the sines of the angles:

$$\sin A = \frac{\sqrt{(1 - \cos^2 a - \cos^2 b - \cos^2 c + 2\cos a \cos b \cos c)}}{\sin b \sin c},$$

$$\sin B = \frac{\sqrt{(1 - \cos^2 a - \cos^2 b - \cos^2 c + 2\cos a \cos b \cos c)}}{\sin a \sin c},$$

$$\sin C = \frac{\sqrt{(1 - \cos^2 a - \cos^2 b - \cos^2 c + 2\cos a \cos b \cos c)}}{\sin a \sin b}.$$

Now we set

$$\sqrt{(1 - \cos^2 a - \cos^2 b - \cos^2 c + 2\cos a \cos b \cos c)} = k$$

and, to simplify the calculation, we introduce in the numerators the notation

$$\cos a = \alpha, \cos b = \beta, \cos c = \gamma,$$

in such a way that we obtain

$$k^2 = 1 - \alpha^2 - \beta^2 - \gamma^2 + 2\alpha\beta\gamma.$$

This being done, we obtain

$$\cos A = \frac{\alpha - \beta\gamma}{\sin b \sin c}; \ \cos B = \frac{\beta - \alpha\gamma}{\sin a \sin c}; \ \cos C = \frac{\gamma - \alpha\beta}{\sin a \sin b},$$

$$\sin A = \frac{k}{\sin b \sin c}; \ \sin B = \frac{k}{\sin a \sin c}; \ \sin C = \frac{k}{\sin a \sin b}.$$

9. Now we add first of all the angles A and B and we find

$$\sin(A+B) = \sin A\cos B + \cos A\sin B = \frac{k(1-\gamma)(\alpha+\beta)}{\sin a\sin b\sin^2 c},$$

$$\cos(A+B) = \cos A\cos B - \sin A\sin B = \frac{(\alpha-\beta\gamma)(\beta-\alpha\gamma)-k^2}{\sin a\sin b\sin^2 c},$$

therefore, if we add now the third angle C, we get

$$\begin{aligned}\sin(A+B+C) &= \sin A\cos B\cos C + \sin B\cos A\cos C\\ &\quad + \sin C\cos A\cos B - \sin A\sin B\sin C,\end{aligned}$$

$$\begin{aligned}\cos(A+B+C) &= \cos A\cos B\cos C - \cos A\sin B\sin C\\ &\quad - \cos B\sin A\sin C - \cos C\sin A\sin B.\end{aligned}$$

Thus, it remains only to substitute in these formulae, instead of the capital A, B, C the values just now assigned.

First Investigation
[A Formula for] $\sin S$
10. Since we have

$$\sin S = -\sin(A+B+C),$$

we get

$$\begin{aligned}\sin S &= \sin A\sin B\sin C - \sin A\cos B\cos C\\ &\quad - \sin B\cos A\cos C - \sin C\cos A\cos B.\end{aligned}$$

This expression consists of four terms that we develop separately. Thus, we obtain

$$\text{I.}\quad \sin A\sin B\sin C = \frac{k^3}{\sin^2 a\sin^2 b\sin^2 c} = \frac{k(1-\alpha^2-\beta^2-\gamma^2+2\alpha\beta\gamma)}{\sin^2 a\sin^2 b\sin^2 c};$$

$$\text{II.}\quad \sin A\cos B\cos C = \frac{k(\beta-\alpha\gamma)(\gamma-\alpha\beta)}{\sin^2 a\sin^2 b\sin^2 c} = \frac{k(\beta\gamma-\alpha\beta^2-\alpha\gamma^2+\alpha^2\beta\gamma)}{\sin^2 a\sin^2 b\sin^2 c};$$

$$\text{III.}\quad \sin B\cos A\, cos C = \frac{k(\alpha-\beta\gamma)(\gamma-\alpha\beta)}{\sin^2 a\sin^2 b\sin^2 c} = \frac{k(\alpha\gamma-\beta\alpha^2-\beta\gamma^2+\beta^2\alpha\gamma)}{\sin^2 a\sin^2 b\sin^2 c};$$

$$\text{IV.}\quad \sin C\cos A\cos B = \frac{k(\alpha-\beta\gamma)(\beta-\alpha\gamma)}{\sin^2 a\sin^2 b\sin^2 c} = \frac{k(\alpha\beta-\gamma\alpha^2-\gamma\beta^2+\gamma^2\alpha\beta)}{\sin^2 a\sin^2 b\sin^2 c}.$$

Therefore, taking into account the fact that everywhere the denominator that appears is the same,

$$\sin^2 a \sin^2 b \sin^2 c = (1-\alpha^2)(1-\beta^2)(1-\gamma^2),$$

adding together the third members [of these equalities], we obtain

$$\frac{k(\alpha\beta+\alpha\gamma+\beta\gamma-\alpha\beta^2-\alpha\gamma^2-\beta\gamma^2-\alpha^2\beta-\alpha^2\gamma-\beta^2\gamma+\alpha\beta\gamma(\alpha+\beta+\gamma)}{(1-\alpha^2)(1-\beta^2)(1-\gamma^2)}.$$

11. To make these formulae more treatable, let us set, for brevity,

$$\alpha+\beta+\gamma=p;\ \alpha\beta+\alpha\gamma+\beta\gamma=q;\ \alpha\beta\gamma=r,$$

whence it follows

$$\alpha^2+\beta^2+\gamma^2=p^2-2q,$$

and consequently

$$k^2=1-p^2+2q+2r.$$

From here, taking into account the identity

$$pq=\alpha^2\beta+\alpha^2\gamma+\beta^2\alpha+\beta^2\gamma+\gamma^2\alpha+\gamma^2\beta+3\alpha\beta\gamma,$$

it follows

$$\alpha^2(\beta+\gamma)+\beta^2(\alpha+\gamma)+\gamma^2(\alpha+\beta)=pq-3r.$$

Once these values are substituted, the last three members combined give

$$\frac{k(q-pq+3r+pr)}{(1-\alpha^2)(1-\beta^2)(1-\gamma^2)}.$$

If we subtract this sum from the first member

$$\frac{k(1-p^2+2q+2r)}{(1-\alpha^2)(1-\beta^2)(1-\gamma^2)},$$

we obtain what we seek for, therefore

$$\sin S=\frac{k(1+q-r-p^2+pq-pr)}{(1-\alpha^2)(1-\beta^2)(1-\gamma^2)}.$$

Here, it is useful to observe that, setting $\alpha = 1$, the denominator vanishes. In this case the numerator also vanishes and this must also be satisfied in the cases $\beta = 1$ and $\gamma = 1$, in such a way that the numerator must necessarily contain the factors

$$1 - \alpha;\ 1 - \beta;\ 1 - \gamma.$$

Their product being equal to $1 - p + q - r$, the numerator will be at the same time divisible by it. Performing the division one finds that the quotient is equal to $1 + p$.

Instead, the denominator, divided by the same divider, gives

$$(1 + \alpha)(1 + \beta)(1 + \gamma) = 1 + p + q + r,$$

and thus the following formula results:

$$\sin S = \frac{k(1 + p)}{1 + p + q + r},$$

namely, reintroducing the initial values,

$$\sin S = \frac{(1 + \alpha + \beta + \gamma)\sqrt{1 - \alpha^2 - \beta^2 - \gamma^2 + 2\alpha\beta\gamma}}{(1 + \alpha)(1 + \beta)(1 + \gamma)},$$

where α, β and γ denote respectively $\cos a$, $\cos b$ and $\sin c$.

And it is worth illustrating the usefulness of this formula with some examples.

First Example

12. Let us assume that the sides b and c are quadrants.[10] In this case, we have $\beta = 0$ and $\gamma = 0$, and get

$$\sin S = \sqrt{(1 - \alpha^2)},$$

namely $\sin S = \sin a$, and, consequently, the area S will be $S = a$. But when both sides AB and AC are quadrants and the side BC is equal to a, then both angles B and C will be right. Having $\cos A = \alpha = \cos a$, the angle A will be equal to a, therefore the sum of all the angles will be equal to $180^{\circ} + a$, that is, the area sought for is $S = a$.

Second Example

13. Let us assume that the spherical triangle ABC is right at A. Then, from the formulae of spherical trigonometry one has $\cos BC = \cos AB \cos AC$; it follows that $cosa = \cos b \cos c$, that is, $\alpha = \beta\gamma$. Substituting these values, we find

$$\sin S = \frac{(1 + \beta + \gamma + \beta\gamma)\sqrt{(1 - \beta^2 - \gamma^2 + \beta^2\gamma^2}}{(1 + \beta)(1 + \gamma)((1 + \beta\gamma)} = \frac{\sqrt{(1 - \beta^2)(1 - \gamma^2)}}{1 + \beta\gamma}.$$

[10] A quadrant is a quarter of a circle.

Therefore, since $\sqrt{(1-\beta^2)} = \sin b$ and $\sqrt{(1-\gamma^2)} = \sin c$, we get, for the area of the right triangle

$$\sin S = \frac{\sin b \sin c}{1+\cos b \cos c} = \frac{\sin b \sin c}{1+\cos a}.$$

Third Example

14. If the triangle is equilateral, that is, if $\alpha = \beta = \gamma$, its area has an expression such that one has

$$\sin S = \frac{(1+3\alpha)\sqrt{(1-3\alpha^2+2\alpha^3)}}{(1+\alpha)^3}.$$

In this formula the term under the square root decomposes into the product $(1-\alpha)^2(1+2\alpha)$, and therefore one obtains

$$\sin S = \frac{(1+3\alpha)(1-\alpha)\sqrt{(1+2\alpha)}}{(1+\alpha)^3}.$$

Therefore, if these three sides are quadrants, implying $\alpha = 0$, one will have $\sin S = 1$ and consequently $S = \frac{\pi}{2}$.

Fourth Example.

15. If all the sides a, b, c of the triangle are infinitely small, the spherical triangle degenerates into a planar triangle, and since we have

$$\alpha = \cos a = 1 - \frac{1}{2}a^2 + \frac{1}{24}a^4 - \dots$$

and, analogously,

$$\beta = \cos b = 1 - \frac{1}{2}b^2 + \frac{1}{24}b^4 - \dots$$

and

$$\gamma = \cos c = 1 - \frac{1}{2}c^2 + \frac{1}{24}c^4 - \dots,$$

the rational factor [that is, the factor without the radical] of our formula will be equal to $\frac{4}{2^3} = \frac{1}{2}$, once the infinitesimal parts are naturally neglected.

Instead, for what concerns the irrational part, among them not only the finite parts cancel out, but also those in which a, b, c appear squared. For this reason it is necessary to develop each of the parts until the forth power.

Thus, one obtains what follows:

$$\alpha^2 = 1 - a^2 + \frac{1}{3}a^4,$$

$$\beta^2 = 1 - b^2 + \frac{1}{3}b^4,$$

$$\gamma^2 = 1 - c^2 + \frac{1}{3}c^4,$$

$$\alpha\beta = 1 - \frac{1}{2}a^2 - \frac{1}{2}b^2 + \frac{1}{24}a^4 + \frac{1}{24}b^4 + \frac{1}{4}a^2b^2,$$

and therefore

$$\alpha\beta\gamma = 1 - \frac{1}{2}a^2 - \frac{1}{2}b^2 - \frac{1}{2}c^2 + \frac{1}{24}a^4 + \frac{1}{24}b^4 + \frac{1}{24}c^4 + \frac{1}{4}a^2b^2 + \frac{1}{4}a^2c^2 + \frac{1}{4}b^2c^2.$$

Then we gather the quantity under the square root sign in the form

$$\begin{aligned} &-2 + a^2 + b^2 + c^2 - \frac{1}{3}a^4 - \frac{1}{3}b^4 - \frac{1}{3}c^4 \\ &+2 - a^2 - b^2 - c^2 + \frac{1}{12}a^4 + \frac{1}{12}b^4 + \frac{1}{12}c^4 \\ &+\frac{1}{2}a^2b^2 + \frac{1}{2}a^2c^2 + \frac{1}{2}b^2c^2, \end{aligned}$$

which, when the terms which cancel out are eliminated, reduces to

$$\frac{1}{2}a^2b^2 + \frac{1}{2}a^2c^2 + \frac{1}{2}b^2c^2 - \frac{1}{4}a^4 - \frac{1}{4}b^4 - \frac{1}{4}c^4.$$

Consequently, since the area of S is extremely small, whence $\sin S = S$, we have the sought for area:

$$S = \frac{1}{2}\sqrt{(\frac{1}{2}a^2b^2 + \frac{1}{2}a^2c^2 + \frac{1}{2}b^2c^2 - \frac{1}{4}a^4 - \frac{1}{4}b^4 - \frac{1}{4}c^4)},$$

namely,

$$S = \frac{1}{4}\sqrt{(2a^2b^2 + 2a^2c^2 + 2b^2c^2 - a^4 - b^4 - c^4)},$$

which is the well-known formula[11] of the area of a planar triangle.

Second Investigation
[A Formula for] $\cos S$
16. Since $\cos S = -\cos(A+B+C)$, one gets

$$\cos S = \cos A \sin B \sin C + \cos B \sin A \sin C + \cos C \sin A \sin B - \cos A \cos B \cos C,$$

and expanding separately each of the summands, we find

$$\text{I.}\quad \cos A \sin B \sin C = \frac{k^2(\alpha - \beta\gamma)}{\sin^2 a \sin^2 b \sin^2 c};$$

$$\text{II.}\quad \cos B \sin A \sin C = \frac{k^2(\beta - \alpha\gamma)}{\sin^2 a \sin^2 b \sin^2 c};$$

$$\text{III.}\quad \cos C \sin A \sin B = \frac{k^2(\gamma - \alpha\beta)}{\sin^2 a \sin^2 b \sin^2 c}.$$

For what regards the last term, one has, first of all,

$$\cos A \cos B = \frac{(\alpha - \beta\gamma)(\beta - \alpha\gamma)}{\sin^2 a \sin^2 b \sin^2 c} = \frac{\alpha\beta - \alpha^2\gamma - \beta^2\gamma + \alpha\beta\gamma^2)}{\sin^2 a \sin^2 b \sin^2 c},$$

and therefore

$$\cos A \cos B \cos C = \frac{(\alpha\beta\gamma - \alpha^2\beta^2 - \alpha^2\gamma^2 - \beta^2\gamma^2 + \alpha\beta\gamma^3 + \alpha\gamma\beta^3 + \beta\gamma\alpha^3 - \alpha^2\beta^2\gamma^2)}{\sin^2 a \sin^2 b \sin^2 c}.$$

17. Now if we set again

$$\alpha + \beta + \gamma = p;\ \alpha\beta + \alpha\gamma + \beta\gamma = q;\ \alpha\beta\gamma = r,$$

[11] Euler refers here to the so-called Heron formula, which gives the area of a planar triangle in terms of its side lengths. According to Heath, the (planar) Heron formula was already known to Archimedes (see T. L. Heath, A History of Greek Mathematics, vol. II, Oxford University Press, 1921, pp. 321–323). Heron of Alexandria lived in the 1st c. AD, whereas Archimedes lived in the 3rd c. BC. Let us also note that Euler also gave an analogue of this formula for spherical triangles; this is the subject of his memoir *Various speculations on the area of spherical triangles*, translated in Chap. 13 of the present volume.

adding the three terms on the right hand side of **I, II** and **III** above, we obtain

$$\frac{k^2(p-q)}{\sin^2 a \sin^2 b \sin^2 c}.$$

If then we put the last summand in the form

$$\frac{(\alpha\beta\gamma - \alpha^2\beta^2 - \alpha^2\gamma^2 - \beta^2\gamma^2) + \alpha\beta\gamma(\alpha^2+\beta^2+\gamma^2) - \alpha^2\beta^2\gamma^2}{\sin^2 a \sin^2 b \sin^2 c},$$

taking into account the identities

$$\alpha^2+\beta^2+\gamma^2 = p^2 - 2q$$

and

$$\alpha^2\beta^2 + \alpha^2\gamma^2 + \beta^2\gamma^2 = q^2 - 2pr,$$

we get

$$\frac{r - q^2 + 2pr + p^2r - 2qr - r^2}{\sin^2 a \sin^2 b \sin^2 c}.$$

Then, since

$$k^2 = 1 - p^2 + 2q + 2r,$$

summing all the parts, we have

$$\cos S = \frac{(p-q)(1-p^2+2q+2r) - r + q^2 - 2pr - p^2r + 2qr + r^2}{\sin^2 a \sin^2 b \sin^2 c},$$

and developing the calculations the formula becomes

$$\cos S = \frac{p - q - r + 2pq + p^2q - p^2r - q^2 + r^2 - p^3}{\sin^2 a \sin^2 b \sin^2 c}.$$

18. Because here again the denominator vanishes in the cases where $\alpha = 1, \beta = 1$ and $\gamma = 1$, in these cases the numerator must necessarily vanish as well, and therefore it contains the factor

$$(1-\alpha)(1-\beta)(1-\gamma) = 1 - p + q - r.$$

Thus, when this division is done, at the numerator we obtain $p-q-r+p^2$, whereas at the denominator the quotient will be

$$(1+\alpha)(1+\beta)(1+\gamma) = 1 + p + q + r,$$

and in this way we attain the following expression:

$$\cos S = \frac{p(1+p) - q - r}{1+p+q+r}.$$

Then, substituting the letters p, q and r by their values, we get

$$\cos S = \frac{(\alpha+\beta+\gamma)(1+\alpha+\beta+\gamma) - \alpha\beta - \alpha\gamma - \beta\gamma - \alpha\beta\gamma}{(1+\alpha)(1+\beta)(1+\gamma)},$$

or also

$$\cos S = \frac{\alpha+\beta+\gamma+\alpha^2+\beta^2+\gamma^2+\alpha\beta+\alpha\gamma+\beta\gamma-\alpha\beta\gamma}{(1+\alpha)(1+\beta)(1+\gamma)}.$$

First Example
19. When the two sides b and c are quadrants, and hence $\beta = 0$ and $\gamma = 0$, it follows that

$$\cos S = \frac{\alpha(1+\alpha)}{1+\alpha} = \alpha = \cos a.$$

Consequently, one will also have, like above, $S = a$.

Second Example
20. Let us consider a spherical triangle with right angle A for which we have, as we saw above, $\cos a = \cos b \cos c$, namely, $\alpha = \beta\gamma$. Substituting, one finds

$$\cos S = \frac{\beta+\gamma+2\beta\gamma+\beta^2+\gamma^2+\beta^2\gamma+\beta\gamma^2}{(1+\beta)(1+\gamma)(1+\beta\gamma)},$$

that is,

$$\cos S = \frac{(\beta+\gamma)(1+\beta+\gamma+\beta\gamma)}{(1+\beta)(1+\gamma)(1+\beta\gamma)} = \frac{\beta+\gamma}{1+\beta\gamma}.$$

But in the same case above, we found

$$\sin S = \frac{\sqrt{(1-\beta^2)(1-\gamma^2)}}{1+\beta\gamma},$$

and this is admirably consistent, since from here we have

$$\sin^2 S + cos^2 S = \frac{1+2\beta\gamma+\beta^2\gamma^2}{(1+\beta\gamma)^2} = 1.$$

Third Example

21. Let us assume that the triangle is equilateral, that is, that one has $\alpha = \beta = \gamma$ and therefore

$$\cos S = \frac{3\alpha + 6\alpha^2 - \alpha^3}{(1+\alpha)^3}.$$

On the other hand, we found, for this case,

$$\sin S = \frac{(1+3\alpha)(1-\alpha)\sqrt{(1+2\alpha)}}{(1+\alpha)^3}.$$

To show that these expressions are consistent, we take the square of both formulae, and we obtain:

$$\cos^2 S = \frac{9\alpha^2 + 36\alpha^3 + 30\alpha^4 - 12\alpha^5 + \alpha^6}{(1+\alpha)^6}$$

and

$$\sin^2 S = \frac{(1+6\alpha+9\alpha^2)(1-3\alpha^2+2\alpha^3)}{(1+\alpha)^6}$$
$$= \frac{1+6\alpha+6\alpha^2-16\alpha^3-15\alpha^4+18\alpha^5}{(1+\alpha)^6}.$$

Adding these two fractions, we obtain

$$\frac{1+6\alpha+15\alpha^2+20\alpha^3+15\alpha^4+6\alpha^5+\alpha^6}{(1+\alpha)^6} = 1.$$

Fourth Example

22.[12] Let us suppose that the sides of the triangle are infinitely small, and therefore that its area is almost zero, whence it follows $\cos S = 1 - \frac{1}{2}S^2$. Therefore, from the formula expressed as a function of the letters p, q, r, one has

$$1 - \frac{1}{2}S^2 = \frac{p(1+p)-q-r}{1+p+q+r},$$

[12] In this section, Euler shows that the formula he obtained for the area of a spherical triangle reduces to the one of a planar (Euclidean) triangle, when the three sides of the spherical triangle become infinitely small. In modern terms, this is a consequence of the fact that the geometry of the sphere, like the geometry of any Riemannian manifold, is infinitesimally Euclidean.

whence we obtain

$$S^2 = \frac{2 + 4q + 4r - 2p^2}{1 + p + q + r},$$

and, substituting p, q, r by their values, we find

$$S^2 = \frac{2k^2}{(1+\alpha)(1+\beta)(1+\gamma)}.$$

If here at the denominator one puts the unity instead of the letters α, β, γ, the denominator becomes equal to 8. For what regards the numerator, as we have seen above, one has

$$\begin{aligned} k &= \sqrt{(1 - \alpha^2 - \beta^2 - \gamma^2 + 2\alpha\beta\gamma)} \\ &= \sqrt{(\frac{1}{2}a^2b^2 + \frac{1}{2}a^2c^2 + \frac{1}{2}b^2c^2 - \frac{1}{4}a^4 - \frac{1}{4}b^4 - \frac{1}{4}c^4)}. \end{aligned}$$

Substituting this value [of k], we find

$$S^2 = \frac{2a^2b^2 + 2a^2c^2 + 2b^2c^2 - a^4 - b^4 - c^4}{16},$$

therefore

$$S = \frac{1}{4}\sqrt{(2a^2b^2 + 2a^2c^2 + 2b^2c^2 - a^4 - b^4 - c^4)}.$$

Third Investigation
[A Formula for] $\tan S$ **and** $\tan \frac{1}{2}S$

23. After having calculated for the area of our spherical triangle $\sin S$ and $\cos S$, the expression of the tangent of this area appears naturally, that is,

$$\tan S = \frac{(1 + \alpha + \beta + \gamma)\sqrt{1 - \alpha^2 - \beta^2 - \gamma^2 + 2\alpha\beta\gamma}}{\alpha + \beta + \gamma + \alpha^2 + \beta^2 + \gamma^2 + \alpha\beta + \alpha\gamma + \beta\gamma - \alpha\beta\gamma},$$

a formula which in general it is not possible to express in a simpler form.

24. However one may represent the tangent of half of the area in a much more symmetric form. In fact, having

$$\tan \frac{1}{2}S = \frac{\sin S}{1 + \cos S},$$

we first keep the letters p, q and r, in such a way that in the numerator we have

$$\sin S = \frac{(1+p)\sqrt{(1-p^2+2q+2r)}}{1+p+q+r},$$

whereas at the place of the denominator, as we have

$$\cos S = \frac{p(1+p)-q-r}{1+p+q+r},$$

we obtain

$$1+\cos S = \frac{(1+p)^2}{1+p+q+r}.$$

Therefore, substituting these values, we find

$$\tan\frac{1}{2}S = \frac{\sqrt{(1-p^2+2q+2r)}}{1+p},$$

and, returning to the initial values,

$$\tan\frac{1}{2}S = \frac{\sqrt{(1-\alpha^2-\beta^2-\gamma^2+2\alpha\beta\gamma)}}{1+\alpha+\beta+\gamma},$$

a formula which anyway turns out to be very useful in practice.

First Example

25. If two sides b and c are quadrants, whence $\beta = 0$ and $\gamma = 0$, it will follow that

$$\tan\frac{1}{2}S = \frac{\sqrt{((1-\alpha^2)}}{1+\alpha} = \frac{\sin a}{1+\cos a}.$$

Consequently, it is evident that one has

$$\tan\frac{1}{2}S = \tan\frac{1}{2}a,$$

therefore $S = a$, as we have already found before.

Second Example

26. Let us suppose that the spherical triangle is right at A, whence we have $\cos a = \cos b\cos c$ and $\alpha = \beta\gamma$. Then, substituting these values, one finds

$$\tan\frac{1}{2}S = \frac{\sqrt{((1-\beta^2-\gamma^2+\beta^2\gamma^2))}}{1+\beta+\gamma+\beta\gamma} = \frac{\sqrt{(1-\beta^2)(1-\gamma^2)}}{(1+\beta)(1+\gamma)},$$

and dividing above and below by $\sqrt{(1+\beta)(1+\gamma)}$, reduces this fraction to the following:

$$\tan\frac{1}{2}S = \sqrt{\frac{(1-\beta)(1-\gamma)}{(1+\beta)(1+\gamma)}}.$$

On the other hand,

$$\sqrt{\frac{(1-\beta)}{(1+\beta)}} = \sqrt{\frac{(1-\cos b)}{(1+\cos b)}} = \tan\frac{1}{2}b$$

and, analogously,

$$\sqrt{\frac{(1-\gamma)}{(1+\gamma)}} = \sqrt{\frac{(1-\cos c)}{(1+\cos c)}} = \tan\frac{1}{2}c.$$

Consequently, the following formula results, which can be very well memorized:

$$\tan\frac{1}{2}S = \tan\frac{1}{2}b\tan\frac{1}{2}c,$$

and one proves without any difficulty that it is in accordance with the results found above.

Third Example
27. If the triangle is equilateral, that is, if $\alpha = \beta = \gamma$, one has

$$\tan\frac{1}{2}S = \frac{\sqrt{((1-3\alpha^2+2\alpha^3)}}{1+3\alpha} = \frac{(1-\alpha)\sqrt{(1+2\alpha)}}{1+3\alpha}.$$

Therefore, in the case where all the sides are quadrants, and therefore $\alpha = 0$, one has $\tan\frac{1}{2}S = 1$, and therefore $\frac{1}{2}S = 45^{\circ}$ and $S = \frac{\pi}{2}$.[13]

Fourth Example
28. Finally, let us suppose that the three sides a, b, c are extremely small. Since

$$\tan\frac{1}{2}S = \frac{1}{2}S,$$

one will get

$$S = 2\frac{\sqrt{(1-\alpha^2-\beta^2-\gamma^2+2\alpha\beta\gamma)}}{1+\alpha+\beta+\gamma}.$$

[13] This is compatible with the Girard (or Harriot) theorem which Euler mentions in the introduction of the present memoir.

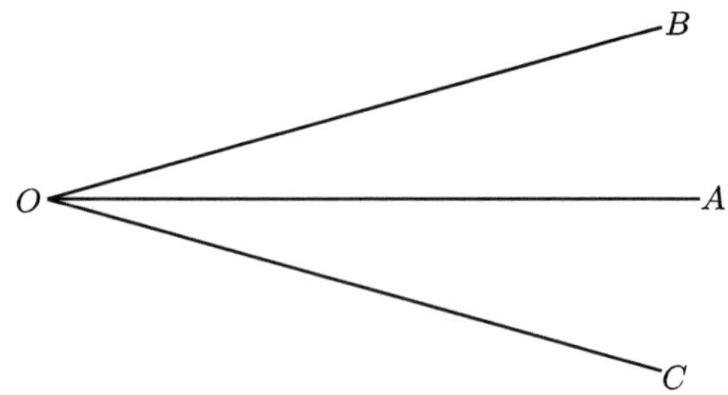

Fig. 14.4 Solution of Problem 1

Now it is therefore sufficient that for the denominator one supposes $\alpha = 1$, $\beta = 1$, $\gamma = 1$, so that the coefficient of the radical in the formula is equal to $\frac{1}{2}$; but we have already seen a few times that the same radical in the formula is

$$\sqrt{(\frac{1}{2}a^2b^2 + \frac{1}{2}a^2c^2 + \frac{1}{2}b^2c^2 - \frac{1}{4}a^4 - \frac{1}{4}b^4 - \frac{1}{4}c^4)}.$$

Therefore the area can be expressed exactly as before.

Problem 1 (Fig. 14.4)
29. Given a solid angle $AOBC$ formed by three planar angles $BOC = a$, $AOC = b$ and $AOB = c$, to determine its actual area.[14]

Solution
Since the measure of this solid angle can be set to be equal to the area of the spherical triangle having sides a, b, c, supposing that the radius of the sphere is equal to 1, from the preceding considerations one may understand that the solid angles, as well as the planar ones, can be expressed either in degrees and in minutes, or using circular arcs. Thus, let us suppose that S expresses the measure of the solid angle considered. Then, having set, for brevity,

$$\cos a = \alpha, \ \cos b = \beta, \ \cos c = \gamma,$$

this measure S can be assigned in three ways; first of all, using the sine

$$\sin S = \frac{(1 + \alpha + \beta + \gamma)\sqrt{1 - \alpha^2 - \beta^2 - \gamma^2 + 2\alpha\beta\gamma}}{(1 + \alpha)(1 + \beta)(1 + \gamma)};$$

and then using the cosine:

$$\cos S = \frac{\alpha + \beta + \gamma + \alpha^2 + \beta^2 + \gamma^2 + \alpha\beta + \alpha\gamma + \beta\gamma - \alpha\beta\gamma}{(1 + \alpha)(1 + \beta)(1 + \gamma)};$$

[14] A literal translation would be: To assign to a given solid angle $AOBC$, formed by three planar angles $BOC = a$, $AOC = b$ and $AOB = c$, its true measure.

and, thirdly, but in a truly convenient way, using the tangent of its half:

$$\tan\frac{1}{2}S = \frac{\sqrt{(1-\alpha^2-\beta^2-\gamma^2+2\alpha\beta\gamma)}}{1+\alpha+\beta+\gamma}.$$

Here, first of all, it is useful to observe that if all the three angles a, b, c are right, then the measure of the solid angle becomes equal to 90°, and this agrees marvelously with the common manner of speaking, since the solid angles of this type are usually called also by the house builders right angles; and from here one may at the same time understand which angles have to be considered as greater or smaller than a right angle.

Scholion I[15]

30. It would be extraordinary if this measure of solid angle could lead also to excellent properties as those that hold for the plane figures, like: that the sum of the planar angles of a triangle is equal to two right angles. But meanwhile, such a property in solid figures is not absolutely verified, because of our measure. And not even in all tetrahedra, which are made up of four solid angles, summing all solid angles results in the same quantity, but according to whether the tetrahedra are constructed more or less oblique, the sum of the four solid angles may become sometimes larger, sometimes smaller. In fact, if we examine a regular tetrahedron, each of whose solid angles is formed by three plane angles of 60°, we get $\alpha = \beta = \gamma = \frac{1}{2}$; therefore the measure of each of the angles is determined in such a way that one has $\tan\frac{1}{2}S = \frac{\sqrt{2}}{5}$, hence from the trigonometric table it results: $\frac{1}{2}S = 15^\circ 48'$, or else $S = 31^\circ 36'$.

Therefore the sum of all four angles of this tetrahedron will be $126^\circ 24'$.

Now consider a triangular pyramid, having also as basis an equilateral triangle, but whose vertex terminates in a very acute cusp, so that its measure tends to zero. Then, for the three solid angles that are at the base, an angle will be $a = 60^\circ$, while for the remaining two, $b = c = 90^\circ$, in such a way that $\alpha = \frac{1}{2}$ and $\beta = \gamma = 0$. Whence, we obtain $\tan\frac{1}{2}S = \frac{1}{\sqrt{3}} = \tan 30^\circ$, so that we have $S = 60^\circ$.

From here, it results that the sum of all the solid corners of this pyramid is 180°, while before, for the tetrahedron, its value was only 126°. But although the sum of the solid angles of any solid does not reveal any significant property, perhaps in other relationships this measure may reveal non-negligible properties.

Scholion II

31. The considerations that have been reported here are related to the measurement of solid angles that are formed only by three planar angles. If, on the contrary, a solid angle is made up of four or more planar angles, its measure will be the area of the spherical quadrilateral or the polygon with several sides, whose sides are equal to the corresponding planar angles that form the solid angle. Then, it suffices

[15] A *scholion* is a commentary or a short note; from the Greek σχόλιον.

to decompose this polygon into spherical triangles and determine the area of each of them, and of course their sum will give the solid angle measure. But in these cases it is not sufficient to know each of the flat planar angles, but in addition it is necessary to know what inclination the planar angles mutually make with each other. Since these facts are quite obvious, here I will only add the size of the angles of the regular solids which, whatever their number is, consist of a certain arbitrary number of planar angles mutually equal and equally inclined.

Problem 2 (Fig. 14.5)

32. *Given a solid angle consisting of n equal planar angles, each one equal to a, and equally inclined among themselves, to find the measure of that solid angle.*

Solution

If we take a sphere of radius equal to one circumscribed to that solid angle (Fig. 14.5), its measure will be the area of the spherical polygon having n sides all equal to a. And since also all the angles are equal among themselves, the polygon will be regular, and therefore its center will be at its median point, which we shall call O.

First, let us consider then an arbitrary side AB of the polygon, which has length a, and let us join its extremities with O using the arcs AO and BO, which are equal among themselves, until we have the triangle AOB. Since the number of triangles is equal to n, it will follow that the value of the angle AOB is $\frac{2\pi}{n}$. If we call then S the area of the whole polygon, which will be at the same time the measure of the given angle, the area of this triangle AOB will be equal to $\frac{S}{n}$. From the point O we drop the perpendicular OP on the side AB, which divides into two equal parts the side AB, hence $AP = \frac{1}{2}a$ and the angle AOP is equal to $\frac{\pi}{n}$. If we call Φ the angle

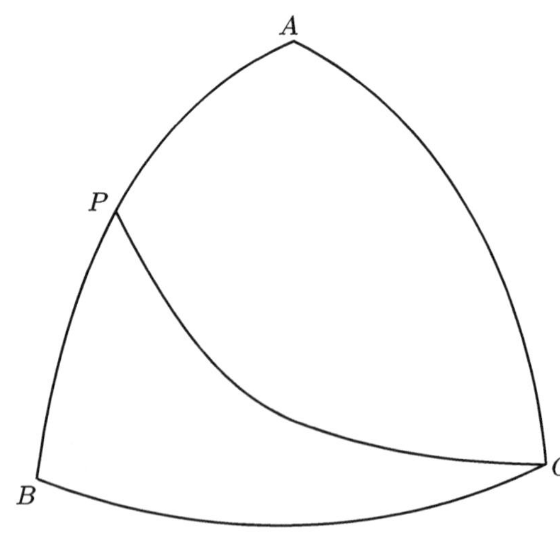

Fig. 14.5 Solution of Problem 2

OAB, recalling the identities of spherical trigonometry,[16] we get

$$\sin\Phi = \frac{\cos\frac{\pi}{n}}{\cos\frac{1}{2}a}.$$

Since the angle OBA is also equal to the angle Φ, the angle sum of the triangle AOB will be equal to $2\Phi + \frac{2\pi}{n}$. Then, subtracting two right angles, we obtain the area of the triangle AOB:

$$\frac{S}{n} = 2\Phi + \frac{2\pi}{n} - \pi,$$

and hence the total area of the polygon is

$$S = 2n\Phi + 2\pi - n\pi = 2n\Phi - (n-2)\pi,$$

which will therefore be the measure of the solid angle considered.

Corollary I

33. Therefore if the solid angle consists of three planar angles equal to a, having $n = 3$, we get

$$\sin\Phi = \frac{\cos 60^{\circ}}{\cos\frac{1}{2}a};$$

and this angle being found, the measure of the solid angle will be

$$S = 6\Phi - \pi = 6\Phi - 180^{\circ}.$$

Corollary II

34. If the solid angle consists of four planar angles which are all equal to a, since in this case $n = 4$, we search for the angle Φ such that

$$\sin\Phi = \frac{\cos 45^{\circ}}{\cos\frac{1}{2}a};$$

then from this we find the measure of the solid angle

$$S = 8\Phi - 2\pi = 8\Phi - 360^{\circ}.$$

[16] In the text: *ex sphaericis*.

Corollary III

35. If the solid angle consists of five planar angles which are all equal to a, since in this case we have $n = 5$, we search for the angle Φ such that

$$\sin \Phi = \frac{\cos 36^{\circ}}{\cos \frac{1}{2}a};$$

and from this we find the measure of the solid angle

$$S = 10\Phi - 3\pi = 10\Phi - 540^{\circ}.$$

Corollary IV

36. If the solid angle consists of six planar angles all equal to a, having $n = 6$, one searches for the angle Φ such that

$$\sin \Phi = \frac{\cos 30^{\circ}}{\cos \frac{1}{2}a}.$$

Then the measure of this solid angle is

$$S = 12\Phi - 4\pi = 12\Phi - 720^{\circ}.$$

Scholion

37. Following these rules, we can calculate the solid angles of the five regular solids, as we can more easily compare them with the right angle, which in solids is also equal to 90°; of course here it will be convenient to call acute the solid angles less than 90°, and, instead, obtuse those that are grater than 90°.

Measures of the Solid Angles in the Tetrahedron

38. Since for the constitution of solid angles three solid angles of 60° contribute, we get $\frac{1}{2}a = 30^{\circ}$ and $n = 3$. Therefore from Corollary I, we can use logarithms to perform the calculation in the following mode:

$$\log \cos 60^{\circ} = 9,6989700$$

$$\log \cos 30^{\circ} = 9,9375306$$

$$\log \sin \Phi = 9,7614394,$$

therefore

$$\Phi = 35^{\circ}15'52'',$$

thus

$$6\Phi = 211^{\circ}35'12''$$

whence we obtain each of the solid angles of the tetrahedron

$$S = 31^{\circ}35'12'',$$

and thus in this case the angle barely exceeds the third of a right angle.

The Measure of the Solid Angles of the Octahedron
39. Here, given that each of the angles is composed of four planar angles of 60°, we get $\frac{1}{2}a = 30^{\circ}$ and $n = 4$. Then, according to what is obtained in Corollary II one performs in the following way the calculation with logarithms:

$$\log\cos 45^{\circ} = 9,8494850,$$

$$\log\cos 30^{\circ} = 9,9375306,$$

$$\log\sin\Phi = 9,9119544,$$

and it follows that

$$\Phi = 54^{\circ}44'8'',$$

whence

$$8\Phi = 437^{\circ}53'4'';$$

therefore the measure of the solid angle of the octahedron will be

$$S = 77^{\circ}53'4'',$$

which therefore is not much smaller than a right angle. After all, this angle Φ is the complement to 90° of the preceding [angle Φ].

Measure of the Solid Angles of the Icosahedron
40. Since now the solid angle is composed of five planar angles $a = 60^{\circ}$, we get $\frac{1}{2}a = 30^{\circ}$ and $n = 5$; therefore, for Corollary III the calculation should be carried out as follows:

$$\log\cos 36^{\circ} = 9,9079576,$$

$$\log\cos 30^{\circ} = 9,9375306,$$

$$\log\sin\Phi = 9,9704270,$$

whence it follows that

$$\Phi = 69^{\circ}5'41\frac{1}{2}''.$$

Therefore,

$$10\Phi = 690^{\circ}56'55''.$$

Thus the measure of the solid angle of the icosahedron will be

$$S = 150^{\circ}56'55'',$$

therefore this angle is very obtuse.

Measure of the Solid Angles of the Hexahedron

41. Since in this case each of the solid angles consists of three planar right angles, we get $a = 90^{\circ}$, $\frac{1}{2}a = 45^{\circ}$ and $n = 3$. Here, using Corollary I, the calculation gives

$$\log \cos 60^{\circ} = 9,6989700,$$

$$\log \cos 45^{\circ} = 9,8494850,$$

$$\log \sin \Phi = 9,8494850.$$

Thus, we get $\Phi = 45^{\circ}$ and therefore $6\Phi = 270^{\circ}$. It follows that the solid angle of the hexahedron will be 90°, and of course in this case this angle is right.

Measure of the Solid Angle of the Dodecahedron

42. Given here that each of the angles is composed of three planar angles of 108°, we get $\frac{1}{2}a = 54^{\circ}$ and $n = 3$. From Corollary I, the calculation must be the following:

$$\log \cos 60^{\circ} = 9,6989700,$$

$$\log \cos 54^{\circ} = 9,7692187,$$

$$\log \sin \Phi = 9,9297513,$$

and from this we obtain the angle

$$\Phi = 58^{\circ}16'57'',$$

therefore

$$6\Phi = 349^{\circ}41'42''.$$

Therefore the measure of the solid angle of the tetrahedron will be $169^{\circ}41'42''$, and thus, in this case, among the solid angles of all the regular bodies, the one of the dodecahedron is the greatest.

Scholion

43. And if the solid angle consists of six planar angles $a = 60^{\circ}$, therefore $\frac{1}{2}a = 30^{\circ}$ and $n = 6$, the resulting solid angle turns out to be flattened to a plane, and it intersects the sphere in a great circle. The latter bounds a hemisphere whose area is equal to 360°, and therefore the measure of each of these solid angles is equal to four right angles. Putting together all the solid angles of this type we obtain a regular body which is nothing else than the sphere itself.

Therefore it is also made clear from the calculation done using Corollary IV that

$$\log\cos 30^{\circ} = 9,9375306,$$

$$\log\cos 30^{\circ} = 9,9375306,$$

$$\log\sin\Phi = 10,000000,$$

therefore $\Phi = 90^{\circ}$ and $12\Phi = 1080^{\circ}$, whence the solid angle will be $S = 360^{\circ}$. The same thing occurs if the solid angle is composed of four planar right angles, whence $\frac{1}{2}a = 45^{\circ}$ and $n = 4$. Thus, in fact, we have

$$\sin\Phi = \frac{\cos 45^{\circ}}{\cos 45^{\circ}} = 1.$$

Therefore, $\Phi = 90^{\circ}$, and for the solid angle one has $S = (8 - 4)90^{\circ} = 360^{\circ}$. Finally, if the solid angle consists of three planar angles, in such a way that we have $a = 120^{\circ}$, then $\frac{1}{2}a = 60^{\circ}$ ed $n = 3$; whence we have again:

$$\sin\Phi = \frac{\cos 60^{\circ}}{\cos 60^{\circ}} = 1,$$

therefore $\Phi = 90^{\circ}$, and the solid angle will be $S = (6 - 2)90^{\circ} = 360^{\circ}$.

Chapter 15
A Construction Relative to a Problem of Pappus of Alexandria

Leonhard Euler

Abstract In this memoir, Euler solves, in the Euclidean and then in the spherical setting, a problem inspired by a construction of Pappus of Alexandria (Proposition 117 of Book VII of the *Collection*). The problem is to construct a triangle which is inscribed in a given circle and whose three sides are contained in three lines that are required to pass through three given points. Pappus gives this construction in the Euclidean setting and only in the special case where the three given points are collinear. Euler solves the general case, in the Euclidean and in the spherical settings.

Keywords Greek mathematics · Pappus problem · Castillon problem · Euclidean geometry · Leonhard Euler · Spherical geometry

AMS Codes 01A50, 01A35, 01A20, 51N20

Theorem 1
If from the extremities A and B of an arbitrary straight line, to any point P of an arbitrary circle are drawn the lines AP and BP, cutting the circle at a and b, and if the points F and G are taken [on this line] in such a way that

$$AF = \frac{AP \cdot Aa}{AB} \quad \text{and} \quad BG = \frac{BP \cdot Bb}{AB},$$

then we always have

$$FP \cdot Ff = GP \cdot Gg = AF \cdot BG,$$

[where f and g are the intersections of FP and GP with the circle].

The author Leonhard Euler (1707–1783).
Translated from the Latin by Renzo Caddeo.
Problematis cuiusdam Pappi Alexandrini constructio, *Acta Academiae Scientiarum Imperialis Petropolitanae*, 4, 1783, pp. 91–96. Opera Omnia: Series 1, Volume 26, pp. 237–242.

R. Caddeo, A. Papadopoulos (eds.), *Spherical Geometry in the Eighteenth Century I*, https://doi.org/10.1007/978-3-032-12466-1_15

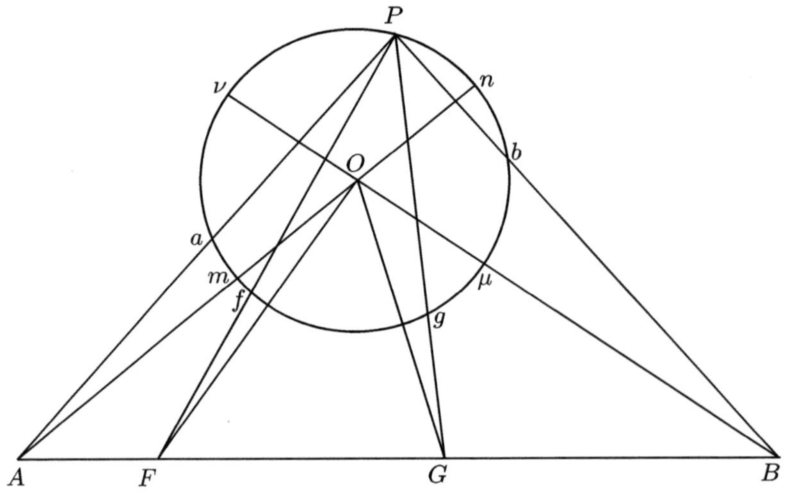

Fig. 15.1 Theorem 1

Proof
We represent the line through A and B with its points F and G on the same side as the center O of the circle,[1] and we set $AO = a$, $BO = b$, $Om = On = r$, the radius of the circle, and $AB = c$, $FO = f$ and $GO = g$, then $AP \cdot Aa = An \cdot Am$. It is true that $An = a + r$ and $Am = a - r$, and therefore

$$AP \cdot Aa = a^2 - r^2.$$

In a similar way, one has

$$BP \cdot Bb = B\nu \cdot B\mu = (b + r)(b - r)\,,$$

or

$$BP \cdot Bb = b^2 - r^2\,.$$

In the same way, it results that

$$FP \cdot Ff = f^2 - r^2 \;\text{ and }\; GP \cdot Gg = g^2 - r^2.$$

Therefore if we suppose that

$$AF = \frac{a^2 - r^2}{c} \;\text{ and }\; BG = \frac{b^2 - r^2}{c},$$

[1] In Fig. 15.1, the line is taken to be in the exterior of the circle.

we need to show that

$$f^2 - r^2 = g^2 - r^2 = (a^2 - r^2)(b^2 - r^2)/c^2\,,$$

and for that purpose we appeal to the following lemma.

Lemma
If from the point O of the triangle AOB we draw onto a point F on its opposite side AB the segment OF, we shall have

$$FO^2 = \frac{AO^2 \cdot BF + BO^2 \cdot AF}{AB} - AF \cdot BF.$$

Proof
Dropping from O[2] onto AB the perpendicular $O\Pi$, we shall have

$$AO^2 = A\Pi^2 + \Pi O^2 = (AF + F\Pi)^2 + FO^2 - F\Pi^2,$$

that is,

$$AO^2 = AF^2 + FO^2 + 2AF \cdot F\Pi;$$

and in the same way we find

$$BO^2 = BF^2 + FO^2 - 2BF \cdot F\Pi.$$

If we add the first of these equations multiplied by BF to the other one multiplied by AF, we obtain

$$AO^2 \cdot BF + BO^2 \cdot AF = BF(AF^2 + FO^2) + AF(BF^2 + FO^2)\,,$$

or

$$AO^2 \cdot BF + BO^2 \cdot AF = FO^2 \cdot AB + BF \cdot AF \cdot AB,$$

therefore

$$FO^2 = \frac{AO^2 \cdot BF + BO^2 \cdot AF}{AB} - AF \cdot BF.$$

Q. E. D.

Continuation of the Proof of the Preceding [Theorem]
If we set (Fig. 15.1)

$$AF = \frac{a^2 - r^2}{c} = \alpha\ ,\ \ BG = (b^2 - r^2)/c = \beta,$$

[2] In the original edition: from E.

we shall have

$$a^2 = \alpha c + r^2 \quad \text{and} \quad b^2 = \beta c + r^2 .$$

Then from the lemma it follows that

$$cf^2 = a^2(c - \alpha) + b^2\alpha - \alpha c(c - \alpha) ,$$

and if instead of a^2 and b^2 one substitutes the values just given, one shall have

$$cf^2 = cr^2 + \alpha\beta c$$

or

$$f^2 - r^2 = \alpha\beta .$$

Since on the other hand we have[3]

$$FO^2 = \frac{AO^2 \cdot BF + BO^2 \cdot AF}{AB} - AF \cdot BF ,$$

in a similar way, one proves that

$$cg^2 = cr^2 + \alpha\beta c ,$$

that is,

$$g^2 - r^2 = \alpha\beta ,$$

and therefore

$$f^2 - r^2 = g^2 - r^2 = (a^2 - r^2)(b^2 - r^2)/c^2 .$$

Q. E. D.

At this moment we can state the following

Theorem 2
If (Fig. 15.2) from a point O of the triangle ABO[4] one draws onto the base AB two segments OF and OG equal among themselves, we shall have

[3] In the next formula, we have corrected a typo which was in the original edition.

[4] In the original publication, " of the triangle ABC".

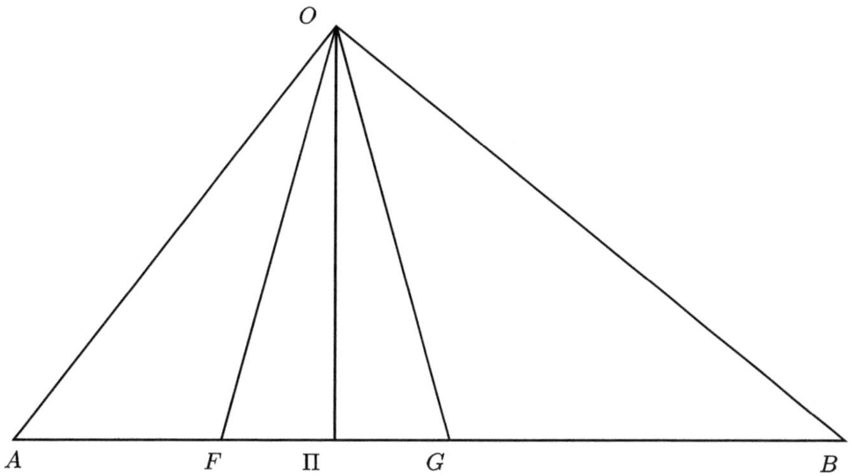

Fig. 15.2 Theorem 2

$$AO^2 - AB \cdot AF = BO^2 - AB \cdot BG\,.$$

Proof
Dropping from the vertex O the perpendicular $O\Pi$, we shall have

$$F\Pi = G\Pi = \tfrac{1}{2} \cdot FG\,.$$

Since we have

$$AO^2 = AF^2 + FO^2 + 2AF \cdot F\Pi\,,$$

that is,

$$AO^2 = AF^2 + FO^2 + AF \cdot FG,$$

then we shall have

$$AO^2 = FO^2 + AF \cdot AG. \tag{15.1}$$

In a similar way one finds

$$BO^2 = FO^2 + BF \cdot BG. \tag{15.2}$$

From (15.1) it is obtained

$$FO^2 = AO^2 - AF \cdot AG,$$

whence

$$FO^2 - AF \cdot BG = AO^2 - AB \cdot AF.$$

From equation (15.2) one obtains

$$FO^2 = BO^2 - BF \cdot BG,$$

and therefore

$$FO^2 - AF \cdot BG = BO^2 - AB \cdot BG,$$

whence it follows

$$AO^2 - AB \cdot AF = BO^2 - AB \cdot BG.$$

Q. E. D.

Corollary
Thus, each time one has

$$AO^2 - AB \cdot AF = BO^2 - AB \cdot BG = \Delta,$$

the two segments FO and GO will be equal to each other, and at the same time we will have

$$FO^2 - AF \cdot BG = \Delta\,.$$

In fact, if one takes

$$AF = \frac{AO^2 - \Delta}{AB} \text{ and } BG = \frac{BO^2 - \Delta}{AB}$$

he finds that $FO = GO$. But in the preceding theorem we had

$$AF = \frac{a^2 - r^2}{c} \text{ and } BG = \frac{b^2 - r^2}{c},$$

whence

$$\Delta = r^2\,,\; AF = \frac{AO^2 - r^2}{AB}\,,\; BG = \frac{BO^2 - r^2}{AB}$$

and

$$FO^2 - r^2 = AF \cdot BG\,.$$

Problem 1
Given a circle of center O, to inscribe a triangle $a\ b\ c$ whose three sides ab, ac, bc, extended, pass through the three given points C, B, A (Fig. 15.3).

Construction
Let A, B and C be the three given points, having distances $AO = a$, $BO = b$, $CO = c$ from the center O of the circle, and we assume that the radius of this circle is equal to 1. Then from the point B we take the segment $BF = \frac{b^2-1}{AB}$ and we get

$$FO^2 - 1 = \frac{BF(a^2-1)}{AB}.$$

Then, we draw the segment FC, and on the same we take the interval

$$FK = \frac{FO^2-1}{FC} = \frac{BF(a^2-1)}{AB \cdot FC},$$

and we get

$$KO^2 - 1 = \frac{FK(c^2-1)}{FC}.$$

Therefore, from the center O one draws the radius Om such that the cosine of the angle KOm is equal to $\dfrac{\cos BFC}{KO}$. Then if we bisect the angle BFC by the ray FS, and we draw from the point m the segment mb parallel to it, the point b will be one of the vertices of the desired triangle. If we join this point with the point A, then the extension of the segment Ab[5] will intersect the circle in c.[6] From this point c[7] one draws onto B the ray cB which intersects the circle in a; but then the extension of the side ba will pass by the third given point C, and abc will be the desired triangle (Fig. 15.3).

Corollary 1
Suppose that A and C, two of the given points, are infinitely distant [from the center O of the circle] and belong to the straight lines *ABA* and *CBC* which intersect in B (Fig. 15.4). From B one draws the straight line Bca, which cuts on the circle the arc ca, on whose boundary are subtended angles which are equal to the angle CBA: then we draw from a the straight lines ab, parallel to the straight line CBC, and bc, parallel to ABA; the desired triangle will be abc.

[5] In the original edition, AB.
[6] In the original edition, C.
[7] In the original edition, C.

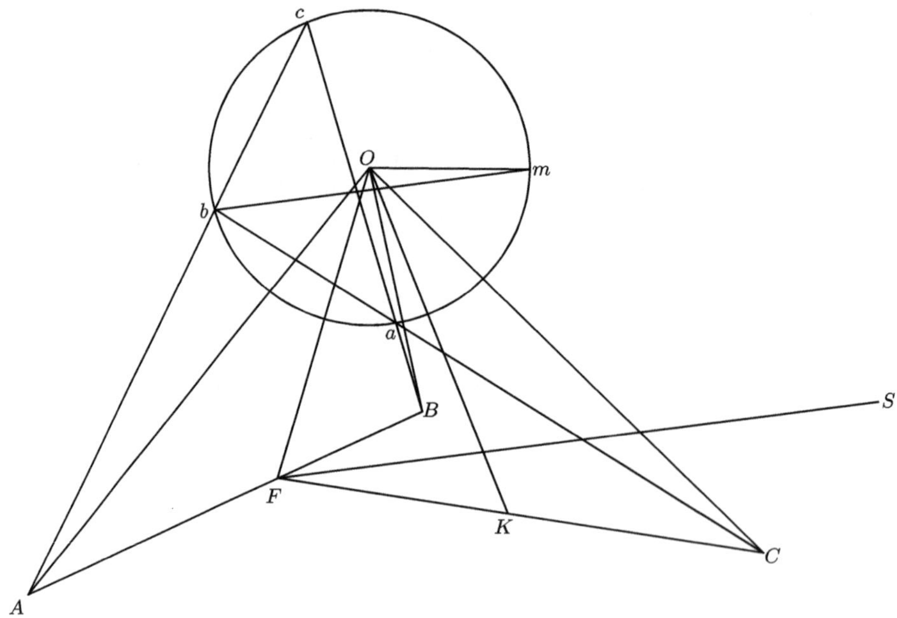

Fig. 15.3 Problem 1

Corollary 2

Let us assume that all the three points lie on the straight lines OA, OB, OC at infinite distances [from the center O of the circle]; then to the straight line OA one draws the parallel line bc at distance $OX = \cos BOC$ from the center; then drawing the straight lines ba parallel to OC and ca parallel to OB, we obtain the desired triangle (Fig. 15.5).

Scholion 1

However it is good to notice here that the construction given above provides two solutions, according to whether the angle[8] KOm is to the right or to the left.[9] Besides, since the three points A, B, C are interchangeable among themselves, the construction given here may be erected in six different modes, therefore all of them must give rise to the same two solutions; however there is no evident reason for this fact.

Scholion 2

This problem can be solved also on the sphere, in order to know how to inscribe in a small circle drawn on a sphere a triangle abc in such a way that the extension of its sides ab, ac, bc pass through three points C, B, A which are fixed on the surface

[8] See Fig. 15.3.

[9] With respect to the origin.

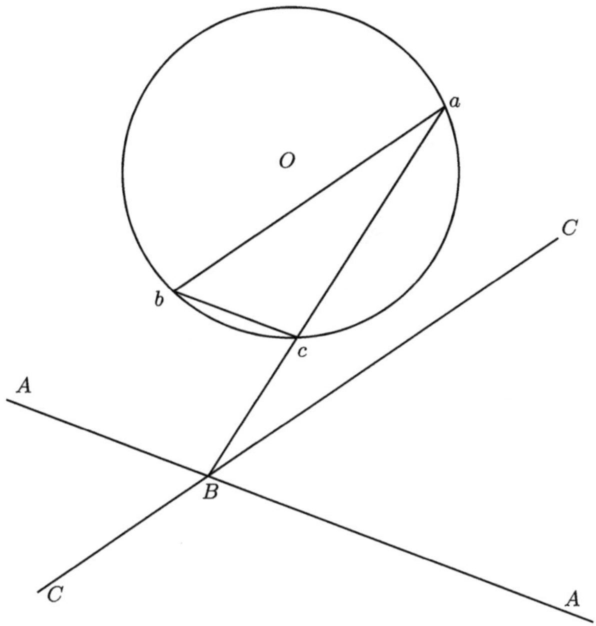

Fig. 15.4 Corollary 1

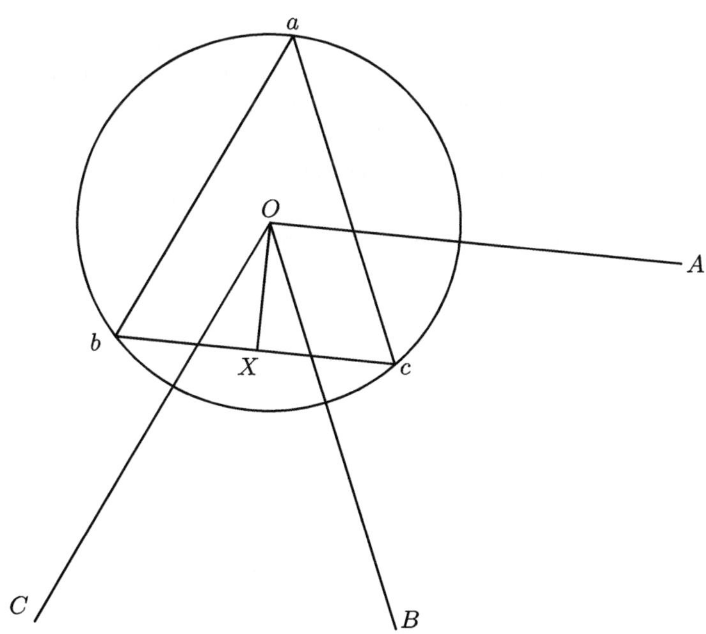

Fig. 15.5 Corollary 2

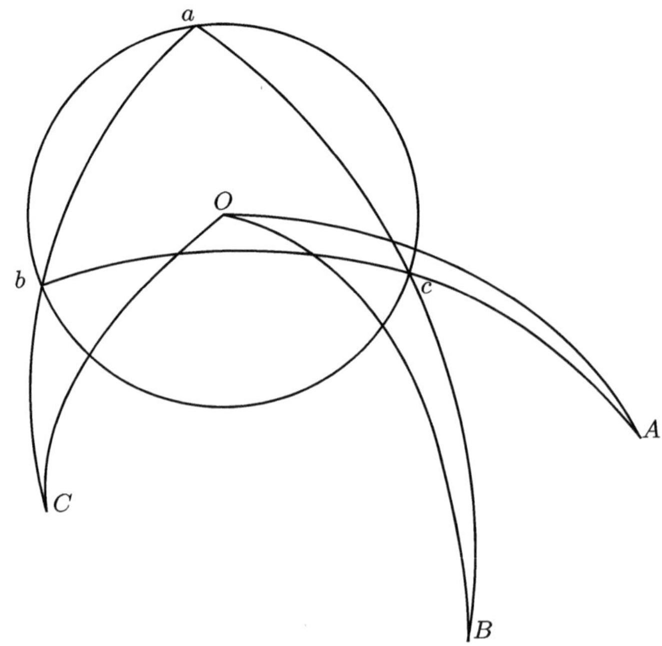

Fig. 15.6 Scholion 2

of the sphere (Fig. 15.6). In fact, one considers the tangent plane to the sphere at the center O of the circle, on which a planar triangle is already constructed in the way we described above; the transfer of this (triangle) on the surface of the sphere will be very easy, since at the center, which is on the surface, all the angles, whether they are on the plane or on the surface, are the same, while the distances of the given points A, B, C and of the vertices a, b, c from the center O are different from those of their corresponding points in the tangent plane.

Chapter 16
Some Questions of the Geometry of the Plane and of the Sphere

Leonhard Euler

Abstract In this memoir, Leonhard Euler starts with a triangle in the Euclidean plane and three segments joining one vertex of the triangle to the opposite side, such that these three segments intersect at a common point. He establishes a relation between the six parts cut on these three segments by the intersection point. He then shows that conversely, if we start with the relation found, then we can construct a triangle and three lines joining the vertices to the opposite sides such that the six segments that arise in this construction satisfy the given relation. After the Euclidean case, Euler proves an analogous result in the spherical case.

Keywords Leonhard Euler · Spherical geometry · Euclidean geometry

AMS Codes 01A50, 97G60, 51M04, 53A35

§1. Recently, while I was reflecting on the case in which, in any triangle ABC (Fig. 16.1), from the angles one leads to the opposite sides, in an arbitrary way, the straight lines Aa, Bb and Cc that intersect each other at a common point O, the following question came to my mind: in what way, given the two parts of each of these straight lines, is it possible to construct the triangle? However, I quickly realized that in general such a question admits of no solution, unless some special condition takes place between those six parts. In this way I came across the following very significant theorem.

Theorem 1

If in any triangle ABC (Fig. 16.1) from the angles are led to the opposite sides, in whatever way, the lines Aa, Bb and Cc that intersect each other at the common point O, then the property expressed by the following equality will always take place

The author Leonhard Euler (1707–1783).
Translated from the Latin by Renzo Caddeo.
Geometrica et sphaerica quaedam, *Mémoires de l'Academie impériale des sciences de Saint-Petersbourg*, 5, 1815, pp. 96–114; *Opera Omnia* Series 1, vol. XXVI, pp. 344–358.

R. Caddeo, A. Papadopoulos (eds.), *Spherical Geometry in the Eighteenth Century I*, https://doi.org/10.1007/978-3-032-12466-1_16

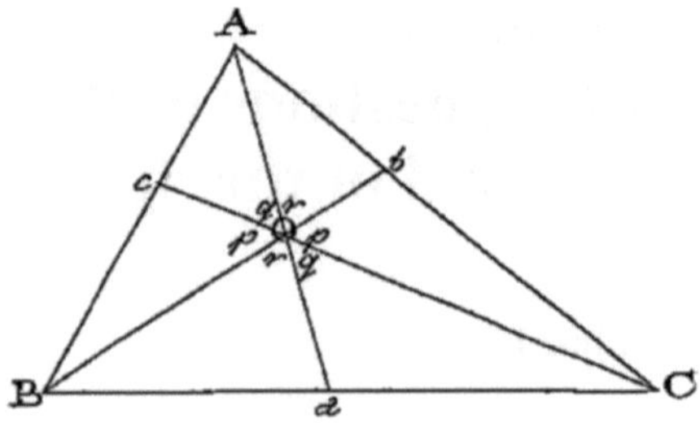

Fig. 16.1 Theorem 1

$$\frac{AO}{Oa}\cdot\frac{BO}{Ob}\cdot\frac{CO}{Oc}=\frac{AO}{Oa}+\frac{BO}{Ob}+\frac{CO}{Oc}+2.$$

Proof

§2. To prove this, we introduce for the described parts the following notation:

$$AO = A;\ BO = B;\ CO = C,$$

$$Oa = a;\ Ob = b;\ Oc = c,$$

and we also denote all six angles formed around the point O as they are shown in Fig. 16.1, where it is immediately evident that $p + q + r = 180^{\circ}$. Now, from the formula by which in a triangle, from two sides and the angle between them we usually determine its area, we will have the area

$$AOc = \frac{1}{2}Ac\sin q,$$

and the area

$$BOc = \frac{1}{2}Bc\sin p,$$

and then therefore we obtain the area

$$AOB = \frac{1}{2}AB\sin(p+q).$$

On the other hand we have

$$\sin(p+q) = \sin r,$$

and from here, this triangle being equal to the sum of the previous two, we then derive the following equation:

$$AB \sin r = Ac \sin q + Bc \sin p.$$

Similarly, the remaining parts give the equations

$$BC \sin p = Ba \sin r + Ca \sin q,$$

$$CA \sin q = Cb \sin p + Ab \sin r.$$

§3. In order to put these equations into a more convenient form for our intended use, we transform them into the following:

$$\frac{\sin r}{c} = \frac{\sin q}{B} + \frac{\sin p}{A},$$

$$\frac{\sin p}{a} = \frac{\sin r}{C} + \frac{\sin q}{B},$$

$$\frac{\sin q}{b} = \frac{\sin p}{A} + \frac{\sin r}{C}.$$

Here it is evident that the three angles p, q, r refer equivalently to the three letters A, B, C, or to a, b, c.

§4. We also set $A = \alpha a$, $B = \beta b$, $C = \gamma c$; then moreover, we introduce the notation

$$\frac{\sin p}{A} = \frac{\sin p}{\alpha a} = P;$$

$$\frac{\sin q}{B} = \frac{\sin q}{\beta b} = Q;$$

$$\frac{\sin r}{C} = \frac{\sin r}{\gamma c} = R.$$

In this way we obtain de facto the following three very simple formulae:

$$\gamma R = P + Q,$$

$$\alpha P = Q + R,$$

$$\beta Q = R + P.$$

But from the differences in these equations[1] we immediately obtain the ratio between each of the pairs of letters P, Q, R; for from them we deduce the following proportions:

$$P : R = (\gamma + 1) : (\alpha + 1),$$

$$Q : P = (\alpha + 1) : (\beta + 1),$$

$$R : Q = (\beta + 1) : (\gamma + 1),$$

from which the following double proportion follows clearly:

$$P : Q : R = \frac{1}{\alpha + 1} : \frac{1}{\beta + 1} : \frac{1}{\gamma + 1}.$$

Naturally, the three letters P, Q, R refer in an orderly way to the letters α, β, γ.
§5. But since from the first of our three equations it follows that $R = \frac{P+Q}{\gamma}$, while from the second it results that $R = \alpha P - Q$, from these two values, equalized with each other, it results the following ratio between P and Q:

$$\frac{P}{Q} = \frac{\gamma + 1}{\alpha\gamma - 1}.$$

Therefore, having

$$\frac{P}{Q} = \frac{\beta + 1}{\alpha + 1},$$

we thus get an equation free from the letters P, Q, R, which will be

$$\alpha\beta\gamma = \alpha + \beta + \gamma + 2,$$

which is manifestly the same property stated in the theorem, since we have

$$\alpha = \frac{AO}{Oa}, \ \beta = \frac{BO}{Ob}, \ \gamma = \frac{CO}{Oc},$$

Hence, having premised this theorem, let us move on to address the issue mentioned at the beginning.

[1] More precisely, subtracting the second from the first, the third from the second and the third from the first.

Problem 1

If the two sides of each of these lines, that is, if

$$AO = A,\ BO = B,\ CO = C,$$

$$Oa = a;\ Ob = b;\ Oc = c,$$

are known, from these six quantities, which are given so that the property proved above applies, study the construction of such a triangle.

Solution

§6. We retain all the notation introduced in the proof of the theorem, viz. $A = \alpha a$, $B = \beta b$, $C = \gamma c$; then,

$$\sin p = \alpha a P;\ \sin q = \beta b Q;\ \sin r = \gamma c R;$$

first of all, among the letters α, β, γ the following relationship must take place:

$$\alpha\beta\gamma = \alpha + \beta + \gamma + 2.$$

§7. Since the letters P, Q, R are in the same ratio to each other as the fractions $\frac{1}{\alpha+1}$, $\frac{1}{\beta+1}$, $\frac{1}{\gamma+1}$, are to each other, we set

$$P = \frac{\Delta}{\alpha + 1};\ Q = \frac{\Delta}{\beta + 1};\ R = \frac{\Delta}{\gamma + 1},$$

and thus in this way the same letters P, Q, R, and consequently also the sines of the angles p, q, r, are completely eliminated from the calculation, and instead, the only new unknown Δ enters and, once its value is found, all the elements that are necessary for the construction of the triangle become known.

§8. But we must try to find this unknown Δ from the condition that the sum $p+q+r$ of the three angles must be equal to two right angles, that is, that we have $\sin r = \sin(p+q)$. But since we have found only the expression of the sines of these angles, it will be convenient to reduce this equation back to the sines only, and in order that this can be done more easily, let us set $\sin p = f$, $\sin q = g$, $\sin r = h$, whereby we must have

$$h = f\sqrt{1 - g^2} + g\sqrt{1 - f^2},$$

where we must eliminate the irrationalities.[2] Taking the squares[3] we then find

$$h^2 = f^2 + g^2 - 2f^2g^2 + 2fg\sqrt{(1 - f^2)(1 - g^2)}.$$

[2] That is, the radicals, the terms under root sign.

[3] That is, of the two members of this equality.

If we now move the rational parts[4] to the left and take the squares again, we arrive at this equation:

$$f^4 + g^4 + h^4 - 2f^2g^2 - 2f^2h^2 - g^2h^2 = 0.$$

§9. Now, since we have $f = \sin p$, taking into account the notation fixed earlier, we will have

$$f = \alpha a P = \frac{\alpha a \Delta}{\alpha + 1};$$

$$g = \frac{\beta b \Delta}{\beta + 1};$$

$$h = \frac{\gamma c \Delta}{\gamma + 1}.$$

For this reason, we set, for [love of] concision,

$$\frac{\alpha a}{\alpha + 1} = F;$$

$$\frac{\beta b}{\beta + 1} = G;$$

$$\frac{\gamma c}{\gamma + 1} = H,$$

in such a way that we have

$$f = F\Delta; \ g = G\Delta; \ h = H\Delta,$$

and if we substitute these values into the equation just found, we arrive at an equation divisible by Δ^4, that is

$$F^4 + G^4 + H^4 - 2F^2G^2 - 2F^2H^2 - 2G^2H^2 + 4F^2G^2H^2\Delta^2 = 0,$$ [5]

and from which we deduce

$$\Delta^2 = \frac{2F^2G^2 + 2F^2H^2 + 2G^2H^2 - F^4 - G^4 - H^4}{4F^2G^2H^2},$$

[4] That is, without the radicals.

[5] In the original, the part " $= 0$" is missing

so that our unknown Δ is now perfectly determined, being

$$\Delta = \frac{\sqrt{(2F^2G^2 + 2F^2H^2 + 2G^2H^2 - F^4 - G^4 - H^4)}}{2FGH},$$

where the second member can also be expressed as a product of factors, that is,

$$\Delta = \frac{\sqrt{(F+G+H)(F+G-H)(F+H-G)(G+H-F)}}{2FGH}.$$

§10. Since, therefore, the letters F, G, H are obtained immediately from the known quantities A, B, C and a, b, c, the expression found, which may seem somewhat tortuous, nevertheless can be constructed very easily by means of the area of the triangle. Indeed, let us construct the triangle whose sides are F, G, H, and look for its area, which we shall call M^2, and which, as it was noted, is

$$M^2 = \frac{1}{4}\sqrt{(F+G+H)(F+G-H)(F+H-G)(G+H-F)};$$

Introducing this notation we will have $\Delta = \frac{2M^2}{FGH}$, and, knowing this value, the whole problem is completely solved. In fact, from here we immediately find the sines of the angles p, q, r, that is,

$$\sin p = \frac{\alpha a}{\alpha+1}\Delta;\ \sin q = \frac{\beta b}{\beta+1}\Delta;\ \sin r = \frac{\gamma c}{\gamma+1}\Delta.$$

Thus, having found only one of these angles, one can immediately construct the triangle, a matter which is easily understood without further explanation.

§11. But, having prepared these things, I have now finally realized that the theorem given above can be stated much more conveniently and much more elegantly in the following way:

Theorem 2

Having conducted in an arbitrary triangle ABC from the angles A, B, C to the opposite sides the straight lines Aa, Bb, Cc which intersect at a common point O, we will always have the property

$$\frac{Oa}{Aa} + \frac{Ob}{Bb} + \frac{Oc}{Cc} = 1;$$

i.e., if the parts Oa, Ob, Oc of those lines are each divided by the entire segment of its line, the sum of the three fractions thus obtained will always be equal to unity.

Proof, Derived from the Preceding One

§12. If, as we have done before, we set $AO = \alpha \cdot Oa$; $BO = \beta \cdot Ob$; $CO = \gamma \cdot Oc$, then, as we have previously proved, we always have

$$\alpha\beta\gamma = \alpha + \beta + \gamma + 2.$$

Now we add to both members the expression

$$\alpha\beta + \alpha\gamma + \beta\gamma + \alpha + \beta + \gamma + 1.$$

Then, from the left hand term we obtain $(\alpha + 1)(\beta + 1)(\gamma + 1)$, whereas from the right hand term we have $\alpha\beta + \alpha\gamma + \beta\gamma + \alpha + \beta + \gamma + 2(\alpha + \beta + \gamma) + 3$, and the latter expression can evidently be rewritten in the form $(\alpha + 1)(\beta + 1) + (\alpha + 1)(\gamma + 1) + (\beta + 1)(\gamma + 1)$. If we now substitute these forms into the equality[6] and divide both members of the equality by the product $(\alpha + 1)(\beta + 1)(\gamma + 1)$, we find

$$1 = \frac{1}{\gamma + 1} + \frac{1}{\beta + 1} + \frac{1}{\alpha + 1}.$$

Q. E. D.

§13. From here we can also deduce the following remarkable identity:

$$\frac{\alpha}{\alpha + 1} + \frac{\beta}{\beta + 1} + \frac{\gamma}{\gamma + 1} = 2.$$

If in fact the latter identity is added to the previous one, we find the equality $1 + 1 + 1 = 3$.

A Very Simple Proof Based on Common Notions

§14. Through the point O (Fig. 16.2) we draw the parallel to each side of the triangle, and in particular the segments $f\zeta$ parallel to BC, $g\eta$ to AC and $h\theta$ to AB, and it is immediately evident that we have

$$\frac{Bf}{AB} + \frac{A\eta}{AB} + \frac{f\eta}{AB} = 1.$$

Now, on the other hand, from the similarity of the triangles ABa and AfO we get

$$Bf : BA = Oa : Aa$$

and that is how the first fraction $\frac{Oa}{Aa} = \frac{Bf}{BA}$ comes out. Then, since ΔBAb and $\Delta B\eta O$ [are also] similar, we get

$$A\eta : AB = Ob : Bb,$$

[6] This is the equality at the beginning of the present paragraph.

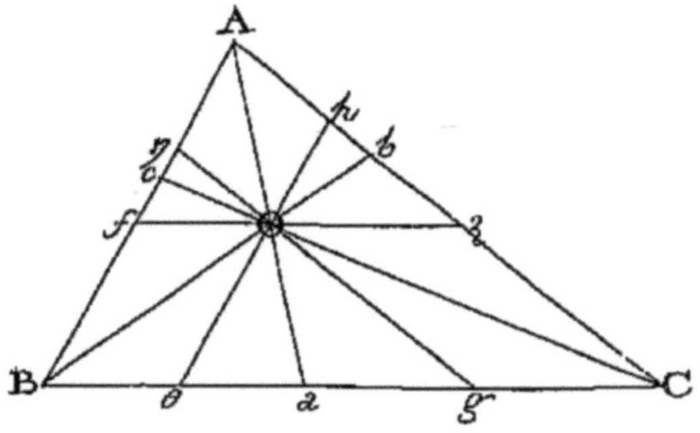

Fig. 16.2 Theorem 2. Case when the point O is inside the triangle

and thus obtain $\frac{A\eta}{AB} = \frac{Ob}{Bb}$. Finally, taking into account the similarity between the triangles $\Delta fO\eta$ and ΔBCA, we have $f\eta : BA = fO : BC$, and thus

$$\frac{f\eta}{AB} = \frac{fO}{BC}.$$

On the other hand, we have $fO = BA$, and therefore, since the triangle ΔBCc is similar to the triangle $\Delta\theta CO$, we will have

$$B\theta : Bc = Oc : Cc,$$

thus

$$\frac{f\eta}{BA} = \frac{Oc}{Cc};$$

if these values are now substituted for it, the identity

$$\frac{Bf + A\eta + f\eta}{AB} = 1$$

takes the form

$$\frac{Oa}{Aa} + \frac{Ob}{Bb} + \frac{Oc}{Cc} = 1,$$

which is precisely the equality that was to be proved.
§15. (Figure 16.3). This remarkable property will always take place wherever the point O is also taken outside the triangle, as in Fig. 16.3, provided that the notation adopted through the letters Aa, Bb, Cc are adopted in the prescribed way. Therefore, in this figure, for the line Aa we will have $AO = A$, $Oa = a$, while in place of the line Bb, we placed $BO = B$, we will have $Ob = -b$, and in place of the line Cc we will have to place $CO = C$ and $OC = -c$.

Fig. 16.3 Theorem 2. Case when the point O is outside the triangle

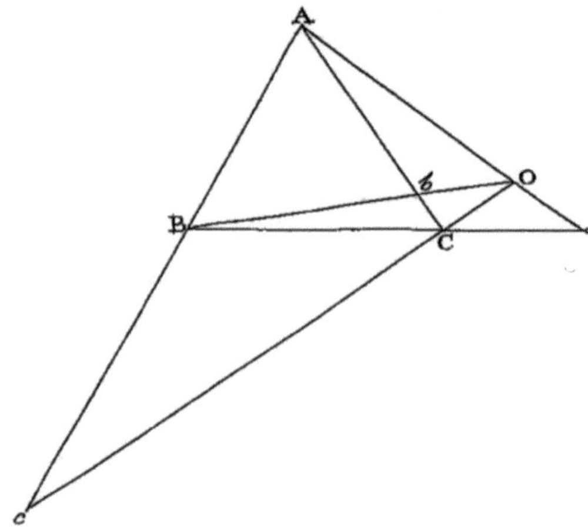

Consequently, we will then have $Aa = A + a$, $Bb = B + b$, $Cc = -(C + c)$. Then, since we always have

$$\frac{a}{A+a} + \frac{b}{B+b} + \frac{c}{C+c} = 1,$$

we will have, for the lines drawn in the figure,

$$\frac{Oa}{Aa} - \frac{Ob}{Bb} + \frac{Oc}{Cc} = 1.$$

§16. On the other hand, once this property is established, it is possible to find the area of the entire triangle ABC quite easily. In fact, since the area of the triangle AOB is equal to $\frac{1}{2}AB \sin r$, as $\sin r = CR$, this area [of the triangle AOB] will be equal to $\frac{1}{2}ABC \cdot R$. Similarly, we find that the area of the triangle AOC is equal to $\frac{1}{2}ABC \cdot Q$ and that the area of the triangle BOC is equal to $\frac{1}{2}ABC \cdot P$, and so the area of the entire triangle will be equal to $\frac{1}{2}ABC(P + Q + R)$.
§17. Then in fact also have set

$$P = \frac{F\Delta}{A}, \; Q = \frac{G\Delta}{B}, \; R = \frac{H\Delta}{C}.$$

But we also had

$$F = \frac{A}{\alpha+1}, \; G = \frac{B}{\beta+1}, \; H = \frac{C}{\gamma+1}$$

(see §9); therefore the area of the triangle will be equal to

$$\frac{1}{2}ABC\Delta(\frac{1}{\alpha+1} + \frac{1}{\beta+1} + \frac{1}{\gamma+1}).$$

However, we have shown that

$$\frac{1}{\alpha+1}+\frac{1}{\beta+1}+\frac{1}{\gamma+1}=1,$$

so the area of our triangle will be equal to $\frac{1}{2}ABC\Delta$. In addition to this we actually considered the triangle formed by the three sides F, G, H, and we have denoted by M^2 its area, which, once determined, allows us to find the value

$$\Delta = 2\frac{M^2}{FGH}.$$

Having substituted this value, the area of our triangle takes on the expression

$$\frac{ABCM^2}{FGH}.$$

Thus, since

$$F=\frac{A}{\alpha+1},\quad G=\frac{B}{\beta+1},\quad H=\frac{C}{\gamma+1},$$

the area will be

$$(\alpha+1)(\beta+1)(\gamma+1)M^2,$$

and in this way the area of the triangle ABC initially considered is in a very simple proportion to the area M^2 of the triangle called for help, i.e. as that between

$$(\alpha+1)(\beta+1)(\gamma+1)$$

and 1, or by means of line segments, the proportion between the areas of the two triangles will be

$$\frac{ABC}{M^2}=\frac{Aa\cdot Bb\cdot Cc}{Oa\cdot Ob\cdot Oc}.$$

Spherical Geometry

§18. These results we have found so far about planar triangles can be adapted to spherical triangles as well. More precisely, let us consider the spherical triangle ABC (Fig. 16.4) in which from the angles are led to the opposite sides arcs intersecting each other at the same point O; first we must certainly seek what relation must exist between the parts of these arcs in order that, from the knowledge

Fig. 16.4 Theorem 3

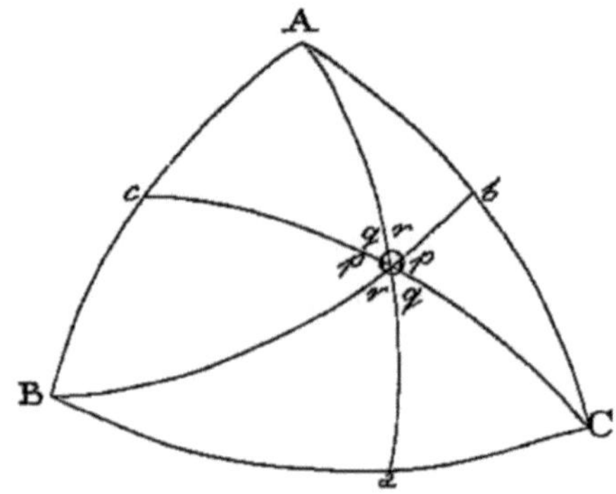

of these parts, we may construct the triangle. For this purpose we must introduce the following theorem:

Theorem 3

If in an arbitrary spherical triangle ABC we draw, from the three angles to the opposite sides, the arcs Aa, Bb, Cc that intersect each other at a common point O, then, setting, for the sake of brevity,

$$\frac{\tan AO}{\tan Oa} = \alpha, \quad \frac{\tan BO}{\tan Ob} = \beta, \quad \frac{\tan CO}{\tan Oc} = \gamma,$$

we always get

$$\alpha\beta\gamma = \alpha + \beta + \gamma + 2,$$

and this property can also be reduced to the form

$$\frac{1}{\alpha + 1} + \frac{1}{\beta + 1} + \frac{1}{\gamma + 1} = 1.$$

Proof

§19. Let us call the angles around the intersection point O by the names by which they are marked in the figure, and for the arcs let us set $AO = A$, $BO = B$, $CO = C$ and $Oa = a$, $Ob = b$, $Oc = c$. In the triangle AOc we will have

$$\tan AcO = \frac{\sin A \sin q}{\cos A \sin c - \sin A \cos c \cos q}.$$

In the triangle BOc we will have

$$\tan BcO = \frac{\sin B \sin p}{\cos B \sin c - \sin B \cos c \cos p}.$$

§20. Now, since these two angles taken together give two right angles, the sum of their tangents must be equal to zero, hence the following equation arises

$$\sin A \cos B \cos c \sin q - \sin A \sin B \cos c \cos p \sin q$$
$$+ \sin B \cos A \sin c \sin p - \sin A \sin B \cos c \cos q \sin p = 0,$$

which can be reduced to the simpler one

$$\sin A \cos B \sin c \sin q + \sin B \cos A \sin c \sin p = \sin A \sin B \sin c \sin r,$$

from which we get

$$\sin r = \frac{\sin A \cos B \sin c \sin q + \sin B \cos A \sin c \sin p}{\sin A \sin B \cos c},$$

and this expression then produces this equation:

$$\frac{\sin r}{\tan c} = \frac{\sin p}{\tan A} + \frac{\sin q}{\tan B}.$$

And in a similar way the other two equations are found:

$$\frac{\sin p}{\tan a} = \frac{\sin q}{\tan B} + \frac{\sin r}{\tan C},$$

$$\frac{\sin q}{\tan b} = \frac{\sin r}{\tan C} + \frac{\sin p}{\tan A}.$$

§21. But since in the theorem we have set $\frac{\tan A}{\tan a} = \alpha$, $\frac{\tan B}{\tan b} = \beta$, $\frac{\tan C}{\tan c} = \gamma$, substituting these values, these three equations take the forms

$$\frac{\sin r}{\tan c} = \frac{\sin p}{\alpha \tan a} + \frac{\sin q}{\beta \tan b},$$

$$\frac{\sin p}{\tan a} = \frac{\sin q}{\beta \tan b} + \frac{\sin r}{\gamma \tan c},$$

$$\frac{\sin q}{\tan b} = \frac{\sin r}{\gamma \tan c} + \frac{\sin p}{\alpha \tan a}.$$

§22. Now let us further set $\frac{\sin p}{\alpha \tan a} = P$, $\frac{\sin q}{\beta \tan b} = Q$, $\frac{\sin r}{\gamma \tan c} = R$, consequently our three equations will be

$$\gamma R = P + Q, \alpha P = Q + R, \beta Q = R + P,$$

from the first of which we get

$$R = \frac{P + Q}{\gamma},$$

and from the second

$$R = \alpha P - Q;$$

and from these equalized values we get $\frac{P}{Q} = \frac{\gamma+1}{\alpha\gamma-1}$. Then indeed, if from the second equation we subtract the third, we find $\alpha P - \beta Q = Q - P$, from which we get $\frac{P}{Q} = \frac{\beta+1}{\alpha+1}$, so we have the equation

$$\frac{\gamma + 1}{\alpha - 1} = \frac{\beta + 1}{\alpha + 1},$$

which developed and rearranged becomes

$$\alpha\beta\gamma = \alpha + \beta + \gamma + 2.$$

Another Proof, Derived from the First Elements of Geometry

§23. Suppose we take the plane tangent to the sphere at the point O (a plane that we do not represent with a figure since [everything we will say] can be easily understood), and toward which from the center of the sphere to the points A, B, C and a, b, c we lead out the straight lines intersecting the plane at points A', B', C' and a', b', c'. If in the [tangent] plane we lead from O to these points the straight lines OA', OB', OC' and Oa', Ob', Oc', it is evident that we have $OA' = \tan A$ and, similarly, $OB' = \tan B$, $OC' = \tan C$; moreover, we also have $Oa' = \tan a$, $Ob' = \tan b$, $Oc' = \tan c$, and in fact the angles denoted by the letters p, q, r will remain the same also in the plane.

§24. Now, in this way we have obtained the planar triangle $A'B'C'$ from the angles of which we lead to the opposite sides the straight lines $A'a'$, $B'b'$, $C'c'$, which intersect each other at the point O, and thus the whole problem is brought back to the case of a planar triangle, so if we set

$$\frac{OA'}{Oa'} = \frac{\tan A}{\tan a} = \alpha;$$

$$\frac{OB'}{Ob'} = \frac{\tan B}{\tan b} = \beta;$$

$$\frac{OC'}{Oc'} = \frac{\tan C}{tanc} = \gamma,$$

it will certainly occur, as has been previously proved, both

$$\alpha\beta\gamma = \alpha + \beta + \gamma + 2,$$

and

$$\frac{1}{\alpha + 1} + \frac{1}{\beta + 1} + \frac{1}{\gamma + 1} = 1.$$

Problem 2

If in an arbitrary spherical triangle ABC from the angles we lead to the opposite sides the arcs Aa, Bb, Cc which intersect each other at a common point O, and if the two sides of each of these arcs are known, that is, if we know

$$AO = A, BO = B, CO = C, Oa = a, Ob = b, Oc = c,$$

and if these six quantities are assumed to be such that the property proved above takes place, investigate this triangle.

Solution

§25. We set, as it has already been done in the theorem, $\frac{\sin p}{\alpha \tan a} = P$, $\frac{\sin q}{\beta \tan b} = Q$, $\frac{\sin r}{\gamma \tan c} = R$, and we obtain three equations:

$$\gamma R = P + Q, \alpha P = Q + R, \beta Q = R + P,$$

from whose differences we immediately derive the following formulae:

$$\gamma R - \alpha P = P - R,$$

from which it follows

$$\frac{P}{R} = \frac{\gamma + 1}{\alpha + 1}.$$

Then,

$$\alpha P - \beta Q = Q - P,$$

from which it follows

$$\frac{Q}{P} = \frac{\alpha + 1}{\beta + 1}.$$

Finally,

$$\gamma R - \beta Q = Q - R,$$

from which it follows

$$\frac{R}{Q} = \frac{\beta + 1}{\gamma + 1},$$

from where it is evident that the letters P, Q, R are in the same ratio with each other as are the fractions $\frac{1}{\alpha+1}$, $\frac{1}{\beta+1}$, $\frac{1}{\gamma+1}$, and for this reason we set $P = \frac{\Delta}{\alpha+1}$, $Q = \frac{\Delta}{\beta+1}$, $R = \frac{\Delta}{\gamma+1}$.

§26. But now from the first equation we derive $R = \frac{P+Q}{\gamma}$, from the second equation [we obtain] $R = \alpha P - Q$, and from their equality we get $\frac{P}{Q} = \frac{\gamma+1}{\alpha\gamma-1}$. On the other hand, a little while ago we found that

$$\frac{P}{Q} = \frac{\beta + 1}{\alpha + 1},$$

and these two values, equalized with each other, give rise to the same condition already proved in the theorem.

§27. Now, having determined the values of the letters P, Q, R, from here we extract for the sines of the angles p, q, r the following values:

$$\sin p = \tan A \cdot \frac{\Delta}{\alpha + 1};$$

$$\sin q = \tan B \cdot \frac{\Delta}{\beta + 1};$$

$$\sin r = \tan C \cdot \frac{\Delta}{\gamma + 1}.$$

If then for brevity, we also set

$$\frac{\tan A}{\alpha + 1} = F, \quad \frac{\tan B}{\beta + 1} = G, \quad \frac{\tan C}{\gamma + 1} = H,$$

the condition that the sum of the angles must be $p + q + r = 180^{\circ}$, gives us for Δ directly, as in the previous problem, the following value :

$$\Delta = \frac{\sqrt{(F+G+H)(F+G-H)(F+H-G)(G+H-F)}}{2FGH}.$$

Therefore, if we consider again another triangle whose sides are F, G, H and whose area we denote by M^2, we will have, as before, directly $\Delta = \frac{2M^2}{FGH}$, and so the [quantity denoted by the] letter Δ is determined by the known quantities only.

§28. Thus, since the value of this Δ is known, we will have

$$\sin p = \frac{2M^2}{GH},$$

$$\sin q = \frac{2M^2}{FH},$$

$$\sin r = \frac{2M^2}{FG};$$

therefore it is possible to determine each of the angles p, q, r; however it is sufficient to know only one of them to construct the triangle. But here it will be useful to observe that the formula we have found gives two values for the angle p, of which one is complementary of the other to two right angles, so that there must be two solutions; and since the cosines of the obtuse angles are negative, this negativity must always be taken into account, so that we have

$$\cos r = -\cos(p+q).$$

So in this way this problem of spherical geometry is also completely solved.

Supplement *(Containing a very simple analysis of both the proof of the theorem [Theorem 3] and the solution of the previously proposed problem [Problem 2]).*

§29. Let (Fig. 16.5) ABC be any triangle. From its angles we lead in an arbitrary way to the opposite sides the lines Aa, Bb, Cc, which intersect each other at the point O, and introduce as before the notation

$$AO = A,\ Oa = a,\ BO = B,\ Ob = b,\ CO = C,\ Oc = c;$$

then from the point O we lead the lines $O\beta$ and $O\gamma$ parallel to the sides AB and AC [respectively], and it is necessary to solve the problem completely only by making use of these lines.

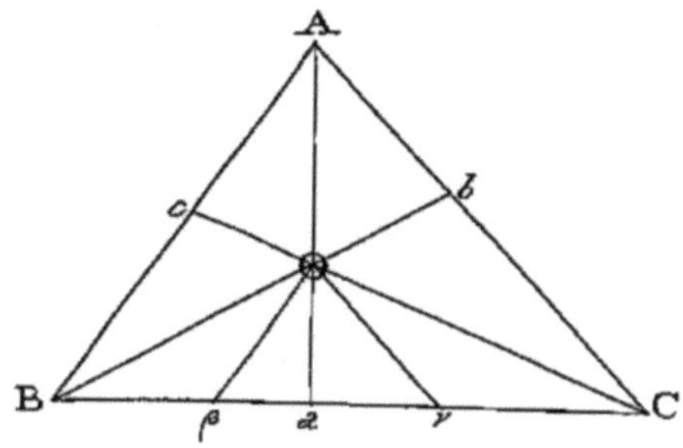

Fig. 16.5 Theorem 3 and Problem 2

§30. Having established these things, the similarity of the triangles BCb and $B\gamma O$ will give us

$$\frac{C\gamma}{BC} = \frac{Ob}{Bb}.$$

Then from the similarity of the triangles CBc and $C\beta O$ it follows that $\frac{B\beta}{BC} = \frac{Oc}{Cc}$. Finally, the triangle $\beta O\gamma$ is similar to the triangle BAC, and in both the one and the other the lines Oa and Aa are drawn similarly, therefore $\frac{\beta\gamma}{BC} = \frac{Oa}{Aa}$, and so we have the following three equations:[7]

$$\frac{Oa}{Aa} = \frac{a}{A+a} = \frac{\beta\gamma}{BC},$$

$$\frac{Ob}{Bb} = \frac{b}{B+b} = \frac{C\gamma}{BC},$$

$$\frac{Oc}{Cc} = \frac{c}{C+c} = \frac{B\beta}{BC}.$$

Thus, the sum of these three fractions will be

$$\frac{\beta\gamma + C\gamma + B\beta}{BC} = 1,$$

and undoubtedly this is a very short proof of the theorem, compared with the previous one, which was derived by following a rather circuitous route.

§31. We set now, for brevity,

$$\frac{Oa}{Aa} = \frac{a}{A+a} = \alpha,$$

$$\frac{Ob}{Bb} = \frac{b}{B+b} = \beta,$$

$$\frac{Oc}{Cc} = \frac{c}{C+a} = \gamma$$

where these letters α, β, γ should not be confused with those used above. From here we will then have

$$a = \frac{\alpha A}{1-\alpha},$$

[7] In the original, the last fraction is (mistakingly) $\frac{B\beta}{AC}$.

or

$$A = \frac{a(1-\alpha)}{\alpha}.$$

In the same way,

$$b = \frac{\beta B}{1-\beta},$$

or

$$B = \frac{b(1-\beta)}{\beta}.$$

$$c = \frac{\gamma C}{1-\gamma},$$

or

$$C = \frac{c(1-\gamma)}{\gamma}.$$

And now the letters α, β, γ are related with each other in such a way that we have $\alpha + \beta + \gamma = 1$.

§32. We now denote by x the entire base BC[8] of the triangle; its three parts $B\beta$, $\beta\gamma$, γC will be determined by the relations $\beta\gamma = \alpha x$, $B\beta = \gamma x$, $C\gamma = \beta x$. Therefore, since the triangle $Oa\beta$ is similar to triangle AaB, we will have

$$\frac{Ba}{B\beta} = \frac{Aa}{AO},$$

and from here we deduce the line segment[9]

$$Ba = \frac{B\beta \cdot Aa}{AO} = \frac{\gamma x(A+a)}{A} = \frac{\gamma x}{1-\alpha}.$$

In a similar way we find

$$Ca = \frac{\beta x}{1-\alpha}.$$

[8] In the original version, this is written bC.

[9] The third term corrects the term $\frac{\gamma x}{A(A+a)}$ which appears in the original.

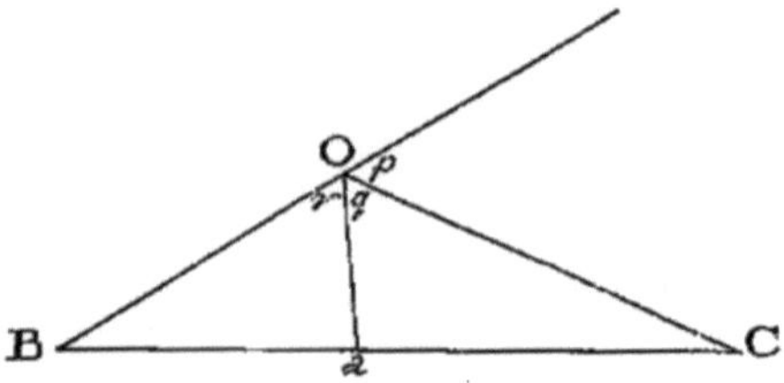

Fig. 16.6 Problem 3

§33. Having found these values, let us consider (Fig. 16.6) only the triangle BOC in which the line Oa is drawn, and let set, as we have already done, $OB = B$, $OC = C$, $Oa = a$; from this line $[Oa]$, the base BC is intersected in a so that it results

$$Ba : Ca = \gamma : \beta,$$

and in this way we are brought back to the following problem:

Problem 3

If we are given the sides OB and OC of the triangle and the line Oa intersecting the base BC in the given ratio, study in what way we can construct the triangle itself.

§34. For this purpose, we denote the angles at the point O as they are marked in the figure, and whose sum $p + q + r$ must equal two right angles. Now, to study the ratio of the sines of these angles, let us consider in the triangle OBa the following proportions

$$\sin B : \sin r = Oa : Ba = a : \frac{\gamma x}{1-\alpha}.$$

But from the whole triangle BOC it follows that

$$\sin p : \sin B = BC : OC,$$

and multiplying these proportions together gives

$$\sin p : \sin r = a : \frac{C\gamma}{1-\alpha},$$

so that we have

$$\sin r = \frac{C\gamma \sin p}{a(1-\alpha)},$$

and in a similar way we find

$$\sin q = \frac{B\beta \sin p}{a(1-\alpha)}.$$

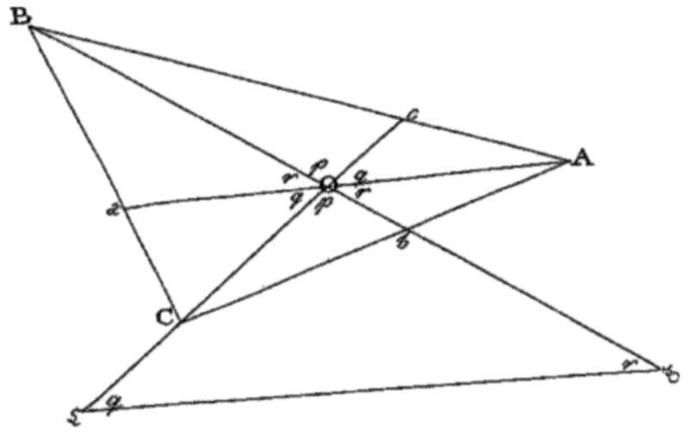

Fig. 16.7 Construction of the triangle OCB[10]

So is clear the ratio between the sines of the angles q and r and the sine of the angle p, since we have assumed that the quantities A, B, C and a, b, c, and therefore also α, β, γ, are known.

§35. Let us now consider more closely these three angles p, q, r; since their sum is equal to two right angles, there exists some triangle whose angles are equal to them. Therefore, since their three sines, i.e., $\sin p$, $\sin q$, $\sin r$, have the same ratio to each other as do the following known three quantities: $a(1-\alpha)$, $B\beta$, $C\gamma$, if we construct a triangle with these three sides, the angles opposite to them will be the same angles p, q, r that are sought.

§36. So our task will be completely accomplished if we construct (Fig. 16.7) the triangle $O\mathfrak{C}\mathfrak{B}$ having as base $\mathfrak{C}\mathfrak{B} = a(1-\alpha) = A\alpha$, and sides $\mathfrak{B}O = B\beta$ and $\mathfrak{C}O = C\gamma$, for the angles of which we will actually have the following: the one in O will be equal to p, the one in $\mathfrak{B}$ equal to r, and the one in $\mathfrak{C}$ equal to q. If we produce the side $\mathfrak{B}O$ to B so that $OB = B$, if in the other side $O\mathfrak{C}$ we cut out the portion $OC = C$, these two lines constitute the angle $q + r$ between them, as postulated by the previous calculation; but then, if from O we lead the line Oa parallel to $\mathfrak{B}\mathfrak{C}$, we will simultaneously have the point a on the line BC, so it will be the same line $Oa = a$, and in this way our problem is completely solved. For if in fact we extend the line BO on the other side as far as b, the line CO as far as c, and the line aO as far as A, the whole figure proposed initially will be complete, and in this way from the quantities A, B, C together with a, b, c, given and such that the condition found is satisfied, by this very simple operation the whole triangle is constructed.

§37. Consequently, the case we have treated here must be considered worthy of more attention, since initially it seemed to require very abstruse and troublesome calculations but, nevertheless, having overcome all the difficulties, we arrived at a solution that is both very simple and very elegant.

[10] The points $\mathfrak{C}$ and $\mathfrak{B}$ (the endpoints of the lowest side in this figure) do not appear clearly in the original Euler figure.

Part III
Translations of Lagrange's Memoirs

Joseph-Louis de Lagrange
Engraving by Robert Hart, nineteenth c., after an nineteenth century portrait

Chapter 17
Algebraic Solution of a Problem in Geometry

Joseph-Louis Lagrange

Abstract In this memoir, Lagrange gives a solution to a problem of plane geometry that Euler calls the "Pappus problem" in his memoir *Problematis cuiusdam Pappi Alexandrini constructio*. The problem is to construct a triangle which is inscribed in a given circle and whose three sides are contained in three lines that are required to pass through three given points.

Keywords Pappus problem · History of geometry · Joseph-Louis Lagrange · Mathematics in the eighteenth century

AMS Classification 01A50, 51-03, 53A05, 53A35

Problem

Given in magnitude and position the circle $RMNP$, to inscribe in this circle a triangle MNP whose three sides NM, PM, PN, produced if necessary, pass by the three given points A, B, C.[1]

---

[1] The problem considered in this memoir is not a problem of spherical geometry. Lagrange deals here with a problem of plane geometry that Euler calls the "Pappus problem" and which he studies in the memoir titled *Problematis cuiusdam Pappi Alexandrini constructio* (A construction relative to a problem of Pappus). Indeed, this is a problem that Pappus solves in the *Collection*. In the same memoir, Euler solves the analogue of Pappus' problem in the spherical case, and it is for this

---

The author Joseph-Louis Lagrange (1736–1813).
Translated from the French by Athanase Papadopoulos.
Solution algébrique d'un problème de géométrie, *Nouveaux mémoires de l'Académie royale des sciences et belles-lettres de Berlin* année 1776, pp. 335–339. [Note by the Journal editor] This solution was published by M. de Castillon at the beginning of a memoir titled *Sur une nouvelle propriété des sections coniques* and which is part of the volume of the Berlin Academy for the year 1776. The author of this memoir expresses himself in the following terms: "The day after I read at the Academy my solution of the problem concerning the circle and the triangle to be inscribed in this circle, in such a way that each side passes by three given points, M. de la Grange sent me the following algebraic solution."

R. Caddeo, A. Papadopoulos (eds.), *Spherical Geometry in the Eighteenth Century I*, https://doi.org/10.1007/978-3-032-12466-1_17

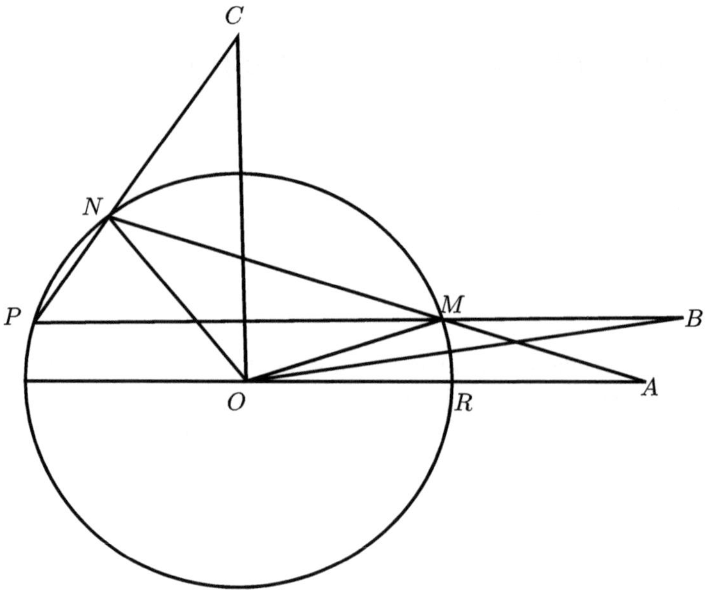

Fig. 17.1 Figure for Problem

Algebraic Solution
I draw from the three given points to the center O of the circle the lines AO, BO, CO; these lines are given in magnitude and position, because they determine the position of three given points A, B, C.

Calling then AO, a; BO, b; CO, c, the angle AOB, m, the angle AOC, n, the five quantities a, b, c, m, n are given and known.

Now I draw from the three points M, N, P of the circumference of the circle, where the angles of the desired triangle MNP lie, the rays OM, ON, OP: it is clear that these three lines are given in magnitude, because the circle is assumed to be given in magnitude; but their position is unknown, and this is what has to be found (Fig. 17.1).

Calling then x the angle AOM, y the angle AON and z the angle AOP, the question will be reduced to finding the values of the three unknowns x, y, z. Furthermore, I call r the radius of the circle.

This being set, I consider first the isosceles triangle NOM, in which we have:

$$\text{the central angle } NOM = y - x,$$

reason that Euler's paper is included in this volume. We have decided to include here Lagrange's solution of the same problem in the Euclidean case, so that the reader can compare Euler's and Lagrange's solutions in this setting. Pappus' solution of the same problem is surveyed in the essay by Guillaume Théret contained in this volume (Chap. 4).

therefore,

$$\text{the angle } ONM = \frac{180^{\circ} - y + x}{2} = 90^{\circ} - \frac{y - x}{2}.$$

Then I consider the angle AON, in which we have:

$$\text{the central angle } AON = y - x$$

and

$$\text{the angle } ONA = 90^{\circ} - \frac{y - x}{2},$$

therefore

$$\text{the angle } OAN = 90^{\circ} - \frac{y + x}{2}.$$

Therefore, by the proportionality of the sides to the sines of the opposite sides, we get, in the same triangle,

$$AO : NO = \sin \widehat{ONA} : \sin \widehat{OAN},$$

that is,

$$a : r = \sin\left(90^{\circ} - \frac{y - x}{2}\right) : \sin\left(90^{\circ} - \frac{y + x}{2}\right),$$

or else,

$$a : r = \cos \frac{y - x}{2} : \cos \frac{y + x}{2},$$

whence we deduce the equation

$$a \cos \frac{y + x}{2} = r \cos \frac{y - x}{2},$$

which, by the well-known theorems, reduces to the following:

$$a\left(\cos \frac{y}{2} \cos \frac{x}{2} - \sin \frac{y}{2} \sin \frac{x}{2}\right) = r\left(\cos \frac{y}{2} \cos \frac{x}{2} + \sin \frac{y}{2} \sin \frac{x}{2}\right),$$

or else

$$(a - r) \cos \frac{y}{2} \cos \frac{x}{2} = (a + r) \sin \frac{y}{2} \sin \frac{x}{2},$$

and, dividing by $(a+r)\cos\frac{y}{2}\cos\frac{x}{2}$,

$$\tan\frac{x}{2}\tan\frac{y}{2}=\frac{a-r}{a+r}. \tag{17.4}$$

We shall find another similar equation by considering first the isosceles triangle POM, and then the whole triangle POB. And without doing any new computation, it will suffice to substitute the line OB instead of the line OA, and the radius OP instead of the radius ON. Thus, instead of a we get b, instead of the angle $\widehat{MOA}$ (x) we get the angle $\widehat{MOB}$ $(x-m)$, and instead of the angle $\widehat{NOA}$ (y) we get the angle POB $(z-m)$.

Therefore the new equation will be

$$\tan\frac{x-m}{2}\tan\frac{z-m}{2}=\frac{b-r}{b+r}. \tag{17.5}$$

Finally, we shall find a third similar equation by the consideration of the isosceles triangle PON and the triangle POC; and for this, it will suffice to insert in the first Eq. (17.4), instead of the line AC (a) the line OC (c), instead of the angle MOA (x) the angle NOC $(y-n)$, and instead of the angle NOA (y) the angle POC $(z-n)$, in such a way that we get

$$\tan\frac{y-n}{2}\tan\frac{z-n}{2}=\frac{c-r}{c+r}, \tag{17.6}$$

and these three equation will be used to determine of three unknown angles x, y, z.

Let us set, for further simplicity,

$$\tan\frac{x}{2}=s,\ \tan\frac{y}{2}=t,\ \tan\frac{z}{2}=u,\ \tan\frac{m}{2}=p,\ \tan\frac{n}{2}=q,$$

$$\frac{a-r}{a+r}=A,\ \frac{b-r}{b+r}=B,\ \frac{c-r}{c+r}=C.$$

The three equations we just found, (17.4), (17.5), (17.6), will become, by the well-known property of tangents,

$$st=A,\ \frac{s-p}{1+ps}\cdot\frac{u-p}{1+pu}=B,\ \frac{t-q}{1+qt}\cdot\frac{u-q}{1+qu}=C.$$

The first gives

$$t=\frac{A}{s};$$

the second gives

$$u = \frac{B - p^2 + (1 + B)ps}{-(1 + B)p + (1 - Bp^2)s};$$

and after these values are substituted in the third, we get

$$\frac{A - qs}{Aq + s} \cdot \frac{B - p^2 + (1 + B + pq + \big((1 + B)p - (1 - Bp^2)q\big)s}{-(1 + B)p + (B - p^2)q + \big(1 - Bp^2 + (1 + B)pq\big)s} = C,$$

an equation which, being ordered in terms of the unknown s, will rise to the second degree, and will therefore be soluble by ruler and compass.

For conciseness again, let

$$B - p^2 + (1 + B)pq = F,$$

$$(1 + B)p - (1 - Bp^2)q = G,$$

$$-(1 + B)p + (B - p^2)q = H,$$

$$1 - Bp^2 + (1 + B)pq = K;$$

we get the equation

$$\frac{A - qs}{Aq + s} \cdot \frac{F + Gs}{H + qs} = C,$$

which reduces to

$$(CK - Gq)s^2 + \big(CH - AG + (CK - F)Aq\big)s = A(F - CHq),$$

whence it is easy to extract s. Then we get t and u by the above formulae. From here, we shall then know the tangents of the angles $\frac{AOM}{2}$, $\frac{AON}{2}$, $\frac{AOP}{2}$; consequently, the points M, N, P will be determined with respect to the line OA.

Chapter 18
Solution of Some Problems Relative to Spherical Triangles with a Complete Analysis of These Triangles

Joseph-Louis Lagrange

Abstract In this memoir, Lagrange establishes a complete system of formulae of spherical trigonometry. Using these formulae, he establishes a formula for the area of a triangle in terms of its sides lengths, that is, a spherical Heron formula. At the end of his memoir, Lagrange makes a comparison between the Euclidean and the spherical formulae and he establishes a relation between them.

Keywords Spherical trigonometry · Joseph-Louis Lagrange · History of spherical geometry · Mathematics in the eighteenth century

AMS Classification 01A50, 51-03, 53A05, 53A35

It is known that in the rectilinear triangles the sides are proportional to the sines of the opposite angles, and we can easily prove that this constant ratio of the sides to the sines is equal to the diameter of the circumscribed circle.

We can also express this same ratio by means of the area of the triangle, and it is easy to prove that it is equal to the product of the three sides divided by the double of the area.

1. But if we wanted to express this ratio only by the sides of the triangle, it would be sufficient to consider that, by calling a, b, c the three sides and A, B, C the angles that are opposite to them, we have, by the well-known theorem,

$$a^2 = b^2 + c^2 - 2bc \cos A,$$

therefore

$$\cos A = \frac{b^2 + c^2 - a^2}{2bc}$$

The author Joseph-Louis Lagrange (1736–1813).

Translated from the French by Vincent Alberge and Athanase Papadopoulos

Solution de quelques problèmes relatifs aux triangles sphériques avec une analyse complète de ces triangles, *Journal de l'École Polythechnique*, 6 (1799), pp. 331–359

R. Caddeo, A. Papadopoulos (eds.), *Spherical Geometry in the Eighteenth Century I*, https://doi.org/10.1007/978-3-032-12466-1_18

and henceforth

$$\sin A = \frac{\sqrt{4b^2c^2 - \left(b^2 + c^2 - a^2\right)^2}}{2bc}.$$

Thus, if, for short, we make

$$d = \sqrt{4b^2c^2 - \left(b^2 + c^2 - a^2\right)^2},$$

we get

$$\sin A = \frac{d}{2bc};$$

therefore

$$\frac{a}{\sin A} = \frac{2abc}{d};$$

this is the desired ratio.

If we call r the radius of the circle circumscribed to the triangle, and s the surface or the area of the same triangle, we get

$$\frac{a}{\sin A} = \frac{2abc}{d} = 2r = \frac{abc}{2s};$$

therefore

$$r = \frac{abc}{d}, \text{ and } s = \frac{abc}{4r} = \frac{d}{4}.$$

2. By expanding the square of $b^2 + c^2 - a^2$ and reducing, we have

$$d = \sqrt{2a^2b^2 + 2a^2c^2 + 2b^2c^2 - a^4 - b^4 - c^4},$$

a formula where we see that the three sides a, b, c are evenly involved, as it should be.

But we can put this formula in a simpler and more convenient form for the logarithmic calculus, by decomposing it into factors. Indeed, we have

$$\begin{aligned} 4b^2c^2 - (b^2 + c^2 - a^2)^2 &= (2bc + b^2 + c^2 - a^2)(2bc - b^2 - c^2 + a^2) \\ &= \left[(b + c)^2 - a^2\right]\left[a^2 - (b - c)^2\right]; \end{aligned}$$

and, decomposing again each of these factors into two, we get

$$d = \sqrt{(a+b+c)(a-b+c)(a+b-c)(-a+b+c)}.$$

These formulae are well known and I am recalling them only for the purpose of an introduction to the investigations that follow.

3. Since in the spherical triangles the sines of the sides are proportional to the sines of the angles opposite to these sides, one may be curious about knowing this constant ratio, and also about seeing whether it also depends, as in the case of rectilinear triangles, on the radius of the circumscribed circle or on the area of the triangle.

Likewise, let us denote by a, b, c the three sides of a spherical triangle, and by A, B, C the three angles opposite to these sides. We get, by the well-known theorem

$$\cos a = \cos b \cos c + \sin b \sin c \cos A;$$

therefore

$$\cos A = \frac{\cos a - \cos b \cos c}{\sin b \sin c},$$

and henceforth

$$\sin A = \frac{\sqrt{\sin^2 b \sin^2 c - (\cos a - \cos b \cos c)^2}}{\sin b \sin c}.$$

Let us set, for short,

$$f = \sqrt{\sin^2 b \sin^2 c - (\cos a - \cos b \cos c)^2}\ ;$$

we get

$$\sin A = \frac{f}{\sin b \sin c};$$

therefore

$$\frac{\sin a}{\sin A} = \frac{\sin a \sin b \sin c}{f},$$

an expression of the desired ratio, analogous to the one we found for rectilinear triangles (§1).

Since the quantity f is expressed by a square root, and may have, consequently, the *plus* and *minus* sign, we will note that, relatively to the spherical triangles, it must always be taken positively, since the sides and the angles of any triangle being always less than two right angles, their sines are necessarily always positive.

4. The square root quantity f is also subject to reductions analogous to those of §2.

Because substituting for $\sin^2 b$ and $\sin^2 c$ their values in cosines, $1 - \cos^2 b$, $1 - \cos^2 c$, and reducing, we get

$$f = \sqrt{1 - \cos^2 a - \cos^2 b - \cos^2 c + 2\cos a \cos b \cos c},$$

where we see that the three sides a, b, c enter evenly [that is, f is invariant for every permutation of a, b, c].

Likewise, one can solve the quantity under the sign into factors. Indeed, we will first have

$$\begin{aligned}\sin^2 b \sin^2 c - (\cos a - \cos b \cos c)^2 &= (\sin b \sin c + \cos a - \cos b \cos c) \\ &\quad (\sin b \sin c - \cos a + \cos b \cos c) \\ &= [\cos a - \cos(b + c)]\,[\cos(b - c) - \cos a]\,.\end{aligned}$$

But we have, in general,

$$\cos a - \cos h = 2\sin\frac{a + h}{2}\sin\frac{h - a}{2},$$

therefore, decomposing in this way the two factors, we get

$$f = 2\sqrt{\sin\frac{a + b + c}{2}\sin\frac{a - b + c}{2}\sin\frac{a + b - c}{2}\sin\frac{-a + b + c}{2}},$$

an expression which is very convenient for the logarithmic calculus.

5. Let us search now for the radius of the circle circumscribed to the spherical triangle. It is evident that this circle can only be a small circle of the sphere, and that it will also be circumscribed to the rectilinear triangle formed by the three chords of the arcs a, b, c. But these chords being expressed by $2\sin\frac{a}{2}$, $2\sin\frac{b}{2}$, $2\sin\frac{c}{2}$, it only remains to substitute them instead of a, b, c in the expression of the radius r (§2), that is, in $\frac{abc}{d}$.

Let us call R the radius of this circumscribed circle, and h the quantity that results from d after these substitutions; we get

$$R = \frac{8\sin\frac{a}{2}\sin\frac{b}{2}\sin\frac{c}{2}}{h}.$$

But we have

$$\begin{aligned}h^2 = 32\left(\sin\frac{a}{2}\sin\frac{b}{2}\right)^2 + 32\left(\sin\frac{a}{2}\sin\frac{c}{2}\right)^2 + 32\left(\sin\frac{b}{2}\sin\frac{c}{2}\right)^2 \\ - 16\left(\sin\frac{a}{2}\right)^4 - 16\left(\sin\frac{b}{2}\right)^4 - 16\left(\sin\frac{c}{2}\right)^4;\end{aligned}$$

and, substituting $2\left(\sin\frac{a}{2}\right)^2, 2\left(\sin\frac{b}{2}\right)^2, 2\left(\sin\frac{c}{2}\right)^2$ by their values $1-\cos a$, $1-\cos b$, $1-\cos c$, we obtain, after the reductions,

$$h^2 = 12 - 8\cos a - 8\cos b - 8\cos c + 8\cos a\cos b + 8\cos a\cos c + 8\cos b\cos c$$
$$- 4\cos^2 a - 4\cos^2 b - 4\cos^2 c,$$

an expression which can be reduced to the following

$$4f^2 + 8(1-\cos a)(1-\cos b)(1-\cos c),$$

that is,

$$4f^2 + 64\left(\sin\frac{a}{2}\sin\frac{b}{2}\sin\frac{c}{2}\right)^2,$$

by substituting the value of f from §4. Thus, we get

$$R = \frac{4\sin\frac{a}{2}\sin\frac{b}{2}\sin\frac{c}{2}}{\sqrt{f^2 + 16\left(\sin\frac{a}{2}\sin\frac{b}{2}\sin\frac{c}{2}\right)^2}}.$$

6. Now, if we consider the radius of the sphere that passes through the center of the small circumscribed circle, we can see that this radius will be perpendicular to the plane of this circle, and that it will terminate at the point of the surface which will be the pole of the same circle. Therefore, calling φ the arc that measures the distance from the pole to the circumference of this circle, we get evidently $R = \sin\varphi$; therefore

$$\sin\varphi = \frac{4\sin\frac{a}{2}\sin\frac{b}{2}\sin\frac{c}{2}}{\sqrt{f^2 + \left(4\sin\frac{a}{2}\sin\frac{b}{2}\sin\frac{c}{2}\right)^2}};$$

whence we deduce

$$\cos\varphi = \frac{f}{\sqrt{f^2 + \left(4\sin\frac{a}{2}\sin\frac{b}{2}\sin\frac{c}{2}\right)^2}},$$

and from here

$$\tan\varphi = \frac{4\sin\frac{a}{2}\sin\frac{b}{2}\sin\frac{c}{2}}{f}.$$

Thus, since $\sin a = 2\sin\frac{a}{2}\cos\frac{a}{2}$, and likewise for the other sines, we get (§3)

$$\frac{\sin a}{\sin A} = 2\tan\varphi\cos\frac{a}{2}\cos\frac{b}{2}\cos\frac{c}{2}.$$

7. If we wanted to have the area of the rectilinear triangle inscribed in the given small circle and formed by the chords of the arcs a, b, c, then, calling S this surface, it would suffice to change in the formula $s = \frac{abc}{4r}$ of §1, s into S, r into R and a, b, c into $2\sin\frac{a}{2}, 2\sin\frac{b}{2}, 2\sin\frac{c}{2}$, which would immediately imply

$$S = \frac{2\sin\frac{a}{2}\sin\frac{b}{2}\sin\frac{c}{2}}{R},$$

or, because $R = \sin\varphi$ (§6),

$$S = \frac{2\sin\frac{a}{2}\sin\frac{b}{2}\sin\frac{c}{2}}{\sin\varphi}.$$

Now if we consider the triangular pyramid having this triangle as base and whose apex is at the center of the sphere, it is clear that the height of this pyramid will be $\cos\varphi$; therefore its volume[1] will be $\frac{S\cos\varphi}{3}$; or, by replacing S by the value we just obtained,

$$\frac{2\sin\frac{a}{2}\sin\frac{b}{2}\sin\frac{c}{2}}{3\tan\varphi},$$

and, furthermore, substituting the value of $\tan\varphi$ found above (§6), we get $\frac{f}{6}$ as the value of the volume of the pyramid.

8. It remains for us to consider the area itself of the spherical triangle formed by the arcs a, b, c.

We know the beautiful theorem according to which the area of a spherical triangle is, to the entire surface of the sphere, like the excess of the three angles of the triangle over two right angles,[2] to eight right angles. It is commonly attributed to Albert Girard, who indeed states it in the work called *Invention nouvelle en Algèbre* and printed in Amsterdam in 1629.[3] But since the proof that he gives is not rigorous[4]

---

[1] Lagrange uses the word "solidity", which we translate by "volume".

[2] This is commonly called the "angle excess", or "angular excess" of the triangle.

[3] Girard's rule is also mentioned and used by Euler in his memoir *Various speculations on the area of spherical triangles*, see Chap. 13 of the present volume. We indicated there in a footnote that the identification of the area of a spherical triangle with its angle excess was discovered by Thomas Harriot, before Girard.

[4] In fact, Girard declares in his book that he is not completely satisfied with his proof and he felt the need for a more rigorous proof.

and cannot even be regarded as an induction, we should rather attribute this theorem to Cavalieri who gave it in the *Directorium generale uranometricum*,[5] printed in Bologna in 1632, with the beautiful proof reported on by Wallis[6] and inserted since then in most of the Trigonometries.

Let us denote by Σ the excess to two right angles of the three angles of the triangle; we shall have, keeping the same names we used up to now, and calling D the right angle,

$$\Sigma = A + B + C - 2D.$$

In this way, the area of the triangle whose sides are a, b, c and opposite angles A, B, C will be the part $\frac{\Sigma}{8D}$ of the entire surface of the sphere; and if we consider this surface as equal to $8D$, we can then take Σ to be the value of the area of the same triangle.

9. If we imagine that the sides b and c which contain the angle A are produced until a quarter of a circle, the angles B and C will become right, and the side a will become equal to the opposite angle A; then the area of the isosceles right triangle will become A. Hence, if we subtract from it the first triangle whose sides around the angle A are b and c, we get the spherical quadrilateral whose base will be A and whose sides perpendicular to this base will be $D - b$ and $D - c$; and the area of this quadrilateral will be expressed by simply $B + C - 2D$.

But by the well-known Neper analogies,[7] we have in any spherical triangle the following equation, which we shall prove below,

---

[5] Bonaventura Cavalieri (1598–1647), *Directorium generale uranometricum, in quo trigonometriae logarithmicae fundamenta ac regulae demonstrantur, astronomicaeque supputationes ad solam fere vulgarem additionem reducuntur* (A general guide for celestial measures, in which the fundamentals and rules of logarithmic trigonometry are demonstrated, and astronomical calculations reduced to almost only common addition), Bologna, 1632.

[6] Wallis' report on Cavalieri's proof is contained in Vol. II of his *Collected works* (Operum Mathematicorum, Oxford, 1693), p. 875.

[7] The name refers to John Napier (1550-1617), the Scottish theologian, physicist, astronomer and mathematician. It seems that the French spelling Neper was invented to imitate the Scottish pronunciation of Napier. The Napier analogies, in spherical geometry, are relations between certain trigonometric functions of the variables $\frac{A+B}{2}$, $\frac{A-B}{2}$ and C involving the angles and $\frac{a+b}{2}$, $\frac{a-b}{2}$ and c involving the sides of a spherical triangle of sides a, b, c and angles A, B, C. The word "analogy" comes from the Greek "analogia" (ἀναλογία), which is a complicated word with several meanings, which was used by Euclid to mean "proportion" in the mathematical sense we use it today, and which is the sense in which it is used here, the Napier analogies being equalities between proportions. The formulae simplify computations in a spherical triangle and are used in astronomy; see e.g. Ossian Bonnet, *Astronomie sphérique*, Premier fascicule : notes sur le cours professé pendant l'année 1887, G. Carré, Paris, 1889, p. 56. The usual proof that is given for the Napier formulae use the so-called Gauss formulae, but Napier lived long before Gauss. For an exposition of Napier's work on spherical trigonometry, see Lambert's memoir *Notes and Comments on Trigonometry* and A. A'Campo's commentary The *pentagrammma mirificum* from Napier to Lambert: Notes on Lambert's memoir on spherical trigonometry, both published in the present volume. In particular, Lambert attributes to Napier the use of logarithms in the trigonometric formulae, see §1 of Lambert's memoir.

$$\tan\frac{B+C}{2} = \frac{\cos\frac{b-c}{2}}{\cos\frac{b+c}{2}}\cot\frac{A}{2}.$$

Thus, if we denote by σ the area or the surface of the quadrilateral we are dealing with, we get

$$\tan\frac{\sigma}{2} = \cot\frac{B+C}{2} = \frac{1}{\tan\frac{B+C}{2}},$$

therefore

$$\tan\frac{\sigma}{2} = \frac{\cos\frac{b+c}{2}}{\cos\frac{b-c}{2}}\tan\frac{A}{2};$$

and if we denote by β and γ the two sides of the quadrilateral that are perpendicular to the base A, in such a way that $\beta = D - b$ and $\gamma = D - c$, we get, for the determination of the area σ, the formula

$$\tan\frac{\sigma}{2} = \frac{\sin\frac{\beta+\gamma}{2}}{\cos\frac{\beta-\gamma}{2}}\tan\frac{A}{2}.$$

This formula is analogous to the well-known formula $\sigma = \frac{\beta+\gamma}{2}A$ for rectilinear quadrilaterals where the base is A, the two vertical sides β, γ, and the area σ. And since the latter is of the greatest usage for measuring planar surfaces bounded by straight lines, the formula that we just gave will also be useful for measuring the spherical surfaces terminated by arcs of great circles. Thus, it can be used with great advantage to determine the extent of a country, when we know the latitudes and the differences of longitudes of several points situated on the circumference, since by connecting these points by arcs of great circles, we get a spherical polygon of which we will easily find the area, by decomposing it into quadrilaterals formed by the circles of latitude and by the arcs of the equator that are intercepted by these circles.

10.[8] But if we wanted to have the value of the area Σ from the three sides a, b, c of the spherical triangle, then it would suffice to consider that, since $\Sigma = A + B + C - 2D$, we get

$$\cot\frac{\Sigma}{2} = -\tan\frac{A+B+C}{2} = -\frac{\tan\frac{A}{2} + \tan\frac{B+C}{2}}{1 - \tan\frac{A}{2}\tan\frac{B+C}{2}}.$$

[8] In this section, Lagrange gives a formula for the value of the area of a spherical triangle in terms of the lengths of its sides. Other formulae for the area in terms of the sides were given by Euler in §7 of his memoir *De mensura angulorum solidorum* (On the measure of solid angles), translated in Chap. 14 of the present volume.

If we substitute $\tan\frac{B+C}{2}$ by its value found above (the preceding §), we get

$$\cot\frac{\Sigma}{2} = -\frac{\tan\frac{A}{2}\cos\frac{b+c}{2} + \cot\frac{A}{2}\cos\frac{b-c}{2}}{\cos\frac{b+c}{2} - \cos\frac{b-c}{2}},$$

a formula that can easily be transformed into the following:

$$\cot\frac{\Sigma}{2} = \frac{\cos\frac{b}{2}\cos\frac{c}{2} + \sin\frac{b}{2}\sin\frac{c}{2}\cos A}{\sin\frac{b}{2}\sin\frac{c}{2}\sin A}.$$

Now, if we substitute in this formula the values of $\sin A$ and $\cos A$ of §3, we get, dividing the top and the bottom by $\sin\frac{b}{2}\sin\frac{c}{2}$,

$$\cot\frac{\Sigma}{2} = \frac{4\left(\cos\frac{b}{2}\cos\frac{c}{2}\right)^2 + \cos a - \cos b\cos c}{f};$$

but

$$2\left(\cos\frac{b}{2}\right)^2 = 1 + \cos b, \text{ and } 2\left(\cos\frac{c}{2}\right)^2 = 1 + \cos c;$$

therefore, making these substitutions and reversing the fraction, we get

$$\tan\frac{\Sigma}{2} = \frac{f}{1 + \cos a + \cos b + \cos c},$$

a formula which is the simplest for the determination of the area Σ of a spherical triangle by means of its three sides a, b, c.

11. We have seen (§7) that $\frac{f}{6}$ is the volume of the triangular pyramid formed by the three radii of the sphere that correspond to the three angles of the spherical triangle.

Let us consider now a triangular pyramid formed by these same radii produced as much as needed, in such a way that they become p, q, r, and that a, b, c be the arcs or angles contained by these lines. To obtain the volume of this pyramid, it will suffice to consider it as lying on one of its faces, for instance, the one whose sides are p and q, and to draw from the extremity of the third line r a perpendicular P onto the plane of the same face. First, it is easy to see that if a is the angle contained by p and q, the area of the face that we regard as the base of the pyramid will be $\frac{pq\sin a}{2}$; therefore the volume of the pyramid will be $\frac{Ppq\sin a}{6}$.

But if we call θ the angle that the line r makes with the plane passing through the lines p and q, it is clear that we get $P = r\sin\theta$; therefore the desired volume will be $\frac{pqr\sin a\sin\theta}{6}$.

The angle θ is nothing but the arc drawn perpendicularly from the angle A of the spherical triangle onto the opposite side a; consequently we can determine the value of $\sin\theta$ by the sines or cosines of the sides a, b, c of the triangle. But, for our purpose, it suffices to consider that this value, along with the one of the sine of a, being independent of the lines p, q, r, if we make $p = 1, q = 1, r = 1$, we get the case of the pyramid that we talked about above, and whose volume is $\frac{f}{6}$.

Whence it follows that we get $\sin a \sin \theta = f$; consequently we get in general $\frac{pqrf}{6}$ for the volume of the triangular pyramid in which the three sides or edges that form any of the solid angles are p, q, r and the angles contained by these three sides are a, b, c.

This expression for the volume of an arbitrary triangular pyramid by means of the three sides and the angles contained is, as we can see, very simple et very convenient for calculations, especially if we use for the value of f the factorized expression of §4, and it can be very useful for the determination of the volume of all the bodies bounded by planes, since we can always resolve them into triangular pyramids, in the same way as we resolve all polygons into triangles.

12. Besides, since have found $\sin a \sin \theta = f$, we get

$$\sin \theta = \frac{f}{\sin a};$$

thus we can determine from this formula the perpendicular θ in any spherical triangle whose base is a and whose other two sides are b, c.

13. I solved the preceding problems only in order to have the occasion to show the origin and the use of some remarkable formulae, and most of all, of the function which I denoted by f and which particularly deserves the attention of the analysts for its various applications. I will pass now to general considerations on spherical trigonometry considered analytically.

The analytic resolutions of spherical triangles have first been only simple applications of algebra to the geometric constructions. After that, people were content with the establishment of some fundamental propositions using geometry, and all the formulae of spherical trigonometry were deduced from the equations given by these propositions. The most elegant in this genre is Euler's memoir titled *Trigonometria sphærica ex primis principiis derivata*,[9] printed in the *Actes de Pétersbourg* for the year 1779, in which we find a complete system of trigonometric formulae, based only on three equations. But can't we simplify further this system, by reducing it to a single fundamental equation?

This reduction could serve to enhance the analytic theory of spherical triangles, since, in analysis, perfection consists in using only the smallest number of principles and of deducing from these principles all the truths that they may contain, by the sole power of analysis; in the synthetic method of lines, it consists on the contrary in proving separately each proposition, in the simplest way, using the propositions that are already proved.

[9] This is Euler's memoir *Trigonometria sphaerica universa, ex primis principiis breviter et dilucide derivata* (General spherical trigonometry, deduced from fundamental principles in a brief and clear manner), translated in Chap. 12 of the present volume.

The late de Gua[10] already had the idea of making all spherical trigonometry dependent on a single general property of spherical triangles; but the Memoir that he gave on this subject in the 1783 volume of the Académie des sciences contains computations that are so complicated that they seem to be more adapted to show the drawbacks of his method than to make it accepted.

I propose to myself here the same goal, and I shall present a brief table of all the formulae of spherical trigonometry, deducing them, by simple transformations, from a single equation given by the nature of the spherical triangles.

14. We shall start, as de Gua did, from the equation (§3)

$$\cos a = \cos b \cos c + \sin b \sin c \cos A,$$

in which a, b, c are the three sides or arcs of the triangle and A is the angle opposite to the side a.

This equation can be easily proved by the only consideration of the two rectilinear triangles that are formed, the one by the two tangents to the two arcs b and c and by the line that joins the extremities of these tangents, and the other by this same line and by the two secants of the same arcs, since it is evident that the two tangents form together the angle A contained by the arcs b and c, and that the two secants form together the angle a which is the side of the spherical triangle opposite to the angle A. Thus, calling h the side common to these two triangles, we get immediately, by the well-known theorem on rectilinear triangles, the equation

$$h^2 = \tan^2 b + \tan^2 c - 2 \tan b \tan c \cos A$$

for the first triangle, and the equation

$$h^2 = \sec^2 b + \sec^2 c - 2 \sec b \sec c \cos a$$

for the second triangle.

Whence we deduce

$$\tan^2 b + \tan^2 c - 2 \tan b \tan c \cos A = \sec^2 b + \sec^2 c - 2 \sec b \sec c \cos a.$$

But we have evidently $\sec^2 b - \tan^2 b = 1$, and likewise $\sec^2 c - \tan^2 c = 1$; therefore the equation will become

$$\sec b \sec c \cos a = 1 + \tan b \tan c \cos A;$$

[10] Jean-Paul de Gua de Malves (1710–1786) is a French mathematician, one of the initiators of the French *Encyclopédie*. Lagrange probably refers to the latter's *Trigonométrie sphérique, déduite très brièvement et complètement de la seule solution algébrique du plus simple de ses problèmes généraux, au moyen de diverses transformations, dont les rapports des sinus et cosinus, tangentes et cotangentes, sécantes et cosécantes d'un même arc ou d'un même angle-plan, rendent cette solution susceptible, et comprenant quelques formules et observations qu'on croit utiles et neuves*, Mémoire de l'Académie royale des sciences (1783).

substituting for $\sec b$, $\sec c$, $\tan b$, $\tan c$ their values $\frac{1}{\cos b}$, $\frac{1}{\cos c}$, $\frac{\sin b}{\cos b}$, $\frac{\sin c}{\cos c}$, and multiplying by $\cos b \cos c$, we get the fundamental equation

$$\cos a = \cos b \cos c + \sin b \sin c \cos A. \tag{A}$$

Since, by hypothesis, there is no other condition among the quantities a, b, c, A than the fact that a, b, c are the three sides of the triangle and A the angle opposite to the side a, it follows that, calling B and C the angles opposite to the sides b and c, we get similar equations relatively to these angles by only changing A into B or C, provided we change at the same time a into b or into c.

15. Now, if we deduce from the preceding equation the value of $\cos A$, and if we form out of it the one of $\sin A$, we get, as we already found in §3,

$$\frac{\sin a}{\sin A} = \frac{\sin a \sin b \sin c}{f}$$

where the quantity f is a function of a, b, c in which these three quantities enter equally and in such a way that it stays unchanged by whatever permutation performed on them.

Thus, changing a into b and A into B, the right-hand side of the equation will not change, and consequently we get the equation

$$\frac{\sin a}{\sin A} = \frac{\sin b}{\sin B}. \tag{B}$$

This is what we call the common analogy[11] of sines; and it is visible that by changing a into c and A into C, we get likewise

$$\frac{\sin a}{\sin A} = \frac{\sin c}{\sin C}.$$

16. Let us take again Eq. (A) of §14,

$$\cos a = \cos b \cos c + \sin b \sin c \cos A;$$

changing a into c and A into C we get likewise

$$\cos c = \cos a \cos b + \sin a \sin b \cos C;$$

substituting this value of $\cos c$ in the first equation, it will become

$$\cos a = \cos a \cos^2 b + \sin a \sin b \cos b \cos C + \sin b \sin c \cos A,$$

[11] The word "analogy" which is used here and elsewhere in this volume comes from the Greek "analogia" (ἀναλογία) which has several meanings and which means here "proportion". See Footnote 7.

that is,

$$\cos a \sin^2 b = \sin a \sin b \cos b \cos C + \sin b \sin c \cos A,$$

and, dividing by $\sin b$,

$$\cos a \sin b = \sin a \cos b \cos C + \sin c \cos A.$$

Let us substitute $\sin c$ by its value $\frac{\sin a \sin C}{\sin A}$, extracted from Eq. (B) of the previous §, changing b into c and B into C; let us divide then by $\sin a$, and let us put $\cot a$ and $\cot A$ instead of $\frac{\cos a}{\sin a}$ and $\frac{\cos A}{\sin A}$. We get the equation

$$\cot a \sin b = \cot A \sin C + \cos b \cos C. \tag{C}$$

17. Finally, the same equation obtained above,

$$\cos a \sin b = \sin a \cos b \cos C + \sin c \cos A,$$

gives, changing a into b and A into B,

$$\sin a \cos b = \cos a \sin b \cos C + \sin c \cos B.$$

Substituting this value of $\sin a \cos b$ in the same equation, we have

$$\cos a \sin b = \cos a \sin b \cos^2 C + \sin c \cos B \cos C + \sin c \cos A,$$

that is,

$$\cos a \sin b \sin^2 C = \sin c \left(\cos B \cos C + \cos A\right).$$

Let us substitute $\sin c$ by its value $\frac{\sin b \sin C}{\sin B}$, extracted from Eq. (B), changing a into c and A into C. Next, let us divide by $\sin b \sin C$ and multiply by $\sin B$; we get

$$\cos a \sin B \sin C = \cos B \cos C + \cos A,$$

that is,

$$\cos A = -\cos B \cos C + \sin B \sin C \cos a. \tag{C}$$

18. We can see that this formula is completely analogous to Formula (A) of §14 with which we started; the angles A, B, C have replaced the sides a, b, c and conversely; and the cosines became negative, the sines remaining positive, which indicates that the sides became the supplements to two right angles of the angles, and the angles the supplements to two right angles of the sides. Thus, all the formulae which result from Formula (A) will also be true, by making there the same changes.

We deduce from here the well-known property of spherical triangles, that every spherical triangle can be changed into another one whose sides and angles are respectively the supplements of the angles and sides of the first one; and we know that this new triangle, which we call *supplementary*, is the one that is formed on the sphere by the three poles of the arcs that constitute the sides of the given triangle, by joining these poles by arcs of great circles; this can easily be proved by a very simple construction.[12]

19. The four equations, (A), (B), (C), (D) that we just found contain the solution of all the questions of spherical trigonometry; because, since in a spherical triangle there are only six elements, the three sides and the three angles, and since three of these elements suffice to determine the triangle, it is clear that the simplest relations can only hold between four elements; but all the different combinations that can be done with the six elements, taken four-by-four, are reduced to the following four:

(1) Between three sides and one angle: this relation is given by Eq. (A);
(2) Between two sides and two angles, which can be opposite respectively to the two sides, or one opposite and the other adjacent to the same side, which makes two cases: the relation between two sides and the two opposite angles is given by Eq. (B);
(3) Between two sides and two angles, one of which is opposite and the other adjacent to the same given side: this relation is contained in Eq. (C);
(4) Between three angles and one side: this relation is given by Equation (D).

20. If we assume that the angle A is right, the preceding equations can be simplified and give the following

$$\cos a = \cos b \cos c,$$

$$\sin a = \frac{\sin b}{\sin B},$$

$$\cot a = \cot b \cos C,$$

$$\cos a = \cot B \cot C;$$

and, if we assume that the angle C is right, Eq. (C) and (D) give also the following two

$$\cot A = \cot a \sin b,$$

$$\cos A = \sin B \cos a.$$

[12] Euler makes a similar remark in §7 of his memoir *Trigonometria sphaerica universa, ex primis principiis breviter et dilucide derivata* (General spherical trigonometry, deduced from fundamental principles in a brief and clear manner), translated in the present volume (Chap. 12). We note here as well that this is explained using polarity theory in spherical geometry, discovered by the Arab mathematicians of the ninth–tenth century; see Chap. 10 of the present volume.

These six equations provide directly the solution of all the cases of right spherical triangles; and since they are under a form which is very useful for the use of logarithms, they are commonly used in Trigonometry, decomposing all triangles into right triangles by drawing a perpendicular. But we can also solve all cases by the four general equations, after factorizing these equations, using the transformations that we shall present.

21. Equation (A) between the three sides a, b, c and an angle A opposite to the side a can be useful to determine: 1° A from a, b, c; 2° a from b, c and A; 3° b from a, c and A.

(1) To determine A from a, b, c we get

$$\cos A = \frac{\cos a - \cos b \cos c}{\sin b \sin c},$$

from which we deduce

$$1 + \cos A = 2\left(\cos\frac{A}{2}\right)^2 = \frac{\cos a - \cos(b+c)}{\sin b \sin c} = \frac{2\sin\frac{b+c-a}{2}\sin\frac{b+c+a}{2}}{\sin b \sin c},$$

$$1 - \cos A = 2\left(\sin\frac{A}{2}\right)^2 = \frac{\cos(b-c) - \cos a}{\sin b \sin c} = \frac{2\sin\frac{a-b+c}{2}\sin\frac{a+b-c}{2}}{\sin b \sin c};$$

therefore

$$\tan\frac{A}{2} = \sqrt{\frac{\sin\frac{a+b-c}{2}\sin\frac{a-b+c}{2}}{\sin\frac{b+c+a}{2}\sin\frac{b+c-a}{2}}}. \tag{a}$$

(2) To determine a from b, c and A we have

$$\cos a = \cos b \cos c + \sin b \sin c \cos A.$$

It does not seem to be possible to factorize immediately this equation for the usage of logarithms; but we can attain this by using a subsidiary angle.

Indeed, if we make

$$\tan c \cos A = \tan\varphi, \tag{b}$$

we get

$$\sin c \cos A = \cos c \tan\varphi,$$

therefore

$$\cos a = \cos c\,(\cos b + \sin b \tan\varphi);$$

but

$$\cos b + \sin b \tan \varphi = \frac{\cos b \cos \varphi + \sin b \sin \varphi}{\cos \varphi} = \frac{\cos (b - \varphi)}{\cos \varphi};$$

thus

$$\cos a = \frac{\cos c \cos (b - \varphi)}{\cos \varphi}. \tag{c}$$

It is not difficult to see that this transformation amounts to the division of the triangle into two right triangles by a perpendicular drawn from the angle B onto the side b, and that φ is the segment of the side adjacent to the angle A.

(3) To determine b by a, c and A, one would have to use, in the main Eq. (A), $\sqrt{1 - \sin^2 b}$ instead of $\cos b$, then take the square in order to cancel the square root, and deduce the value of $\sin b$ by solving a quadratic equation, which would give for $\sin b$ a formula which is complicated and which would not be fit for logarithmic calculus. But the transformation used above also serves to solve this case, since, having determined the angle φ by Eqs. (b) and (c) will give

$$\cos (b - \varphi) = \frac{\cos a \cos \varphi}{\cos c}. \tag{d}$$

22. Equation (B) between two sides a and b and the opposite angles A and B can be useful: 1° to determine A from a, b, B and 2° to find a from b, A, B.

(1) To determine A from a, b, B, we get

$$\sin A = \frac{\sin a \sin B}{\sin b}. \tag{e}$$

(2) To determine a from b, A, B we get

$$\sin a = \frac{\sin b \sin A}{\sin B}. \tag{f}$$

These formulae do not require any transformation for the application of logarithms.

23. Equation (C) between two sides a and b and two angles A, C, the first opposite and the second adjacent to the side a, may be useful: 1° to determine A from a, b, C; 2° to determine a from b, A, C; 3° to determine C from a, b, A; 4° to determine b from a, A, C.

(1) To determine A from a, b, C, we get the equation

$$\cot A = \frac{\cot a \sin b}{\sin c} - \cot C \cos b;$$

and to factorize it, we make $\frac{\cot a}{\cos C} = \cot \varphi$, or

$$\tan a \cos C = \tan \varphi; \tag{g}$$

therefore $\cot a = \cos C \cot \varphi$; and, substituting this value, we get

$$\begin{aligned} \cot A &= \cot C\,(\cot \varphi \sin b - \cos b) \\ &= \frac{\cot C}{\sin \varphi}(\cos \varphi \sin b - \sin \varphi \cos b) = \frac{\cot C \sin (b - \varphi)}{\sin \varphi}; \end{aligned}$$

therefore

$$\cot A = \frac{\cot C \sin (b - \varphi)}{\sin \varphi}. \tag{h}$$

This factorization also amounts to dividing the triangle into two right ones by a perpendicular drawn from the angle B onto the opposite side b; and the subsidiary angle φ is the segment of this side adjacent to the angle C.

(2) To determine a by b, A, C, we get the equation

$$\cot a = \frac{\cot A \sin C}{\sin b} + \cot b \cos C,$$

and to factorize it we shall make

$$\frac{\cot A}{\cos b} = \tan \varphi; \tag{i}$$

therefore, substituting in the equation $\cot A$ by its value $\tan \varphi \cos b$, we get

$$\begin{aligned} \cot a &= \cot b\,(\sin C \tan \varphi + \cos C) \\ &= \frac{\cot b}{\cos \varphi}(\sin C \sin \varphi + \cos C \cos \varphi) = \frac{\cot b \cos (C - \varphi)}{\cos \varphi}; \end{aligned}$$

therefore

$$\cot a = \frac{\cot b \cos (C - \varphi)}{\cos \varphi}. \tag{k}$$

This factorization amounts again to dividing the triangle into two right ones by drawing a perpendicular from the angle C onto the side c, and the subsidiary angle φ is the segment of the angle C adjacent to the side b.

(3) To determine C from a, b, A, one must deduce from Eq. (C) the value of $\sin C$ or $\cos C$ by the resolution of a quadratic equation, and we would get an expression which would contain a square root. But the preceding transformation

is also useful to solve these cases, since, having found the angle φ by Eqs. (i) and (h) will give

$$\cos(C-\varphi)=\frac{\cot a\cos\varphi}{\cot b}. \tag{l}$$

(4) Finally, to determine b from a, A, C, one also ought to deduce from the same Eq. (C) the value of $\sin b$ or $\cos b$ by the resolution of a quadratic equation. But the transformation used for the first case will also be useful to solve this one, since, having found the subsidiary angle φ by Eqs. (i) and (h) will give

$$\sin(b-\varphi)=\frac{\cot A\sin\varphi}{\cot C}. \tag{m}$$

24. Finally, Equation (D) between the three angles A, B, C and a side a will be useful for the determination of: 1° a from A, B, C; 2° A from a, B, C; 3° B from a, A, C. Since these three cases agree with those that depend on Eq. (A), of which Equation (D) is only a transformed version, one can immediately apply the formulae that we gave in §21, substituting, instead of the sides a, b, c, the supplements to two right angles of the angles A, B, C, and instead of these angles the supplements to two right angles of the opposite sides a, b, c (§18).

Thus:

(1) To determine a from A, B, C, Eq. (a) of §21 will give the transformed one:

$$\cot\frac{a}{2}=\sqrt{\frac{\cos\frac{A+B-C}{2}\cos\frac{A-B+C}{2}}{-\cos\frac{B+C+A}{2}\cos\frac{B+C-A}{2}}}. \tag{n}$$

(2) To determine A from a, B, C, we shall make the same substitutions in Formulae (b) and (c) of the same §; and, taking instead of the angle φ its complement to a right angle, we get the following two equations

$$\tan C\cos a=\cot\varphi, \tag{o}$$

$$\cos A=\frac{\cos C\sin(B-\varphi)}{\sin\varphi}. \tag{p}$$

(3) The same equations will serve to determine B from a, A, C, since, having found φ by Eqs. (o) and (p) will give

$$\sin(B-\varphi)=\frac{\cos A\sin\varphi}{\cos C}. \tag{q}$$

25. Thus, by these formulae, we can find directly an arbitrary side or angle by three given parts, either sides or angles; this contains all the theory of spherical triangles.

But, when one has to search at the same time for two sides from the third side and the two opposite angles, or for two angles from the third angle and the two opposite sides, it is more convenient to use the formulae found by Neper between these five parts. Here is the simplest way to reach these formulae.

We found in §21

$$\tan\frac{A}{2} = \sqrt{\frac{\sin\frac{a+b-c}{2}\sin\frac{a-b+c}{2}}{\sin\frac{b+c+a}{2}\sin\frac{b+c-a}{2}}}.$$

Likewise we get for the angle B, by changing A into B and a into b,

$$\tan\frac{B}{2} = \sqrt{\frac{\sin\frac{a+b-c}{2}\sin\frac{b-a+c}{2}}{\sin\frac{a+b+c}{2}\sin\frac{a-b+c}{2}}},$$

therefore, multiplying together,

$$\tan\frac{A}{2}\tan\frac{B}{2} = \frac{\sin\frac{a+b-c}{2}}{\sin\frac{a+b+c}{2}};$$

if we change in this equation b into c and B into C, we get

$$\tan\frac{A}{2}\tan\frac{C}{2} = \frac{\sin\frac{a-b+c}{2}}{\sin\frac{a+b+c}{2}},$$

and changing in the same equation a into b and A into B, we get

$$\tan\frac{B}{2}\tan\frac{C}{2} = \frac{\sin\frac{b+c-a}{2}}{\sin\frac{a+b+c}{2}},$$

therefore, adding together, we get

$$\left(\tan\frac{A}{2} + \tan\frac{B}{2}\right)\tan\frac{C}{2} = \frac{\sin\frac{a-b+c}{2} + \sin\frac{b+c-a}{2}}{\sin\frac{a+b+c}{2}} = \frac{2\sin\frac{c}{2}\cos\frac{a-b}{2}}{\sin\frac{a+b+c}{2}},$$

and, subtracting the same equations from each other, we get

$$\left(\tan\frac{A}{2} - \tan\frac{B}{2}\right)\tan\frac{C}{2} = \frac{\sin\frac{a-b+c}{2} - \sin\frac{b+c-a}{2}}{\sin\frac{a+b+c}{2}} = \frac{2\cos\frac{c}{2}\sin\frac{a-b}{2}}{\sin\frac{a+b+c}{2}}.$$

On the other hand, we have

$$1+\tan\frac{A}{2}\tan\frac{B}{2}=\frac{\sin\frac{a+b+c}{2}+\sin\frac{a+b-c}{2}}{\sin\frac{a+b+c}{2}}=\frac{2\sin\frac{a+b}{2}\cos\frac{c}{2}}{\sin\frac{a+b+c}{2}},$$

$$1-\tan\frac{A}{2}\tan\frac{B}{2}=\frac{\sin\frac{a+b+c}{2}-\sin\frac{a+b-c}{2}}{\sin\frac{a+b+c}{2}}=\frac{2\sin\frac{c}{2}\cos\frac{a+b}{2}}{\sin\frac{a+b+c}{2}}.$$

Therefore, since

$$\tan\frac{A\pm B}{2}=\frac{\tan\frac{A}{2}\pm\tan\frac{B}{2}}{1\mp\tan\frac{A}{2}\tan\frac{B}{2}},$$

we get the following two equations

$$\tan\frac{A+B}{2}\tan\frac{C}{2}=\frac{\cos\frac{a-b}{2}}{\cos\frac{a+b}{2}} \tag{r}$$

$$\tan\frac{A-B}{2}\tan\frac{C}{2}=\frac{\sin\frac{a-b}{2}}{\sin\frac{a+b}{2}}. \tag{s}$$

We can deduce similar formulae from Eq. (n) of §24, and, without doing a new computation, it will suffice to change the sides a, b, c into the supplements to two right angles of the angles A, B, C, and to change these angles into the supplements to two right angles of the same sides. In this manner, we get right away these two other equations

$$\tan\frac{a+b}{2}\cot\frac{c}{2}=\frac{\cos\frac{A-B}{2}}{\cos\frac{A+B}{2}} \tag{t}$$

$$\tan\frac{a-b}{2}\cot\frac{c}{2}=\frac{\sin\frac{A-B}{2}}{\sin\frac{A+B}{2}}. \tag{u}$$

Thus, these four equations will be used to find directly, and without the help of any auxiliary angle, the two angles A and B from the two opposite sides a and b, together with the intercepted angle C, or these two sides from the opposite angles and the third side.

26. Before ending this memoir, I think it is my duty to say two words about the comparison between spherical and rectilinear triangles. By taking, as we usually do, the radius of the sphere as the unit, it is clear that the arcs that form a spherical triangle naturally express angles whose sines and cosines are found in the tables; if the radius of the sphere is not the unit, then, in order to have the angular value of the sides of the triangle, we must divide their absolute value by the radius. Thus, if α,

β, γ are the absolute lengths of the arcs that form a spherical triangle on the surface of a sphere whose radius is r, we get $\frac{\alpha}{r}$, $\frac{\beta}{r}$, $\frac{\gamma}{r}$ for the angles corresponding to these arcs, and these are the quantities that one must take for the sides that we denoted by a, b, c, assuming that A, B, C are the angles opposite to the arcs α, β, γ in the given triangle.

Now if the radius of the sphere becomes infinitely large, its surface changes into a plane, and the spherical triangle becomes rectilinear; from here it follows that if in the formulae of the triangles, we substitute everywhere $\frac{\alpha}{r}$, $\frac{\beta}{r}$, $\frac{\gamma}{r}$ instead of a, b, c and if we further assume that r is infinitely large, and that having developed in series the sines and cosines of these angles, we reject the terms that vanish under the assumption $\frac{1}{r} = 0$, we get the case of rectilinear triangles, in which α, β, γ are the sides and A, B, C the opposite angles.

Thus, if $V = 0$ is an equation between the sines and cosines of a, b, c and A, B, C, we shall substitute, for $\sin a$, $\frac{\alpha}{r} - \frac{\alpha^3}{2{\cdot}3r^3} + \cdots$; for $\cos a$, $1 - \frac{\alpha^2}{2r^2} + \frac{\alpha^4}{2{\cdot}3{\cdot}4r^4} \cdots$; and for $\sin b$, $\cos b$, $\sin c$, $\cos c$ similar expressions; changing α into β, γ and transforming into series according to the decreasing powers of r, which will give

$$V = \frac{P}{r^m} + \frac{Q}{r^{m+1}} + \frac{R}{r^{m+2}} + \cdots,$$

we get for the rectilinear triangle the equation $P = 0$, since from the equation $V = 0$, multiplying by r^m,

$$P + \frac{Q}{r} + \frac{R}{r^2} + \cdots = 0,$$

and, making $\frac{1}{r} = 0$, we have $P = 0$.

In this way, we can deduce the rules of rectilinear trigonometry from the fundamental equations of spherical trigonometry; but this would be useful only as an interesting computational exercise, since it would be proving the simple by the composite: it will suffice for us to remark that Equation (D) of §17 gives immediately the following one

$$\cos A = \sin B \sin C - \cos B \cos C,$$

that is,

$$\cos A = -\cos(B - C),$$

whence we deduce

$$A = 2D - B - C, \text{ that is, } A + B + C = 2D,$$

D being the right angle; which is the well-known property of rectilinear triangles.

27. Now, if the radius r of the sphere, instead of being infinitely large, is only very large, the spherical triangle will not become rectilinear, but will be very close to being so; and in this case, since the angles a, b, c that correspond to the sides become very small, the ordinary trigonometric tables will not offer enough precision for the computation of the sides and the angles. Therefore there is an advantage in treating the spherical triangles as rectilinear, taking into account the small correction which results from their difference.

The case considered takes place mostly in the computation of the triangles that are formed on the surface of the Earth for measuring an arc of meridian; in these triangles, the quantities α, β, γ are the very lengths of the sides, and r is the radius of the Earth.

To determine the correction that we just talked about, we shall take Eq. (A) of §14, which serves as a foundation for all spherical trigonometry, and which gives

$$\cos A = \frac{\cos a - \cos b \cos c}{\sin b \sin c}.$$

Let us perform in the right-hand side the substitutions indicated above, stopping to the terms divided by r^4. We will first have

$$\cos A = \frac{\frac{\beta^2+\gamma^2-\alpha^2}{2r^2} + \frac{\alpha^4-\beta^4-\gamma^4}{2\cdot 3\cdot 4r^4} - \frac{\beta^2\gamma^2}{4r^4}}{\frac{\beta\gamma}{r^2}\left(1 - \frac{\beta^2+\gamma^2}{2\cdot 3r^2}\right)};$$

multiplying the top and the bottom of the fraction by r^2 and substituting the factor $1 + \frac{\beta^2+\gamma^2}{2\cdot 3r^2}$ instead of the divisor $1 - \frac{\beta^2+\gamma^2}{2\cdot 3r^2}$, we get, neglecting the terms divided by powers higher than r^2,

$$\begin{aligned}\cos A &= \frac{\beta^2+\gamma^2-\alpha^2}{2\beta\gamma} + \frac{\alpha^4-\beta^4-\gamma^4}{2\cdot 3\cdot 4\beta\gamma r^2} - \frac{\beta^2\gamma^2}{4\beta\gamma r^2} + \frac{(\beta^2+\gamma^2-\alpha^2)(\beta^2+\gamma^2)}{3\cdot 4\beta\gamma r^2}\\ &= \frac{\beta^2+\gamma^2-\alpha^2}{2\beta\gamma} + \frac{\alpha^4+\beta^4+\gamma^4-2\alpha^2\beta^2-2\alpha^2\gamma^2-2\beta^2\gamma^2}{2\cdot 3\cdot 4\beta\gamma r^2}.\end{aligned}$$

Making $\frac{1}{r} = 0$, the angle A becomes the angle opposite to the side α in the rectilinear triangle whose sides would be α, β, γ.

Let us denote this angle by A'; then we get

$$\cos A' = \frac{\beta^2+\gamma^2-\alpha^2}{2\beta\gamma},$$

whence

$$\sin^2 A' = \frac{2\alpha^2\beta^2+2\alpha^2\gamma^2+2\beta^2\gamma^2-\alpha^4-\beta^4-\gamma^4}{4\beta^2\gamma^2},$$

as we saw above (§1 and 2); thus, substituting these values in the preceding equation, it becomes

$$\cos A = \cos A' - \frac{\beta\gamma \sin^2 A'}{2 \cdot 3r^2};$$

but in the rectilinear triangle whose sides are α, β, γ, it is clear that $\frac{\beta\gamma \sin A'}{2}$ expresses its area. Therefore, if we denote this area by θ, we get

$$\cos A = \cos A' - \frac{\theta \sin A'}{3r^2},$$

whence it follows that we get, up to the quantities of order $\frac{1}{r^4}$,

$$A = A' + \frac{\theta}{3r^2}.$$

And, since in changing the side α into β or γ, the angle A is changed into B or C, if we denote likewise by B' and C' the angles opposite to the sides β and γ in the rectilinear triangle, we also get

$$B = B' + \frac{\theta}{3r^2},$$

$$C = C' + \frac{\theta}{3r^2},$$

since the quantity θ, which is equal to the area of the rectilinear triangle, is a function which depends evenly on the three sides α, β, γ, in such a way that it does not change by making among these quantities any exchange whatsoever.

Therefore, when we have a spherical triangle drawn on the surface of a sphere whose radius r is very large, if we form a rectilinear triangle whose sides have the same length as those of the spherical triangle, the angles of the latter will be equal to the corresponding angles of the rectilinear triangle, each one increased by the quantity $\frac{\theta}{3r^2}$, θ being the area of the rectilinear triangle, making sure of reducing the value of this quantity into angles, that is, taking as unit the angle that corresponds to the arc which is equal to the radius.

28. If we add together the three equations

$$A = A' + \frac{\theta}{3r^2}, \quad B = B' + \frac{\theta}{3r^2}, \quad C = C' + \frac{\theta}{3r^2},$$

we have

$$A + B + C = A' + B' + C' + \frac{\theta}{r^2};$$

but we know that

$$A' + B' + C' = 2D,$$

D being the right angle; therefore we get

$$\frac{\theta}{r^2} = A + B + C - 2D.$$

Whence we can conclude that by subtracting from each angle of the spherical triangle the third of the excess of the sum of its three angles over two right angles, we get the three angles of a rectilinear triangle whose sides will be equal in length to those of the spherical triangle. Thus, we can treat the latter as a rectilinear triangle, and the results will be exact up to the quantities of the order $\frac{1}{r^4}$.

This beautiful theorem is due to Legendre, who first stated it without proof in the *Mémoires de l'Académie des Sciences* for the year 1789,[13] and who just proved it in a way which is slightly different from the previous one, in a Memoir on the method for determining the length of the quarter of the meridian. Since it can be of great use in all the cases where we have to compute spherical triangles that differ slightly from rectilinear triangles, we thought that it would be convenient to find it here.

[13] The year should be 1787. This subject is treated by Legendre in his *Méthode pour déterminer la longeur exacte du quart du Méridien d'après les observations faites pour la mesure de l'arc compris entre Dunkerque et Barcelone* (A method for determining the exact length of the quarter of the meridian, after the observations made for the measure of the arc comprised between Dunkerque and Barcelona). Note III is titled *Résolution des triangles sphériques dont les côtés sont très petits par rapport au rayon de la sphère* (Resolution of spherical triangles whose sides are very small compared to the radius of the sphere). Legendre writes, about this result: "This is the proposition which I had given without proof in the *Memoirs of the Academy*, year 1787, p. 338. It immediately reduces to rectilinear trigonometry the resolution of spherical triangles that are very slightly curved, or whose sides are very small compared to the radius of the sphere." Legendre's memoir appeared as the first article in the book *Méthodes analytiques pour la détermination d'un arc du méridien. Précédées d'un mémoire sur le même sujet, par A.-M. Legendre* (Analytical methods for the determination of an arc of the meridian, preceded by a memoir on the same subject by A.-M. Legendre) by J.-B.-J. Delambre, Paris, Duprat, 1798.

Part IV
Translation of Lambert's Memoir

Johann Heinrich Lambert,
Godefroy Engelmann (1788–1839), drawing and lithography, 1829

Chapter 19
Notes and Comments on Trigonometry

Johann Heinrich Lambert

Abstract In this memoir, Lambert establishes a complete system of formulae of spherical trigonometry, which he reduces then to a set of four formulae, and he makes them fit to the use of computations with logarithms. He also gives a geometric interpretation of the so-called Napier rules which summarize the relations between any three parameters (angle or side length) of a right-angled spherical triangle, using cyclic transformations of a given triangle.

Keywords Spherical trigonometry · Johann Heinrich Lambert · History of spherical geometry · Mathematics in the eighteenth century

AMS Classification 01A50, 51-03, 53A05, 53A35

From the Preface to the First Volume of the Contributions

Here[1] I provide some treatises with the purpose of extending mathematical knowledge in its own right, and furthermore to make it applicable both in common life as well as in studying nature and for experiments.[2]

[1] These are the relevant passages of Lambert's preface to Volume 1 of his *Contributions to mathematics and its applications*, Berlin, 1765, concerning Treatise III on trigonometry.

[2] The following discussion of the first two treatises is omitted here.

The author Johann Heinrich Lambert (1728–1777).
Translated from the German by Annette A'Campo Neuen.
In this chapter, the footnotes are due to the translator.
Johann Heinrich Lambert, Anmerkungen und Zusätze zur Trigonometrie, in: Beyträge zum Gebrauche der Mathematik und deren Anwendung. Vol. **1**, Treatise III, Berlin 1765, pp. 369–424, in: Johann Heinrich Lambert, Opera Mathematica I, ed. Andreas Speiser, Orell Füssli Verlag Zürich, 1946, pp. 52–84.

R. Caddeo, A. Papadopoulos (eds.), *Spherical Geometry in the Eighteenth Century I*, https://doi.org/10.1007/978-3-032-12466-1_19

The third treatise concerns trigonometry, and it provides notes and comments on it. The two rules of *Neper*[3] for right-angled spherical triangles are well-known.[4] They refer to 10 different cases. I have found that on the sphere one can place five right-angled triangles next to each other in a circle such that their hypotenuses lie next to each other, and that each hypotenuse by prolongation to 90 degrees[5] on both sides provides a cathete to both adjacent triangles.[6] Hereby, it was possible to present all 10 cases of Neper's rules in a simple, nice and evident way.[7] This figure has some more beautiful properties, e.g. that the area of the five triangles together with twice the area of the pentagon that they enclose, always add up to $3/8$ of the entire area of the sphere. Therefore I have taken more time for its consideration.

If one wants to describe spherical trigonometry with general formulae,[8] one can reduce all 60 possible cases to 4 cases. So in the following, I chose suitable notation for these 4 cases, and then presented each of them by formulae that I proved in the shortest way. Moreover, I showed how to avoid hereby occurring irrationalities,[9] and how the formulae could be made convenient for using logarithms in explicit calculations.[10] For example, I explained how one could find the height of the sun for every moment from the length of a day by the simple addition of three logarithms.[11]

Notes and Comments on Trigonometry

§1. One may consider trigonometry as a science serving as a model for the other sciences with respect to its completeness. It lists all possible cases that can occur considering the solution of triangles. The cases are divided into classes, and by proving that there are neither more nor less cases possible, it gives a solution for each case. This enables the user in each case to find any part of the triangle from the

---

[3] John Napier, or latinized Neper, 1550–1617, was a Scottish mathematician and theologian who became known as one of the inventors of logarithms. In fact, he developed logarithms as a tool for calculations in spherical trigonometry (see [John Napier, Mirifici logarithmorum canonis descriptio ejusque usus in utraque trigonometria etc. Edinburgh 1614] and [English translation of Mirifici logarithmorum canonis descriptio etc., Ian Bruce, http://www.17centurymaths.com/contents/napiercontents.html].

[4] Napier explains his mnemonic for setting up identities between any 3 chosen parts of a right-angled spherical triangle in Book IIa, Chapter III of the work quoted in Footnote 3, but without going into the details of a proof.

[5] Here it is assumed that sides measured by their angles are less than 90°.

[6] See Fig. 19.1.

[7] In fact, a similar figure already appears in Napier's text, but there is no explicit explanation on how it is related to his mnemonic rule.

[8] Now Lambert addresses the general case of oblique triangles with no right angle.

[9] Here Lambert refers to the occurence of square roots in his formulae.

[10] In order to benefit from logarithms, there should only appear quotients or products of terms.

[11] Following, there is a description of the fourth treatise, and we omit it here.

ones that are necessary and sufficient to know. If one tries to contribute something to this science, the aim is not to increase the number of possible cases since this would be impossible, but the goal is to make the solution of problems more beautiful, shorter and more convenient. In that respect there is still a lot to be thought about.

§2. The shortening of solutions depends on the purpose for which trigonometric problems are dealt with. We can distinguish three kinds of purposes. Firstly, there are a lot of cases where the known parts of a triangle are given as numbers and also the unknown ones are asked for as numbers. For this aim there are trigonometric tables, and one considers the solution as most elegant if it shows how to find the desired values using the tables in the shortest possible way and without unnecessary calculations. Even before inventing logarithms, one has tried to put all the solutions into mere proportions, and since the eminent Neper introduced logarithms, this has become even more important.

§3. Furthermore, one started with great success to use trigonometry for algebraic problems and vice-versa algebra for trigonometrical problems. This happened when in the calculation of general theorems one did not only use lines and Pythagoras' theorem but also angles. Thereby one could eliminate many roots from the equations, the terms became more elegant, the irrationalities were transformed into simple proportions, several relations could be expressed by just one angle, and the theorems that in order to transform the irrational values had to be deduced from Pythagoras' theorem and its consequences, became trigonometrical theorems that one could use to complete calculations that otherwise would have been long and complicated.

§4. These advantages made it even more important to apply algebra to trigonometry, and for all the occurring cases one had to find general formulae suitable for the chosen purpose. The rules given by these formulae are almost necessarily different from the ones that are most convenient for explicit calculations with given numbers. They only coincide if the entire solution consists in one proportion. But if one has to draw perpendiculars, to divide angles and sides, determine one part by another, and other than proportions has to add or subtract angles or sides, then the algebraic solutions use different tricks and therefore divert necessarily from the rules used with numbers.

§5. The third useful application of trigonometry occurs in calculus. When integrating one often ends up with arcs and their sines and tangents even though initially one did not think of any figure at all. Integral calculus would have taught us to consider angles, arcs and their secants as proportions if we had not thought about it otherwise. It is easy to see that here again one had to find new formulae in trigonometry different from the two kinds described previously, since here the focus lies on differential equations and their integration. And since this part of trigonometry is completely new, it is not surprising that neither the number of possible cases nor their corresponding classes have yet been determined. Anyway this will not be easy since the number of parts that can occur in a differential formula is necessarily indetermined. But there would be means to look at the simpler classes and to give complete solutions for them.

§6. We can consider these three purposes as the current state of trigonometry which might still develop further with time. It is easy to see that the task here is not just a simple gleaning but that in particular with respect to the last two purposes there is a lot to do and the field is still open. In this treatise I will present what came to my mind lately, and for the sake of the reader, this will be described together with well-known parts such that one obtains a unified picture. For trigonometry this is both necessary and useful.

§7. Up to now, in spherical trigonometry one has used an induction[12] which of course is admissible even according to the strictest rules of arguing. But one might still ask for a proper proof since this is better than even the strictest principle of induction. More precisely, one shows that for right-angled triangles there are exactly 30 cases occurring. Leaving aside the right angle, there are 5 parts remaining. Each of these 5 parts can be the one we ask for, and to find it we need to know two of the other four parts. These can be combined two-by-two in six different ways. So each of the five parts can be found in six different ways. This yields $5 \cdot 6 = 30$ cases.

§8. These 30 cases can be reduced to 10, if one does not distinguish between the given and the desired parts, but only considers the relation between any three of the 5 parts. These 10 cases can be further reduced to 6, if one does not look at the actual position of the parts but interchanges the two catheti and their angles with the hypotenuse. But we will stay with the 10 cases in order to take into account the difference in position. Now the induction is the following.

§9. One proves the relation for each of these 10 (or possibly 6) cases separately. This was until Neper, to whom we owe such a lot in trigonometry, had the idea to compare the 10 relationships with each other and observed a general principle. Namely, in five cases he found the relationship expressed with sines and in the other five cases with sines and tangents. Introducing a clever notation,[13] he was able to reduce all 10 relations and hence all 30 cases occurring with right-angled spherical triangles, to two general rules that are now included in all introductions to spherical geometry and are used with profit.

§10. This induction is as strict as one might require. We know all possible cases. We divide them into two classes. For each type belonging to the same class we employ the same rule. And we conclude rightfully: *What is true for each case in particular is true for all of them, and hence for the whole class.* If only all our general theorems in science were as true as this one!

---

[12] Lambert here does not refer to mathematical induction in the modern sense of this expression. By induction, he means a method of case-by-case analysis.

[13] Lambert implicitly uses the following complementary version of Napier's rules (compare Fig. 19.3a, b and §11). Let us denote the three sides of the spherical triangle by a, b, c and the opposite angles by A,B,C, and assume that $C = 90°$. Write down $A, c, B, 90° - a, 90° - b$ in this order on the margin of a circle divided into 5 equal sectors. Choosing one of these parts as a middle one, it is adjacent to two neighboring parts and opposite to the two others. Now the identities are of the following form. The cosine of the middle part equals the product of the cotangents of the adjacent parts, and moreover the product of the sines of the opposite parts.

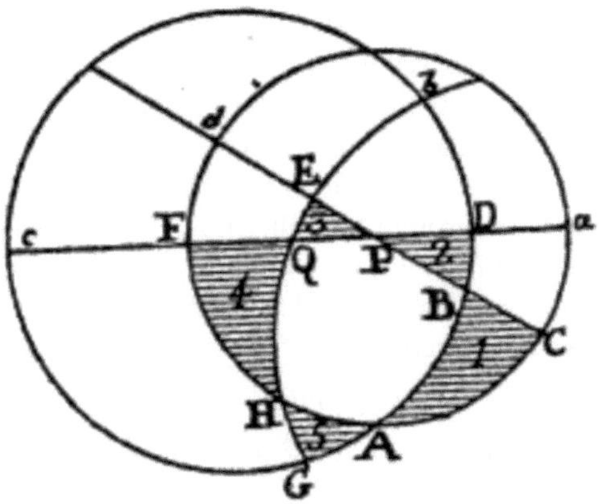

Fig. 19.1 Figure showing the chain of five triangles

§11. However, could it be that in the art of measurement a true general rule exists which cannot be proved other than by induction from proving case by case? Should not the proof be as general as the rule itself? Or is it necessary to split the proof into five parts and to prove each of them separately? The latter seems to be necessary since Neper's rule without making a difference between angles and sides *refers to three chosen parts of the triangle as a middle one and two adjacent or two opposite ones.* But in the cases themselves there is a difference between angles and sides, and in order to apply the rule one has to take the complements of the catheti.

§12. Thus, a general proof of this rule must be based on the transformation of sides into angles and angles into sides. It is easy to see that one needs five such transformations. Let A, B denote the two angles, let H be the hypotenuse, let C be the cathetus at angle A and let K be the cathetus at angle B. Then for the middle and the adjacent parts we have

$$HAB\,,\quad ACH\,,\quad CKA\,,\quad BKH\,,\quad KCB\,,$$

and for the two separate parts we have

$$HCK\,,\quad ABK\,,\quad CHB\,,\quad BAC\,,\quad KAH\,.$$

So there will arise five triangles if the sides and angles of each are transformed into the sides and angles of the next one or into their complements, and it is easy to see that the complements need to occur whenever the transformation concerns a cathetus. This follows from the above note to the use of Neper's rule.

§13. I found out that these five triangles on the sphere do not just lie next to each other but that they form a chain around the sphere, and the fifth touches the first just as the first touches the second, and this the third etc. In Fig. 19.1 you see these five triangles drawn according to the same projection that is used for astrolabia, if the eye is placed in the pole and the figure is in the equator plane.

§14. Let $AadF$ and $ADcG$ be two great circles of the sphere, and let P and Q be their respective poles. Let $cQPa$ be a great circle which appears in the projection as a straight line since the eye is placed in the antipode of pole P. Moreover, let dPC be another circle[14] through the pole P with a right angle to $AadF$ at C and forming the triangle ABC that we call the first triangle. Finally, draw the circle $GHQb$ through the other pole Q perpendicular to dC. Then these circles cut out the five triangles that we have shaded in Fig. 19.1, and we have to prove that these are the desired triangles.[15] For the proof we need the following propositions that can be deduced easily from basic facts in spherical geometry.

(1) At C, D, E, F, G are the five right angles of the five triangles.
(2) The arcs Aa, AD, Ac, AF, PC, Pd, Pa, PF, QD, Qc, QG, Qb are quadrants.[16]
(3) The points A, B, H are poles of the circles cPa, GQb, dPC respectively.
(4) Hence the arcs Bb, BG, HC, Hd, HE are also quadrants, and it is easy to see which of the arcs are complements of the others.
(5) The arcs $Da = cF$, Eb, $aC = Fd$, $Db = cG$ and dE measure the angles A, B, P, Q, H respectively, since their legs are quadrants perpendicular to these arcs. *This observation provides the transformation of angles into sides and sides into angles.*

§15. If for example we follow the angle A in all its transformations in the five triangles, we find

$$A = aD = 90^\circ - DP = QP = 90^\circ - FQ = cF = A\,.$$

So in the first and in the fifth triangle, A occurs as an angle, in the third one it becomes the hypotenuse, and in the second and fourth its complement appears as cathetus. The angle B passes through the same transformations if we start at the second triangle or if we go backwards from the first towards the fifth triangle.

§16. More generally, each hypotenuse has its two complements next to itself which are catheti in the adjacent triangles, and in the two remoter triangles it becomes the angle which lies on the opposite side in the pentagon $ABPQH$.

§17. If we enclose the catheti by brackets () to denote their complements, we can present all the transformations in the following table.

[14] From now on all the circles are great circles.
[15] providing the transformations described in §12.
[16] Their length is 90°.

$$\begin{array}{ccccccccc}
1 & & 2 & & 3 & & 4 & & 5 \\
A & = & (PD) & = & QP & = & (FQ) & = & A \\
B & = & B & = & (QE) & = & QH & = & (HG) \\
(AC) & = & P & = & P & = & (FH) & = & AH \\
(BC) & = & PB & = & (EP) & = & H & = & H \\
AB & = & (BD) & = & Q & = & Q & = & (AG) \\
C & = & D & = & E & = & F & = & G
\end{array}$$

§18. This shows that each side and angle of the first triangle has a special position in each of the other triangles, and that hence these five triangles represent all possible transformations of them. Moreover, remarkably those parts which according to Neper's rule are called the middle and the adjacent ones, or the middle and the opposite ones respectively, do not change their nature through all five transformations. This is because all five parts occur in all five triangles in the same order if we omit the right angle, since in each triangle the right angle moves between two other places of the five ones and hence goes to all five positions.

§19. To make this amazing transformation even more apparent, in Fig. 19.2, I have depicted these triangles with their sides and angles and denoted the catheti for which one has to take the complement by small letters.

§20. All five triangles are determined by five circles. On each of these circles there are two catheti and one hypotenuse, and both catheti are complements of the hypotenuse which lies in between them. The five hypotenuses enclose the pentagon which lies in the interior of the chain of triangles, and the five right angles lie on the outside.

§21. Moreover, the point where two hypotenuses come together is a pole of the circle containing the opposite hypotenuse in the pentagon. It is equal to the angles that the triangles at the pole present to each other. Therefore they are denoted by the same letter in Fig. 19.2.

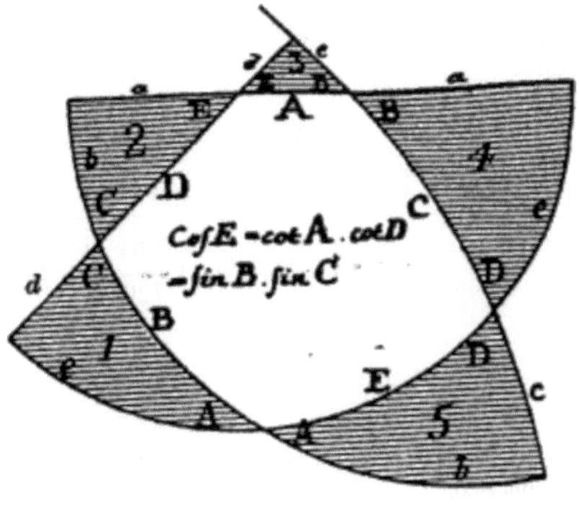

Fig. 19.2 Figure with Neper's rule

§22. Also each of these poles is at distance 90° from the two opposite ones since they lie on the circle corresponding to the pole.
§23. The five poles have their antipode on the other hemisphere where the five circles also intersect each other. *Hence, the pentagon and the five triangles occur twice on the sphere, and each triangle is antipodal to a triangle which is equal and similar in all parts.*
§24. Finally, the whole system has the property that *the area of the five triangles plus twice the area of the pentagon equals* $3/8$ *of the area of the entire sphere or* $3/4$ *of the area of the hemisphere.*
§25. In order to prove this we want to decompose the entire sphere into 720 pieces which we call degrees. Each of these degrees is a triangle whose longer sides are quadrants enclosing an angle of one degree. This seems to be the most convenient scale to determine the parts of a sphere in relation with the whole sphere. Then one can easily find the area of any spherical triangle whose sides are arcs of great circles from its angles by adding them up and subtracting 180° being the angle sum of a planar triangle. The excess gives the area of the spherical triangle in terms of the number of such pieces of which the entire sphere has 720.[17]
§26. Analogously, one finds the area of any polygon on the sphere in terms of such pieces by forming the sum of its angles and subtracting the sum of the angles of a corresponding planar polygon with the same number of sides.

Hence if we denote the area of the pentagon $= V$, the sum of the areas of the five triangles $= \Delta$, then

$$V = 5 \cdot 180^\circ - A - B - C - D - E - 3 \cdot 180^\circ\,.$$

Thus

$$V = 360^\circ - (A + B + C + D + E)\,.$$

Moreover,

$$\Delta = 2(A + B + C + D + E) + 5 \cdot 90^\circ - 5 \cdot 180^\circ\,,$$

hence

$$\Delta = 2(A + B + C + D + E) - 450^\circ\,,$$

Therefore

$$2V + \Delta = 270^\circ \text{ and } \frac{270}{720} = \frac{3}{8} \text{ of the area of the sphere.}$$[18]

[17] Lambert does not use arc measure for angles, and in terms of degrees it becomes more complicated.
[18] The fraction $\frac{270}{720}$ refers to the decomposition of the sphere into special triangles as described by Lambert in §25.

§28.[19] So one finds the area of the pentagon by subtracting the sum of its sides from 360°. The equation

$$\Delta = 2(A + B + C + D + E) - 450^\circ$$

can easily be transformed into

$$\Delta = A + B + C + D + E - a - b - c - d - e\,,$$

hence one finds the area of all five triangles by adding up the hypotenuses and subtracting the sum of its complements.

§29. The five circles defining the whole system intersect each other so that each of them is divided into eight parts where each is the complement of the following one. Moreover, on the sphere, apart from the two pentagons and the 10 triangles there are 10 quadrangles whose sides are known. Each of them has three right angles and may therefore be cut into two right-angled triangles with given catheti. Namely, they are one side of the pentagon and the complement of the adjacent one. Hence the entire sphere is divided into 2 pentagons and 30 right-angled triangles. The sides of the pentagons are A, B, C, D, E and the catheti of the triangles are

$$\begin{gathered} ab,\ bc,\ cd,\ de,\ ea. \\ aB,\ bC,\ cD,\ dE,\ eA. \\ Ab,\ Bc,\ Cd,\ De,\ Ea. \end{gathered}$$

§30. In addition to the observed similarities we note that since each of the five circles cuts the sphere into two equal parts, this yields 10 hemispheres, and on each of them there are the same kind of parts, namely, the pentagon and 15 right-angled triangles of which we listed the catheti.

§31. Though I assume that among so many coincidences there are more general laws hidden, I could not find anything else but the proof of Neper's rule. For example, from the fact that the five triangles lie around the pentagon and form a closed chain, that any three of the triangles determine the other two and that their area plus twice the area of the pentagon has a fixed proportion to the area of the entire sphere, one may assume that any one of the five parts of a right-angled spherical triangle might be determined by the other four in a different way than by the well-known trigonometric rules leaving aside the fact that from two given parts one can find the three others with these rules.

§32. Now the proof of both Neper's rules does not present any difficulties any more. Figure 19.3a, b show the entire trigonometry of spherical right-angled triangles, and the two inscribed equations give an interpretation that does not need further explanation. This is because Fig. 19.3a contains all the five cases for the middle and

[19] There is no Paragraph 27 in Lambert's original memoir.

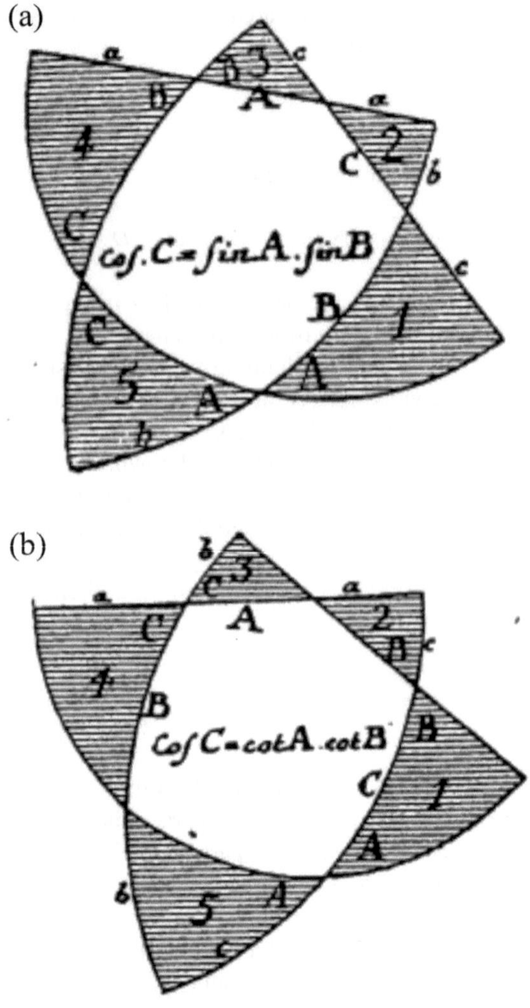

Fig. 19.3 Figure for the proof of both Neper's rules in §32

the opposite parts, and Fig. 19.3b the five cases for the middle and the adjacent parts (see §12, §18).

§33. If one has shown the relationship inscribed into the figure for one of the five triangles, it is clear that the proof also holds for the other four, and Neper's rule will automatically be valid for all five cases. Hence it is not necessary to prove each case separately, and the commonly used induction is obsolete.[20]

§34. Since all cases for right-angled triangles can be solved by a single analogy, we can also use logarithms here, and the algebraic solution is not different from the

[20] This ends Lambert's proof of Neper's rules. Amazingly, he never bothers to write down the rule or the identities explicitly.

solution in numbers. But for triangles with only oblique angles, in most cases there is a notable difference, and one cannot reduce the solution to such simple rules due to several features since mostly one needs two analogies and some preparation. We want to assume some well-known formulae that are trigonometric theorems and are useful in all calculations involving angles.

§35. Let y, z be two circular arcs or angles and $y > z$. Then

$$\begin{aligned}
&(1)\ \sin(y+z) = \sin(y)\cos(z) + \cos(y)\sin(z)\\
&(2)\ \cos(y+z) = \cos(y)\cos(z) - \sin(y)\sin(z)\\
&(3)\ \sin(y-z) = \sin(y)\cos(z) - \cos(y)\sin(z)\\
&(4)\ \cos(y-z) = \cos(y)\cos(z) + \sin(y)\sin(z)
\end{aligned}$$

$$\begin{aligned}
&(5)\ 2\sin(y)\cos(z) = \sin(y+z) + \sin(y-z)\\
&(6)\ 2\cos(y)\sin(z) = \sin(y+z) - \sin(y-z)\\
&(7)\ 2\cos(y)\cos(z) = \cos(y-z) + \cos(y+z)\\
&(8)\ 2\sin(y)\sin(z) = \cos(y-z) - \cos(y+z)
\end{aligned}$$

$$\begin{aligned}
&(9)\ \sin(y) + \sin(z) = 2\sin\left(\frac{y+z}{2}\right)\cos\left(\frac{y-z}{2}\right)\\
&(10)\ \sin(y) - \sin(z) = 2\cos\left(\frac{y+z}{2}\right)\sin\left(\frac{y-z}{2}\right)\\
&(11)\ \cos(y) + \cos(z) = 2\cos\left(\frac{y+z}{2}\right)\cos\left(\frac{y-z}{2}\right)\\
&(12)\ \cos(y) - \cos(z) = -2\sin\left(\frac{y+z}{2}\right)\sin\left(\frac{y-z}{2}\right)
\end{aligned}$$

$$\begin{aligned}
&(13)\ \tan(y+z) = (\tan(y)+\tan(z)) : (1-\tan(y)\tan(z))\\
&(14)\ \cot(y+z) = (1-\tan(y)\tan(z)) : (\tan(y)+\tan(z))
\end{aligned}$$

$$
\begin{aligned}
&(15)\ (\sin(y)+\sin(z)) : (\sin(y)-\sin(z)) &&= \tan\left(\frac{y+z}{2}\right) : \tan\left(\frac{y-z}{2}\right)\\
&(16)\ (\cos(y)+\cos(z)) : (\cos(z)-\cos(y)) &&= \cot\left(\frac{y+z}{2}\right) : \tan\left(\frac{y-z}{2}\right)\\
&(17)\ (\sin(y)+\sin(z)) : (\cos(z)+\cos(y)) &&= \tan\left(\frac{y+z}{2}\right) : 1\\
&(18)\ (\sin(y)+\sin(z)) : (\cos(z)-\cos(y)) &&= 1 : \tan\left(\frac{y-z}{2}\right)\\
&(19)\ (\sin(y)-\sin(z)) : (\cos(z)+\cos(y)) &&= \tan\left(\frac{y-z}{2}\right) : 1\\
&(20)\ (\sin(y)-\sin(z)) : (\cos(z)-\cos(y)) &&= 1 : \tan\left(\frac{y+z}{2}\right)
\end{aligned}
$$

$$
\begin{aligned}
&(21)\ (\sin(y)+\sin(z)) : (\cos(z)+\cos(y)) = (\cos(z)-\cos(y)) : (\sin(y)-\sin(z))\\
&(22)\ (\sin(y)+\sin(z))\cdot(\sin(y)-\sin(z)) \ = \sin(y+z)\cdot\sin(y-z) : 1
\end{aligned}
$$

$$
\begin{aligned}
&(23)\ (\tan(y)+\tan(z)) : (\tan(y)-\tan(z)) = \sin(y+z) : \sin(y-z)\\
&(24)\ (\cot(y)+\cot(z)) : (\cot(y)-\cot(z)) = \sin(y+z) : \sin(y-z)\\
&(25)\ (\tan(y)+\tan(z)) : (\cot(z)+\cot(y)) = (\sin(y)\sin(z)) : (\cos(y)\cos(z))\\
&(26)\ (\tan(y)-\tan(z)) : (\cot(z)-\cot(y)) = (\sin(y)\sin(z)) : (\cos(y)\cos(z))
\end{aligned}
$$

Moreover, in order to remove an irrationality in the various functions of the same angle, we have

$$
\begin{aligned}
&(27)\ \sin(2y) &&= 2\tan(y) : (1+\tan^2(y))\\
&(28)\ \cos(2y) &&= (1-\tan^2(y)) : (1+\tan^2(y))\\
&(29)\ \tan(2y) &&= 2\tan(y) : (1-\tan^2(y))\\
&(30)\ \cot(2y) &&= 2\,(1-\tan^2(y)) : \tan(y)\\
&(31)\ \sec(2y) &&= (1+\tan^2(y)) : (1-\tan^2(y))\\
&(32)\ \mathrm{cosec}(2y) &&= (1+\tan^2(y)) : (2\tan(y))
\end{aligned}
$$

Finally, the proportions of the corresponding parts of an oblique-angled triangle as in Fig. 19.4 on the two sides of the vertical perpendicular are presented in the following formulae.

$$
\begin{aligned}
&(33)\ \cos(A) : \cos(B) &&= \cos(a) : \cos(b)\\
&(34)\ \cos(C) : \sin(D) &&= \cos(c) : \sin(d)\\
&(35)\ \sin(A) : \sin(a) &&= \sin(c) : \sin(C)\\
&(36)\ \cos(D) : \cot(A) &&= \cos(d) : \cot(a)\\
&(37)\ \sin(B) : \cot(C) &&= \sin(b) : \cot(c)\\
&(38)\ \tan(B) : \tan(b) &&= \cot(d) : \cot(D)
\end{aligned}
$$

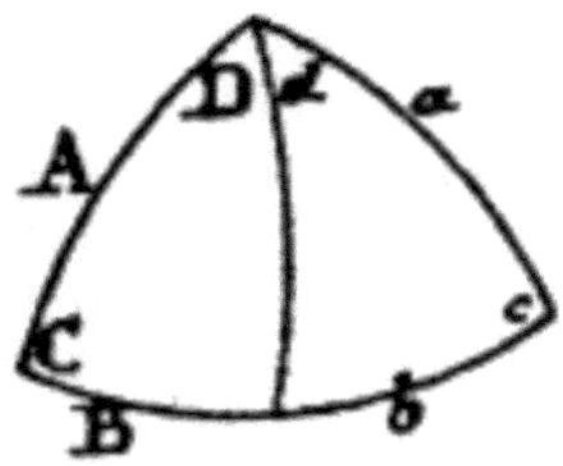

Fig. 19.4 Figure for §35

The last six formulae follow almost directly from Neper's rules. The others are consecutively derived from the first four formulae. In all the formulae we assume the radius of the sphere to be $= 1$.

§36. In every spherical triangle it is sufficient to know three parts to determine each of the other three parts.[21] Thus in all the solutions appear four parts, and there are as many relations as there are possibilities to choose 4 of 6 parts. Since there are 15 such combinations, there are 15 relations and hence all together 60 cases since each of the four parts in the relation can be considered as unknown.

§37. But these 60 cases can be reduced to much fewer. If we ignore the difference between given and unknown parts, only 15 cases remain. If moreover, we only consider the order in which the four parts of the triangle follow each other, but keep the difference between angles and sides, then there are not more than four cases. Namely, among the four parts there is either

(1) just one side, and all three angles,
(2) or all three sides and one angle,
(3) or there are two sides and those two angles that lie opposite to the sides,
(4) or there are two sides and two angles where one of them is enclosed between the two sides.

This is because if there are just two angles chosen, one of them is missing. This missing angle can be the one between the two chosen sides or it lies at the end of one of the sides.

§38. Since each triangle can be transformed into another triangle where the sides become the angles and the angles become sides,[22] the first and the second of these four cases have analogous solutions. This means that we can reduce all 60 cases of oblique-angled triangles to the following 3 cases, where the four parts are

(1) either opposite to each other: *partes oppositae*,
(2) or they follow each other: *partes continuae*,
(3) or one is separated from the others: *partes seiunctae*.

[21] This is true if all parts are assumed to be less than 90°. Lambert comes back to ambiguities in the general case later, see §48.

[22] Namely the polar triangle.

In the first case, two by two parts lie next to each other and two by two lie opposite to each other. In the third case, three parts follow each other, and the fourth part is opposite to the middle one. In the second case, all 4 parts follow each other and the first and last part are opposite to each other.

So one could also distinguish the cases by the way an opposite position is occurring. *oppositio reciproca, media, extrema*, since the four parts are either alternatingly opposite (Case 1) or only the middle ones are opposite (Case 2), or the extremal ones are opposite (Case 3).

§39. However, we will not use this last reduction since it is based on the duality of angles and sides which is not in our focus, and we want their difference to be easily perceived from the formulae themselves. So we go back to the four general cases listed in §37. For these cases the general algebraic formulae are well-known and make it possible to sum up the entire spherical trigonometry in 4 equations. These equations are very useful for analytic solutions of spherical problems. However, they are not all equally suitable for working with numbers and in particular with logarithms. Since they follow easily from the above formulae (see §35), we will present them here, using a simple way of drawing all the cases.

§40. For preparation,

(1) Let me denote the three sides of the triangle with the first three letters of the alphabet in capital form, and the opposite angles with the same letters in small form, so that A is opposite to a, B opposite to b and C opposite to c.
(2) I will arrange the formulae such that A and a always occur. Since in all cases necessarily there is at least one opposition of two parts (see §38), I will start with that pair and denote the side A and the angle a. Then the other two parts are already clear. If there is another side, it is called B, and for the other cases and parts there is no more choice of notation.

§41. This enables us to draw any possible case and to recognise the corresponding formula without referring to any other figure than the one following from the above choices. Namely, we get

3 sides and one angle	$A,\ a,\ B,\ C,$
3 angles and one side	$a,\ A,\ b,\ c,$
four consecutive parts	$A,\ a,\ B,\ c,$
four opposite parts	$A,\ a,\ B,\ b,$

and hence we can easily recognise the corresponding formula.

§42. These are all 4 possible cases which are depicted in Fig. 19.5a–d.

The letters denote the values of the total[23] angles and sides. It goes without saying that the arc going through each triangle is the perpendicular decomposing it into two right-angled triangles. This arc either divides one of the sides or one of the angles occurring in the calculation. One part is denoted by x and the other one by the

[23] Undivided.

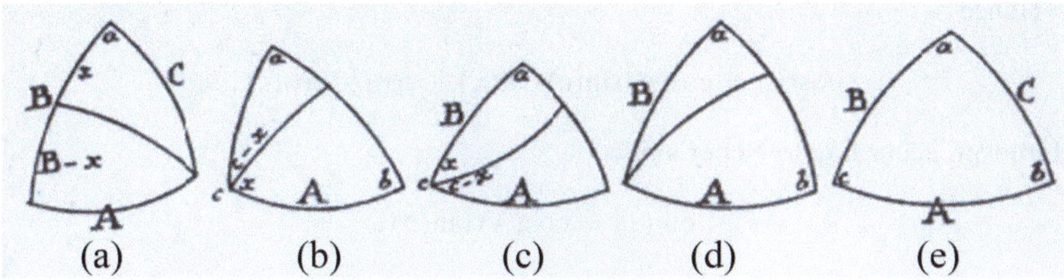

Fig. 19.5 Figures for Case I-IV below

difference. After this preparation, the proof of each of the desired four equations consists of an application of Neper's rule and one of the formulae given in §35.

Case I

§43. Relation between three sides and one angle: Here we have (see Fig. 19.5a):

$$\begin{aligned} \cos(A) : \cos(C) &= \cos(B-x) : \cos(x) && (\S 35, (33)) \\ \cos(B-x) &= \cos(B)\cos(x) + \sin(B)\sin(x) && (\S 35, (4)), \end{aligned}$$

and hence

$$\cos(A) = \cos(B)\cos(C) + \sin(B)\cos(C)\tan(x).$$

Moreover, according to Neper's rule

$$\tan(x) = \tan(C)\cos(a).$$

This implies the desired formula

$$\cos(A) = \cos(B)\cos(C) + \sin(B)\sin(C)\cos(a).$$

Case II

§44. Relation between three angles and one side: Here we have (see Fig. 19.5b):

$$\begin{aligned} \cos(a) : \cos(b) &= \sin(c-x) : \sin(x) && (\S 35, (34)) \\ \sin(c-x) &= \sin(c)\cos(x) - \cos(c)\sin(x) && (\S 35, (3)). \end{aligned}$$

Hence

$$\cos(a) = \cos(b)\sin(c)\cot(x) - \cos(b)\cos(c)\,.$$

Moreover, according to Neper's rule

$$\cot(x) = \cos(A)\tan(b)\,.$$

This implies the desired formula

$$\cos(a) = \sin(c)\sin(b)\cos(A) - \cos(b)\cos(c)\,.$$

Case III

§45. Relation between four consecutive parts. Here we have (see Fig. 19.5c):

$$\begin{aligned} \cot(A) : \cot(B) &= \cos(c-x) : \cos(x) && (\S 35, (36)) \\ \cos(c-x) &= \cos(c)\cos(x) + \sin(c)\sin(x) && (\S 35, (4))\,. \end{aligned}$$

Hence

$$\cot(A)\tan(B) = \cos(c) + \sin(c)\tan(x)\,.$$

Moreover, according to Neper's rule

$$\tan(x) = \cot(a) : \cos(B)\,.$$

This implies the desired formula

$$\sin(B)\cot(A) = \cos(c)\cos(B) + \sin(c)\cot(a)\,.$$

Case IV

§46. Relation between four opposite parts. Here (see Fig. 19.5d and §35,(35)) we have

$$\sin(A) : \sin(a) = \sin(B) : \sin(b)\,.$$

§47. So the four general formulae for all cases in skew-angled spherical triangles are the following (see Fig. 19.5e):

I. $\cos(A) = \sin(B)\sin(C)\cos(a) + \cos(B)\cos(C)$.

II. $\cos(a) = \sin(b)\sin(c)\cos(A) - \cos(b)\cos(c)$.
III. $\sin(B)\cot(A) = \cos(c)\cos(B) + \sin(c)\cot(a)$.
IV. $\sin(A)\sin(b) = \sin(B)\sin(a)$.

§48. These four formulae are sufficient if one is only looking for the relationship between any 4 given parts of a triangle. Here we have assumed all sides and angles to be less than 90° so that the cosines, tangents and cotangents remain positive as one usually supposes. However, if an angle or a side is larger than 90°, then if it is given, one has to change the sign of its cosine, tangent or cotangent. And if it is the unknown then the sign changes by itself. This follows automatically from the mechanical procedure of calculating with symbols.[24] It is easy to see that the first and the second of the above formulae only differ in the sign which has to change when interchanging angles and sides.

§49. In all 4 formulae, the opposite parts are easier to find than the other two, and the simplicity doubles if the other two are also opposite to each other as in the fourth formula. This is because they occur just once, whereas the other two parts in the first three formulae occur twice and in an irrational manner since they are expressed by their sine or cosine.

§50. Furthermore, in the first formula, the sides B, C and in the second the angles b, c are involved in a similar manner. Thus one finds the first in the same way as the latter by interchanging them. If one leaves aside the difference in notation, one might omit one of the two formulae. But for the sake of clarity, we will keep both since the interchanging of angles and sides determining the choice of notation leads too far.

§51. If we distinguish between given and unknown parts, then each of the first two formulae give three and the third one yields four equations. We leave the fourth formula as it is since it cannot be simplified.

§52. The irrationality occurring in the first three cases can be expressed in different ways. The simplest way is to use the tangent of half the angle or half the side. We will explain that in one case, and the other cases are dealt with similarly.

§53. Suppose that in the first formula

$$\cos(A) = \cos(B)\cos(C) + \sin(B)\sin(C)\cos(a)$$

we look for the side B. Set the tangent of its half $= z$, i.e. $\tan(\frac{1}{2}B) = z$. Then by §35, (27),

$$\begin{aligned} \sin(B) &= 2z : (1 + z^2) \\ \cos(B) &= (1 - z^2) : (1 + z^2), \end{aligned}$$

and therefore

$$\cos(A) = \frac{1 - z^2}{1 + z^2}\cos(C) + \frac{2z}{1 + z^2}\sin(C)\cos(a),$$

[24] Lambert speaks of calculation with letters, in German: *Buchstabenrechnung*.

which implies

$$\cos(A) + z^2\cos(A) = \cos(C) - z^2\cos(C) + 2z\sin(C)\cos(a)\,.$$

Hence

$$z = \frac{\sin(C)\cos(a) + \sqrt{\sin^2(C)\cos^2(a) + \cos^2(C) - \cos^2(A)}}{\cos(C) + \cos(A)}\,,$$

or in shorter form

$$\tan(\frac{1}{2}B) = \frac{\sin(C)\cos(a) + \sqrt{\sin^2(A) - \sin^2(a)\sin^2(C)}}{\cos(C) + \cos(A)}\,.$$

§54. If similarly, we look for the tangent of half of b, B, c in the second and the third formula, then for any choice of given or unknown values we obtain the following 11 formulae.

Case I: Three Sides and One Angle

1. $\cos(A) = \sin(B)\sin(C)\cos(a) + \cos(B)\cos(C)$
2. $\cos(a) = (\cos(A) - \cos(B)\cos(C) : (\sin(B)\sin(C))$
3. $\tan(\frac{1}{2}B) = \dfrac{+\sin(C)\cos(a) + \sqrt{\sin^2(A) - \sin^2(C)\sin^2(a)}}{\cos(C) + \cos(A)}$

Case II: Three Angles and One Side

4. $\cos(a) = \sin(b)\sin(c)\cos(A) - \cos(b)\cos(c)$
5. $\cos(A) = (\cos(a) + \cos(b)\cos(c)) : (\sin(b)\sin(c))$
6. $\tan(\frac{1}{2}b) = \dfrac{-\sin(c)\cos(A) + \sqrt{\sin^2(a) - \sin^2(c)\sin^2(A)}}{\cos(c) - \cos(a)}$

Case III: Four Consecutive Parts

7. $\cot(A) = (\cos(c)\cos(B) + \sin(c)\cot(a)) : \sin(B)$
8. $\cot(a) = (\sin(B)\cot(A) - \cos(c)\cos(B)) : \sin(c)$
9. $\tan(\frac{1}{2}B) = \dfrac{-\cos(A)\sin(a) + \sqrt{\sin^2(a) - \sin^2(c)\sin^2(A)}}{\sin(A)\sin(a - c)}$

10. $\tan(\frac{1}{2}c) = \dfrac{+\sin(A)\cos(a) + \sqrt{\sin^2(A) - \sin^2(B)\sin^2(a)}}{\sin(a)\sin(B + A)}$

Case IV: Four Opposite Parts

11. $\sin(A)\sin(b) = \sin(B)\sin(a)$

These are the simplest formulae that one can find without changing the notation for the four parts. Since the same notation is common for general spherical theorems and differential equations where usually angles and arcs have to be kept separately, we have not changed it. The two expressions $\sin(a - c)$ and $\sin(A + B)$ occurring in formula 9 and 10, are easily solved into simpler terms using the above theorems (see §35) if needed.

§55. Since the square root can be taken both as positive as well as negative, the formulae containing a square root yield two values that are both equally possible, and they do not only arise from a mixture of acute and obtuse angles that one might have avoided. So one has to conclude from other circumstances which of these two angles is desired or will be taken. Since in all spherical triangles the tangent of half an arc or half an angle is necessarily positive, Formulae 3, 6, 9 and 10 in §54 must yield two positive values. Should one of the values be negative then the corresponding side or angle does not meet the requirements and has a totally opposite position. But we will not discuss this further since anyway the formulae might change their appearance in calculations for solving problems due to the associated conditions. For finding the numbers for a triangle there are other ways (see §4).

§56. Note however that the irrational terms in these formulae are all similar to each other. They contain two opposite parts, and a third one. Looking for the part opposite to this third part and using the corresponding Formula 11 of §54, it is possible to reduce the irrational term such that it disappears completely. For example, to simplify the third formula of §54, use

$$\sin(C)\sin(a) = \sin(A)\sin(c)\,,$$

and hence

$$\sqrt{\sin^2(A) - \sin^2(C)\sin^2(a)} = \sqrt{\sin^2(A) - \sin^2(A)\sin^2(c)} = \sin(A)\cos(c)\,.$$

§57. With this reduction we have

3. $\tan(\frac{1}{2}B) = (\sin(C)\cos(a) + \sin(A)\cos(c)) : (\cos(C) + \cos(A).$
6. $\tan(\frac{1}{2}b) = (-\sin(c)\cos(A) + \sin(a)\cos(C)) : (\cos(c) - \cos(a).$
9. $\tan(\frac{1}{2}B) = (-\cos(A)\sin(a) + \sin(a)\cos(C)) : (\sin(A)\sin(a - c)).$
10. $\tan(\frac{1}{2}c) = (\sin(A)\cos(a) + \sin(A)\cos(b)) : (\sin(a)\sin(B + A)).$

§58. This shows that the irrational term contains the cosine of a side or an angle which was not included in the four parts of the formula and which could be both greater or less than 90°.

§59. These formulae expressing the desired part by the tangent of its half are the simplest since the square root term has no other coefficients. If one uses the sine or cosine instead of the tangent then one finds similar formulae but with more extra factors.

For example, we find

$$3.\ \sin(B) = \frac{\cos(A)\sin(C)\cos(a) + \cos(C)\sqrt{\sin^2(A) - \sin^2(a)\sin^2(C)}}{1 - \sin^2(C)\sin^2(a)}.$$

$$3.\ \cos(B) = \frac{\cos(A)\sin(C) + \sin(C)\cos(a)\sqrt{\sin^2(A) - \sin^2(a)\sin^2(C)}}{1 - \sin^2(C)\sin^2(a)}.$$

$$6.\ \sin(b) = \frac{\cos(a)\sin(c)\cos(A) + \cos(c)\sqrt{\sin^2(a) - \sin^2(c)\sin^2(A)}}{1 - \sin^2(c)\sin^2(A)}$$

$$6.\ \cos(b) = \frac{-\cos(a)\cos(c) + \sin(c)\cos(A)\sqrt{\sin^2(a) - \sin^2(c)\sin^2(A)}}{1 - \sin^2(c)\sin^2(A)}$$

$$9.\ \sin(B) = \sin(A)\frac{\cos(A)\cos(a)\sin(c) + \cos(c)\sqrt{\sin^2(a) - \sin^2(c)\sin^2(A)}}{\sin(a)(1 - \sin^2(A)\sin^2(c))}$$

$$9.\ \cos(B) = \frac{-\cos(c)\sin(c)\sin^2(A)\cos(a) + \cos(A)\sqrt{\sin^2(a) - \sin^2(c)\sin^2(A)}}{\sin(a)(1 - \sin^2(A)\sin^2(c))}$$

§60. The relation between the sine of each side and its opposite angle can be found from the three sides or the three angles in the following way.

Case I: From the Sides

We have (§54,(1)):

$$\cos(A) = \sin(B)\sin(C)\cos(a) + \cos(B)\cos(C).$$

Moreover (§35,(1),(2))[25]

$$\cos(a) = 1 - 2\sin^2(\frac{1}{2}a) = 2\cos^2(\frac{1}{2}a) - 1,$$

hence

$$\cos(A) = \cos(B)\cos(C) + \sin(B)\sin(C) - 2\sin(B)\sin(C)\sin^2(\frac{1}{2}a)$$

[25] There is a typo in the original edition, referring to §38.

$$= \cos(B)\cos(C) - \sin(B)\sin(C) + 2\sin(B)\sin(C)\cos^2(\frac{1}{2}\,a)\,.$$

Furthermore (§35,(4),(2))

$$\begin{aligned}\cos(B)\cos(C) + \sin(B)\sin(C) &= \cos(B - C)\\ \cos(B)\cos(C) - \sin(B)\sin(C) &= \cos(B + C)\,.\end{aligned}$$

This implies

$$\begin{aligned}2\sin(B)\sin(C)\sin^2(\frac{1}{2}\,a) &= \cos(B - C) - \cos(A)\\ 2\sin(B)\sin(C)\cos^2(\frac{1}{2}\,a) &= \cos(A) - \cos(B + C)\,,\end{aligned}$$

and (§35,(12))

$$\begin{aligned}2\,\sin(B)\sin(C)\sin^2(\frac{1}{2}\,a) &= 2\,\sin\left(\frac{A+B-C}{2}\right)\sin\left(\frac{A-B+C}{2}\right)\\ 2\,\sin(B)\sin(C)\cos^2(\frac{1}{2}\,a) &= 2\,\sin\left(\frac{B+C+A}{2}\right)\sin\left(\frac{B+C-A}{2}\right).\end{aligned}$$

Therefore

$$\begin{aligned}\sin(\frac{1}{2}\,a) &= \frac{\sqrt{\sin\left(\frac{A+B-C}{2}\right)\sin\left(\frac{A-B+C}{2}\right)}}{\sqrt{\sin(B)\sin(C)}}\\ \cos(\frac{1}{2}\,a) &= \frac{\sqrt{\sin\left(\frac{A+B+C}{2}\right)\sin\left(\frac{B+C-A}{2}\right)}}{\sqrt{\sin(B)\sin(C)}}\,.\end{aligned}$$

§61. Thus we found the sine and the cosine of half the arc. Now

$$\sin(\frac{1}{2}\,a)\cos(\frac{1}{2}\,a) = \frac{1}{2}\sin(a)\,,$$

hence

$$\sin(a) = 2\cdot\frac{\sqrt{\sin\left(\frac{A+B+C}{2}\right)\sin\left(\frac{A+B-C}{2}\right)\sin\left(\frac{A-B+C}{2}\right)\sin\left(\frac{B+C-A}{2}\right)}}{\sin(B)\sin(C)}\,,$$

therefore

$$\sin(a) : \sin(A) = 2 \cdot \frac{\sqrt{\sin\left(\frac{A+B+C}{2}\right)\sin\left(\frac{A+B-C}{2}\right)\sin\left(\frac{A-B+C}{2}\right)\sin\left(\frac{B+C-A}{2}\right)}}{\sin(A)\sin(B)\sin(C)}.$$

Case II: From the Angles

Similarly, we find

$$\sin(\frac{1}{2}A) = \sqrt{\frac{-\cos\left(\frac{a+b+c}{2}\right)\cos\left(\frac{c+b-a}{2}\right)}{\sin(b)\sin(c)}}$$

$$\cos(\frac{1}{2}A) = \sqrt{\frac{\cos\left(\frac{a+c-b}{2}\right)\cos\left(\frac{a-c+b}{2}\right)}{\sin(b)\sin(c)}},$$

hence

$$\sin(A) = 2 \cdot \sqrt{\frac{-\cos\left(\frac{a+b+c}{2}\right)\cos\left(\frac{a+b-c}{2}\right)\cos\left(\frac{a+c-b}{2}\right)\cos\left(\frac{b+c-a}{2}\right)}{\sin(b)\sin(c)}},$$

therefore

$$\sin(A) : \sin(a) = 2 \cdot \sqrt{\frac{-\cos\left(\frac{a+b+c}{2}\right)\cos\left(\frac{a+b-c}{2}\right)\cos\left(\frac{a+c-b}{2}\right)\cos\left(\frac{b+c-a}{2}\right)}{\sin(a)\sin(b)\sin(c)}}.$$

§62. The term under the square root in the last two formulae has a negative sign, and hence it seems to give an impossible value. But since the sum of the three angles in a spherical triangle is necessarily greater than 180°, half the sum is greater than 90°. Therefore the *cos* $\frac{(a+b+c)}{2}$ is negative, and the seeming impossibility disappears.

§63. These formulae are written in such a way that one can find the desired proportion with logarithms. The four factors under the square root are half of the sum or their excess over any of the three sides or the three angles. Hence if we add the logarithms of their sine or cosine, we have to take half of the sum and then subtract half the sum of the logarithms of the sines of all three sides or angles. The result is the logarithm of the desired proportion divided by 2.

§64. Let us now consider the triangle as infinitely small. Then we can replace the sine of the sides by the sides themselves, and their cosine becomes $= 1$, if they do not occur alone in the formula. If they occur alone or together with squares of the sine, then we set $\cos(A) = 1 - \frac{1}{2}A^2$, $\cos(B) = 1 - \frac{1}{2}B^2$ etc., and higher powers are omitted. In this way, the above formulae[26] are transformed into those that hold for planar triangles as one can also find directly from the properties of such triangles. The result is the following:

Case I: Three Sides and One Angle

1. $A^2 = B^2 + C^2 - 2BC \cdot \cos(a)$
2. $\cos(a) = (B^2 + C^2 - A^2) : 2BC$
3. $B = C \cdot \cos(a) + \sqrt{A^2 - C^2 \sin^2(a)}$
4. $\sin(\frac{1}{2}a) = \sqrt{\frac{\frac{A+B-C}{2} \cdot \frac{A-B+C}{2}}{BC}}$
5. $\cos(\frac{1}{2}a) = \sqrt{\frac{\frac{A+B+C}{2} \cdot \frac{B+C-A}{2}}{BC}}$
6. $\sin(a) = 2\sqrt{\frac{\frac{A+B+C}{2} \cdot \frac{A+B-C}{2} \cdot \frac{A-B+C}{2} \cdot \frac{B+C-A}{2}}{BC}}$

Case II: Three Angles and One Side

Is undetermined.

Case III: Four Consecutive Parts

7. $B = A(\cos(c) + \sin(c)\cot(a)) = A\sin(a+c) : \sin(a)$
8. $\cot(a) = (B - A\cos(c)) : (\sin(c)A)$
9. $A = B : (\cos(c) + \sin(c)\cot(a)) = \sin(a)B : \sin(a+c)$
10. $\tan(\frac{1}{2}c) = \frac{A\cos(a) + \sqrt{A^2 - B^2\sin^2(a)}}{(A+B)\sin(a)}$

Case IV: Four Opposite Parts

11. $A\sin(b) = B\sin(a)$

[26] See §54 and §60,§61.

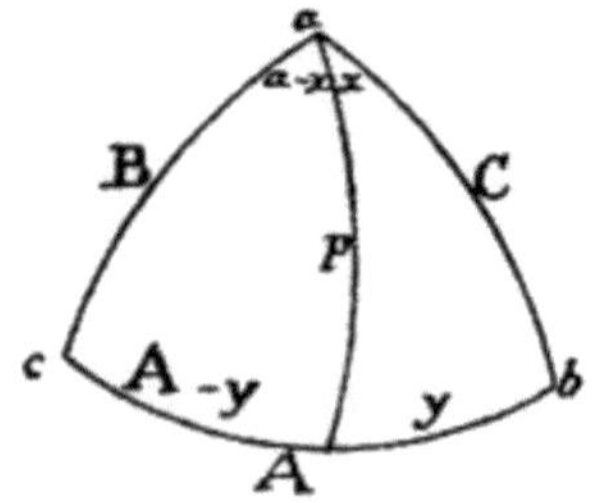

Fig. 19.6 Figure for §66

§65. Usually perpendiculars are drawn just as a means to solve a given triangle, however it can happen that one needs to find the perpendiculars themselves. Here the most difficult cases are the ones where only the three sides or three angles of the triangle are given, or the two sides and an angle from where the perpendicular originates. We want to describe two ways to proceed in this case. The first one is useful for examples calculated with given numbers and the second one is more appropriate for algebraic calculations.

§66. Let us find the perpendicular p from given sides A, B, C. Here is (see Fig. 19.6 and §35,(33))

$$\cos(B) : \cos(C) = \cos(A - y) : \cos(y)\,.$$

Hence by a simple transformation

$$\begin{aligned}(\cos(B) + \cos(C)) &: (\cos(C) - \cos(B)) \\ &= (\cos(A - y) + \cos(y)) : (\cos(y) - \cos(A - y))\,.\end{aligned}$$

Therefore (§35,(16))

$$\tan(\frac{B + C}{2}) \cdot \tan(\frac{B - C}{2}) = \tan(\frac{1}{2}A) \cdot \tan(\frac{1}{2}A - y)$$

and thus

$$\tan(\frac{1}{2}A - y) = \frac{\tan(\frac{1}{2}(B + C)) \cdot \tan(\frac{1}{2}(B - C))}{\tan(\frac{1}{2}A)}\,.$$

Since with this proportion one finds $\frac{1}{2}A - y$ and thereby also y, one obtains

$$\cos(p) = \cos(C) : \cos(y)\,.$$

§67. The algebraic formula follows from the equations

$$\cos(B) : \cos(C) = \cos(A - y) : \cos(y)$$

$$\cos(p) = \cos(C) : \cos(y)\,.$$

This is because we have (§35,(4))

$$\cos(A - y) = \cos(A)\cos(y) + \sin(A)\sin(y)\,,$$

hence[27]

$$\cos(B) = \cos(C)\cos(A) + \cos(C)\sin(A)\tan(y)$$

or

$$\cot(y) = \cos(C)\sin(A) : (\cos(B) - \cos(C)\cos(A))\,.$$

Since moreover

$$\cot^2(y) + 1 = \mathrm{cosec}^2(y) = 1 : \sin^2(y)$$

and

$$1 : \sin(y) = \cos(p) : \sqrt{\cos^2(p) - \cos^2(C)}\,,$$

we find

$$\cos(p) = \frac{\sqrt{\cos^2(B) + \cos^2(C) - 2\cos(B)\cos(C)\cos(A)}}{\sin(A)}\,.$$

§68. Similarly, given B, C, a, we find

$$\cot(p) = \frac{\sqrt{\cos^2(B) + \cot^2(C) - 2\cot(B)\cot(C)\cos(a)}}{\sin(a)}\,.$$

This formula can be compared with the first one given for planar triangles in §64.[28] Since if we make $\cot(B)$ and $\cot(B)$ into the sides of a triangle enclosing the angle a then the third side will be $=\cot(p)\sin(a)$.

[27] There was a mistake in the following formula, noted by the editor Andreas Speiser, namely in the original edition, Lambert wrote $cot(y)$ instead of $tan(y)$ which implied further mistakes in the sequel. Here we reproduce the corrected version of §67, presented by Speiser in [Johann Heinrich Lambert, Anmerkungen und Zusätze zur Trigonometrie, in: Beyträge zum Gebrauche der Mathematik und deren Anwendung. Vol. **1**, Treatise III, Berlin 1765, pp. 369–424, in: Johann Heinrich Lambert, Opera Mathematica I, ed. Andreas Speiser, Orell Füssli Verlag Zürich, 1946, pp. 52–84].

[28] Planar cosine law.

§69. Now suppose we have to calculate with numbers for an occurring example. Among the formulae given so far there are only a few that are suitable for the use of logarithms. Namely, for opposite parts these are the fourth formula of §47, and for the case where one has to find an angle from given sides or a side from given angles, each of the formulae of §60 and §61. These cases have something special in the use of logarithms, and they differ from the other cases which can be brought to a general rule. But we have to consider the problem as a whole since the use of logarithms requires proportions,[29] and we also have to distinguish between given and desired parts. Due to this distinction, it is not sufficient to consider 4 general cases but there are 12 cases that can be derived from the 4 general ones (see §41) in the following way:

I. For three sides and one angle A, a, B, C we have 3 cases, since either the angle a is desired or the opposite side A or one of the adjacent sides B, C.
II. Similarly, there are three cases for three angles and one side. This is because one is either asking for A or a, or for one of the angles b, c.
III. For four consecutive parts A, a, B, c there are 4 cases since here there is no similarity of position and hence each part has something special.
IV. For the opposite parts A, a, B, b there are two cases, either one is asking for a side or for an angle.

§70. Among these 12 cases, those of opposite parts and those where from the sides an angle and from the angles a side is desired, have their own solution that we have already described above (see §60, §61, §47). In the other eight cases, one has to draw a perpendicular in order to find the desired part with two proportions. Here one can describe a general method and with suitable notation treat all eight cases in the same way and present it in general formulae.

§71. Having chosen the four parts of the triangle involved in the calculation, one should draw the perpendicular such that it divides only one of these parts and leaves two other given ones on one side. These two parts will then consecutively lie next to the divided one. Let us call the divided part C, the adjacent or middle one A, the following or outer one B, then the fourth part will be either a or b, depending on whether it is opposite to A or to B. Moreover, call the two parts into which C is divided by the perpendicular, x and y so that x lies next to A.

§72. Now one obtains the following formulae:

$$x + y = C$$

$$\tan(x) = \cos(A)\tan(B)$$

$$\cos(x)\cos(a) = \cos(y)\cos(B)$$

$$\sin(x)\tan(A) = \sin(y)\tan(B)\,.$$

[29] See Footnote 3.

Note that if C is an angle, then for x and y you have to take its complement to 90°.
§73. From these four formulae, in each case one can leave aside either the third or the fourth one depending on whether a or b is missing in the chosen 4 parts. The desired part will always be C or a or b, and the cases will be the following:

$$\begin{array}{llll} \text{Given} & A, B, a & \text{find} & C \\ \text{Given} & A, B, b & \text{find} & C \\ \text{Given} & A, B, C & \text{find} & a \\ \text{Given} & A, B, C & \text{find} & b\,. \end{array}$$

§74. The order of the steps of calculation is obvious. One always starts with the second formula to find x. If C is given then y can be found from the first formula and then a or b via the third or the fourth. But if C is desired, then y can be found from the third or fourth formula and then C from the first one. All possible cases are presented in Fig. 19.7a–d

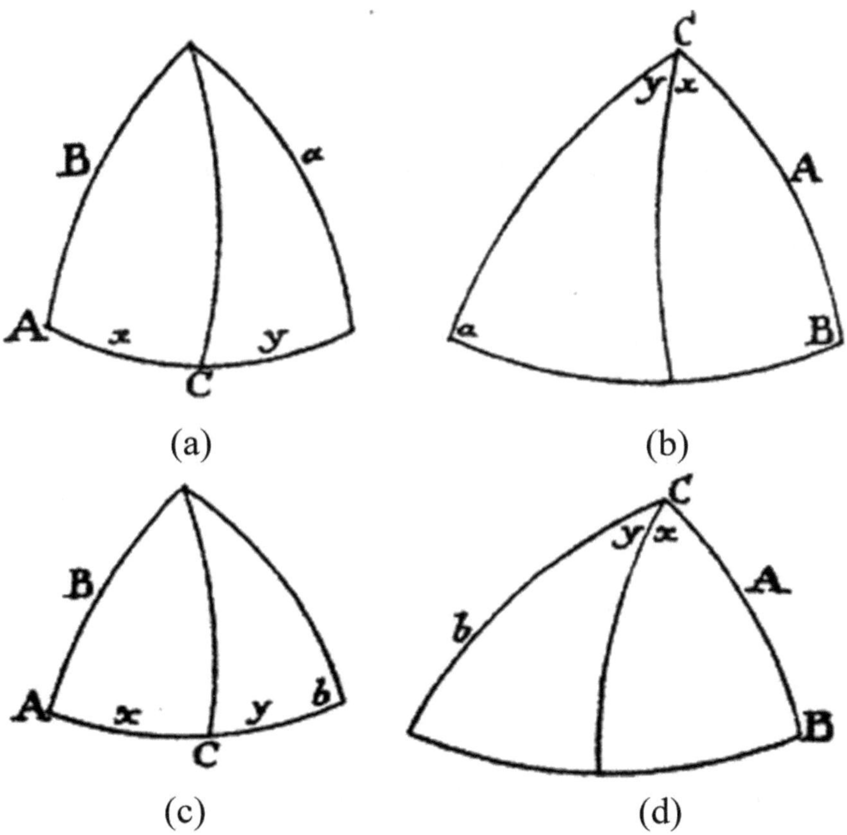

Fig. 19.7 Figures for §74

§75. Since in all these cases one needs two proportions, they are particularly tedious for the calculation of the entire tables. Therefore it is worthwhile to see if by suitable preparation one might spare one proportion. It is clear that the work will then be reduced to half of it. I found that this is possible in some cases and I will now present the method that I used.

§76. Suppose that given an angle and its two adjacent sides, we want to find the opposite side (see Fig. 19.7a). This occurs when calculating tables of solar altitude, and is therefore a frequent problem. Here the given angle is the hours angle, the two adjacent sides are the distance between the celestial pole and the zenith and the distance between the pole and the sun, and the third side opposite to the angle is the distance between the sun and the zenith.[30] Since we have three sides and an angle, the first formula from §47 applies:[31]

$$\cos(A) = \cos(B)\cos(C) + \sin(B)\sin(C)\cos(a)\,.$$

We will transform this in the following way. Firstly

$$\cos(a) = 1 - 2\sin^2(\frac{1}{2}a)\,,$$

and hence

$$\cos(A) = \cos(B)\cos(C) + \sin(B)\sin(C)(1 - 2\sin^2(\frac{1}{2}a))\,.$$

This implies

$$\frac{\cos(A)}{2\sin(B)\sin(C)} = \frac{\cos(B)\cos(C) + \sin(B)\sin(C)}{2\sin(B)\sin(C)} - \sin^2(\frac{1}{2}a)\,.$$

Moreover (§35,(4))

$$\cos(B)\cos(C) + \sin(B)\sin(C) = \cos(B - C)\,,$$

thus

$$\frac{\cos(A)}{2\sin(B)\sin(C)} = \frac{\cos(B - C)}{2\sin(B)\sin(C)} - \sin^2(\frac{1}{2}a)\,.$$

---

[30] See §77, Lambert considers the triangle formed by the celestial pole, the sun and the zenith. Then the angle in this triangle at the pole is the hours angle.

[31] Note that here the notation differs from Fig. 19.7a. The letters a and A are interchanged. In the formula, a denotes the angle between the sides B, C and A is the opposite side. Compare § 77.

Now set

$$\frac{\cos(B-C)}{2\sin(B)\sin(C)} = \sin^2(\frac{1}{2}\varphi)\,,$$

then we have

$$\begin{aligned}\frac{\cos(A)}{2\sin(B)\sin(C)} &= \sin^2(\frac{1}{2}\varphi) - \sin^2(\frac{1}{2}a)\\ &= (\sin(\frac{1}{2}\varphi) + \sin(\frac{1}{2}a))(\sin(\frac{1}{2}\varphi) - \sin(\frac{1}{2}a))\\ &= \sin(\frac{1}{2}\varphi + \frac{1}{2}a)\sin(\frac{1}{2}\varphi - \frac{1}{2}a) \qquad (\S 35,\ (22))\,,\end{aligned}$$

hence

$$\cos(A) = 2\sin(B)\sin(C)\sin\left(\frac{\varphi+a}{2}\right)\sin\left(\frac{\varphi-a}{2}\right),$$

which is the desired formula.

§77. If we apply this formula to the calculation of the solar altitude then a is the hour angle, C is the distance of the celestial pole to the zenith, D is the distance of the sun to the celestial pole and $90° - A$ is the solar altitude. For the same day, both C and B and thus φ is stable. Having found φ, one has to look for the logarithm of $2\sin(B)\sin(C)$, and then the last formula will give the logarithm of the sine of the solar altitude by the simple addition of two logarithms to this value.

§78. It is not difficult to find what in this case the arc φ that we introduced to simplify the calculation means. Namely, if in the equation

$$\frac{\cos(A)}{2\sin(B)\sin(C)} = \sin^2(\frac{1}{2}\varphi) - \sin^2(\frac{1}{2}a)$$

we set $\cos(A) = 0$ and hence $A = 90°$, we obtain

$$0 = \sin^2(\frac{1}{2}\varphi) - \sin^2(\frac{1}{2}a)\,,$$

therefore

$$\sin(\frac{1}{2}\varphi) = \sin(\frac{1}{2}a)\,, \quad \text{which implies} \quad \varphi = a\,.$$

So φ is the angle a if the arc A becomes 90°.

§79. Applying this to the calculation of the solar altitude, then $A = 90°$ if the sun is at the horizon. In that case a is half the day arc and therefore also φ. In the deduced formula,[32] we have $\frac{1}{2}(\varphi + a)$ and $\frac{1}{2}(\varphi - a)$. Hence these two arcs represent the distance of the sun between sunrise and sunset in degrees of the hour angle, and together they give the day arc.

§80. *If we know the length of the day, we can find the solar altitude for each moment by a simple addition of three logarithms.* Namely, transforming *half* of the time from sunrise to the given moment, and from this moment to the sunset into degrees, one gets $\frac{1}{2}(\varphi + a)$ and $\frac{1}{2}(\varphi - a)$. Adding to the logarithms of the sines of these arcs the logarithm of $2\sin(B)\sin(C)$, one obtains the logarithm of the sine of the solar altitude.

§81. At places and days where the sun does not set, this rule cannot be applied since then the expression

$$\frac{\cos(B - C)}{2\sin(B)\sin(C)} > 1\,,$$

and hence cannot be expressed by a sine. Yet in these cases set

$$\frac{\cos(B - C)}{2\sin(B)\sin(C)} = \tan^2(\psi)$$

and

$$\sin^2(\frac{1}{2}a) = \tan^2(\rho)\,.$$

Then

$$\frac{\cos(A)}{2\sin(B)\sin(C)} = \tan^2(\psi) - \tan^2(\rho) =$$

$$(\tan(\psi) + \tan(\rho))(\tan(\psi) - \tan(\rho)) = \frac{\sin(\psi + \rho)\sin(\psi - \rho)}{\cos^2(\psi)\cos^2(\rho)}\,,$$

therefore

$$\cos(A) = \frac{2\sin(B)\sin(C)\sin(\psi + \rho)\sin(\psi - \rho)}{\cos^2(\psi)\cos^2(\rho)}\,.$$

This formula can also be calculated with logarithms. However, it is much more complicated than the first one, but it can be applied in all cases.

§82. *In a similar manner, from a given side and the two adjacent angles, one can find the third angle which is opposite to the side.* This case arises if we want to find the angle between the ecliptic and the horizon at a given time. Then the horizon, the celestial equator and the ecliptic form a triangle in which the two angles between the equator and the horizon or the ecliptic respectively are given as well as the arc of

[32] See §76.

the equator lying between the horizon and the point of the equinox. For each latitude one calculates tables presenting the angle of the ecliptic with the horizon. It is clear that the calculation of such tables is simplified considerably if it can be reduced to the addition of three logarithms. Now this can be done as follows.

§83. Since here occur three angles and one side, we take the second formula from §47:

$$\cos(a) = \sin(b)\sin(c)\cos(A) - \cos(b)\cos(c)\,.$$

We would like to express the angle a by b, c, A just with a multiplication. Now

$$\cos(A) = -1 + 2\cos^2(\frac{1}{2}A)\,,$$

hence

$$\cos(a) = 2\sin(b)\sin(c)\cos^2(\frac{1}{2}A) - \sin(b)\sin(c) - \cos(b)\cos(c)\,.$$

Moreover,

$$\cos(b)\cos(c) + \sin(b)\sin(c) = \cos(b-c)\,,$$

therefore

$$\cos(a) = 2\sin(b)\sin(c)\cos^2(\frac{1}{2}A) - \cos(b-c)\,,$$

and

$$\frac{\cos(a)}{2\sin(b)\sin(c)} = \cos^2(\frac{1}{2}A) - \frac{\cos(b-c)}{2\sin(b)\sin(c)}\,.$$

Now set

$$\frac{\cos(b-c)}{2\sin(b)\sin(c)} = \cos^2(\frac{1}{2}\varphi)\,.$$

Then we have

$$\frac{\cos(a)}{2\sin(b)\sin(c)} = \cos^2(\frac{1}{2}A) - \cos^2(\frac{1}{2}\varphi)$$
$$= (\cos(\frac{1}{2}A) + \cos(\frac{1}{2}\varphi))(\cos(\frac{1}{2}A) - \cos(\frac{1}{2}\varphi)) =$$

$$\sin(\frac{1}{2}A + \frac{1}{2}\varphi)) \cdot \sin(\frac{1}{2}A - \frac{1}{2}\varphi))\,,$$

and thus finally

$$\cos(a) = 2\sin(b)\sin(c)\sin(\frac{1}{2}(A+\varphi)) \cdot \sin(\frac{1}{2}(A-\varphi)).$$

In this formula, $\varphi = A$ if $a = 90°$. So if a denotes the angle between the ecliptic and the horizon, then it is clear that the formula can only be used between solstice circles since nowhere else, a becomes $90°$.

§84. For the other cases, set

$$\frac{\cos(b-c)}{2\sin(b)\sin(c)} = \tan^2(\psi),$$

and

$$\cos^2(\frac{1}{2}A) = \tan^2(\rho).$$

Then we have

$$\frac{\cos(a)}{2\sin(b)\sin(c)} = \tan^2(\rho) - \tan^2(\psi) = (\tan(\rho)+\tan(\psi))(\tan(\rho)-\tan(\psi)),$$

hence

$$\cos(a) = \frac{2\sin(b)\sin(c)\sin(\rho+\psi)\sin(\rho-\psi)}{\cos^2(\rho)\cos^2(\psi)}.$$

This formula is valid more generally, but is again more complicated than the previous one.

§85. Let us consider a third example: *Given four consecutive parts, one shall find the outmost angle.* This situation arises in the calculation of solar azimuth tables where one wants to find the solar azimuth angle from the distance of the sun to the celestial pole, the hour angle, and the distance of the celestial pole to the zenith. Since here are four consecutive parts, we take the third formula from §47

$$\sin(B)\cot(A) = \cos(c)\cos(B) + \sin(c)\cot(a).$$

Now, since

$$\cot(A) = \cos(A) : \sin(A),$$

we have

$$\sin(B)\cos(A) = \cos(c)\cos(B)\sin(A) + \sin(c)\cot(a)\sin(A),$$

thus

$$\sin(c)\sin(A)\cot(a) = \sin(B)\cos(A) - \cos(B)\sin(A)\cos(c)\,.$$

Now

$$\cos(c) = 2\cos^2(\frac{1}{2}c) - 1\,,$$

therefore

$$\sin(c)\sin(A)\cot(a) = \sin(B)\cos(A) + \cos(B)\sin(A)(1 - 2\cos^2(\frac{1}{2}c))\,.$$

Furthermore,

$$\sin(B)\cos(A) + \sin(B)\cos(A) = \sin(B + A)\,,$$

hence

$$\sin(c)\sin(A)\cot(a) = \sin(B + A) - 2\cos(B)\sin(A)\cos^2(\frac{1}{2}c)\,,$$

or

$$\frac{\sin(c)\cot(a)}{2\cos(B)} = \frac{\sin(B + A)}{2\cos(B)\sin(A)} - \cos^2(\frac{1}{2}c)\,.$$

Now set

$$\frac{\sin(B + A)}{2\cos(B)\sin(A)} = \cos^2(\frac{1}{2}\varphi)\,.$$

Then

$$\frac{\sin(c)\cot(a)}{2\cos(B)} = \cos^2(\frac{1}{2}\varphi) - \cos^2(\frac{1}{2}c) = \sin(\frac{1}{2}(\varphi + c)) \cdot \sin(\frac{1}{2}(\varphi - c))\,,$$

and finally

$$\cot(a) = \frac{2\sin(\frac{1}{2}(\varphi + c))\sin(\frac{1}{2}(\varphi - c))\cos(B)}{\sin(c)}\,.$$

Here $\varphi = c$ if $\cot(a) = 0$ or $a = 90°$. If a represents the solar azimuth angle taken from midnight on, it cannot become 90° unless $A < B$, i.e. if the distance of the sun to the pole is greater than the distance of the pole to the zenith. In this case, φ represents the hour angle if the solar azimuth angle is 90°.

Index

R. Caddeo, A. Papadopoulos (eds.), *Spherical Geometry in the Eighteenth Century I*,
https://doi.org/10.1007/978-3-032-12466-1

MIX
Papier aus verantwortungsvollen Quellen
Paper from responsible sources
FSC® C105338

If you have any concerns about our products,
you can contact us on
ProductSafety@springernature.com

In case Publisher is established outside the EU,
the EU authorized representative is:
Springer Nature Customer Service Center GmbH
Europaplatz 3, 69115 Heidelberg, Germany

Printed by Libri Plureos GmbH
in Hamburg, Germany